MEDICAL MYCOLOGY

GIANT COLONIES OF RINGWORM FUNGI

(Sabouraud's Maltose Agar)

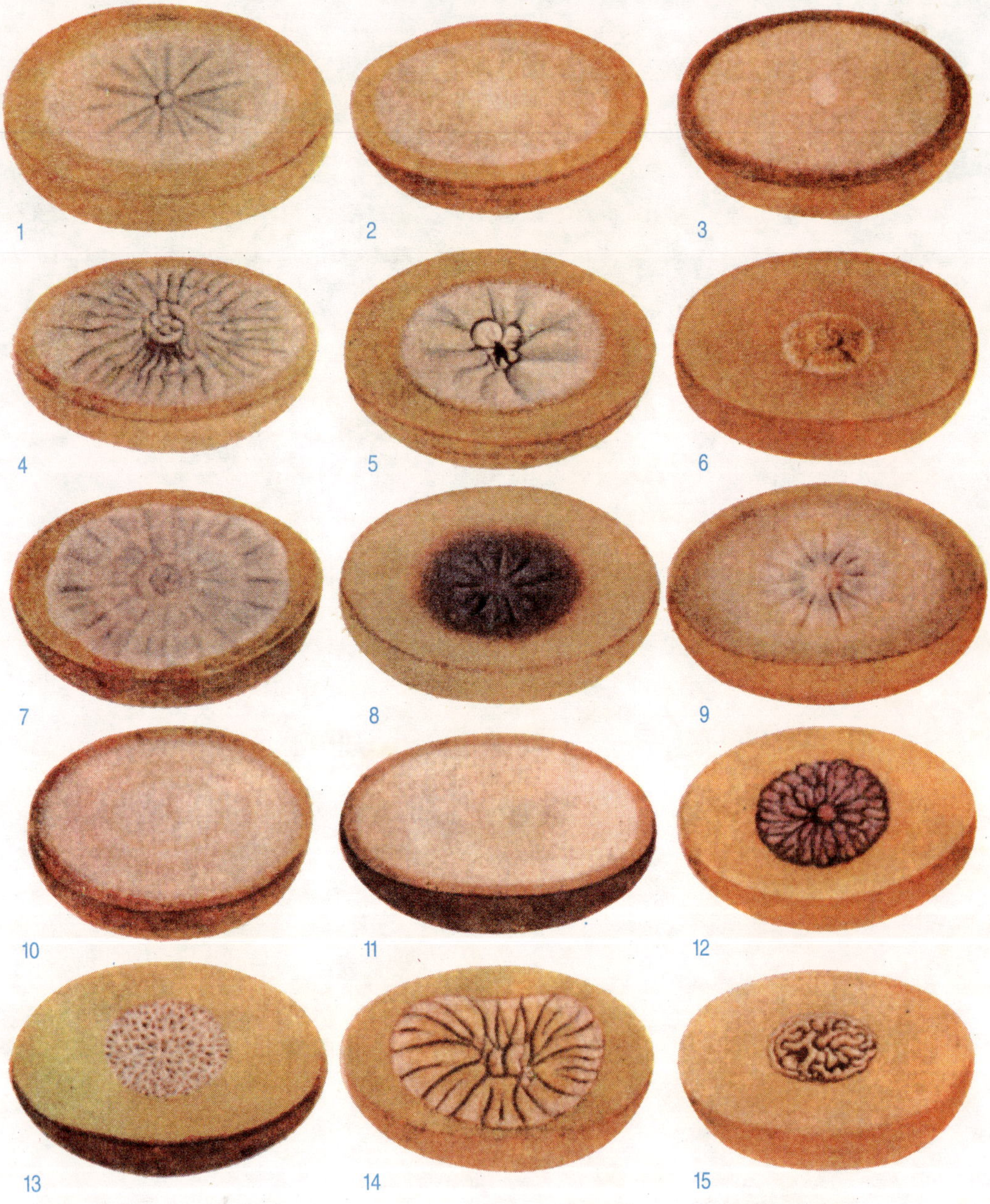

(1) *Microsporum audouini*, (2) *M. canis*, (3) *M. gypseum*, (4) *Epidermophyton floccosum*, (5) *Trichophyton tonsurans*, (6) *T. sulphureum*, (7) and (9) *T. purpureum*, (8) *T. purpureum*, (Glucose agar), (10) *T. mentagrophytes* (Gypseum), (11) *T. interdigitale*, (12) *T. violaceum*, (13) *T. schoenleinii*, (14) *T. ferrugineum*, (15) *T. concentricum*.

(Frontispiece)

This book is meant for undergraduate medical students, BSc Botany students, BSc Microbiology students, Postgraduate Medical (MD, PhD) students and Postgraduate Microbiology students

MEDICAL MYCOLOGY

Dr N C Dey

BSc, MD, FDS, FRSM (Lond.), DBact (Manchester), FISD (New York)

Dr H L E Grueber MSc, MC Path, MD

Formerly Professor of Pathology
Christian Mecical College, Ludhiana

Dr T K Dey

MD, DBact (Manchester), FISD (New York), FRSM (London)
Professor and Head, Department of Pathology
Janaki Medical College, Janakpur, Nepal

New Central Book Agency (P) Ltd

LONDON

HYDERABAD ERNAKULAM BHUBANESWAR

DELHI KOLKATA PUNE GUWAHATI

NCBA

REGD OFFICE

8/1 Chintamoni Das Lane, Kolkata 700 009, India
email: ncbapvtltd@eth.net

OVERSEAS

NCBA (UK) Ltd, 149 Park Avenue North, Northampton, NN3 2HY, UK
email: ncbauk@yahoo.com

EXPORT

NCBA Exports Pvt. Ltd, 212 Shahpur Jat, New Delhi 110 049
email: ncbaexp@ncbapvtltd.com

BRANCHES

208B Peacock Lane, Shahpur Jat
New Delhi 110 049
email: ncbadel@ncbapvtltd.com

House No. 3-1-315, 1st Floor
Nimboliadda, Kachiguda, **Hyderabad** 500 027
email: ncbahydb@ncbapvtltd.com

Shop Nos. 3 & 4, Vinayak Towers
681/B Budhwar Peth, Appa Balwant Chowk
Pune 411 002
email: ncbapune@ncbapvtltd.com

GSS Shopping Complex, 1st Floor
Opposite College Post Office, Convent Road
Ernakulam 682 035
email: ncbaernk@ncbapvtltd.com

Shop No. 15, Bharati Towers, Ground Floor
Forest Park, **Bhubaneswar** 751 001
email: ncbabub@ncbapvtltd.com

Radhamayee Complex (Second Floor)
Opp. S.B.I. Panbazar, Jaswanta Road
Guwahati 781 001

MEDICAL MYCOLOGY

First Published: 1957
First Central Edition: 2006
Reprinted: 2012

PUBLISHER AND TYPESETTER
New Central Book Agency (P) Ltd
8/1 Chintamoni Das Lane, Kolkata 700 009

PRINTER
New Central Book Agency (P) Ltd
Web-Offset Division, Dhulagarh, Sankrail, Howrah

COVER DESIGNER
Soumen Paul

COVER PRINTER
Biswanath Studio, Kolkata

TECHNICAL EDITOR
Dr Mita Sen (MBBS)

PROJECT TEAM
Sukumar Patra, Ananta Bhowmik and Arun Kabri

ISBN 978 81 7381 495 2

Price ₹ 465.00

He who is worried cannot smile, only through wisdom can his sorrow be wiped out. You will be a man of bad luck if you are jealous of others' good luck. You should establish yourself in virtue, if you want to be successful in your life. Good luck is always with you, if you pay attention to it. Bodily, vocal and mental purification is the essence of life. If you want to be happy in this long life, first you should purify yourself. Poverty does not matter, matter generosity and hospitality. One who has self-purification, is best among men and deity.

DEDICATED TO
THE MEMORY OF
MY WIFE

N. C. DEY

" Ring out old shapes of foul disease
Ring out the narrowing lust of gold
Ring out the thousand wars of old
Ring in the thousand years of peace."

TENNYSON

Contents

SECTION II SYSTEMATIC MYCOLOGY

FOREWORD

I have great pleasure in wirting a foreword to the book on Medical Mycology by Professor N. C. Dey of the Assam Medical College. The author was a colleague of mine in Calcutta School of Tropical Medicine for many years. He devoted himself whole-heartedly for several years in the study of Dermatology and isolation of pathogenic fungi from skin diseases and published a large number of original articles. Having thus gained considerable experience, he has fully utilised his knowledge in the production of this book. The book has been amply illustrated with plates, text figures, coloured plates and diagrams, collected and prepared during the long period of his work in the School as well as in the Assam Medical College. The speciality of the book is that the author has followed a natural classification of fungi pathogenic to man. The laboratory methods mentioned in the book are those which were found to be useful by the author, while working in the School of Tropical Medicine, Calcutta.

I congratulate the author for sparing no pains in the production of such a valuable book. I am confident that his book will be a very useful guide to postgraduate students and practitioners for whom it is meant.

G. PANGA
Late of the Department of Dermatology and
Professor of Bacteriology and Pathology,
Calcutta School of Tropical Medicine

30 December 1957
117 Vivekananda Road
Kolkata 700 006

Preface to the First Central Edition

It is a matter of great pleasure for the authors as well as for the readers that New Central has agreed to publish the third edition of "Medical Mycology".

It is also a very hopeful and pleasant aspect that Medical Mycology has been generously considered to be accepted in the under-graduate syllabus of the Indian universities. It would hopefully allow the basically qualified doctors to understand the meaning, diagnosis as well as prognosis of fungal diseases.

As regards antifungal therapy, the late 1990s are a few remarkable years when antifungal chemotherapeutics have been thoroughly revolutionised with the further advancement in the field of Azole compounds. There have been a lot of therapeutic improvements to cure most of the fungal diseases. Every one of them has been incorporated in this edition.

Some new chapters have been added. The other chapters have been thoroughly revised and rewritten wherever found necessary. A large number of illustrations and a large number of both coloured and black and white plates have been incorporated to make the book self-explanatory. As such, a good textbook on Mycology is rarely available. It is expected that this book will be beneficial to all concerned and be considered as "The" Standard Textbook of Medical Mycology.

T. K. Dey

Professor and Head
Department of Pathology
Janaki Medical College
Janakpur, Nepal
November 2005

Preface to the First Edition

Little importance has been laid on the study of Medical Mycology in the undergraduate course of the Indian universities. The subject has, however, been recently included as a part of the post-graduate curricula of Dermatology and Tropical Medicine of the Calcutta University. This has been judiciously done, as diseases caused by fungi are not uncommon in India, a tropical country.

Since the time of the late Lt. Col. H. W. Acton, then Director, The School of Tropical Medicine, Calcutta, 1932, I have had the opportunity of studying Medical Mycology as an Assistant Research Worker under him and others following him, in the Department of Dermatology and Medical Mycology, under the Indian Research Fund Association (present I.C.M.R.) and also in other places in India including the Assam Medical College, Dibrugarh, as the Professor of Pathology. The articles on these works were published from time to time in Medical Journals in India and the present volume is the result of compilation of those articles. There are certain diseases like North and South American blastomycoses, coccidioidomycosis, etc., which have not been authentically recorded in India and these diseases have been compiled from American literatures to make the book complete for the students and practitioners. In this volume common diseases caused by fungi in man and animals have been recorded. Common laboratory methods have been included in the appendix including certain recipes which might be of help to the medical practitioners. The author will consider his endeavour to be amply rewarded if the book becomes helpful to the students and practitioners in India.

The author is thankful to Sri H. C. Sarkar, Manager of Pressagents Private Limited, Calcutta, for the keen interest he has taken in the production of this volume.

Dibrugarh
15th December 1957

N. C. D.

1

Introduction

The science that deals with the study of fungi is called *mycology*. Fungi are **heterotrophic microorganisms** with simple plant body called *thallus* **without anatomical differentiation** into roots, stems and leaves. It is composed of ***interwined filaments, mycelia***, or ***individual filaments, hyphae*** and show various types of ***reproductive bodies*** or ***spores***. Fungi are differentiated from Algae by **lack of chlorophyll** which acts as photodynamic pigment. Thus, algae are autotrophic; **chlorophyll** enables the algae to prepare food from water, CO_2 and minerals in presence of sunlight. Being heterotrophic, fungi require prepared food and thus live as saprophytes or parasites with nutritional requirements like those of bacteria. The taxonomic position of fungi in the plant kingdom has been reviewed. Fungi are now considered *pari passu* with algae (containing chlorophyll). The algae are now classified under a **large number of phyla**. On the contrary **bacteria, slime molds and fungi have been placed in separate subdivisions and classes as in Table 1.1**.

Table 1.1

KINGDOM : PLANTAE

DIVISION :	**SCHIZOMYCOTA**	**MYCOTA**	
SUBDIVISION :	**SCHIZOMYCOTINA**	**MYXOMYCOTINA**	**EUMYCOTINA**
CLASS :	**Schizomycetes Fission fungi, or bacteria including *Actinomyces and Nocardia***	**Myxomycetes Slime mold with multinucleated *coenocytic aseptate hyphae***	**Eumycetes**

Eumycetes:

ZYGOMYCETES	***ASCOMYCETES***	***BASIDIOMYCETES***	***DEUTEROMYCETES OR HYPHOMYCETES***
Fungi are aseptate coenocytic; reproduce by various motile and non-motile sporangiospores; rarely pathogenic to man	Reproduce by ascospores : rarely pathogenic to man	Reproduce by Basidiospores; mushrooms, smut fungi, etc. Produces mycotoxins which causes human pathogenicity	(Fungi imperfecti) The sexual stage is imperfector not known; most of the human pathogenic fungi belong to this group

Schizomycetes

They fission fungi or bacteria are which are the lowest form of organisms in the vegetable kingdom (devoid of chlorophyll). They are 1µ or less in diameter without a well defined nuclear membrane and hence called *prokaryotic nuclei*.

Eumycetes (true fungi) unlike bacteria show definite nuclei with definite nuclear membrane and hence called *eukaryotic* nuclei.

They have **plant volutin** in the cytoplasm. They **reproduce by sexual or asexual spores**. Fungi differ from bacteria morphologically; it is filamentous and showing lateral branching. Actinomyces and Nocardia are **higher bacteria** being 1µ or less in diameter; **unlike bacteria** they show branching. Modern trend is to include them under the class Schizomycetes (Bergey, 1948).

ACTINOMYCETALES is Order between Eubacteriales (true bacteria) and Eumycetes (true fungi). They are higher bacteria as they are slender organisms with being 1µ or less in diameter. Actinomyces fragment readily into **bacillary or coccoid elements**; in **Streptomycetaceae** they produce spores in the filament like the sporulating bacilli. They are sensitive to **penicillin** and **sulphonamides** respectively; true fungi are not affected by these drugs. Because of these peculiarities, the **genera Actinomyces, Nocardia** and **Streptomyces** are placed in the class Schizomycetes, position between Eubacteriales and Eumycetes.

Eumycetes. Fungi resemble bacteria in their mode of isolation and cultivation, capacity to produce infections and their fermentative power for which they are utilised in industrial processes.

While they differ from bacteria in the mode of **reproduction (budding or branching filaments): greater size of cell, etc.**

Fungi resemble higher plants in morphological complexity of their cells. Their nuclei are *eukaryotic* with several different chromosomes and a well-defined nuclear membrane (Eukaryote). The cells possess *'mitochondria, endoplasmic reticulum'*, cell membranes contain sterols (as in higher plants). They are absent in bacteria.

CYTOLOGY

Cell Wall. The cell wall of the fungus lies external to the cell membrane. In **cryptococcus** the cell wall is surrounded by an external *capsular polysaccharide*. In bacteria, cell wall is made of brick-like structural units; in fungus, it appears to be thatched. ***Chemically***, it consists of glucose, mannans, or chitins made of N-acetyl glucosamine, forming protein complexes with them; bacterial cell wall is formed by muramic acid, polypeptides, teichoic acid in 'G' - ves but some have diamino pimelic acid residues (in 'G' - ves).

Nucleus. Fungi have *eukaryotic nuclei* (well-defined nuclear membrane); bacterial nuclei are *prokaryotic* (no well-defined nuclear membrane).

Cytoplasm of the fungus has **mitocondria** and **endoplasmic reticulum**; bacteria have no mitochondria or endoplasmic reticulum. The *cytoplasmic membrane* in fungi contains *sterols* that is absent in bacterial cell membrane except mycoplasma (Mendosicutes) which is grown on sterols containing medium.

Protoplasts. As in bacteria, viable protoplasts are left, when the wall of yeasts and molds are digested in hypertonic solution. They are also formed in *media that inhibit cell wall synthesis*. Yeast protoplasts are deficient in some hydrolases the *invertase* and *fructosidase* in bacteria, secreted enzymes are located in the cell wall of fungus, like or between the cell wall and cell bacteria membrane.

Metabolism. All fungi are *heterotrophic* requiring prepared organic food and most of them are *obligate aerobes*. Some are *facultative aerobes* but **none are *obligate anaerobes*.** Fungi exhibit a great diversity of metabolic variations like bacteria. Most of the fungi can grow in minimal media containing organic carbon, inorganic NH_4^+ or even NO_3^-.

Some of the fungi are *thermophilic*; grow at 56ºC or higher temperature. Some fungi can hydrolyse organic substances (wood, bone, tanned leather, chitin, waxes or even synthetic plastics). Fungi can be induced by appropriate substrates to form adaptive enzymes like bacteria.

Nutritional requirements. When vitamins are added to a **basal medium (casein or ammonium nitrate)** the particular species or strain of trichophyton showed enhanced growth. **Nutritional agar** media for different trichophyton species are available from Difco Laboratories and they cater seven trichophyton agar of different permutations and combinations to differentiate the species on their growth character. These **trichophyton agars** are: (1) Casein basal medium; (2) Casein basal medium with inositol; (3) Casein basal medium with inositol and thiamine; (4) Casein basal medium with thiamine; (5) Casein basal medium with nicotinic acid; (6) Ammonium nitrate basal medium; (7) Ammonium nitrate basal medium with histidine. The growth and development of fungus occurs according to the requirement of vitamins incorporated with the basal medium. For details *vide* Chapter on dermatophytes.

Fungi are not susceptible to virus infection; unlike bacteriophages. Dimorphic characters **(yeast, and mold or mycelial phases)** are peculiar to some fungi causing systemic mycosis.

Drug sensitivity. Dermatophytes are sensititve to griseofulvin but some other fungi causing internal mycosis are sensitive to polyene, but not to sulphonamides, penicillin, streptomycin or broad spectrum antibacterial antibiotics; they are also susceptible to Azole compounds, and broad spectrum antifungal (see) antibiotics Ch. 8 (and the tables).

Genetics, biochemical and antigenic differences have been studied much less among fungi.

Reproduction. It occurs (1) **sexual,** (2) **asexual** and (3) **parasexual processes.** Vast majority of human pathogens reproduce by asexual and parasexual processes.

ASEXUAL REPRODUCTION. The **vegetative growth** of a coenocytic mycelium involves nuclear division without septa formation. Process of mitosis of nucleus causes transmission of full complement of chromosomes to each of the daughter nuclei. Further division of a nuclear body to asexual reproduction forms **few clones** without gametes formation.

ASEXUAL SPORE formation is characteristically seen in **Deuteromycetes** (Hyphomycete). They lack sexuality; they show three modes of reproduction **(1) Thallospores; (2) Conidiospores and (3) Sporangiospores.** They also reproduce by fragmentation of hyphae. The details of these spores are illustrated (Plate 1.1) and given in definitions.

Thallospores : They reproduce by budding, typically seen in ***blastospores.*** The bud bulges out from the mother cell, the nucleus of which divides and one of them passes into the bud. The cell-wall material is laid down between the

Table 1.2

ASEXUAL SPORES

- **1.** THALLOSPORES
  - BLASTOSPORES — (By budding). In Yeasts and Candida
  - ARTHROSPORES — (By segmentation of mycelia). In Dermatophytes and Geotrichum
  - CHLAMYDOSPORES — (By mycelial thickening). In Dermatophytes, Candida, etc.
- **2.** CONIDIOSPORES — Conidia. True conidia in Penicillium and Aspergillus. Macro- and Micro-conidia in Dermatophytes
- **3.** SPORANGIOSPORES — (As in sporangia) In Phycomycetes (Zygomycetes)

daughter and mother cells; the bud breaks away. A birth-scar on the daughter cell wall and scar on the mother cell wall are visible under the electron micrography.

Arthrospores are formed by double septation and fragmentation of mycelia or their swelling (**chlamydospores**). ***Conidiospores or conidia*** **(conis = dust)** are borne on *conidiophores* on their sides, or terminally. They are small in size and called ***conidia vera*** (true conidia) as in Penicillium or Aspergillus. In dermatophytes there are two types of conidia-*microconidia* on the side of mycelia or terminally in clusters, and *macroconidia* (large fusiform spores with septa). **Sporangiospores** form in **sporangia (spore case)** as in Phycomycetes (Zygomycetes).

Asexual spores are tabulated in Table 1.2.

Sexual reproduction. Sexual reproduction occurs as follows :

(1) A haploid nucleus of a donor **(male +)** cell penetrates the cytoplasm of a recipient **(female –)**.

(2) The male and female nuclei fuse to form a diploid zygotic nucleus.

(3) By a process of meiosis, the diploid nucleus gives 4 haploid nuclei some of which may be **genetic recombinants**. In most species **haploid cells form** the most part of the **vegetative growth**, and **diploid stage is transient**; in higher animals the opposite is true.

In **HOMOTHALLIC SPECIES the cells of a single colony** (arising from a single nucleus) can **participate in sexual reproduction** and in some of these species **male and female cells are hermaphrodites** (anatomically differentiated); in others they are indistinguishable.

In **HETEROTHALLIC SPECIES sexual cells** for reproduction arise from two different colonies of opposite mating types. The sexual cells of heterothallic species may be distinguished as males and **females (diecious)** and the anatomy of sex organs and mating processes is highly diversified and characteristic for a given species.

SEXUAL REPRODUCTION OF NEUROSPORA CRASSA

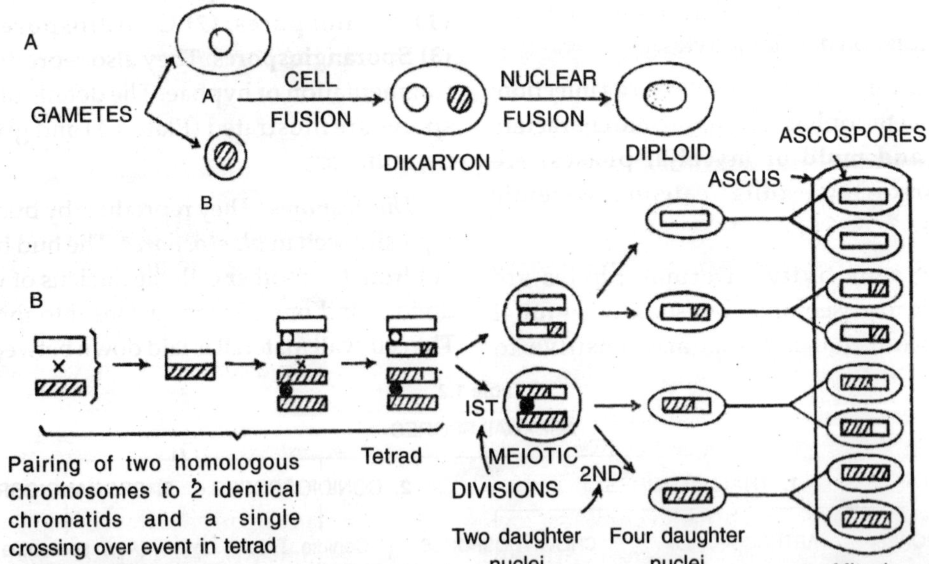

Fig. 1.1 — A. Mating leads to dikaryon, **fusion of nuclei** and diploid cell
B. Meiotic division (recombination) followed by mitosis, producing **eight haploid ascospores**, arranged in an ascus

SEXUAL CYCLE OF NEUROSPORA. Classic studies show sexual process has played a vital role in the development of biochemical genetics as in ***Neurospora crassa***. Neurospora contains seven different types of chromosomes each being a single body. The vegetative organism is haploid showing conidia formation; and asexual reproduction but in the stage of sexual reproduction, two cells and, hyphae (or conidia) of different mating types fuse to form a ***dikaryotic*** cell. The two kinds of haploid nuclei coexist in the same cytoplasm and divide in synchrony for a time. At the initiation of an ascus, **two different haploid nuclei fuse** and form a diploid nucleus, containing pairs of *homologous* chromosomes.

The **diploid cells** then initiate the process of mitosis. The homologous chromosomes pair with each other (synapse) and assume parallel, closely adjacent positions and each chromosome divides without duplication of centromere, whose attachment to spindle fibres guides its migration to one or the other pole during the stage of **anaphase**. Each chromosome thus becomes a pair of identical *Chromatids* connected by a centromere (bivalent) and the homologous bivalents form a ***tetrad*** by a single crossing over event. One or more exchange of parts (crossing over) may then occur at random among the 4 chromatids of the tetrad resulting in **genetic recombination**.

Two meiotic divisions then follow. In the **first meiotic division** there are no divided centromere separates; instead one member of each pair of homologous chromosomes (in the form of a bivalent) is drawn after its centromere to each of the daughter nucleus.

In the **second meiotic division** the chromosome has already divided into chromatids and only the **centromere divides**; one product then migrates to each pole drawing its chromosomes with it. Thus four chromatids of each tetrad are distributed to 4 different cells, each with **haploid chromosomes**. The individual members of each set are thus divided at random from either parent with genetic recombination at the tetrad stage. Here 4 **haploid cells resulting from mitosis** remain together **in the same sac (ascus)** and in many ascomycetes as in neurospora before the ascus is fully mature, each cell divides mitotically in two identical spores. Asci are shaped as narrow pods, i.e., ascospores in an ascus held in a linear array, the order reflecting the mitotic segregation of their chromosomes.

The analysis of genetic combination of four pairs of spores in the same ascus is called ***tetrad analysis*** which gives a possible description of the genetic events occurring during meiosis, and the products of reciprocal recombination can be identified from the distribution study of genetic markers amongst the progeny. Haploid nature of the vegetative phase eliminates the complicating effect of dominance on the relation between genotype and phenotype.

PARASEXUAL CYCLE (mitotic recombination). In some fungi a cycle is seen that imparts some biological advantages of sexuality in the form of recombination of parental DNA without involvement of specialised gametes or mating types. This process is known as **PARASEXUALITY**, which was first demonstrated by Pontecorvo in Aspergillus. The process involves the following steps.

(1) **HYPHAL FUSION** in which different dividing haploid nuclei coexist in a common cytoplasm. A *heterokaryon* is thus formed and 2 sets of nuclei divide at about the same rate.

(2) **RARE NUCLEAR FUSION** may produce *heterozygous* diploid nuclei. These are generally outnumbered by haploid nuclei but once formed they tend to multiply at about the same rate in the haploid nuclei and diploid strains may be isolated.

(3) Though **HOMOLOGOUS CHROMOSOMES** are usually arranged independently on the equatorial plate in the mitosis of a diploid cell, rarely (about 10 per mitosis) sufficient ***somatic pairing*** **occurs to permit crossing over** as in meiosis.

PLATE 1.1
MYCELIAL FORMS

Fig. 1 : Hypha

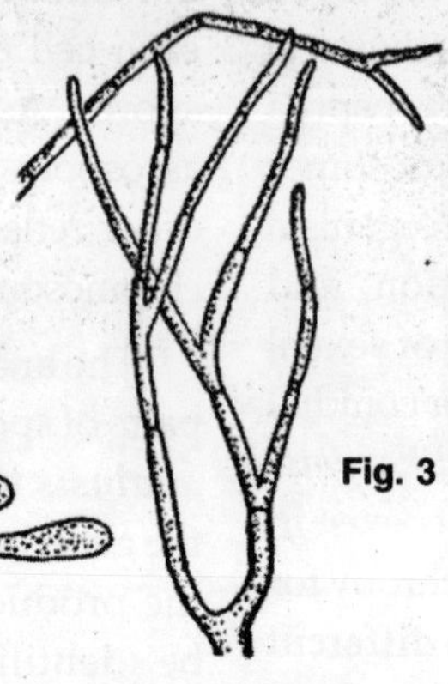

Fig. 2 : Mycelium

Fig. 3 : Coenocyte

Fig. 4 : Rhizoids

Fig. 5 : Clubs

Fig. 6 : Racquets

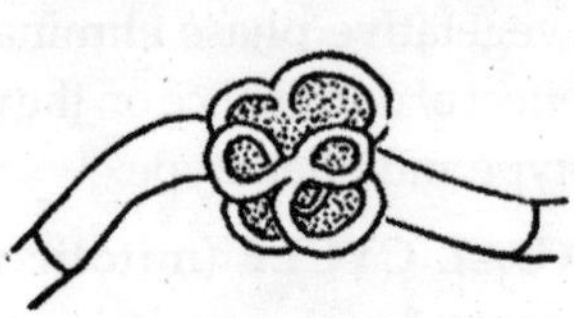

Fig. 7 : Nodular organ

Fig. 8 : Pectinate body

Fig. 9 : Spiral

SEXUAL SPORES

Fig. 10 : Zygospore

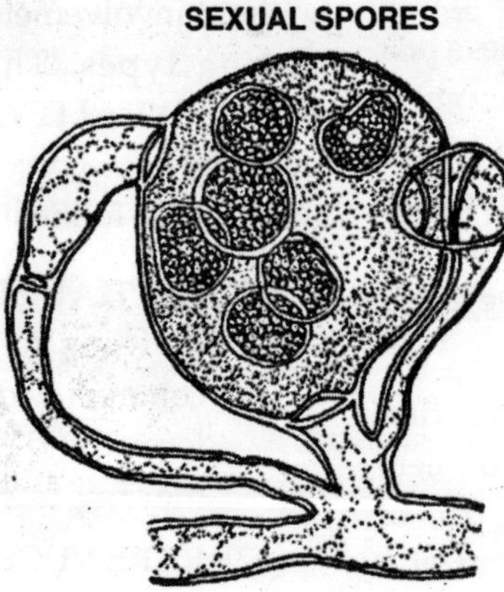

Fig. 11 : Oospore

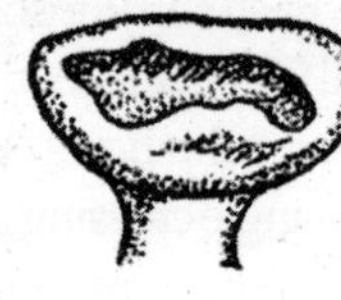

Fig. 12 : Ascocarp

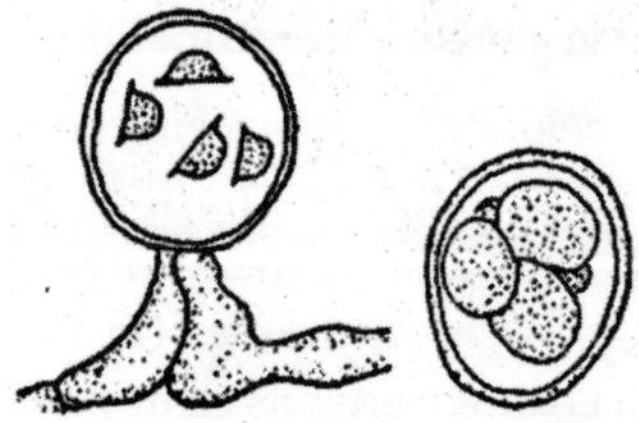

Fig. 13 : Ascospores in asci

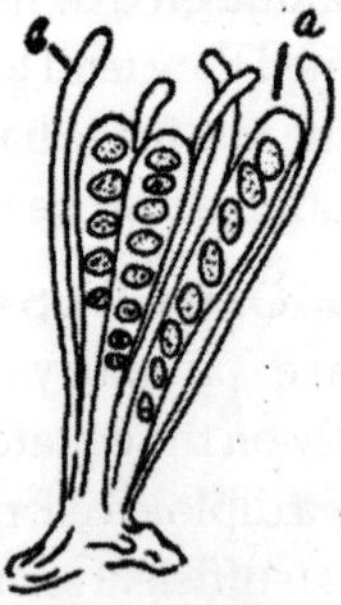

Fig. 14 : (a) Ascospore (b) Hymenium

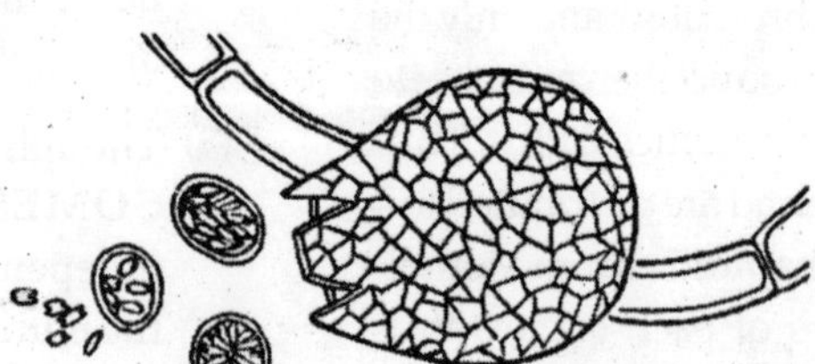

Fig. 15 : Perithecium

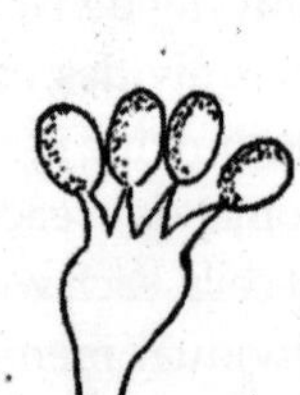

Fig. 16 : Basidiospores

PLATE 1.II
ASEXUAL SPORES

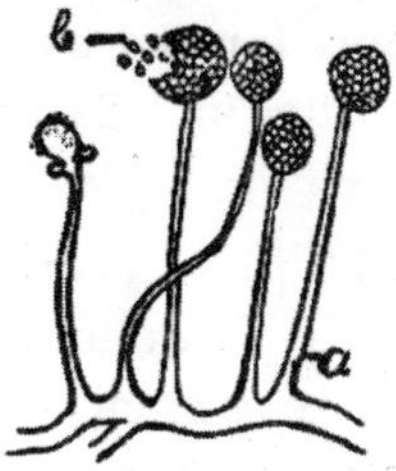

Fig. 17
(a) Sporangiophore
(b) Sporangiospore

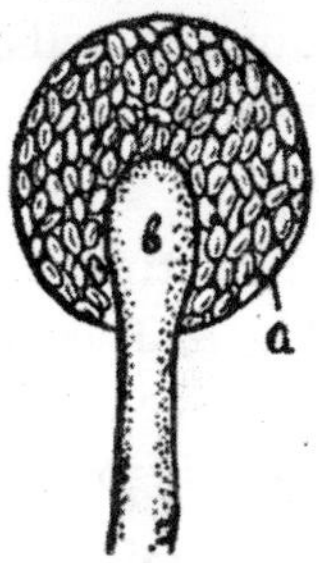

Fig. 18
(a) Sporangiospore
(b) Columella

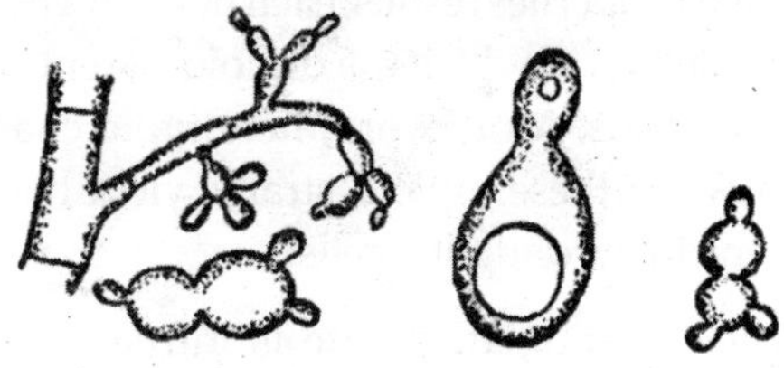

Fig. 19 : Blastospores

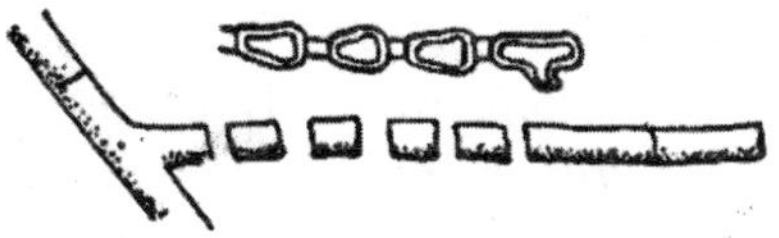

Fig. 20 : Arthrospore

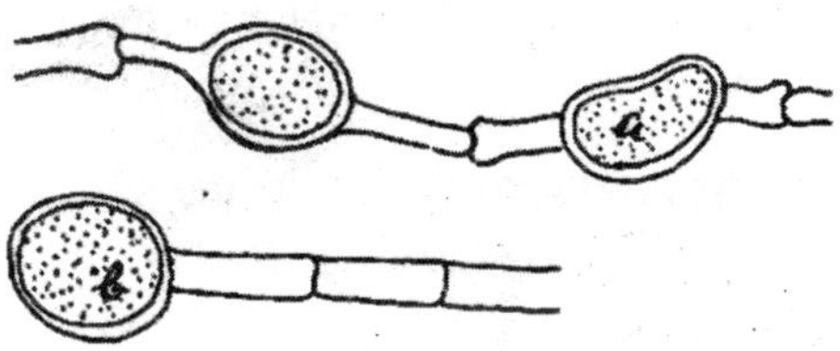

Fig. 21
(a) Intercalary chlamydospore
(b) Terminal chlamydospore

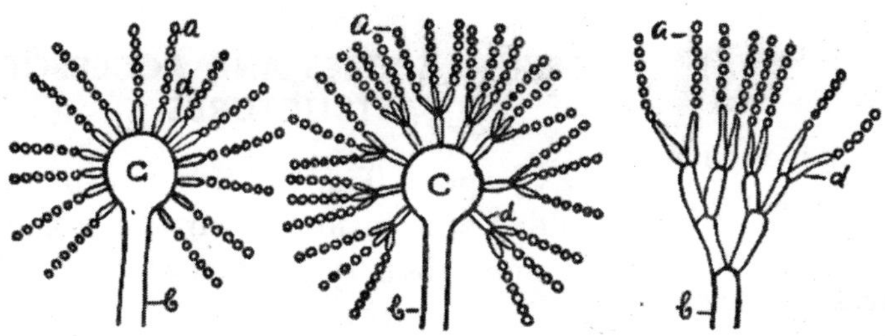

Fig. 22
(a) Conidia vera (b) Conidiophore
(c) Columella (d) Sterigma

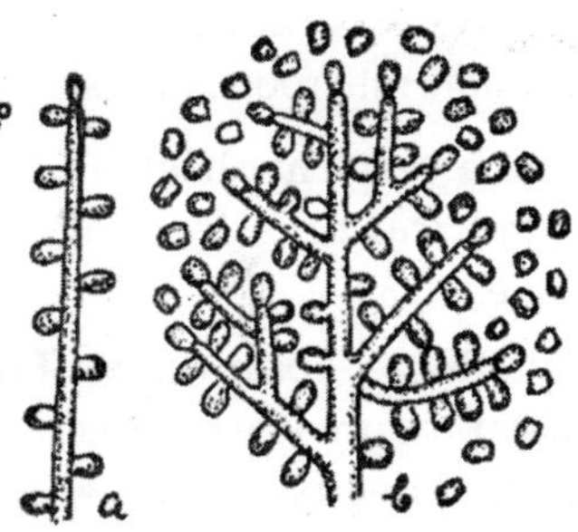

Fig. 23 : Microconidia
(a) Thyrse (b) En grappe

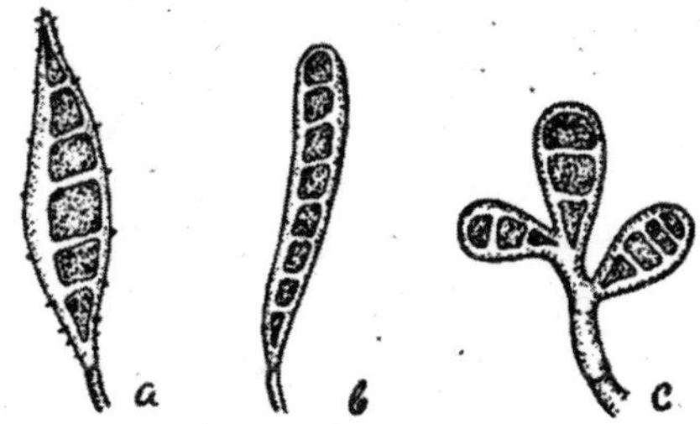

Fig. 24: Macroconidia
(a) Pointed (b) Cylindrical
(c) Blunt-end

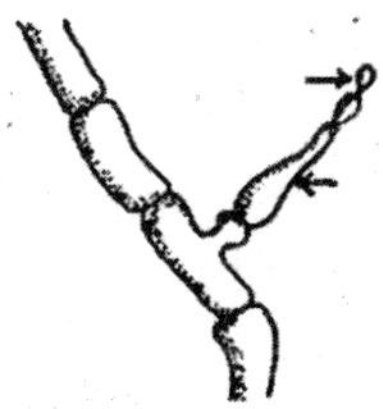

Fig. 25: Phialide
with phialospore

The result of mitotic recombination between heterozygous homologous chromosomes makes the products *homozygous for genes* distal to the point of exchange. Thus two diploid daughter cells with different properties results, each homozygous for some alleles for which the diploid parent cell is heterozygous. Moreover, phenotypic changes occur when these diploid strains yield haploid progeny in the daughter cells.

The recombination rate is infrequent in a parasexual cycle compared with sexual cycle but with a limited range of markers available, the process may be significant for the evolution of asexual fungi.

DEFINITIONS OF TERMS

(*For figures refer to plate 1.I*)

Mycetes — A fungus.

Mycosis — (pl. Mycoses). It is a disease caused by fungus.

Thallus — The entire vegetative apparatus of the fungus is called a thallus. It does not differentiate into roots, stems or leaves.

Hypha — (pl. hyphae). (**Fig. 1**). It is a chain of cylindrical cells forming a filamentous structure.

Mycelium — (pl. mycelia). (**Fig. 2**). It is a felt-like mass produced by a collection of intertwined hyphae, collectively known as mycelium. Mycelia may be surface runners or they may be erect, forming aerial hyphae which often bear spores (sporophores).

Coenocyte — (**Fig. 3**). Multinucleated mass of protoplasm formed by division of nuclei but without any division of cytoplasm; e.g., Phycomycetes.

Rhizoids. (**Fig. 4**). Mycelium, in certain cases, is provided with root-like structures called rhizoids; e.g., Rhizopus.

Club. (**Fig. 5**). The swollen portion of a hypha is called a club.

Mycelial racquets. (**Fig. 6**). These are racket-shaped cells along the course of a hypha, in which the clubbed end of one cell is attached to the thin end of the next cell. Mycelial racquets are often arranged in series.

Nodular organs. (**Fig. 7**). These are large, rounded cells, like knots formed by a few chlamydospores with radiating mycelia from the knot.

Pectinate bodies (Hyphae pectinate). (**Fig. 8**). These are swollen hyphae often curved and give off a row of abortive branches from one side roughly resembling a comb.

Spirals. (**Fig. 9**). These are simple hyphae which develop into closely set spirals or coils. They simulate tendril like apparatus of higher plants.

Spores. Reproductive body of a cryptogam is called a spore. The spores may be sexual or asexual. Sexual spores are derived from the union of two gametes (Perfect stage) but asexual spores arise from fertile hyphae (Imperfect stage).

SEXUAL SPORES AND ACCESSORY STRUCTURES

Zygospore. (**Fig. 10**). It is a sexual spore resulting from fusion of two similar gametes (isogamous) as in zygomycetes.

Oospore. (**Fig. 11**). It is a sexual spore arising from fertilisation of a female cell by a male cell (two dissimilar gametes), antheridium and oogonium, e.g., Oomycetes in Phycomycetes.

Ascocarp. (**Fig. 12**). Any fruit body containing asci is known as ascocarp.

Ascospores. (**Figs. 13, 14**). These are sexual spores 4 or 8 in number enclosed in a membrane known as *ascus* (a *cask* or *case*). The ascus is usually cylindrical but in yeasts it is oval and is formed by fusion of the nuclei of two contiguous cells. The fusion is followed by two or three successive divisions resulting in 4 or 8 nuclei which subsequently form 4 or 8 spores by cutting out cytoplasm around the nuclei. Ascospores are characteristic of the class Ascomycetes.

Hymenium. (**Fig. 14b**). It is a tissue formed by a layer of spore-producing structure found in the fruit bodies of Ascomycetes. The sterile hyphae interposed in the hymenium are known as *paraphyses*.

Perithecium. (**Fig. 15**). It is a globose fruit body containing organs of fructification, interposed with fertile hyphae. A perithecium is enclosed in a loosely woven hymenium and shows asci and ascospores in it.

Apothecium. When the fruit body of an Ascomycetes is wide open like a cup, it is called apothecium, e.g., Dyscomycetes.

Sclerotia. These are composed of compact, hard tissue formed by a close network of hyphae and capable of remaining dormant for a long period under unfavourable conditions. From this a new mycelium germinates under favourable conditions and bears an organ of reproduction, e.g., sclerotia of *Claviceps purpurea*.

Basidiospores. (**Fig. 16**). These are sexual spores 4 in number, held by peg-like structures at the end of a club-shaped spore-bearing organ known as *basidium* (Pl. *basidia*). These are characteristic spores of the class Basidiomycetes.

ASEXUAL SPORES AND THE SPORES-BEARING APPARATUS

Asexual spores may be endogenous, e.g., sporangiospores, or exogenous, e.g., conidia.

Sporangiophore. (**Fig. 17**). The fertile hypha which bears a sporangium (spore-case) is called a sporangiophore.

Sporangiospore. (**Fig. 18**). These are asexual spores formed endogenously in a spore-case, called sporangium, e.g., Mucor.

Columella (small pillar; **Fig. 18b**). It is the terminal dome-shaped swollen part of a sporangiophore bearing a sporangium.

The system of classification of spores by Saccardo is not mentioned here. It is preferable to follow the system of Vuillemin as modified here and mentioned below.

Thallospores. Spores arising from *thallus* (vegetative part of a fungus) are called thallospores. These are subdivided into the following types :

(a) **Blastospores.** (**Fig. 19**). These are spores arising from the vegetative hyphae by a process of budding as in budding yeasts.

(b) **Arthrospores.** (**Fig. 20**). Mycelia when old, show segmentations by transverse double septa at close intervals. These segmented mycelia are either cubical or rounded and spores form in them without swelling of the mycelium.

(c) **Chlamydospores.** (**Fig. 21**) They are formed by swelling of **mycelial filaments**. The process is associated with condensation of the protoplasmic substance often at the cost of the neighbouring cells with formation of a thick cell-wall with double contour. They are often brown and may be *intercalary* along the course of the hypha (**Fig. 21a**); or *terminal* at the end of a hypha (**Fig. 21b**).

Conidiophore. (**Fig. 22b**). It is a fertile hypha bearing conidia.

Conidia. There are several types of conidia. True *conidia* or *conidia vera* (**Fig. 22a**)as a contrast to thallospores are *borne at the terminal* part of a *fertile hypha* - the conidiophore. These are produced by *abstriction of the fertile hyphae*. The conidia may be *basipetal* or *acropetal*. In the *basipetal type*, as in the case of *Penicillium*, the *oldest conidia are at the terminal part* and become detached at the slightest disturbance; in the *acropetal type*, the oldest cell is in the proximal part of the fertile hypha.

Aleuriospores (**microconidia**, conidia). These are small spores about 3 to 4 μ in diameter, borne of fertile hyphae. These are sessile. *Unlike conidia*, they are *liberated only after disintegration of mycelia*. According to their arrangements they are said to form thyrse, i.e., *arranged laterally* (**Fig. 23a**) as in ***Trichophyton rubrum***, or ***en grappe***

Table 1.3 : Classes of Eumycota

Class	Asexual spores	Sexual spores	Mycelia	Representative genera
I. Zygomycetes	Endogenous (in sacs)	Variable	Nonseptate	Rhizopus, mucor, watermolds
II. Ascomycetes	Exogenous	Ascospores (within sacs)	Septate	Neurospora, true yeast, some penicillium and Aspergillus
III. Basidiomycetes	Exogenous	Basidiospores	Septate	Mushrooms, rusts, smuts
IV. Deuteromycetes (Fungi Imperfecti)	Exogenous Macro-and micro-conidia in Dermatophytes	ABSENT	Septate	Most human pathogens

(**Fig. 23b**) as in *T. interdigitale* and *T. mentagrophytes*. These microconidia may be spherical, subspherical, or ovoid.

Macroconidia (Fuseaux). These are spindle-shaped, large conidia, either terminal or lateral. They are nonseptate when young, but pluriseptate when old. They may be pointed, cylindrical or with blunt ends; (**Fig. 24a, 24b, 24c**). These are characteristic of *dermatophytes*.

Sterigma. (**Fig. 22d**). These are small stalks, bearing conidia, often in chains.

Phialide. (**Fig. 25**). These are flask-shaped stalks, borne on sporophores and bear conidia.

Phialospores. (**Fig. 25**). Conidia borne on phialides are called phialospores.

Pycnidia. These are minute, globose or flask-shaped fruit body with apical osteoles, internally lined by conidiophores as in Sphaeropsidales. Conidia are discharged through the osteole in the plant body which they parasitise.

Corremia. These are stalked bodies of cylindrical shape, formed by a cluster of conidiophores which run parallel in a fascicular form, e.g., Stilbaceae.

Sporodochium. It is a sessile, globose fruit body without stalks and contains a collection of conidiophores.

Acervuli. These are fructifying bodies containing conidiophores, packed together, occurring in clusters or heaps and form a saucer-like mass on the plant host; e.g., Melanchoniales.

CLASSIFICATION OF FUNGI

(See also Chapter 9)

I. Mycelium broad, nonseptate, coenocytic, spores usually borne in a sporangium. Sexual fusion results in gametes, gametangia: oospore or a thick walled zygospore. Class: **ZYGOMYCETES**.

II. Mycelium septate, dimension of mycelia more than 1μ.

A. Reproduce by Ascospores, which are sexual spores. The sexual fusion results in formation of an ascus. The ascus contains 4 or 8 ascospores. Class: **ASCOMYCETES**.

B. Reproduce by basidiospore, i.e., sexual fusion results in formation of a club shaped organ called basidium on the surface of which 4 basidiospores are borne by peg-like structures attached to the basidium. Class : **BASIDIOMYCETES**.

C. Reproduce by asexual spores; sexual spores unknown. Class: **DEUTEROMYCETES (Fungi Imperfecti); Hyphomycetes**.

III. Vegetative body 1μ or less in diameter and multiplies by binary fission.

They do not usually form mycelia. Class: SCHIZOMYCETES. (Bacteria, Actinomyces, Nocardia, Streptomyces)

Classification of Eumycotina with their spore characters and reproductive organs are shown in Table 1.3.

SUBCLASSES OF ASCOMYCETES

1. **Asci produced from oogonium** or **by multiplication of diploid cells.**
   Ascogenous hyphae absent — Subclass : **PROTO ASCOMYCETES**
2. **Asci borne on fertile ascogenous hyphae.** (Euascomycetes)
   A. **Asci in a closed perithecium** — Subclass : **PLECTOMYCETES**
      Order : **PLECTASCALES***
   B. **Asci in a hymenium** :
      (i) in an **apothecium** — Subclass : **DISCOMYCETES**
      (ii) in a **perithecium** — Subclass : **PYRENOMYCETES**

CLASS I. (PHYCO) ZYGOMYCETES

Phycomycetes. These show nonseptate, coenocytic vegetative mycelia and reproduce by both sexual and asexual spores. They are divided into three subclasses. The sexual reproduction is by means of gametes, gametangia, oospores or zygospores.

(i) **ArChimycetes**. These are primitive forms and found as parasites of water plants and animals but may sometimes cause diseases of land plants. They mainly reproduce by motile, flagellate spores called *zoospores*. In some cases they form oospores.

(ii)` **Oomycetes**. This subclass is characterised by formation of two dissimilar gametes in their sexual life. The oogonium is fertilised either by a motile, male sperm or by a specialised nonmotile male germ tube. The oogonium or the female organ, when ripe, shows aggregation of protoplasm into one or more *oospheres*. After fertilisation the oospheres round up to form *oospores* which often lie free in the oogonium (**Fig. 11**).

(iii) **Zygomycetes**. The sexuality of this group of fungi is characterised by the formation of similar gametes and is called *isogamous*; the process is known as *conjugation* and the product of union is called a *zygospore* (**Fig. 10**). In the union the ends of two mycelia become swollen, come in contact with each other and the terminal parts of the hyphae are segmented by a transverse wall. The similar terminal cells are the conjugating cells (gametes). The walls between the conjugating cells disappear and the protoplasmic masses of both the cells unite forming a spore called *zygospore*. A zygospore has a double wall which is thick and rough. Under favourable conditions a germinating tube called a *promycelium* develops from a zygospore bearing a sporangium, containing spores. This sporangium is similar to one found in the asexual spores. The classification of Phycomycetes is shown below :

SUBCLASSES OF ZYGOMYCETES

Mycelium absent Subclass : ARCHIMY-CETES Mycelium nonseptate (coenocytec)

A. Sexual reproduction heterogamous

Subclass : **OOMYCETES**

B. Sexual reproduction isogamous

Subclass : **ZYGOMYCETES** (Phycomycetes)

*The family Gymnoascaceae fall in this Order and the dermatophytes fall into this group (see chapter 9 also).

CLASS II. ASCOMYCETES

Ascomycetes. The vegetative body of Ascomycetes may be single-celled (**yeasts**) **or both** yeastlike bodies and hyphae may be present. The hyphae are septate. Asexual spores or conidia are produced exogenously either from conidiophores or directly from hyphae (**blastospores; Pl. I, Fig. 19**).

The sexual life in Ascomycetes is characterised by formation of specialised, endogenous spores called ***ascospores***, borne on ***asci*** or *sacs*. This is the product of fusion of like or unlike sexual cells. The sexual process may not be hidden but naked during development; e.g., true *yeasts*. But in other cases they may be protected and hidden in a specialised structure called ***perithecium***. The class is divided into several subclasses as follows :

CLASS III. BASIDIOMYCETES

Basidiomycetes. The vegetative growths of these fungi show septate filamentous hyphae. They reproduce by basidiospores which are sexual spores. These are borne exogenously on special club-shaped organs called ***basidia*** (**Fig. 16**). In a typical case, a basidium bears four spores held by peg like structures attached to the basidium. Mushrooms (toadstools) are fungi of this group. Lower fungi of this group are **rust and smut fungi**.

CLASS IV. DEUTEROMYCETES (Fungi Imperfecti)

Fungi imperfecti. The fungi of this group do not show any sexual spore, i.e., the life cycle is imperfect or better called imperfectly known. The sexual life of these fungi has not been discovered and they are therefore called Fungi Imperfecti. In the words of Dodge "The Fungi Imperfecti are conceived as a dumping ground until the gaps in our knowledge have fulfilled." In the sexual life they appear similar to Ascomycetes and at least some of them have been found to produce ascospores under special conditions of growth. In such a case the fungus should be transferred as Ascomycetes when it would show formation of asci. These fungi are therefore classified on the basis of asexual spores or conidia which are relatively specific for a particular group of organisms. The spores may be hyaline or dark, unicellular or multicellular. The wall of the spore may be smooth or rough, thin or thick. According to the system of Saccardo (1884) Deuteromycetes (Fungi Imperfecti) are divided into three orders as follows :

Order : **Melanchoniales**. Conidiophores form *acervuli* in the substrate.

Order : **Sphaeropsidales (Phomales)**. Conidiophores form cup-shaped receptacles called *pycnidia*.

Order : **Hyphomycetales (Moniliales)**. Conidiophores are free and do not form fruit bodies.

Order : **Pseudomycetales**. These are yeast-like fungi with or without pseudo or true mycelium.

Order : **Mycelia sterilia**. The members of this group do not produce any characteristic spore. These are often considered pleomorphic.

Fungi producing diseases in man and animals belong to the group : Hyphomycetales and a medical mycologist is specially interested in this particular group of fungi.

Moniliales or Hyphomycetales have been further divided into the following families :

(1) **Moniliaceae** or Mucedinaceae-conidiophores and **conidia are hyaline** and bright in colour.

(2) **Dematiaceae** – mycelia and **conidia are dark** but they may sometimes be hyaline on dark conidiophores.

(3) **Stilbaceae – conidiophores are fasciculate**, parallel and form ***corremia***.

(4) **Tuberculariaceae**. Conidiophores form a globous, sessile fruit-body called ***sporodochium***.

Fungi imperfecti that are of medical interest belong to two families, viz., Moniliaceae and Dematiaceae (dark spores) or the order Moniliales.

Further classification of Saccardo is based on the character of spores wherether the spores are one-celled or two-celled or whether the spores show cross septa or both longitudinal and cross septa (muriform), etc.

SPORE GROUPS IN MONILIALES

(After Saccardo)

The genera and species of Moniliales are classified on the basis of *shape, colour* and *septation* of spores which are as follows :

Amerosporeae, globose, ovate or fusiform (**one celled spores**)	(a) *Hyalosporeae :* hyaline conidia
	(b) *Phaeosporeae :* dark conidia
Didymosporeae	(a) *Hyalodidymae*
(*two celled spores*)	(b) *Pheodydidymae*
Phragmosporeae	(a) *Hyalophragmeae*
(*two or more septa*)	(b) *Phaeophragmeae*
Dictyospoeae (Spores muriform, with both ***longitudinal and cross septa***)	(a) *Hyalodictyosporeae* (b) *Phaeodictyosporeae*
Scolecosporeae (***vermiform***)	
Helicosporeae (***Spirally curved septate spore***)	*Hyalohelicosporeae*
Staurosporeae ***spore forked or shaped***)	*Hyalostaurosporeae*

Dematiaceae (Spores dark). The classification is like that of the Mucedinaceae : but the various names of sections having the **prefix "phaeo" instead of "hyalo"**.

Each of the spore section is then split into two groups :

1. **Conidiospores short** or obsolete, hardly distinguished from mycelia : *Micronemeae.*
2. **Conidiophores elongate** and distinct-*Macronemeae.*

Within these groups different Sub-families and genera are classified according to the arrangement of spores on the conidiophores, whether solitary or in heads, produced directly on the conidiophores or on specialised branches, arranged in cluster or in chains and so on.

For a medical mycologist, it is convenient to classify **Fungi Imperfecti** according to the classification of Vuillemin instead of one followed by Saccardo. Vuillemin (1910) divided Fungi Imperfecti into three orders on the basis of character of the spores and the fourth order **Microsiphonales** was later included (1912). According to a later classification by Vuillemin (1925), the order **Microsiphonales (Actinomycetales)** was included as a family. According to the Bergey (1948), the order Microsiphonales had been included in the order Actinomycetales under Schizomycetes. Vuillemin divided **asexual spores** into two main divisions namely **thallospores** and **conidia**. Conidia may be conidia vera and aleuriospores. Conidia vera are subdivided according to whether they are borne on sterigma or on well-defined conidiophores. The conidiophores have further been differentiated into phialides. The spores borne on phialides have been called *phialospores*. He also recognised *hemispores* which structurally occupy a position between arthrospores and true conidia. Aleuriospores are differentiated from true conidia by the fact that unlike true conidia they are not set free when mature but are liberated only after disintegration of mycelia. These are also called *microconidia* as differentiated from septate *macroconidia* which are larger and pluri-septate. **American workers** are in favour of **calling aleuriospores as *microconidia*** and it is better to call them microconidia **to differentiate them from conidia (conidia vera)** according to the original classification of Vuillemin so that one could be distinguished from the other. Spores arising from thallospores have already been described.

The classification of Fungi Imperfecti on the basis of Vuillemin's classification (1925) is as follows :

CLASSIFICATION

(After Vuillemin modified)

[See "Chapter 9" for detailed Taxonomy and systematic Classification.]

ORDER THALLOSPORALES; reproducing by thallospores.

Suborder *Blastosporineae*; reproducing by blastospores. This includes nonsporing yeasts like **cryptococcus (torula)**. Budding mycelia or pseudomycelia are seen in the genus Candida.

Suborder *Arthrosporineae*; reproducing by arthrospores; e.g., *Geotrichum candidum*, Madurella, Indiella.

Family *Mycodermaceae* : with usual septate mycelia showing elongated budding forms.

ORDER HEMISPORALES; reproducing by hemi-spores. Genus **Hemispora**; ***H. stellata***.

ORDER CONIDIOSPORALES; reproducing by conidia.

Suborder *Aleuriosporineae*; reproducing by aleuriospores. Ringworm fungi according to Ota and Langeron (1923) fall in this group.

Suborder *Sporotrichineae*; reproducing by true conidia but not borne by conidiophores e.g., **Sporotrichum** — ***S. Schenckii***.

Suborder *Sporophorineae* reproducing by true conidia, borne on conidiophores. Most of the common molds of the class **Fungi Imperfecti** belong to this group.

Suborder *Phialidineae*; reproducing by true conidia, borne on phialides — (flask shaped sterigma); e.g., **Aspergillus, Penicillium**, etc.

NOMENCLATURE

Each phylum is subdivided in a descending order into **classes, orders, suborders, families, tribes, genera** and **species**. Species have further been subdivided into strains. Note that **orders end in *ales*. suborders in *ineae*, families in *aceae*, subfamilies in *oideae*, tribes in *eae* and subtribes in *inae*.**

The name of an organism consists of two parts, i.e., the generic part which is a Latin substantive and always written with a capital letter; the second part is the specific name of the fungus which is usually adjective and is not written with capital letter; e.g., ***Penicillium purpurogenum***. In this Penicillium is the generic name and expresses **"broom-like" appearance**, but the specific name purpurogenum indicates a pigment which is purple in colour. Another example is ***Aspergillus flavus***. The world Aspergillus is the generic name. Aspergillum which is a kind of mip used for distributing holy water (Latin Aspergere-sprinkle); Flavus indicates the yellow colour. The species is also named after the name of the discoverer, e.g., ***Trichophyton schoenleinii***. While citing the name of a species, the name of the author who described it originally is often appended to the name of the species. But the name of the author is not written in italics. When the name is amended and a second name is introduced, the name of the original discoverer is put in parenthesis; e.g., *Trichophyton mentagrophytes*. It is amended from the name *Microsporum mentagrophytes* C. Robin, 1853 and expressed as *Trichophyton mentagrophytes* (C. Robin) Blanchard, 1896.

The sequential order of taxonomy is as follows :

Kingdom : In this animal or plant groups are divided, e.g., plant kingdom or **Plantae**.

Phylum : Main groups in which plant or animal kingdom is divided.

Class : Members having certain arbitrarily chosen characters in common are included under a phylum.

Order : Name applied to a closely related group of organisms, under a class.

Family : A group of organisms having similar characters under an order.

Genus : Related group of similar organism under a family.

Species : A basic unit in taxonomy, differentiating members of a genus.

SUMMARY

Cell Structure

Fungal cells are eukaryotic.

Fungal cells contain cells wall and cytoplasmic membrane, mitochondria Eukaryotic (true) nuclei, nuclear membrane. They divide asexually and or sexually.

Cell wall

Fungal cells possess a **rigid cell wall** comprising **chitin**, **glucans**, **mannans**, and **complex polysaccharides**.

Cell membranes

Cell membrane contains **ergosterol** that encloses cytoplasm mitochondria, vacuoles, reticular endothelium, etc. The nuclear membrane surrounds the eukaryotic (true) nucleus.

Fungi (Eumycetes) are classified in 2 ways as follow 3 : See also Chapter 9.

(A) Morphologically and (B) According to nature of sexual spores :

(A) Morphologically

1. **YEASTS :** Unicellular round or oval bodies, **reproduce by budding**, on artificial media form creamy, mucoid, compact colonies : e.g. : *Cryptococcus neoformans*.
2. **YEAST LIKE FUNGI :** Grow partly as yeasts and partly as pseudomycelia (long chains of elongated filamentous cells joined end to end) which give rise to yeast cells ***by budding*** : e.g. : ***Candida albicans***, and **Trichosporon**.
3. **FILAMENTOUS OR MYCELIAL FUNGI :** Grow as long filaments which **interlace, reproduce both by sexual and asexual spores**, and in artificial media they grow as filamentous colony with a

ORIGIN OF FUNGI

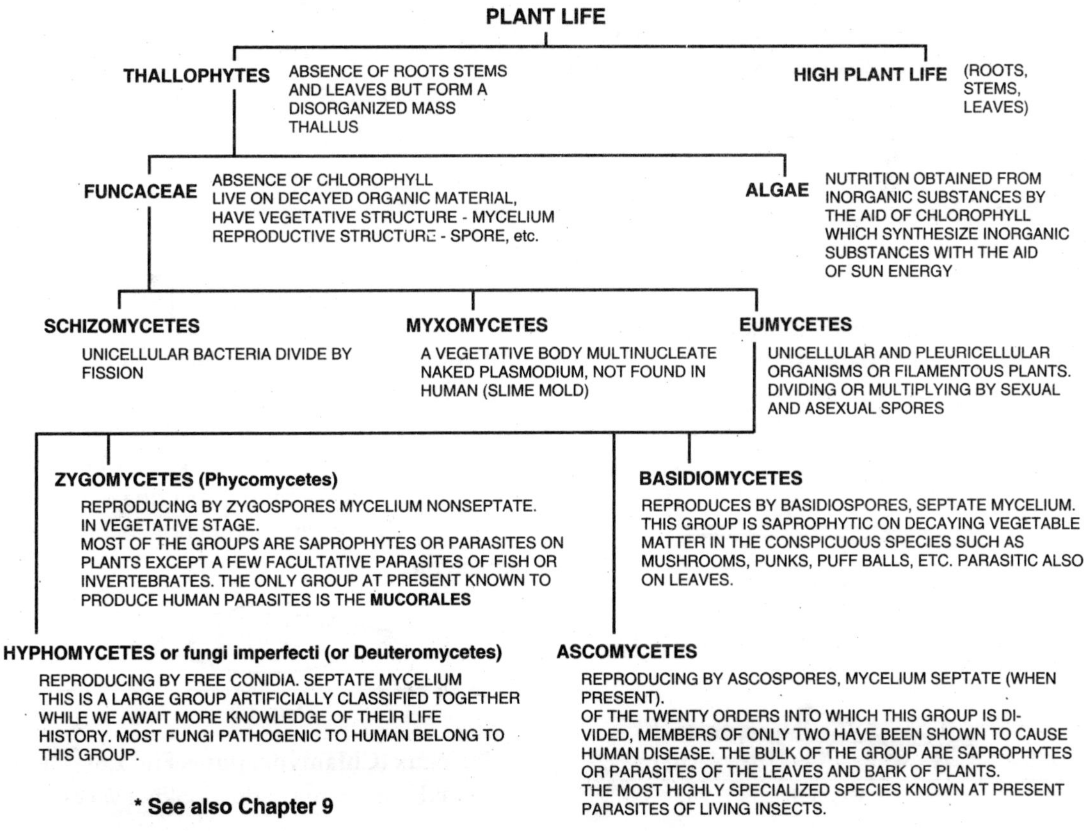

*** See also Chapter 9**

granular surface due to spores : e.g. : **Ringworm fungi, Aspergillus, Penicillium** and **Mucor**.

4. **DIMORPHIC FUNGI :** Grow **both** as **filamentous and yeast forms;** depending on environmental conditions : e.g., **Histoplasma; Blastomyces, Coccidioides and Sporothrix.**

(B) According to sexual spores; (also see later)

1. **Zygomycetes :** With non-septate hyphae; reproduce **SEXUALLY BY 'ZYGOSPORES'**, and **ASEXUALLY BY SPORANGIOSPORES;** e.g. **Rhizopus** and **Mucor (Phycomycetes).**
2. **Ascomycetes :** Septate hyphae, sexual **'ascospores' formed in a sac (ascus), ASEXUALLY REPRODUCE** by various methods e.g. : **Penicillium, Aspergillus.**
3. **Basidiomycetes : Septate hyphae, SEXUAL BASIDIOSPORES** with club shaped basidia; and various **ASEXUAL SPORES;** e.g., **Mushrooms.**
4. **Deuteromycetes (Fungi imperfecti) :** Septate hyphae, **sexual form** of reproduction **not known (no perfect state), REPRODUCE BY ASEXUAL SPORES.**

DEFINITIONS

Thallus : The entire vegetative portion of the fungus.

Hypha : A chain of cylindric or club-shaped cells forming a filament. And

Mycelium and (Fig. 1.2) : A mass or collection of hyphae.

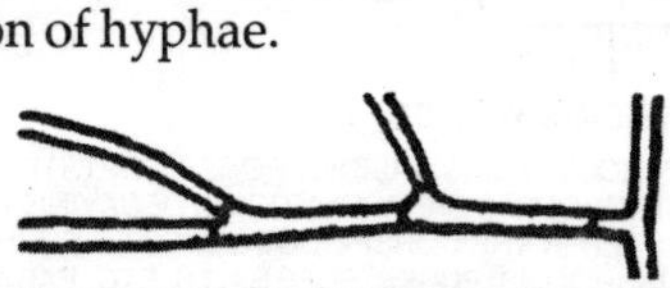

Fig. 1.2

Sporophore (Fig. 1.3) : A specialized hypha to which spores are attached.

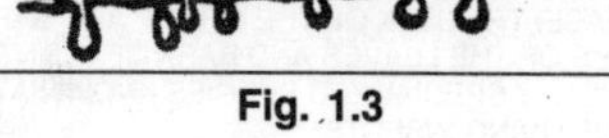

Fig. 1.3

Strerigma (Fig. 1.4) : A specialized hyphal cell to which spores are attached.

Fig. 1.4

Clubs : Swollen hyphal tips, varying greatly in size.

Fig. 1.5

Spirals (Fig. 1.5) : Simple convoluted hyphae that may take all the forms of a tendril : from a spirillum-like form to a closely set coil.

Fig. 1.6A

Fig. 1.6B

Racquet Mycelium (Fig. 1.6A, Fig. 1.6B and Fig. 1.15) : Club-shaped cells, the clubbed end of one cell being attached to the small end of an adjacent cell.

Fig. 1.7

Chlamydospores (Fig. 1.7) : Thallospores formed by the concentration of the protoplasm of a hypha within a swollen portion of the filament, the membrane of which becomes thickened. They are purely resting spores and are closedly analogous in function to the spores of bacteria.

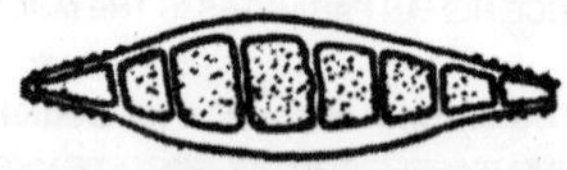

Fig. 1.8

Fuseaux (Chlamydospores Fig. 1.8) : Large elongated chambered bodies looking very much like chlamydospores and divided into segments by parallel transverse septa.

Endospore : Any spore formed within a membrane of the parent cell.

Fig. 1.9

Ascospores (Fig. 1.9) : A special class of endospores formed in a membrane known as the ascus. The number of spores in the sac is limited to two, four, or eight, and is constant for the particular species producing them.

Fig. 1.10

Basidiospores (Fig. 1.10) : An exospore on a special type of sporophore known as a basidium.

And

Thallospore : Cells that form a part of the vegetative portion of the fungus.

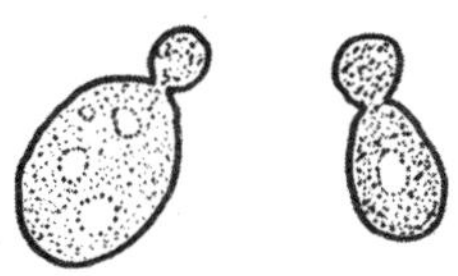

Fig. 1.11

Blastospore (Fig. 1.11) : A thallospore that develops by budding from the end or side of the parent cell and that may in turn throw out another bud or a mycelial filament, without becoming detached, and without any period of latency. The buds of yeast cells are familiar examples.

Fig. 1.12

Arthrospores (Fig. 1.12) : Structures formed by the segmentation of a hypha into a chain of cells, at first cubical, square or rectangular and later rounded.

Conidia : Cells of irregular shape and size, borne free and originating asexually from the mycelial threads by a process of budding, septation, or abstriction. They may be pedunculated, nonpedunculated, lateral, or terminal.

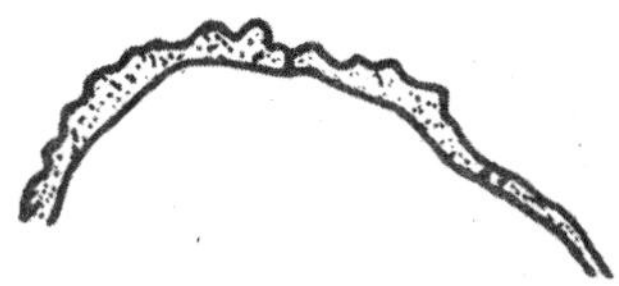

Fig. 1.13

Pectinate Bodies (Fig. 1.13) : Swollen and frequently curved ends of hyphae that give off a row of abortive branches from one side, the structure vaguely resembling a comb.

Fig. 1.14

Nodular Bodies (Fig. 1.14) : Large rounded cells knotted together, forming dense masses.

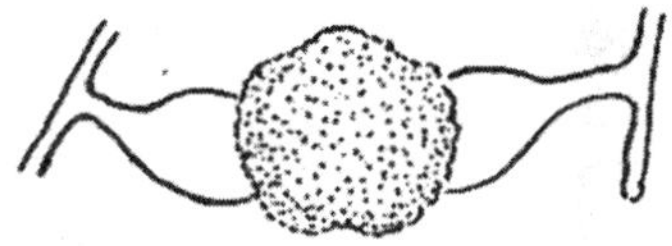

Fig. 1.15

Zygospore (Fig. 1.15) : A sexual cell produced by the fusion of two undifferentiated cells.

REFERENCES

Breed, R. S., Murray, E. G. D., and Hitchens, A. P. Bergey's Manual of Determinative Bacteriology, Sixth Edition, 1948. William & Wilkins Co., Baltimore.

Breed, R. S., Murray, E. G. D. and Smith, N. R. Bergey's Mannual of Determinative Bacteriology, Seventh Edition, 1957. William & Wilkins Co., Baltimore.

Dodge, C. W. — Medical Mycology, Mosby, St. Louis, 1935.

Georg, L. K. and Camp, L. B. - Routine nutritional tests for the identification of dermatophytes. *J. Bact.* ***74***; 113-121. 1957.

Vuillemin, P.; Mate'riaux pour une classification rationelle des Fungi Imperfecti. *Comp Rend., Acad. Sci.* **150** : 882-4, 1910.

Vuillemin, P; Les Champignons. Essai de classification. O. Doin et Fils, Paris, p. 236, 1912.

Vuillemin, P.; Classificatione normale, classement auxiliaire, et groupement pratique des champignons. *Comp. Rend., Acad. Sci.* **180** : 102, 1925.

2

Reproductive Biology of Fungi (Spores)

It occurs by spore formation. So, sporulations and the types of spores are equally important and associated with reproduction. Spores as dispersed for dissemination and propagation of a particular fungus. The hyphae continues to grow and branch and develops into a mat of growth, the mycelium (pleural = mycelia). Part of mycelia that penetrates into the nutrient is called a **vegetative mycelium** Mycelial part that projects up in the air is called an **Aerial Mycelium**. When the **Aerial mycelia develop** various **Reproductive** Spores it is called **a Reproductive Mycelia**.

Fungi reproduce both Asexually and Sexually. In the class Deuteromycetes (hyphomycetes, i.e., fungi imperfecti where the sexual stage is not known in many or undiscovered as yet) only exclusively asexual reproduction occurs. Apart from this, in many fungi where sexual reproduction occurs, asexual reproduction also can occur within the same genera but in different species.

ASEXUAL REPRODUCTION occurs by simple fission (division) of spores and involve mitotic process. There are chiefly **2 main types of Asexual Spores :**

I. *AERIAL SPORES*

(1) **Sporangiospores**, (2) **Conidiospores**, (3) **Phialospores (or Phialide)**

II. *VEGETATIVE SPORES*

(1) **Blastospores** and (2) **Arthrospores** together are called **THALLOSPORES**. (3) **Chlamy**dospores.

THE AERIAL SPORES

1. SPORANGIOSPORES (Fig. 2.1 & 2.2)

These spores are formed within a spherical sporangium. The sporangium is the spores case is borne by the sporangiophore. The sporangiophore is a mycelial structure that is erected up more or less at about 90° angle from the ends of mycelia.

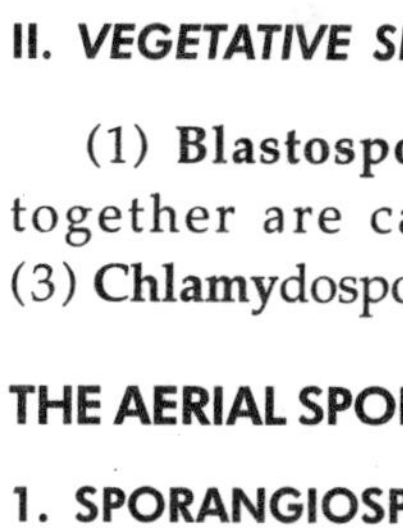

Fig. 2.1 :
(a) Sporangiophore
(b) Sporangium

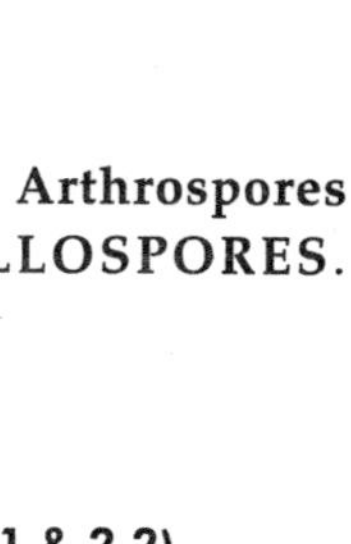

Fig. 2.2 :
(a) Sporangiosphore
(b) Columella

The classical example of sporangio-Spores-forming fungi are :

(a) Rhizopus and

(b) Mucor–Zygomycetes Class

(c) Coccidioidomycetes – Deuteromycetes Class.

(d) Rhinosporidium seeberi.

AT PRESENT IT IS NOT CONSIDERED AS A FUNGUS

2. CONIDIOSPORES

These spores are borne on Conidiospores or Conidia. Condia are specialised hyphal branches which form a swollen part. The comidio are also called Aleuriospores :

ALEURIOSPORES WHEN SMALL **are called** Micro-Aleuriospores or **Microconidia** that represents the **male** (gamet-like) **structure**. They are relatively smaller and made of single cells (Fig. 2.3).

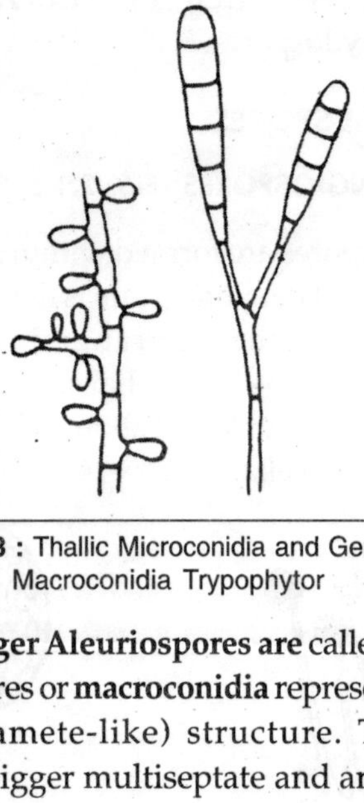

Fig. 2.3 : Thallic Microconidia and Genus : Macroconidia Trypophytor

The **bigger Aleuriospores are** called Macro-Aleuriospores or **macroconidia** representing the female (gamete-like) structure. They are relatively bigger multiseptate and are usually multicellular.

Examples of Conidiospores or Conidia belong to the class **Deuteromyces (Ringworm fungi)** :

(a) Microsporum Genus, (b) Trichophyton Geneus, (c) Epidermophyton Genus (d) Keratinomyces Genus (Epidermophyton) genus contains only Macroconidia **(Fig. 2.4)**. **Microconidia is always absent in the genus Epidermophyton** and hence Epidermophyton genus always shall remain compulsorily in the class Deuteromycetes whereby it will always reproduce by Asexual method of reproduction. The genus Epidermophyton contains only one pathogenic species, which is *E.floccosum*.

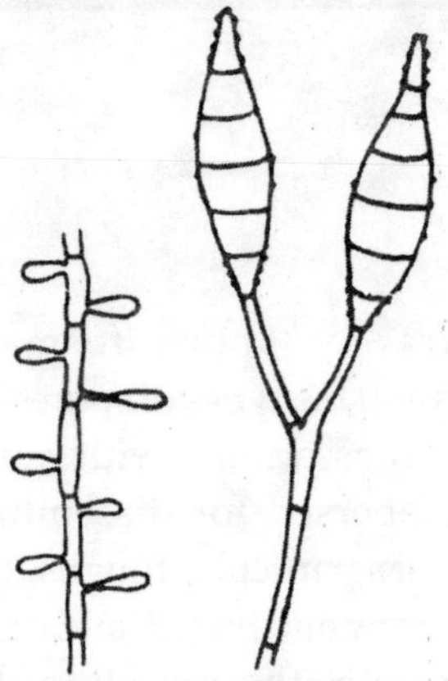

Fig. 2.4 : Micro and Macroconidia Genus : Microsporium

3. PHIALOSPORES

The **phialospores** are borne by the special structure **Phialoconidia** which is called a **Phialide**.

There are 3 chief morphological types of distribution of **PHIALOSPORES**; those are found in the **Genus Phialophora** (Formerly, Genus ***Hormodendrum*** and/or ***Cladosporium***). Most of the species of Phialophora cause subcutaneous mycosis under two divisions :

A.*EUMYCETOMA GROUPS :*

Phialmophora (formerly Genus Torula Torulopsis) *jeanselmei*.

B.**CHROMO BLASTOMYCOSIS** (Chromomycosis) **GROUP** produced by majority of the species.

The following **Table 2.1** shows the three **morphological types of Phialospores** in different species :

Table 2.1 : (Fig. 2.8, 2.9)

TYPICAL PHIALOPHORA	CLADOSPORIUM	ACROTHECA
Type (Sporulation) of Phialospores in organisms	Type (Sporulation) of Phialospores in organisms	Type (Sporulation) of Phialospores in organisms
1. *Phialophora dermatitides*	1. *Phialophora dermatitides*	1. *Phialophora dermatitides*
2. *P. verrucosa*	2. *P. (Cladosporium) Carionii*	
	3. *P. compacta*	
	4. *P. pedrosoi*	

The other organisms of **EUMYCETOMA** bearing conidia are :

(a) *Cephalosporium falciforme*

(b) *Phialophora jeanselmei*

(c) *Psedo-Alescheria* (*Pseudalescheria*) boydii, etc.

CONIDIA

Conidia are basic instruments **for Asexual propagation of conidiospores.** Conidia are made of **specialized conidiogenous cells.** Ontogenically, there are 2 basic types of conidia : (i) **THALLIC CONIDIA**, (ii) **BLASTIC CONIDIA**.

THE THALLIC CONIDIA

The Thallus (the fungal body) gives rise to **Thallic conidia (form the mycelial structure)**, e.g., (a) An **Arthroconidium** (a unicellular thallic conidium) **formed by the condensation of cytoplasm with the thickening of the hyphal cell wall.**

THE ARTHROSPORES

Particularly, in the case of *Coccidioides immitis* the Alternate Mycelial cells are transformed into highly condensed, small **rectangular conidia** called **Arthro(jointed) Conidia**. They often become dried and are ultimately detached from the mother (main) filamentous mycelial (hyphal) structure. These isolated dried pieces under suitable condition of moisture (humidity), temperature and necessary nutrient **Germinate** to form the thallus (Fungal body) from arthroconidia, i.e., **Arthrospores (Fig. 2.5)**.

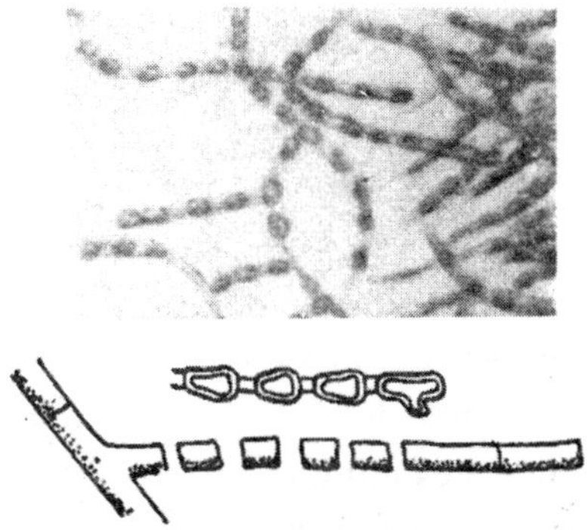

Fig. 2.5 : Arthrospore (Arthroconidia

In the case of Dermatophytes (Ringworm fungi) the **Thallic conidia develop** from the Terminal Hyphal cells which when is **unicellular** (made up of a single cell) is called a **Microconidia** (mentioned before); when the Thallic conidia is **multicellular, it is called Macroconidia.**

MICROCONIDIA IS ABSENT IN EPIDERMOPHYTON GENUS

So far Thallic conidia is discussed.

THE CHLAMYDOSPORES (Fig. 2.6)

The other Thallic vegetative spores are **Chlamydospores** which is an example of a **Thallic Conidia**. A chlamydospore may be (i) **Intercallary**, (ii) **Terminal** and (iii) **Lateral chlamydospores (thallic conidia)**.

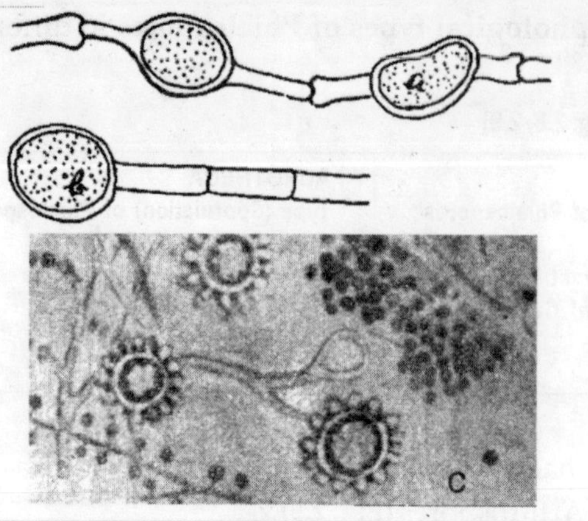

Fig. 2.6 :

(a) Intercalary chlamydospore
(b) Terminal chlamydospore
(c) Conidia and Lateral chlamydospores in culture

THE BLASTIC CONIDIA

A part of the **Conidiogenous cell** develops into **Conidia.** The blastic conidia may form into:
A. Holoblastic Conidia:
B. Enteroblastic Conidia.

Holoblastic Conidia in chains are produced by ***Cladosporium* (*! tachioides or bantianum*). *In ACROPETAL conidia (in chains) SPORULATION occurs as in Cladosporium.***

In case of **Enteroblastic Conidia** the inner layers of the cell wall of the **Conidiogenous** cells take part in the formation of **Enteroblastic Conidia**. In the species of the genus.

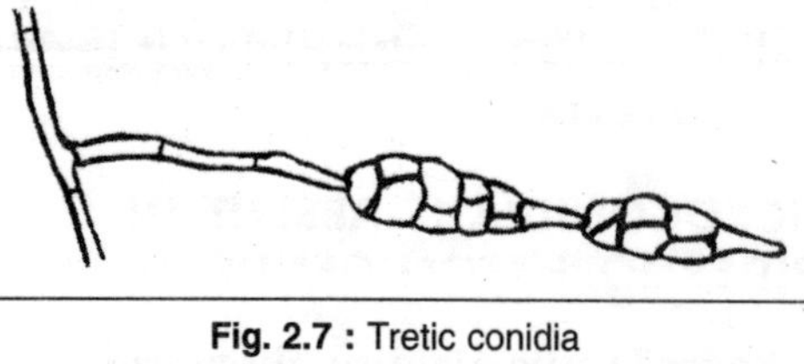

Fig. 2.7 : Tretic conidia

Alternaria a special type of **Conidia,** called **TRETIC CONIDIA** develops (**Fig. 2.7**). Another types of **Enteroblastic Conidia** and **Conidiphores** and **Chlamydospores** are formed in *candida* (**see figure in the Ch. CANDIDA see Fig. 13.1**). **Phialoconidia is an Enteroconidia. Phialides are flask shaped** condiogenous cells producing conidia through a distal opening in the cell wall. The **Phialide cell wall does not take part** in the formation of the Phialoconidia.

3. Phialoconidia : (PHIALOSPORES); Fig. 2.8 & 2.9, 2.10 and 2.11.

These are seen in the genera of (i) **Phialophora,** (ii) **Aspergillus,** and (iii) **Penicillium.**

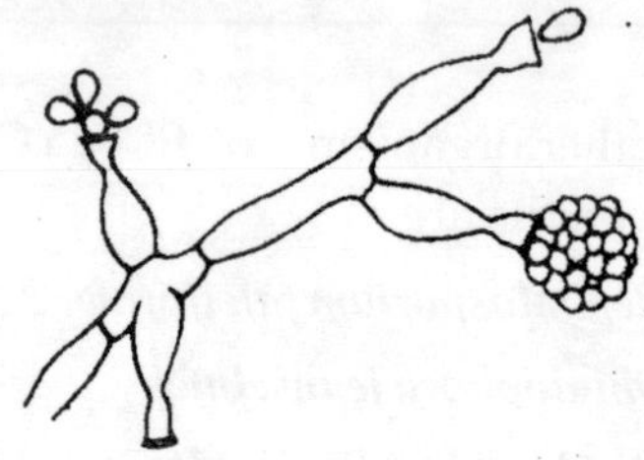

Fig. 2.8 : Phialide

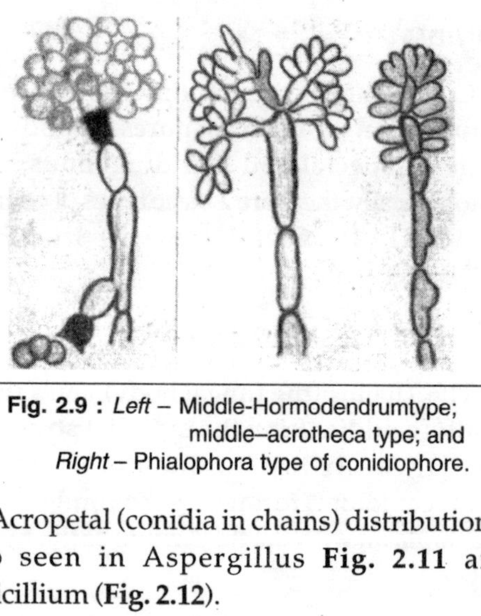

Fig. 2.9 : *Left* – Middle-Hormodendrumtype; middle–acrotheca type; and *Right* – Phialophora type of conidiophore.

Acropetal (conidia in chains) distribution is also seen in Aspergillus **Fig. 2.11** and Penicillium (**Fig. 2.12**).

4. Annello conidia

The name suggests that Ring-shaped scaus (Annelations) develop at junction of the conidiophores and conidia when the Enteroblastic conidiogenous cells elong etc.

Typical examples are:

(i) **Exophiala** and

(ii) **Scopulariopsis**

The simplest and most primitive types of colony and spore formation are found in the

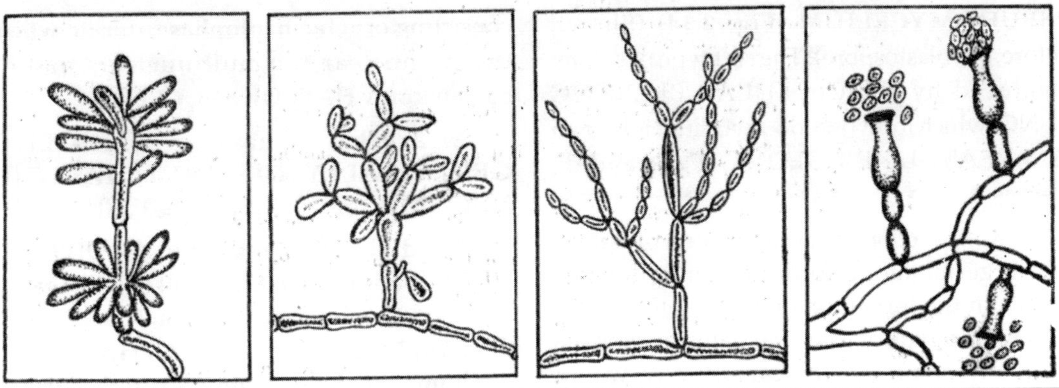

Fig. 2.10 : (a) Acrotheca type of sporulation (b) *Cladosporium* manner or sporulation in *Phialophora compacta* (c) *Cladosporium* type of sporulation in *Cladosporium carrionii.* (d) *Phialophora* type of sporulation

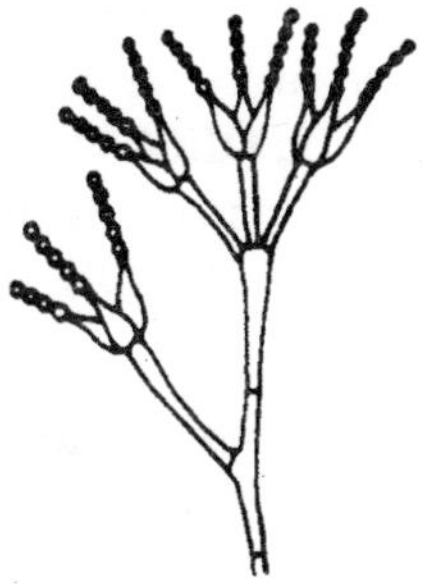

Fig. 2.11 : Aspergillus

Fig. 2.12 : Penicillium

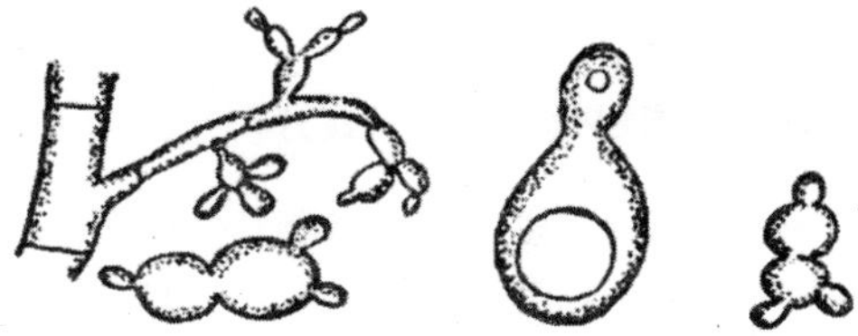

Fig. 2.13 : (a) Blastospores

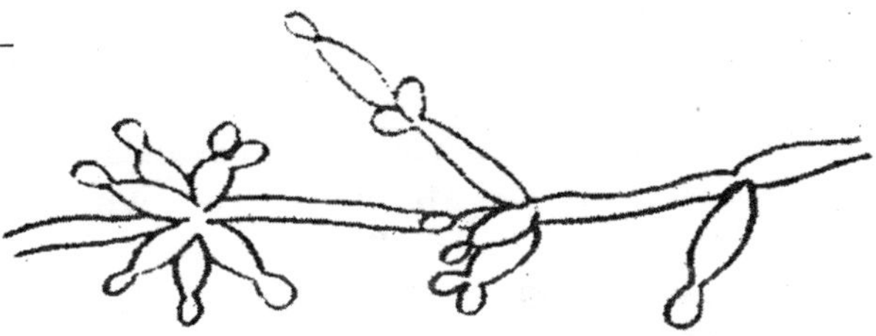

Fit. 2.14 : (b) Blastospores and Pseudomycelium

unicellular fungi, **CRYPTOCOCCUS** and **SACCHAROMYCES**. Both the fungi, when grown on glucose agar at room temperature, develop soft colonies The Cryptococcal cells are round and oval thin-walled cells (3 to 6 n in diameter). The **Cryptococcal cells** do not germinate (to form **germ tubes**and **mycelium**) but they reproduce by **BUDDING**. As the buds break off the parent cell, they enlarge and propagate by budding only. This is an Asexual type of **SPORE FORMATION BY BUDDING (NO FUSION OF NUCLEI** takes place). Such an Asexual spore is called a **BLASTOSPORE (Fig. 2.13)**.

In case of the genus Candida (producing soft, white colonies on the surface of glucose agar at room temperature) comprises oval, **THIN-WALLED CELLS WHICH ALSO REPRODUCE BY BUDDING**. The buds elongate and do not detach from the parent cells. By repeated budding form a branching network consisting of long chains of the attached cells, it is seen that the filaments are not of equal diameter along their length but are **CONSTRICTED AT THE POINTS OF SEPTATION**. The filaments are composed of **ELONGATED BUDDING CELLS** that fail to detach. Such hyphae are called **PSEUDO-HYPHAE** and the corresponding mycelium resulting from their formation is called a

PSEUDOMYCELIUM (Fig. 2.14) (Shown before, see blastospore). The soft type of colony produced by UNICELLULAR BUDDING FUNGI which form pseudo-mycelium is known as a YEAST-LIKE COLONY, e.g., candida species.

The production of true mycelium by spores which germinate to form branching hyphae results in the formation of a FILAMENTOUS colony. This type of colony usually develops as aerial mycelium that appears cottony, wooly, powdery, granular, and so on, and is familiar to most people as the type of growth associated with "molds".

SEXUAL REPRODUCTION

Sexual reproduction in fungi follows the same pattern as in higher eucaryotes. The Sexual process of reproduction starts with **(1) plasmogamy**, which means **two compatible haploid nuclei** within the cell are brought together.

(2) **Karyogamy** is the **fusion of these two nuclei** to form a ***diploid nucleus***. So, Karyogamy follows plasmogamy in a mycelium possessing binucleate cells. Then **MEIOSIS** occurs, with the process of genetic exchange, then reduction division and then mitotic or simple division occurs to yield four haploid progeny nuclei within that particular single cell. Some fungi produce distinct sex organs and gametes. In others, the somatic nuclei perform the sexual function. In some species, sexual compatibility is genetically determined, and compatible thalli are necessary for mating. Nuclei from two compatible thalli can also be brought together by **specialized gametangia** and/or also by anastomosis of the hyphae and the exchange of nuclei by migration from the one hypha to the other (another). Sequentially, hyphal fusion and migration of nuclei as well as cytoplasmic contents from one filament to another occur among the same species (of fungi).

PLASMOGAMY, KARYOGAMY, AND MEIOSIS may not be a single continuous event for all fungi, the life cycle of a fungus as consisting of : (1) a **haplophase,** during which the uninuclear (or multinuclear) thallus contains only **HAPLOID NUCLEI**;

(2) **a dikaryophase**, in which **TWO GENETICALLY DISTINCT HAPLOID NUCLEI** occupy each cell of the thallus;

(3) **a diplophase**, related to the **DIPLOID NUCLEUS** is formed as a result of karyogamy. Most of the **lower fungi lack a distinct dikaryophase**, Where **HAPLOPHASE** is the predominant state. In **complex higher fungi**, have **HAPLOPHASE** (of greater significance), e.g., the *binucleate mycelium* of the dikaryophase in mushrooms (Basidiomycetes) has a macroscopic structure. In some species **HAPLOPHASE** nd **DIKARYOPHASE** occupy significant portions of the life cycle, and **ASEXUAL SPORULATION** propagate both phases.

After meiotic division, the progeny nuclei develop into sexual spores.

Sexual Reproduction (Classical example)

Members of thr tre Saccharomyces species consists of round and oval cells and reproduce by budding. When the growth conditions are favorable they develop spores inside the enlarged cells (**PLASMOGAMY**). Such spores are 8 in number and are called **SEXUAL SPORES** because they develop as a result of a **PRIMARY NUCLEAR FUSION** with **REDUCTION IN CHROMOSOME NUMBER** during their formation (**KARYOGAMY**). The cell or sac which contains the spores is called an **ASCUS** and the spores are referred to as **ASCOSPORES** (Fig.). Such spores allow this fungus to be placed in the Ascomycetes which are characterized by the formation of **ASCOSPORES** (a sexual endospores in a sac). **Saccharomyces** is a primitive member of this class **is a UNICELLULAR FUNGUS AND DOES NOT PRODUCE MYCELIUM**. The pasty type of colony produced by unicellular fungi which **lack mycelium and reproduce by BLASTOSPORES** (and **occasionally by ASCOSPORES**) is known as a **YEAST COLONY**.

The type of mycelium found in the **FILAMENTOUS COLONY** is of great importance in distinguishing the **ZYGOMYCETES** (PHYCOMYCETES) from all other classes of fungi.

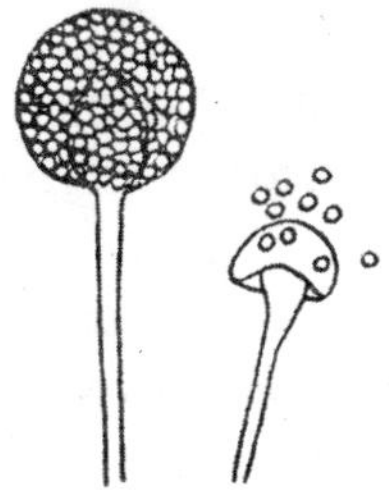

Fig. 2.15

In Zygomycetes class, the hyphae are NON-SEPTATE and COENOCYTIC for **Nuclear fusion (Karyogamy)**. Mycelial septa may occur in old cultures where the spore-producing structures are cut off. **ZYGOMYCETES** PHYCOMYCETES (are further differentiated) from the other classes by producing ASEXUAL SPORES along the inner wall of the swollen structures that develop on the ends of branches. The terminal swollen structure is called a SPORANGIUM (*pl.* SPORANGIA) and the spores contained inside the sporangium are called SPORANGIOSPORES (Fig. 2.16). The branch or hypha from which the SPORANGIUM develop is called a SPORANGIOPHORE (phore = to bear 'sporangium and sporangiospore'). For propagation, (sporangium and sporangiospores) the sporangial wall breaks and the spores are freed to find suitable substrates where they germinate (or carried by air to some distance, where they might germinate) to form new colonies.

Two types of **SEXUAL SPORES** distinguish the genus and species within the ZYGOMYCETES class.

(a) When the tips of approximating hyphae come together and their contents fuse with the development of large, thick-walled bodies, the resulting structure is called (a) **ZYGOSPORE (Fig. 2.15) Mucor and Rhizopus** are notable fungi which develop Zygospores (Phycomycetes).

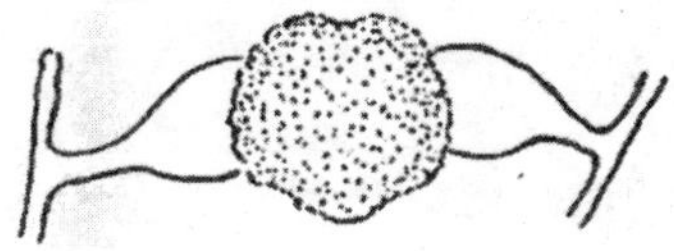

Fig. 2.16 : Zygospore in Phycomycetes

(b) **The second type of sexual spore** develop by the fertilization of the contents of a **SPECIAL FEMALE STRUCTURE ON THE MYCELIUM** by the nucleus of a male structure that develop close to the female structure. Such a male and a female **fused (fertilised) structure** forms into an **OOSPORES (Fig. 2.17)**. An oospore is engaged to from an OOSPHERE (**Fig. 2.16**). So, the members of ZYGOMYCETES class are distinguished from the members of other classes by : (a) Non-septate hyphae. (2) Development of ASEXUAL SPORES, SPORANGIOSPORES (on sporangia) and (3) development of SEXUAL SPORES known as **Zygospores and Oospores**. All other FILAMENTOUS FUNGI are characterized by septate mycelium, and the various classes are distinguished from each other by the **type of spores** which they produce.

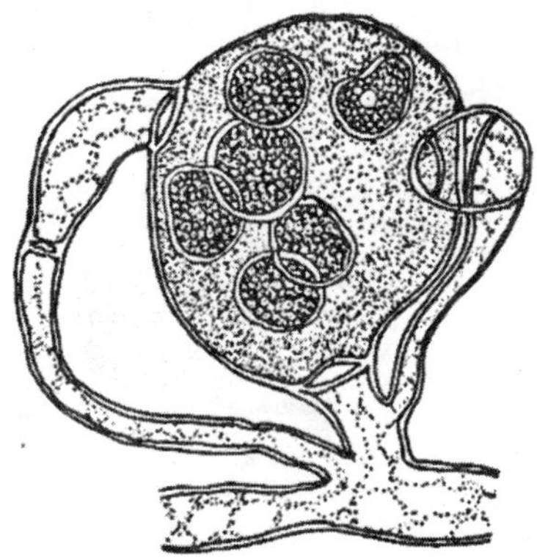

Fig. 2.17 : Oospore

The non-pathogenic fungal class ASCOMYCETES are characterized by the formation of **SEXUAL SPORES** (ASCOSPORES; **Fig. 2.18**) within a sac (ascus). SACCHAROMYCES, as we have already mentioned, is the simplest member of

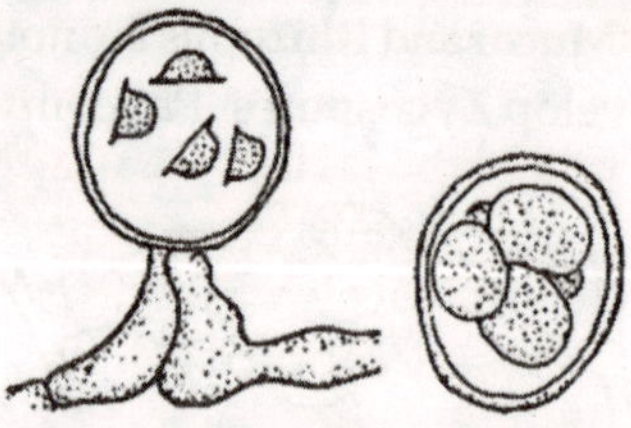

Fig. 2.19 : Ascospores in asci (ascus)

Aseomycetes. It is a unicellular fungus, LACKS MYCELIUM and is placed where only because of the development of 1 to 8 sexual ascospores within a single, round, exposed ASCUS. In the majority of fungi belonging to this class, however, the organisms develop septate mycelia, are MULTICELLULAR, and several club-shaped asci are enclosed in a definite fruiting body called an **ASCOCARP (Fig. 2.19)**. The ascocarp varies greatly in size and shape. A closed ascocarp which is **spherical or flash-shaped** when mature is called a PERITHECIUM; an ascocarp which is OPEN AND saucer or CUP SHAPED when mature is called an **APOTHECIUM** (Fig. 2.20). The morphologic character of these ascocarps serves to divide the ASCOMYCETES into well-defined groups; (i) the PLECTOMYCETES, which LACK AN ASCOCARP, the asci being irregularly arranged in the mycelial mass; (ii) the DISCOMYCETES, which DEVELOP AN **APOTHECIUM**; (**Fig. 2.20**)and (iii) the PYRENOMYCETES, which DEVELOP A **PERITHECIUM (Fig. 2.21)**.

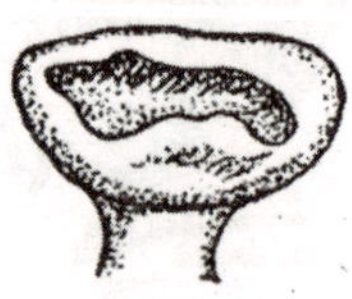

Fig. 2.19 : Ascocarp

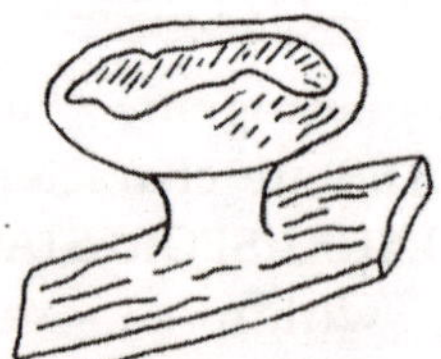

Fig. 2.20 : Apothecilim Higher Ascomycetes

Many members of ASCOMYCETES, IN ADDITION TO THE SEXUAL PHASE is typified by the ascospore, which have an ASEXUAL PHASE and is typified by ASEXUAL SPORES. The asexual spores (of various shapes and sizes) develop from the mycelium. Such a spore is called a CONIDIUM (*pl.* CONIDIA), and the SPECIALISED STRUCTURE which bear them are referred to as **CONIDIOPHORES** or **Sporangiophores, phialophores,** etc. The members of ASCOMYCETES are distinguished from other fungi which develop a) Sexual Spores (ascospores) in the ascus (sac-like structure) and (b) the **ASEXUAL SPORES (CONDIA)** from the mycelium on specialized structure (**conidiophores, sporangiophore, phialophore,** etc.).

THE **BASIDIOMYCETES** develop typical spores, numbering 4 from the ends of club-shaped structure (**columella/basidium Fig. 2.22**). Such a structure, because of its shape, is called a Basidium (**Fig. 2.22**) Basidiospores are SEXUAL SPORES. They are the bracket fungi of the trunks of trees, the mushrooms, and

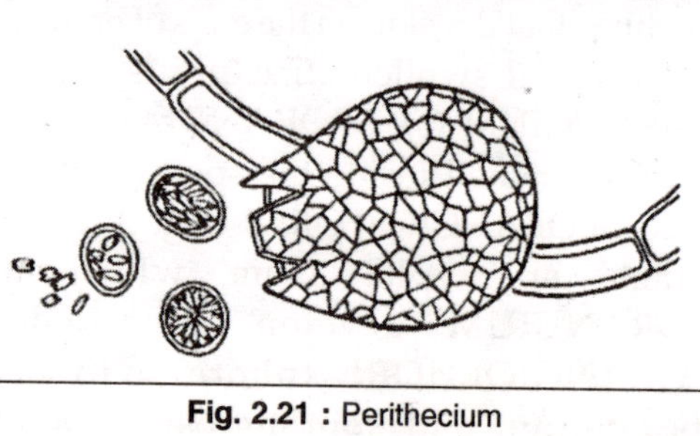

Fig. 2.21 : Perithecium

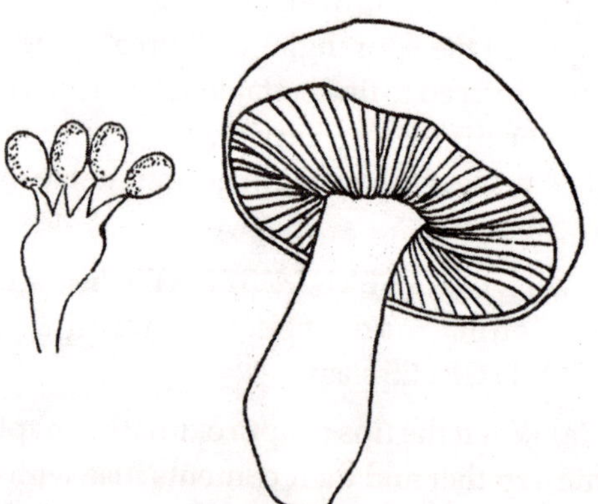

Fig. 2.22 : Basidiospores

toadstools, etc. The mycelia are organized into dense layers which is responsible for the large built body of the fungus. Rusts and smuts, damaging the cereal crops belong to BASIDIOMYCETES. There are many poisonous species which are mistaken for edible mushrooms that causes serious fatal poisoning problem in human. In the club-shaped structure the exogenous sexual spores, the BASIDIOMYCETES, BASIDIO-SPORES develop on the end of basidium.

PARASEXUAL REPRODUCTION

Paresexuality is the eventual effect of **genetic exchange with mitotic recombination**. First described by Pontecorvo, parasexual reproduction has become a laboratory tool for the genetic analysis of many imperfect fungi. It occurs in some species of Ascomycetes, Basidiomycetes, and Deuteromycetes.

In parasexuality there is formation of a **HETEROKARYON**, a **thallus** with **haploid nuclei of** two **different genotypes**. Heterokaryons are most commonly formed by **hyphal anastomosis and nuclear exchange** between genetically different strains of the same species. Rarely, nuclear fusion occurs with the formation of a **stable diploid heterozygote** that multiplies with **two haploid nuclei**. Homologous chromosomes pair up during mitosis of the diploid nuclei enables somatic recombination to occur. The cycle is completed when the diploid state undergoes haploidization (by meiotic division) to the original chromosome number, and the haploid recombinant is eventually isolated.

The parasexual process **mitotic recombination** provides a natural mechanism for genetic exchange among imperfect fungi Deuleromycetes = Hyphomycetes.

By Genetic manipulation mammalian tissue cultures show parasexual process.

3

Mycoses and Pathogenic Fungi

General characteristics of fungal diseases. The estimated species of fungi are about 50,000 to 200,000 and of these, about 100 are known to cause infectious diseases in human. Some have a predilection for epidermal structures and derive their nutrition from their parasitic growth from animal tissues, but for pathogenic species, human infections and other animals is only accidental.

Terminology. Diseases caused by fungi are called mycoses (sing-mycosis). The terminology is based on the site affected or the fungus that produces disease.

(1) **Anatomical terminology** is used according to the site involved by the fungus; viz., dermatomycosis, onychomycosis, otomycosis, pulmonary mycosis etc.

(2) **Mycological terminology** is based after the name of the fungus causing the lesion, e.g., actinomycosis, candidiasis, sporotrichosis, trichophytosis, epidermo-phytosis, etc. In certain cases it is a combination of the above two; viz., bronchomoniliasis, pulmonary aspergillosis, etc.

Classification. Human mycosis is clinically divided into different useful groups differing according to the tissue level at which the infection is localised. It is more or less topographical classification affecting the layers of the skin or deep mycosis affecting the viscera.

1. **Superficial mycosis**.
2. **Cutaneous mycosis**, the fungi of which are known as **dermatophytes** and the diseases are called **Dermatophytosis or Dermycosis**.
3. **Subcutaneous mycosis**.
4. **Systemic or deep mycosis**.
5. **Opportunistic fungi**.

SUPERFICIAL MYCOSIS

These fungal infections are confined to the uppermost horny layer of skin and hair and sometimes the nails, e.g., **tinea versicolor, tinea nigra** and **piedra**. The streptomyces species responsible for affecting the palms and soles in the tropics is ***Streptomyces keratolytica***.

The mycotic nature of tinea nigra is established by microscopic examination of skin scraping obtained from the lesions; from the scraping of tinea versicolor, the fungi and spores of the organism are best demonstrated by **Mc Guire's stain**. Culture of the organism is made in oily glucose agar (**see appendix**).

In piedra, a mycotic infection of the hair, scalp and beard areas, the fungus is demonstrated in the hair by using a keratolytic solution 10 40% KoH or by culturing in Sabouraud's medium.

Streptomyces keratolytica can be cultured in **Norris' medium** for primary isolation.

Erythrasma and trichomyosis nodosa are considered to be caused by bacteria ***Corynebacterium minutissima*** and ***C. tenuis*** respectively.

CUTANEOUS MYCOSIS

In this the **whole skin may be affected**. **Ringworm group of fungi** have a predilection for growth in the **epidermis, hair and nail**. Only a few pathogenic species are found in the **soil**. The commonest one found in the soil is *Microsporum gypseum*. The large number of other dermatophytes are found in the mammalian skin which are obligate parasites of human and other animals; they are transmitted by direct contact with infected individuals, and by bits of infected hair, and desquamated epidermal scales. The diseases are chronic, animal strains (**zoophilic**) often produce acute inflammatory response with formation of **inflammation of the hair follicles (kerion)**, e.g., *Trichophyton interdigitale, M. canis*, etc., causing pustular lesion. But the indigenous dermatophytes in human (anthropophilic), e.g., ***Trichophyton rubrum, Microsporum audouinii***, etc. are obligatory parasites and do not produce inflammatory lesion. Thus human is tolerant to the homologous species of these genera. **Geophilic species** are, ***Microsporum gypseum, Keratinomyces ajelloi, M. cookei, Trichophyton terrestre***, etc., and show **perfect (sexual) stage**.

Ringworm fungi causing ringworm of the skin, hair or nails and fungi causing **candidiasis** of the mucous surfaces causing **thrush, paronychia** and **vaginitis** fall in **CUTANEOUS group**.

Skin or **nail scrapings** or **hairs** should be examined under a keratolytic solutions 10 -40% KoH or cultivated in suitable media in a sterile condition.

In candidiasis, swabs should be examined after staining smears or by wet coverslip preparation. The material should also be cultivated in Sabouraud's dextrose agar (SDA) for isolation of fungi.

SUBCUTANEOUS MYCOSIS

They are saprophytes of the soil, or vegetations. Infection occurs by **direct implantation of spores or mycelial fragments into wounds** (through the skin wounds by thorn pricks). They are prevalent in rural tropical regions amongst **gardeners** and **cultivators**. The disease begins insidiously and progresses slowly. It is a **localised subcutaneous abscess; granulomatous spread occurs by direct extension** breaking through the skin surface forming **draining, ulcerated and crusted lesions**. Extensions through lymphatics lead to **suppurative granulomatous lesions** in the **regional chain of lymph nodes** and metastatic dissemination in the viscera may be fatal.

Soft tissues, fascia and bones are involved. Tortuous sinus tract open on the skin surface. The purulent discharge contains '**granules**' which are **bits of colonies of the causative microorganism**. Indistinguishable abscesses are caused by Schizomycetes: **(Actinomyces, Nocardia and Streptomyces)** known as the **schizomycotic** or **actinomycotic mycetoma** to be distinguished from **eumycotic (fungal) mycetomas** referred to as the **maduramycotic group**. The name Madura foot is after Madura, a place in South India (described in the nineteenth century).

ACTINOMYCOTIC MYCETOMA are caused by ***Actinomyces israeli, A. bovis, Streptomyces madurae, S. pelletieri, Streptomyces somaliensis, Nocardia brasiliensis*** **and** ***N. asteroides*, etc**.

Other subcutaneous mycoses are mycetomas (**Maduramycosis**) caused by various species of filamentous members of higher fungi the commoner being *Pseudo-Allescheria* (***Allescheria***)

boydii, Phialophora jeanselmei, Madurella mycetomii, M. grisea, Cephalosporium falciforme, Leptosphaeria senegalensis and others. The imperfect form of ***Pseudo alescheria (A. boydii)*** is classified as ***Monosporium apiospermum.***

Sporotrichosis and **chromoblastomycosis** are caused by different groups of fungi; rhinosporidiosis affects the nasal mucosa caused by ***Rhinosporidium seeberi***. These are ***subcutaneous mycoses.***

SYSTEMIC MYCOSIS

The fungi causing these diseases are saprophytes of the soil and **inhalation of spores** initiate primary infection in human and animals. The *primary* lesions are pulmonary, occurring as a self-limited pneumonitis; which is often overlooked. Subsequent *chronic form* starts insidiously and progresses slowly to granulomatous or necrotic lesions, forming cavities.

SPREAD BY DIRECT EXTENSION

Sometimes the initial *primary lesion* may be subcutaneous and the fungi spreads by blood stream causing metastatic lesions and granulomas in many organs. They are not transmitted from man or from animal to human. They are not contagious. The *primary visceral lesion* may also involve the skin and spread by metastasis in different viscera. These disseminations were almost always fatal before new drugs appeared.

The viscera involved by systemic mycosis are lungs, liver, spleen and kidneys. The **SYSTEMIC MYCOSES** are **ACTINOMYCOSIS** caused by *Actinomyces israeli* in human and *A. bovis* in cattle, belonging to class Schizomycetes, and **nocardiasis** caused by *Nocardia asteroides* and *N. braziliensis*.

Cryptococcosis, candidiasis, geotrichosis and **histoplasmosis** are seen in India. **North American blastomycosis, South American balstomycosis** and **coccidioidomycosis** are rarely been reported in India. The materials in systemic mycosis are obtained from as **blood, cerebro-spinal fluid, sputum, exudates from draining sinuses and abscesses, bone marrow, scrapings from ulcers, aspirates from abscesses**, and from tissues taken for biopsy (see **Table 3.2**). Biopsy or any other material **should not be fixed in formalin** before culture of the material. They should be sent in containers in sterile saline. Another piece of material can be collected separately in formal saline for **histopathological examination.**

Proper examination of smears directly under the **coverslip preparation** with **potassium hydroxide** or **lactophenol cotton blue** or **India ink preparation** for ***Cryptococcus neoformans*** should be undertaken. **Gram** and **Ziehl-Neelsen** stains of smears are helpful.

OPPORTUNISTIC FUNGI

Certain saprophytic fungi and their spores can infect persons who are sick and run down with an unrelated disease, e.g., **severe diabetes, malignant tumours** (lymphoid tissue) leukaemias, lymphosarcoma, Hodgkin's disease, etc. These saprophytic fungi possess pathogenic potentialities and are designated as ***opportunistic fungi***. Those who receive **excessive treatment with broad spectrum antibiotics**, (their bacterial flora may be reduced) **that make the subject susceptible to these opportunistic fungal infections. Corticosteroids, radiation, cytotoxic drugs** might make the subject susceptible to opportunistic infection. Opportunistic fungi are: ***Candida albicans, Aspergillus fumigatus*** and **Phycomycetes (Mucormycosis fungi).**

Diseases caused by these fungi may be widespread or localised in the respiratory tract (aspergillosis in lungs); bronchial or pulmonary mucous membrane involvement (**broncho-pulmonary candidiasis**); chronic granulomas (**Phycomycosis, affecting the brain as well**).

The conditions of the host conducive to fungus invasion are :

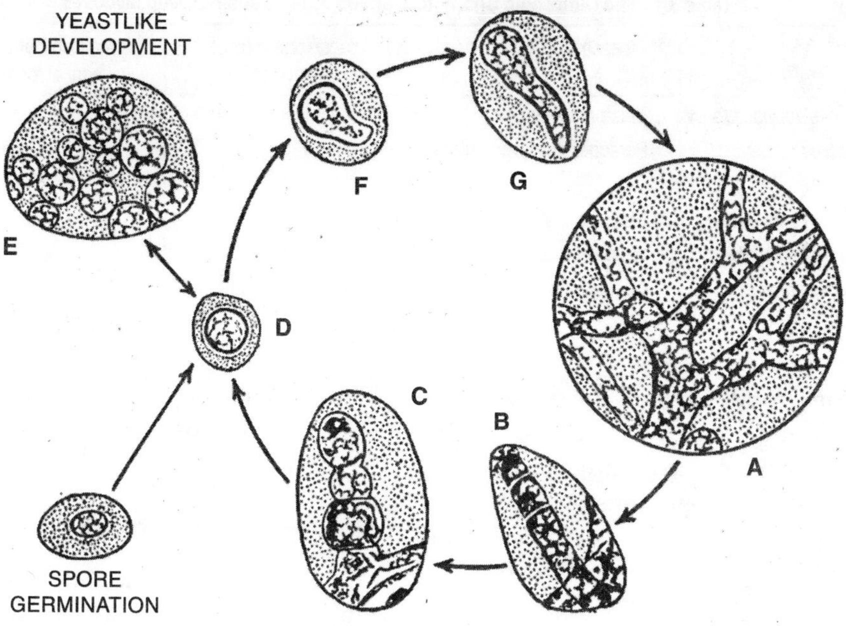

Fig. 3.1 : Yeast-mold dimorphism in *Mucor rouxii.* A—Note mycelial phase with absence of septa typical of phycomycetes in the mold stage. B—and C—Arthrospores are being formed. D—An isolated arthrospore is shown developing into yeast cells (E), or into mold (A) by overgrowth of a filamentous tube (F and G). Modified from S. Bartnicki Garcia. *Bact. Rev.* **27** : 293, 1963.

(1) Impairment of host's **resistance** by chronic diseases;

(2) Administration **of therapeutic**, and

(3) Injury (assault) to host tissue.

The parasite must possess genetic characters to attack or invade the human or animal tissue.

Certain points arise to decide the question of opportunistic fungus as follows :

(1) Does the isolate represent a contaminant?

(2) Is it a normal component of the flora of the body?

(3) Or is the fungus responsible indeed for the pathologic condition, invading fungus can be decided by :

(a) demonstration of fungus cells in the clinical materials.

(b) repeated recovery of the same material from the patient.

(c) correlation of basic features of the isolate with the fungus elements found in the clinical specimen (filamentous or yeast form or spores). Very often, alleged opportunistic fungi have been diagnosed in the isolation tubes **without the demonstration of fungus cells in smears and histologic sections.** Fulfilment of the most fundamentals of **Koch's postulate**, including reproduction of the disease in susceptible animals should be attempted also.

DIMORPHISM

According to the environmental conditions some fungi can grow either as mold (*mycelial*) or *yeast forms*. The capacity for growth in these alternative forms is known as ***dimorphism.*** It is clinically important, since most of the pathogenic fungi human causing systemic mycosis appear in infected tissues as **yeast-*tissue* phase** when cultivated under

Table 3.1 : The Dimorphic Characters of The Fungi Causing Deep Mycoses

Organism	Disease	Yeast form in blood agar at 37°C	Mycelia in Sabouraud's glucose agar at room temperature
Blastomyces dermatitidis	Blastomycosis	+	+
B. brasiliensis	Paracoccidioidal granuloma	+	+
Sporothrix *(Sporotrichum) schenckii*	Sporotrichosis	+	+
Histoplasma capsulatum	Histoplasmosis	+	+
Coccidioides immitis	Coccidioidomycosis	**Spherule seen**	+

+ = Present

conventional conditions they appear as typical **molds *(mycelial phase)***. Dimorphism can be experimentally controlled in artificial media by modifying cultural conditions. A single factor sometimes may be decisive, e.g., *Blastomyces dermatitidis* grows as a **mycelium (*Mold*) at 25°C** but shows **yeast cells at 37°C**.

So is the case with *Histoplasma capsulatum* and *Sporotrichum schenckii*. **Anaerobiasis promotes yeast form** in many fungi. The effect of anaerobiasis in dimorphism is known for about a century. Thus by Pasteur's study (in variation of microbial physiology in absence of oxygen), it was found out that mucor (rarely pathogenic for human) may grow on the surface of the broth in the mycelial form (**Fig. 3.1**). But at the **depth** they grow as **yeast cells**. It is found that a **difference in the CO_2 tension may help dimorphism. With CO_2 tension increases**, the **growth switches** on from the **mycelial to yeast form** and the cell wall is increased in mannan-protein complexes. Thus it forms a longer fraction of total cell weight: **30 per cent of yeast cells** and about **10 per cent of hyphae**.

For most dimorphic fungi, parasitism is a rare event, but it exerts. Dimorphism is a feature peculiar to fungi of many systemic mycoses. The fungus shows a morphological **adaptation to** a simple unicellular **yeast form**, showing rapid reproduction and invasion of the human tissues. In artificial cultures in SDA at room temperature, they show mycelial forms with conidia formation. When the cultures are grown on **blood agar at 37°C, the yeast forms appear** (as in the human body).

The characters of **dimorphic fungi causing subcutaneous and deep mycoses** are shown in **Table 3.1**.

Natural classification

It is apparent that a clinical classification is quite unsatisfactory on a topographical-clinical basis owing to the fact that a fungus which may cause granulomatous lesions of the skin or mucous membrane may often cause a systemic mycosis. For instance, *B. dermatitidis* is said to cause systemic mycosis but it may often produce an affection of the skin and remain so for many years without producing metastatic systemic (North American blastomycosis) one of systemic mycoses is rather too far fetched as it may be localised to the skin for a pretty long time or without causation of systemic mycosis at all. Again, in clinical classification of mycetoma it is found that a number of fungi of different genera or classes may be responsible for the same pathological entity.

Candida albicans may remain as a commensal of the mucous membrane with or without causing any pathogenic changes to the deeper tissues or the same fungus may cause pathological lesion of the skin. Such a fungus under favourable conditions can cause superficial, intermediate or deep mycoses.

All these point out that classification of fungi and a natural classification of mycotic diseases based on the **natural classification of pathogenic fungi** is desirable. Which may be improved with the progress of time.

Epidemiology

The epidemiological factors of internal mycoses are still unexplored. Light has been thrown on the subject of systemic mycoses especially in relation to their epidemiological aspects. In this connection it should be mentioned that systemic mycoses are specially limited to certain areas of the world with their peculiar materiological and other environmental conditions. Gilehrist's blastomycosis is peculiar to North America but only a few cases have been reported from Canada and England, although stray unauthenticated cases have been reported from India and other parts of the world. Paracoccidioidosis is particularly limited to South America; Coccidioidomycosis is a disease of southern parts of United States but both Paracoccidioidomycosis as well as Coccidioidomycosis have been authentically reported by the junior author in Nepal (2003).

Source of infection

Fungus infection often occurs from extraneous sources, particularly from the environment; e.g., **wood, soil**, etc. Incidence of Madura foot has often been found as a result of trauma caused by **splinters of wood** or as a result of **thorn prick**.

Sporotrichoisis is often seen in gardeners and wood-cutters. The parasite *Sporotri schenckii* has been found as a saprophyte in various vegetable substrates, nests of insects, hide of horses which suffered from naturally acquired sporotrichosis. The fungus has a saprophytic existence in timbers in mines, and infection was contracted by mine workers from infected wood.

CHROMOBLASTOMYCOSIS is caused by *Phialophora verrucosa*, ***Phialophora** (Cladosporium or Hormodendrum)* ***pedrosoi*** and ***P. compactum. P. verrucosa*** *(Cadophora americana)* occurs as saprophytes in decaying wood. ***P. pedrosoi*** has close resemblance with saprophytic species of Cladosporium present in the soil. Hence that this infection occurs as a result of contamination of the skin by soil or decaying vegetations after a trauma.

Blastomyces dermatitidis grow saprophytically in the timber of North America and the pulmonary infection occurs by inhalation of spores from the saprophytic sources; inoculation of the fungus into the skin produces skin lesions.

Cryptococcus neoformans, the causative organism of cryptococcosis, has been found in milk products and other food staffs. The infection in these cases has been suggested to occur through the oral mucous membrane or the skin with special relation to ecologic condition, exogenous cryptococcus differs from the endogenous one, which is responsible for producing cryptococcosis.

Reservoir

Coccidioidomycosis is highly infectious and common in its mild *primary stage* in the desert areas of South-Western States of North America. It is shown that rodents harbour the infection. They act as animal reservoir of the disease and the desert soil is contaminated by the reservoir hosts. The spores of the fungus inhaled by man cause primary **coccidioidomycosis of the lung**; but infection is rarely acquired through the skin.

Histoplasma capsulatum produces **systemic mycosis**, but also affects the skin. It is naturally acquired in man and dogs; there is no definite evidence that the dog is the reservoir host.

HOST SPECIFICITY

Ringworm of the hair, **microsporosis and endothrix trichophytosis** are limited to children; spontaneously cured at the attainment of puberty there are instances where endothrix trichophyton persists in some cases after puberty.

Infection with *Microsporum audouinii* is not found in an Indian child in the plane; it is quite, common in Anglo-Indian and European children in India. ***Trichophyton violaceum*** infection is common in Muslim children; no case was found amongst the Hindus.

MICROSPOROSIS and some cases of **endothrix trichophytosis** are self-terminating and disappears at the advent of puberty. This may be due to endocrine secretion at puberty and this makes the hair resistant to those infections as well.

Epidermophytosis is a common disease at the advent of puberty; but it is rare in childhood.

Microsporum canis, a **zoophilic species** causes infection of cats and dogs; human subjects when (children and adults) get infected, it produces intense inflammatory reaction (**kerion**).

ENDOTHRIX TRICHOPHYTA, viz., ***T. violaceum, T. tonsurans***, etc. are pathogenic to children spread; from one child to another. ***T. verrucosum*** is a pathogen of cattle, whereas ***T. mentagrophytes*** (granular) is a pathogen of horses and animals (zoophilic). Its growth may also infect human causing **TINEA BARBAE** and **TINEA CAPITIS**. A variety of the same fungus is seen well adapted in cases of **interdigital ringworm**, viz., ***T. interdigitale*** but produce inflammatory lesions (blisters on the insteps of the foot). On the other hand the same fungus may show a fairly moderate degree of adaptation to human hosts. Infection may persist without any inflammatory change, with recrudescence under favourable conditions of moisture and temperature.

THE HOST SPECIFICITY

Microsporosis caused by *Microsporum audouinii* infect only growing hairs. Hairs in the **resting phase called 'club hairs'** are not penetrated by the fungus. When hairs are highly metabolically active, they become susceptible to infection. When 95 per cent of scalp hairs show active growth phase, they are susceptible to infection (before the age of puberty even).

VIRULENCE OF THE FUNGUS

Soil isolates of ***Microsporum gypseum have low virulence***, but virulence becomes enhanced when the strains are isolated from animals or human.

ENZYMIC ACTIVITY of dermatophytes (**keratinophilic properties**) are responsible for ***hair infection*** as in ***Microsporum gypseum, M. fulvum, M. nanum, M. cooki, Trichophyton terrestre and T. mentagrophytes*** etc.

RELATION OF HOST TO PARASITE

The fungi that are well adapted to human (**anthropophilic**) viz., ***M. audouinii*** may cause spread of ringworm in an ***epidemic form*** in boarding schools but ***M. canis, a zoophilic fungus***, is ill-adapted to human hosts; reaction of the host is very severe as shown by a process of inflammatory reaction producing **KERION**. A **geophilic saprophytic dermatophyte rarely infects human**.

HISTOPATHOLOGY

This is studied in systemic mycosis where it essentially forms an infectious granuloma. In the immediate vicinity of the parasite there is necrosis of the tissue with abscess formation. Many of these abscesses become surrounded by neutrophils and granulation tissue showing mononuclear cells with giant cell formation and a dense layer of fibrous tissue. The reaction varies according to whether the lesion is primary or a secondary one.

In **primary blastomycosis** of the skin, the lesion is essentially a **granuloma** with giant cell formation; whereas, a metastatic lesion shows **abscess formation**. However, it all depends with the **dosage, route of inoculation**, etc.

ANTIBODIES

Unlike bacterial infection in mycoses CMI (cellular reaction) is characteristic and the **antibodies IgE which are cell-fixed, are responsible for sensitisation of the individual and produce allergic reactions**. The cell-wall of the fungus is thick and its intracellular protein hardly diffuses in the tissues. The reactions are due to **liberation** of substances owing to **autolysis of the fungus**. The **serological tests**

positive in fungus infections are: agglutination, precipitation, complement fixation and **fluorescent antibody (FA).**

AGGLUTINATION

Agglutination reaction in sporotrichosis differentiates one strain of *Sporothrix schenckii* from another. Patients serum agglutinates spores of ***S. schenckii*** in a high titre of **1 in 300 to 1 in 400.** But it is seen that, the sera of candida (thrush) cases agglutinate spores of *S. schenckii* even in a higher titre (than they would do with the suspension of *Candida albicans*. Sera of patients suffering from **actinomycosis** also ***agglutinate spores of S. schenckii.* Agglutination reactions are really unsatisfactory** for most of the mycoses.

PRECIPITATION

Precipitin develops in severe coccidioido-mycosis while it is **negative in mild cases.** This test is of little use in mycoses **(except in gel diffusion).**

COMPLEMENT FIXATION TEST (CFT)

The test was found to be positive in sporotrichosis; but it is unreliable. Saline suspension of the yeast form of ***B. dematitidis*** used as an antigen is claimed to show a positive CFT with diluted sera in **North American blastomycosis** and there is no cross reaction with other fungi. While it is negative in mild pulmonary and cutaneous cases even with undiluted sera. CFT is positive in extensive lesions of coccidioidomycosis but the test does not confirm South American blastomycosis.

ALLERGY AND SKIN HYPERSENSITIVITY

Fungi have little invasive power as they often attack traumatised or keratinised tissues. To be invasive, fungi should either increase in their virulence, or there should be some environmental host tissue changes. Increase of virulence could not be done in experimental animals; hence alteration in host tissue is more likely to occur.

Local changes in infective tissues occur due to **ENDOTOXIN** which causes tissue destruction resulting in extension of the lesion. More important changes by fungus infections are the **altered reactivity of the tissue leading to allergy and skin hypersensitivity.** In this latter condition, the fungus, due to its continued existence at the site of primary infection, sensitises the cells of the body and thus the host becomes susceptible to an **accelerated tissue reaction** whereby an allergic reaction is liable to occur with rapid spread of the disease. In dermatophytosis the hypersensitiveness **occurs** when the spores or mycelia enter the circulation (from the site of primary infection) showing of skin eruptions which are also known as **"id" REACTION.** Sensitisation is detected by intradermal injections of extracts of fungi. The **cutireactions** may be positive in cases with previous infection with the same fungus (or an antigenically allied fungus). Cutireaction indicates the presence of **sensitisation.** Fungal antigen (allergen) is of the nature of a **group antigen** (rather than a specific one) and so an antigen which is common in more than one fungus which produces a cross reaction. Hence **a negative reaction is more important than a positive one.** The following skin reactions are practised in various fungal diseases.

TRICHOPHYTIN

Experimentally it has been repeatedly demonstrated sensitisation occurs in guinea pigs only when a **living culture of ringworm fungus is injected** into the skin. Injection of **dead culture** or the fungus extracts, **(trichophytin) does not provoke dermal sensitivity; nor injection of living culture** in tissues **other than skin produces this sensitivity.** The degree of sensitivity depends upon the species that infects the skin. Sensitivity is very much marked with animal species causing deep-seated lesions, like. ***T. mentagrophytes,***

T. quinckaenum and *M. canis*. The **reaction is minimum with human species, like *M. audouinii, T. tonsurans* and *T. schoenleinii*.** By the skin test and by sensitisation and quantitative desensitisation of the guinea pig's uterus by **Schultz-Dale technique**, it was demonstrated that trichophytin contained **a group antigen** and some **minor specific factors**. So, **trichophytin test** is not be considered as a specific but **a group test**. This should not therefore form the sole basis of species specific diagnosis.

Various **preparations of trichophytin** are available and any of the standard preparations may be used for the test [**Metz 1:100; Lederle 1:30 or dermotrichophytin; Muller and Phipps, (India) Ltd.**] It is difficult to standardise a home-made trichophytin. The reaction is judged by **erythema and induration** produced by the **intradermal injection of trichophytin**.

VALUE OF THE TEST. Trichophytin has been used diagnostically specially in suspected cases of **'id' reactions** and also **in deep-seated ringworm infections**. The **skin reaction becomes positive in the majority of deep-seated ringworm infections**. It has been concluded that **negative reaction is of greater value to exclude an inflammatory lesion of dermatophytic origin than a positive trichophytin test.** (If the lesion is of long duration, the fungus cannot be detected microscopically and culturally and the trichophytin test is negative, the inflammatory lesion may be declared not to be of dermatophytic origin.)

COCCIDIOIDIN

Dr. C. E. Smith prepared coccidioidin from strains of *Coccidiodes immitis* in a synthetic liquid medium. **In systemic involvement of coccidioidomycosis marked skin hypersensitiveness develops**. The immunologically active material is **polysaccharide (mannose and galactose)** with small amount of nitrogen (Pappagianis *et. al.*, 1961). The reaction is highly specific but a small number of persons, who had **no history of disease or contact or no history of residing in an endemic area, were found to react to coccidioidin**. This has no explanation. Therefore, the negative test is valuable in the diagnosis as it excludes the infection in cases other than those who are in the **pre-allergic stage of the disease** or in a state of **anergy** resulting from overwhelming infection.

Blastomycin. (Extract of *B. dermatitidis)* The test is specific. **Mild cases with skin lesion often give negative results**. Normal people usually do not show the positive skin reaction. **Hypersensitiveness to it** is **a contra-indication for iodide therapy**. (Cases of blastomycosis should be desensitised before they are treated with X-ray or iodide).

HISTOPLASMIN

It is filtrate of **synthetic broth culture of the fungus *(H. capsulatum)***. Histoplasmin has been used in epidemiological studies and found to be useful for detecting benign type of self-limiting pulmonary histoplasmosis in endemic areas. Histoplasmin **skin test** becomes **positive in the endemic areas** of histoplasmosis, in cases of non-tuberculous pulmonary calcification. **Cross sensitisation reactions** may be noted **for BLASTOMYCIN and HISTOPLASMIN. Patients of coccidioidomycosis often show a positive skin reaction to histoplasmin.**

Reinfection in the sensitive but immunologically deficient patient **results in cavitation due to necrotising pneumonitis.** Such lesions could occur by either exogenous or endogenous reinfection (Schwartz and Baum, 1963).

SPOROTRICHIN

Intracutaneous test with an extract of *S. schenckii* is useful when the test is positive; a **negative result rules out the diagnosis of sporotrichosis**. Occasionally only there may be a **false positive reaction**. According to Du Toit

a positive reaction always occurs in cases of sporotrichosis. The appearance of sensitivity has been found to be on the **fifth day** after inoculation to a volunteer.

OIDIOMYCIN

This is a **commercial vaccine** prepared from *Candida albicans*. According to Lewis and Hopper, **skin sensitivity test with oidiomycin is of little value**. Sensitisation of the skin is possible from a focus in the intestinal tract or from any other lesion which had subsequently resolved. (He mentioned that out of 42 cases of cutaneous candidiasis a positive response was obtained in 57 per cent of cases. Again, out of 91 patients with an infection due to a fungus other than *C. albicans*, 45 per cent showed a positive reaction; whereas out of 192 patients without any evidence of fungus infection, 40 per cent reacted to oidiomycin. It can therefore be concluded that the test has little practical value in the diagnosis of candidiasis.

Table 3.2 : Table showing Source of Clinical Material And Methods of Study of Fungi

Disease/organism	Name of species	Source and type of specimen (and processing)	Media for isolition etc.
SUPERFICIAL MYCOSES			
Tinea versicolor	*Malassazia furfur*	Skin scraping (10% KOH/Hypo-Na-sulphide)	Sabouraud's dextrose agar with oil (for details see appendix)
Tinea nigra (Pheohypha mycosis)	***Exophiala*** *(Cladosporium)* ***werneckii***	Skin scraping (KOH)	Sabouraud's chloramphenicol cycloheximide agar
Chaluni	*Streptomyces keratolytica*	Skin scraping (40% KOH)	Norris' medium
Trichomycosis axillaris (nodosa) (Bacterial)	*Corynebacterium tenui*	Plucked hair by forceps	Ascitic fluid agar, Dorset's egg or Petroff's media
Piedra black Piedra white	*Piedraia hortae, Trichosporon beigelii (cutaneum)*	Clipped hair (10% KOH)	Sabouraud's dextrose agar (SDA)
Candidiasis	*Candida albicans* and *other species*	Scraping from skin and nail by blade and mucocutaneous areas (vagina) by wooden spatula (10% KOH)	Sabouraud's dextrose agar
CUTANEOUS MYCOSES			
Tinea corporis	Various *Trichophyton, Epidermophyton and Microsporum.*	Skin scraping (10% KOH)	Sabouraud's dextrose (BHI) agar
Tinea capitis	*Same*	Plucked hair by forceps (10% - 40% KOH)	Same
Tinea barbae	*Same*	Plucked beard by forceps (10% - 40% KOH)	Same
Tinea unguium (Onychomycosis)	*Same*	Scraping from nail by blade (40% KOH)	Same
SUBCUTANEOUS MYCOSIS			
Mycetoma **SCHIZOMYCOTIC** (Actinomycotic)	*A. israeli, A. bovis*	Pus from draining sinuses, aspirated exudate from sinus tracks and biopsy specimens	Brain-heart infusion agar (anaerobic, or **shake culture**)

Contd...

Disease organism	Name of species	Source and type of specimen (and processing)	Media for isolation, etc.
	Nocardia and Strepto-myces species	Same	Sabouraud's dextrose agar
EUMYCOTIC	Various organism, e.g.; ***Madurella mycotomatis***	Granules in pus of draining sinuses	Sabouraud's dextrose agar
Rhinosporidiosis : NOT CONSIDERED AS FUNGUS	*Rhinosporidium seeberi*	Biopsied nasal and ocular polyps	Not cultivated as yet:
Sporotrichosis	*Sporothrix Sporotrichum (schenekii)*	Pus from ulcers, aspirated fluid from subcutaneous abscesses or lymph nodes	Sabouraud's dextrose agar or blood agar with cycloheximide
Chromoblastomycosis	Various organisms	Crusts or scrapings from warty outgrowth, exudate from lesions and biopsy material	Same
SYSTEMIC MYCOSIS			
Actinomycosis	*Actinomyces israeli*	Pus from draining sinuses, aspirated fluid from subcutaneous abscesses, sputum, bronchial washings, spinal fluid	Brain-heart-infusion (BHI) agar plates (anaerobic, and shake culture)
Nocardiosis	*Nocardia asteroides*	Same	Sabouraud's dextrose agar at room temperature and at 37°C
Candidiasis	*Candida albicans*	Sputum, bronchial washings, spinal fluid, urine, faeces,	Sabouraud's dextrose with cycloheximide which inhibits some species
Cryptococcosis	*Cryptococcus neoformans*	Spinal fluid, sputum, pus from abscesses and sinus tracks, urine	Sabouraud's dextrose agar with chloramphenicol
Histoplasmosis	*H. capsulatum*	Blood, bone marrow, biopsy material, sputum, exudates from ulcers and granulomas, lymph node puncture, scrapings from skin lesions, etc.	Brain-heart-infuslon (BHI) agar with or without chloramphenicol, at 37°C Sabouraud's dextrose agar with cycloheximide
Geotrichosis	*Geotrichum candidum*	Sputum, bronchial washings, faeces	Sabouraud's dextrose agar with or without chloramphenicol
BLASTOMYCOSIS :			
(a) North American	*Blastomyces dermatitidis*	Scrapings from edge of skin lesions, pus from microabs-cesses, sinus tracks, urine, sputum and bronchial washings	Brain-heart-infusion (BHI) agar with chloramphenicol and cyclohexamide (or without antibiotics) at 37°C

Contd...

Disease organism	Name of species	Source and type of specimen (and processing)	Media for isolation etc.
(b) South American (Paracoccidioido-mycosis)	*Paracoccidioides brasiliensis*	From edge of skin lesions, mucous membrane, lymph node biopsy, spinal fluid, urine, faeces	Same
Coccidioidomycosis	*Coccidioides immitis*	Sputum, bronchial washings, spinal fluid, scraping from skin lesions, pus from draining sinuses or abscesses	Sabouraud's dextrose agar with cyclohexamide medium
OPPORTUNISTIC FUNGI			
Aspergillosis	*Aspergillus fumigatus*	Sputum, bronchial washings or biopsy	Sabouraud's dextrose agar with chloramphenicol
Mucormycosis & Phycomycosis and Rhino-entomophthomycosis	Various *Mucors :* Absidia, corymbifera, Mucor, Rhizopus, etc.	Sputum or endoscopic biopsy material	Same
Cladosporiosis	*Cladosporium (Trichoidis) bantianum*	Bronchial washing, Biopsy material	Sabouraud's dextrose agar
PNEUMOCYSIS	*CARINII*	Sputum	Not available
OPPORTUNISTIC FUNGI ARCHI-ASCOMYCETES, e.g., ***Pneumocystosis***	***carinni***	Sputum, Broncho-alveolar lavage	**Cannot be cultivated in artificial media**
Other opportunists :			
Alternaria	- -	Related Specimens	SDA
Biplolaries	-	Related Specimens	SDA
Cladosporium (Torula)	-	Related Specimens	SDA
Curvularia	-	Related Specimens	SDA
Fusarium	-	Related Specimens	SDA
Paecilomyces	-	Related Specimens	SDA
Penicillium marnfeei	-	Related Specimens	SDA
Torulopsis (Torula)	-	Related Specimens	SDA

Table 3.3 : Principal Fungi Pathogenic To Human And Animals

Name of the Fungus	Disease
SCHIZOMYCETES (Higher bacteria) : **(SCHIZOMYCETOMA)**	
ACTINOMYCETOMA **ACTINOMYCETES GROUP**	
ACTINOMYCETES (NON-ACID-FAST), ANAEROBIC	**ACTINOMYCOTIC MYCETOMA**
Actinomyces bovis	**(Subcutaneous Mycosis)** Actinomycosis in cattle
A. israeli	**(Subcutaneous** and **Systemic Mycosis)** Actinomycosis and actinomycetoma in human
AEROBIC, ACID FAST	
Nocardia asteroides *N. brasiliensis* *N. caviae*	Pulmonary nocardiosis and **ACTINOMYCOTIC MYCETOMA**
AEROBIC, NON-ACID FAST	
Streptomyces keratolytica	Superficial Mycosis in cultivators of Bengal
S. madurae *S. pelletieri* *S. somaliensis* *S. paraguyanensis*	**STREPTOMYCOTIC MYCETOMA IN HUMAN**
EUMYCETOMA **EUMYCETES GROUP**	
Madurella mycetomatis *M. grisei and other species* *Phialophora jeanselmei* *Cephalosporium falciforme and other species* *Pyrenochaeta romeroi*	**EUMYCOTIC MYCETOMA** (Subcutaneous mycosis)
Glenospora semoni	Mycetoma (in India)
ASCOMYCETES	
Peudo-allescheria boydii (Perfect form) (Imperfect form is *Monosporium apiospermum*). *Leptosphaeria senegalensis* *Aspergillus glaucus (amstelodami)* *A. nidulans, A. terrens and others*	**EUMYCOTIC MYCETOMA**
ARCHI-ASCOMYCETES :	
PNEUMOCYSTIS CARINII	Systemic Mycosis (in Immunocompromised Host)
Piedraia hortae	Black piedra (Superficial mycosis)
For Gymnoascaceae — see the list of perfect fungi of dermatophytes. (Ch. 28)	
DEUTEROMYCETES (Fungi Imperfecti)	
YEASTS AND YEAST-LIKE FUNGI (*B. dermatitides*)	
North American Blastomycosis (**Systemic Mycosis**)	
Cryptococcus neoformans	Cryptococcosis (Cutaneous, Deep or Systemic Mycosis)
Candida albicans and occasional other species	Candidiasis (**Cutaneous and Deep or Systemic Mycosis**)
Malassezia furfur	Tinea versicolor (**Superficial Mycosis**)

Contd...

Name of the Fungus	Disease
Pityrosporum ovale	Seborrhoea capitis (**Superficial Mycosis**)
Geotrichum candidum	Intestinal and pulmonary geotrichosis (**Systemic Mycosis**)
Trichosporon beigelii (cutaneum)	White piedra (**Superficial Mycosis**)
BLACK YEASTS	
Exophiala (Cladosporium) werneckii	Tinea nigra (Palmaris)
Phialophora (Hormodendrum) pedrosoi *Cladosporium carionii* *Phialophora verrucosa* *Phialophora dermatitidis* *Phialophora (Hormodendrum) compacta*	Chromoblastomycosis (**Subcutaneous Mycosis**)
Cladosporium (trichlodes) bantianum	Cladosporiosis in brain (**Systemic Mycosis**)
DIMORPHIC FUNGI (Yeast and Mycelial Forms) *Blastomyces dermatitidis (North American)* *Paracoccidioides brasiliensis and P. loboi (South American)* *Coccidioides immitis* *Histoplasma capsulatum* *Sporothrix (Sporotrichum) schenckii* (***Subcutaneous and Systemic Mycosis***)	**SYSTEMIC MYCOSIS**
ALEURIOSPORINAE (OPPORTUNIST) *Aleurismae* (tribe) *Acladium castellani*	Gummatous ulcer (**Subcutaneous Mycosis**)
ASPERGILLUS *Aspergillus fumigatus* and other species, like *A. flavus-oryzae, A. niger*	Pulmonary, Systemic and other (Invasive) aspergillosis.
ZYGOMYCETES (OPPORTUNISTS) ***Families :*** ***(a) Mucoraceae*** *Absidia corymbifera* *Mucor* *Rhizopus* ***(b) Entomophthoraceae*** *Basidiobolus* *Entomophthora* (c) *Mortierellaceae* *Mortierella*	Zygomycosis (Phycomycosis) (**Subcutaneous and Systemic Mycosis**)
(d) *Hyphomyces destruens*	Polyps, hyphomycosis in horses
UNDETERMINED FUNGUS *Rhinosporidium seeberi* (**At present it is not considered as a fungus**)	Rhinosporidiosis (**supposed to cause Subcutaneous Mycosis**)

Table 3.4 : Showing Perfect state of Dermatophytes

Perfect State	Imperfect State
Nannizzia	**Microsporum**
Nannizzia cajetana Ajello, 1931	*Microsporum cookei*
N. obtusa Dawson & Gentles, 1961	*M. nanum*
N. incurvata Stockdale, 1961	*M. gypseum complex*
N. grubyia George, Ajello, Friedman and Brinkaman, 1962	*M. vanbreuseghemii*
N. fulva Stockdale, 1963	*M. fulvum*
N. gypsea Stockdale, 1963	*M. gypseum complex**
N. persicolor Stockdale, 1967	*M. persicolor*
Arthroderma	**Trichophyton**
A. quadrifidum Dawson & Gentles, 1961	*T. terrestre*
A. uncinatum Dawson & Gentles, 1961	*Keratinomyces (K.) ajelloi*
A. ciferrii Varsavsky & Ajello, 1964	*T. georgiae*
A. lenticularum Pore, Tsao and Plunkett, 1965	*T. terrestre**
A. simii Stockdale, Mackenzie and Austwick, 1965	*T. simii*
A. gloriae Ajello, 1957a	*T. gloriae*
Arthroderma benhamiae Ajello and Cheng, 1967b	*T. vanbreuseghemii*
A. gertleri Boehme, 1967	*T. vanbreuseghemii*
Arthroderma state Young, 1968	*T. rubrum (T. purpureum)*

*Two imperfect forms are considered as genetical complexes.

N. B. **For references on Nannizzia and Arthroderma, see chapter on Dermatophytes.**

Asthma

Various spores of common molds have been **incriminated as aetiological agent in asthma.** These spores, when inhaled with dust, sensitise the subject and the **sensitisation** can be **detected by cutireaction**. Spores that might cause asthma are those of *Puccinia graminis* (wheat rust fungus), various species of (Invasive) **ASPERGILLUS**, Alternaria and others. Similar fungi might produce **allergic eczema in** those cases of **asthma**.

Farmer's lung

This results from **hypersensitivity reaction in the lungs to antigens from moldy hay** (in dust and occurs in persons exposed to **repeated inhalation of such dust**).

Pepys *et. al.,* (1963) have found thermophilic actinomycetes ***Thermopolyspora polyspora*** and ***Micromonospora vulgaris*** grown in sterilised hay produce antigens reacting in **gel diffusion tests** and in **immuno-electrophoresis tests** with 90 per cent sera from patients with farmer's lung.

FARMER'S LUNG CAN BE DIAGNOSED by **agar gel precipitation** with patient's serum and an extract of *T. polyspora* with few false positive and false negative results.

Dermatophytes are enlisted in Gymnoascaceae family under **three imperfect** and two **perfect genera** :

GENUS : ***Epidermophyton;*** *E. floccosum*

GENUS : *Microsporum*

Microsporum amazonicum, M. audouinii (*), *M. boullardii, M. canis, M. distortum, M. ferrugineum, M. racemosum.*

(*) The doubtful species like *M. langeronii* Vanbreuseghem, 1950b and *M. revalieri* Vanbreuseghem, 1963 are considered synonymous with *M. audouinii* Ajello, 1968.

GENUS : *Trichophyton*

Trichophyton ajelloi, T. concentricum, T. equinum, T. longifusus, T. megninii, T. phaseoliforme, T. schoenleinii, T. soudanense (2), *T. tonsurans, T. verrucosum, T. violaceum.*

(2) *T. gourvilii* Catanei, 1933 and *T. yaoundei* are considered synonymous with *T. soudanense* (2) Joyeaux, 1912, giving the effect of priority.

SUMMARY (TABLE)

Table : 3.5 : The major types of mycoses producing disease & organisms

Classical Zonal Type of Mycosis	Mycosis	Organisms
Superficial	Seborrhoic dermatitis and Dandruff	***Pityro sporum ovale***
	Pityriasis versicolor	*Malassezia furfur*
	Tinea nigra	*Exophiala werneckii*
	White piedra	*Trichosporon beigelii* (*Cutaneum*)
	Black piedra	*Piedraia hortae*
Cutaneous	**Dermatophytosis**	*Microsporum* species, *Trichophyton* species, and *Epidermophyton floccosum*
	Candidiasis of skin, mucosa, or nails	*Candida albicans* and other *Candida* species
Subcutaneous	**Sporotrichosis**	*Sporothrix schenckii*
	Chromoblastomycosis	*Phialophora verrucosa, Fonsecaea pedrosoi,* and toehrs
	Mycetoma	*Pseudallescheria boydii, Madurella mycetomatis,* and others
Systemic (primary, endemic)	**Coccidioidomycosis**	*Coccidioides immitis*
	Histoplasmosis	*Histoplasma capsulatum*
	Blastomycosis	*Blastomyces dermatitidis*
	Paracoccidioidomycosis	*Paracoccidioides brasiliensis*
Opportunistic	**Systemic candidiasis**	*Candida albicans* and aother *Candida* species
	Cryptococcosis	*Cryptococcus neoformans*
	Aspergillosis	*Aspergillus furigatus* and other *Aspergillus* species
	Mucormycosis (zygomycosis; Phycomycosis)	*Species of Rhizopus, Absidia, Mucor,* and *other zygomycetes*

REFERENCES

Ajello, L. (1968). A taxonomic review of dermatophytes and related species. *Sabouraudia* **6** (2), 147.

Conant, N. F. (1937). The occurrence of a human pathogenic fungus as a saprophyte in nature. *Mycologia* **29**, 597.

Du Toit, C. J. Sporotrichosis on the Witwatersrand. *Proc. Transv. Mine Med. Officers' Ass.* **22** : June, 1942.

Jadassohn, W., Schaaf, F. and Wholer, G. (1937). Analyses of composite antigens by Schultz-Dale Technique : Further experimental analyses of Trichophytins. *J. Immunol.* **32** : 203.

Kligman, A. M. (1955). The pathogenesis of tinea capitis due to *Microsporum audouinii* and *Microsporum canis*. II. Dynamics of host parasite relationship. *Arch. Dermat.* **71**, 313-337.

Lewis G. M. and Hopper, M. E. An Introduction to Medical Mycology, Third Edition, 1948. The year Book Publishers. Inc. Chicago.

Pepys, J.; Jenkins, P. A.; Festenstein, G. N. Gregory, P. H.; Lacey, M. and Skinner, F. A. (1963). Farmer's lung, thermophilic actinomycetes as a source of "Farmer's lung hay" antigen. *Lancet* **2**, 607.

Pappagianis, D., Putnam, E. W. and Kobayoshi, G. S. (1961). Polysaccharide of *Coccidioides immitis. J. Bact.* **82**, 714.

Schwartz, J. and Baum, G. L. (1963). Reinfection histoplasmosis. *Arch. Path.* **75**, 475.

Widal, F., Abrami, P., Brissaud, E. and Weill, A. (1910). Serodiagnostique Mycosique. *Ann. Inst. Pasteur* **24**, 1-133.

4

Study of Pathogenic Fungi (Including Histopathology)

Infection with dermatophytes is seen on the surface of the skin. But subcutaneous and systemic mycoses are suspected on the basis of their clinical behaviour in patients in whom the characteristic lesions are formed. From these lesions **fungi are to be isolated and identified for further study. The immunological reactions** like **serological** and **dermal reactions** with antigens of the suspected fungus and **delayed allergic reactions to intra-dermal injection of fungal antigen** may provide indirect evidences for diagnosis but it is not always conclusive. Histologic picture of some of internal mycoses have been worked out and they show characteristic tissue changes in many cases. The most convincing diagnostic evidence is the detection of fungus in lesions and exudates by direct microscopic examination, or their isolation by cultivation. This may be confirmed in some cases by animal inoculation in which a characteristic lesion may develop, e.g., ***Histoplasma capsulatum*** in mice, ***Coccidioidis immitis*** in mice and guinea pigs and ***S. schenckii*** in male rats and guinea pigs. In addition, this animal inoculation also serves to distinguish pathogenic from non-pathogenic fungi. The non-pathogenic saprophytic strains are usually inoccuous in test animals.

Pathogenic fungi are studied under the following heads:

1. **Filtered ultraviolet (Wood's) light** (see dermatophytes)
2. **Direct microscopic examination**
3. **Culture for the isolation and study of GIANT COLONIES**
4. **MICROCULTURE**
5. **Hyphal fusion**
6. **Biochemical reactions and nutritional** (physiological) **requirements** for **dermatophytes** (see chapter on Dermatophytes)
7. **Histopathology for tissue reaction**
8. **Animal inoculation**
9. **Serological reactions.**

Direct microscopic examination : Direct microscopic examination of the suspected material is one of the most important methods available for investigation of fungal diseases. The simplest procedure is to put the scraping of the lesion or bits of exudate, e.g., sputum or pus in a drop of **10 per cent KOH solution. After an hour** or so, the **KOH solution hydrolyses proteins, fats** and many **polysaccharides** and the tissues become optically clear. The **fungal mycelia are visualised with their cell walls intact, because of their alkali resistant glucans.** With some fungi like small yeast cells visualisation of **alkali-resistant cell wall is helped by warming the tissue in 10 per cent KOH.**

For *DERMATOPHYTES* examinations of the **hair, scraping from the glabrous skin or nails** show presence of the fungus in the ringworm lesion. Any keratolytic solution which dissolves and clarifies hairs, scales and nails may be utilised for this purpose. Usually **10 per cent aqueous solution of potassium hydroxide** is used as a keratolytic agent but the author uses **SODIUM SULPHIDE*** as a **ROUTINE KERATOLYTIC AGENT**. In a coverslip preparation with this solution, scales or hairs, when heated for a minute, are quickly keratolysed in about 5 minutes but for nails it takes about 3 hours to dissolve it completely. This solution has the advantage over that of caustic potash in that (i) it clarifies the cells without destroying the fungus; (ii) it shows the fungus clearly and ringworm fungus may be spotted out under the low power and then confirmed by the high power.

STAINING *IN VITRO*

Coverslip preparations for direct examination may be done **with a lactophenol cotton blue mount**. As an alternative method, the fungus can be detected after staining with **Mc Guire's stain.*** By this method the fungi are stained purple. It is very easy to stain fungi by this method when they are present in the thin scales of tinea imbricata, tinea versicolor, seborrhoea capitis, or in granules of actinomycotic mycetoma crushed under a cover slip with a mixture of a drop of glycerine and a drop of the stain.

PUS OR OTHER EXUDATES

Pus from an abscess, exudate from an ulcer or sputum may be examined in the above way but they may be examined satisfactorily by a cover slip preparation with Mc Guier's stain which stains the fungus purple. Granules in actinomycosis and nocardiosis are easily examined by crushing a granule in a drop of stain under the cover slip. It stains the fungus beautifully and the leucocytes as well.

Cerebrospinal fluid : The **cerebrospinal fluid is centrifuged** and examined in any of the above methods or by fixing and staining smears with **polychrome methylene blue** or Leishman's stain. Capsules of ***Cryptococcus neoformans*** can be demonstrated very well by Indian ink cover slip preparation.

Culture for isolation of fungi : Even under optimal conditions the fungi grow much more slowly than bacteria, and it is essential therefore to keep the inoculated media under observation for a prolonged period of at least 3 weeks. But at the same time it is also essential to inhibit the bacterial contaminants by drugs by maintaining the *pH* and incubation at temperature of 25ºC which are helpful for inhibiting bacteria more than fungi. Because of the relatively better growth at low temperature, contaminating fungi may often overrun bacterial cultures stored in the refrigerator.

For isolation of fungus from grossly contaminated material (**ringworm of the nail**), it is better to use **Sabouraud's proof medium containing 1 : 400,000 of gentian violet**. The method is very satisfactory and secondary contamination is practically absent when this method is adopted. **Giant colonies are studied in Sabouraud's proof medium after primary isolation and subculture in Erlenmeyer flasks.**

ANTIBIOTICS IN MEDIA

Dimorphic fungi were studied by Mc Donough *et. al.* (1960) in cycloheximide (0.5%) in Sabouraud's dextrose agar. They cultivated ***Blastomyces dermatitidis, Histoplasma capsulatum, Paracoccidioides brasiliensis*** and ***Sporothrix schenckii*** and found that **yeast phase was more sensitive than mycelial phase** and experimental results supported the general practice of using 0.5 mg/ml of **cycloheximide** and 0.05 mg/ml of **chloramphenicol** in media for selective isolation of many fungi and dermatophytes in grossly contaminated specimen.

The results indicate that some strains could not be recovered at 37ºC with the above concentration of antibiotics. It was also **recommended by the authors that in general practice a concentration of not more than 0.2 mg/**

---

*See appendix 'A'

ml chloramphenicol should be used to preserve sputum, subsequently cultured for *B. dermatitidis, Histoplasma capsulatum, Paracoccidioides brasiliensis* or *Sporothrix schenckii.*

Media : Usually blood agar and Sabouraud's agar are used for cultivating fungi, Sabouraud's medium was introduced by Sabouraud, the distinguished French dermatologist of nineteenth century for dermatophytes and it is virtually useful in the isolation of all pathogenic and nonpathogenic fungi. Glucose and peptone are the sole nutrient of Sabouraud's medium and at pH 5.6 to discourage bacterial growths. At present this medium is adjusted to neutral pH for more favourable growth of most fungi.

Selective Media

Antibacterial agents are added to the medium usually chloramphenicol 40 mg per ml or cycloheximide 500 mg per ml. Cycloheximide has the advantage that it not only reduces the growth of many saprophytic fungi but at the same time it is toxic for some pathogens as well, e.g., *Cryptococcus neoformans* and Candida Sp. **Chloramphenicol** on the other hand **inhibits the yeast phase of some dimorphic fungi** when grown **at 37°C**. Therefore, cultures should be prepared both with and without these drugs and incubated at room temperature and at 37ºC.

BRAIN-HEART-INFUSION CYCLOHEXIMIDE-CHLORAMPHENICOL AGAR

This is an enriched medium of brain heart infusion with **cycloheximide** 0.5 mg and **chloramphenicol** 0.05 mg per ml and agar 2 g per 1000 ml. On this medium the pathogenic, fastidious fungi like ***Histoplasma capsulatum, Blastomyces dermatitidis*** grow at 25ºC or room temperature but yeast phase of dimorphic fungi are inhibited by cycloheximide at 37ºC.

It is recommended for the selective isolation of most fungi pathogenic to human excepting those that are sensitive to cycloheximide, e.g., *Cryptococcus neoformans*, candida species except *C. albicans, Pseudallescheria boydii* **(partially sensitive to cycloheximide),** ***Aspergillus fumigatus*** and ***Nocardia asteroides*** **(partially sensitive to chloramphenicol).**

Selective media for Actinomyces (Anaerobic) : Actinomyces grow well in **primary isolation on Brain Heart Infusion (BHI) agar** which is an enriched medium for Actinomyces. Anaerobiosis is to be maintained. Some strains require blood.

Lenze (1938) recommended **beef bouillion agar** containing 30 per cent ascitic fluid. **Agar - 20 g per litre may be added to the maintenance medium as isolation medium.**

Single colonies may be inoculated in freshly boiled and cooled **thioglycollate (Anaerobic) broth** after removing oxygen, heart infusion broth (sugar free medium) or in **Actinomyces maintenance medium**. When sugar fermentation is studied, inoculum should be taken from a sugar free medium like heart infusion broth.

Fungi isolated should be examined for their identification, and their spores should be clearly visualised since the morphology of spores and the manner in which they are produced are characteristics for many species.

Once the fungus has been recovered, procedures should be followed for the identification of the isolated organism depending on the nature of the organism. If necessary **well-slide culture of the organism** should be prepared **(Fig. 4.1a)** to determine the morphology, nature of spore formation, etc.

For this purpose, fungi are often cultivated under conditions which stimulate sporulation, i.e., in media where they show less (than optimal) growth, e.g., **rice medium** and ***cornmeal agar*** for production of **CHLAMYDOSPORES** in case of ***C. albicans.*** The mycelia are usually white and fluffy in many dermatophytes in the earlier stage but they develop pigmented spores with the age.

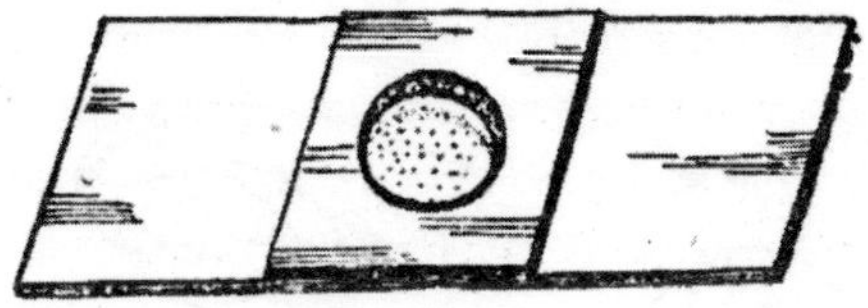

Fig. 4.1a : Well-slide

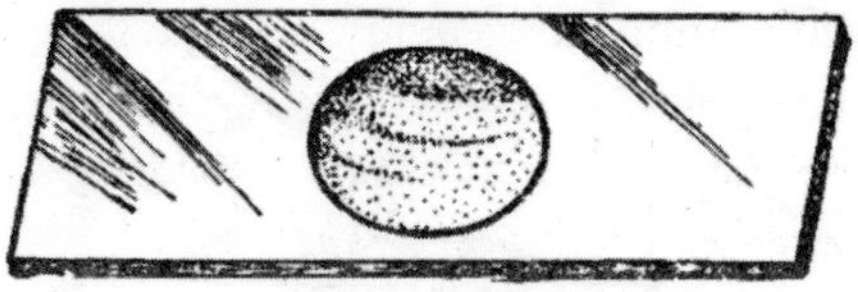

Fig. 4.1b : Hollow ground slide

Fig. 4.1c : Van Tieghem cell

Fig. 4.1d : Microculture in well-slide

MICROCULTURE

For this hanging drop cultures are studied in well-slides containing square slabs of stone about 2.5 cm square and 5 mm in thickness with a central circular hole of 1.5 cm in diameter. The slab is cemented on a microscopic slide (**Fig. 4.1a**). The slide and the coverslip should be sterile and grease-free. Both can be wrapped in craft paper and autoclaved. Sabouraud's semisolid agar (0.5 per cent agar) or any other suitable medium may be used for the study of the morphology of the fungus. The agar is melted in a boiling water bath and a drop of the medium is spread in the centre of the grease-free coverslip with the help of a platinum loop. As soon as the medium is cool, the agar sets on the coverslip and a tiny bit of inoculum of the fungus to be studied, is placed on agar. The coverslip is turned upside down so that the inoculum is in the centre of the well, hanging down from the coverslip (**Fig. 4.1d**). The edge of the coverslip is quickly sealed with a mixture of hard and soft paraffin in a proportion of 1 : 2. The mixture is melted and spread at the edge of the coverslip with the help of a camel hair brush. The fungus grows in the medium and is studied for about 6 weeks. Instead of media, thick wheat or starch solution, scales, hair, shavings from horn, or hoops may be inoculated with the fungus with a drop of sterile water in the chamber. Instead of well-slides, a hollow-ground slide (**Fig. 4.1b**) or a **Van Tieghem cell** (**Fig. 1c**) can be used for the same purpose. The progress of the growth is studied from time to time for reproductive organs.

CULTURE MOUNTS are prepared from the hanging drop culture or by taking a bit of growth from agar. The agar is teased, separated from the growth and gently washed away. The growth is then mounted in lactophenol-cotton blue and sealed. Another useful method is to spread a drop of melted agar under a coverslip. The preparation is preserved in an improvised, sterile, moist chamber. From time to time the reproductive organs are studied. When these reproductive organs are fully developed the coverslip is taken out and a culture mount is prepared with lactophenol blue.

RIDDEL'S AGAR BLOCK METHOD (permanent preparation) : In this about 10 ml of any chosen agar (1.5%) is poured into a sterile 9 cm Petri dish forming a layer of 2 mm deep. When the medium solidifies, 1 cm sq. block is cut with a sterile blade or needle. The dishful of blocks may be stored at 4ºC until required. With the help of sterile forceps transfer the cut agar block on a sterile slide. A needle point inoculum from the tube or plate culture is placed at the mid point of each edge of the agar block over the whole 2 mm depth on four sides of the square. The cover slip is placed centrally upon the surface of the agar square. The slide is then transferred to a moist chamber and incubated at room temperature or at 28ºC. Care is taken

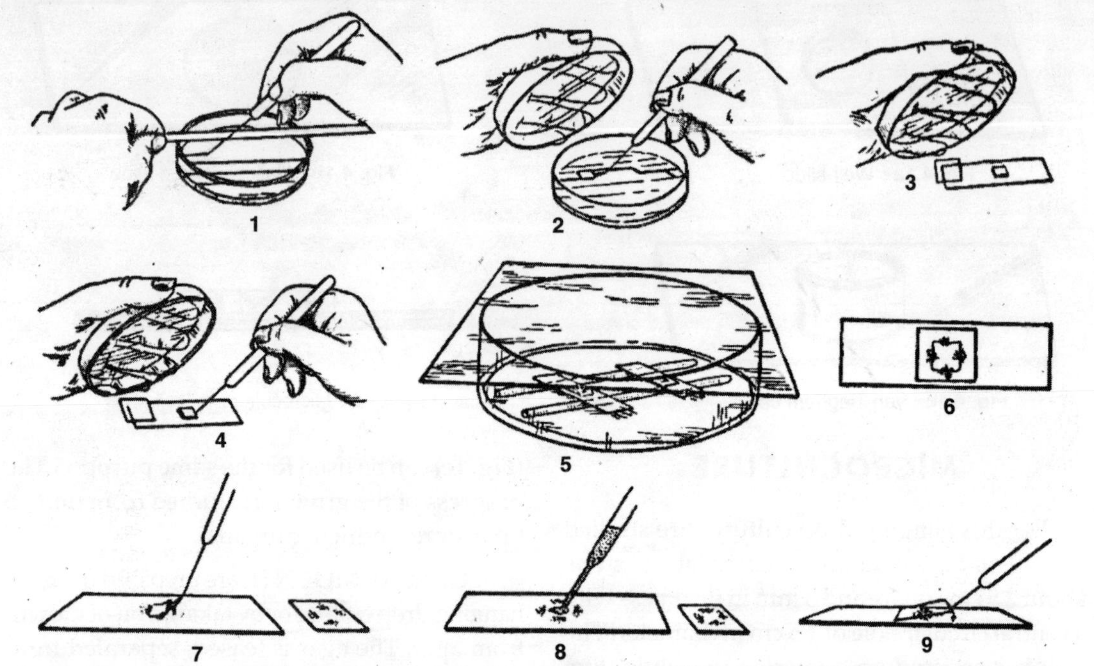

Fig. 4.2 : (1) Flamed glass rod on Petri dish and a sterile dissecting needle. (2) An agar square block is being lifted out of Petri dish. (3) The agar block is transferred rapidly to the centre of the cooled slides. (4) Needle-point inoculum of spore or mycelium is being inoculated at the edge of the agar block. (5) With sterile forceps the inoculated slides are transferred to a moist chamber, care is taken that the coverslip is not moved during growth. (6) Alow to grow from 1-3 mm off the agar block. (7) The lifter is gently inserted under the agar to lift the coverslip completely off the slide. (8) Agar with most of the fungus growths are discarded. Put a drop of ethyl alcohol at the centre of the growth to wet the fungus growth on slide or coverslip. (9) Before evaporation of alcohol put a drop of lactophenol and put a clean coverslip of bigger size on a slide or on the coverslip. Modified from R. Riddel (1950). Permanent mycological preparations. *Mycologia* **42**, 265-270.

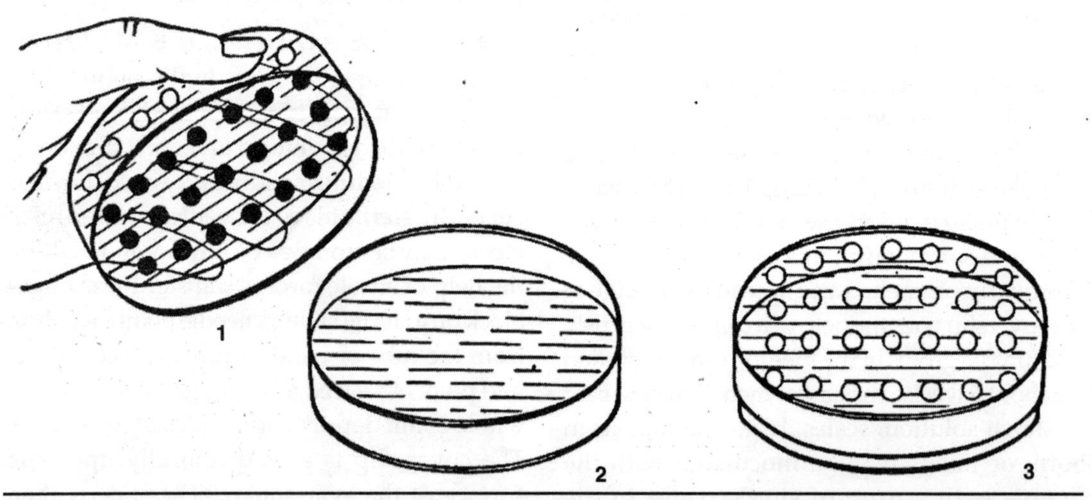

Fig. 4.3 : (1) Sterile Petri dish showing drops of melted Sabouraud's agar on the cover. (2) Petri dish showing sterile distilled water at the bottom. (3) Petri dish showing Sabouraud's agar drops inside the cover of the plate (Dr. Curry's method).

Fig. 4.4 : "Hair on soil" culture showing perforation on the surface of hair (Keratinolysis). This shows Keratinolytic property of the fungus

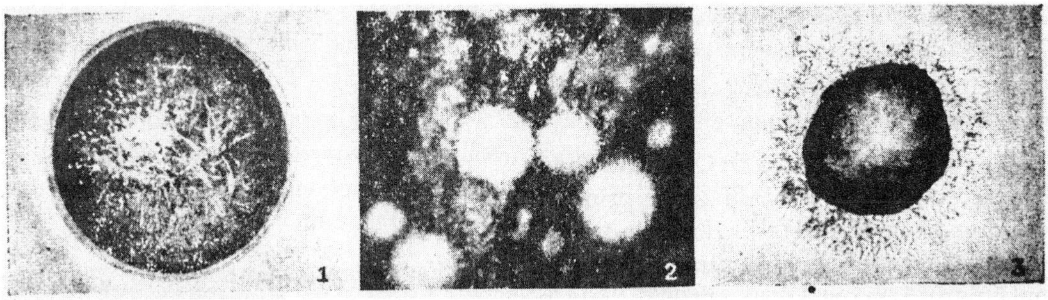

Fig. 4.5 : "Hair on soil" culture : 2 and 3 showing formation of cleistothecia. Modified from Ajello, L. (1961). Ascigerous state of *Microsporum cookei. Sabouraudia* **1**(3), 173-177

that the cover slip is not moved during the growth of the fungus while taking out of the chamber and examining every 3 days or so with low power or changing to high power. After the full growth the coverslip is lifted vertically from the agar block without twisting, (when full growth has occurred). This gives growth of the fungus underneath the coverslip. All that remains on both slide and coverslip. All that remains on both slide and coverslip after discarding agar are 4 areas of growth joined by scanty line of mycelium.

A drop of 95 per cent ethyl alcohol is put in the centre of each of the rings of the growth and is allowed to flow outward to wet the fungus. Just before it has completely evaporated, mount it in lactophenol cotton blue. A clean coverslip is lowered on the slide to spread the solution outward over the damp mycelia. The slide is left overnight and excess of the solution is blotted out with strips of blotting paper applied to the edge of the coverslip. Seal the edge of the coverslip with nail polish. The coverslip over the agar block can be similarly mounted on a clean slide. Thus 2 permanent preparations may be made from one agar block preparation (**Fig. 4.2**).

The method provides a simple means of observing aerial and submerged mycelial growth (of fungus) and the developing **fruit bodies**, like **macro** – or **micro-conidia** *in situ*, and also the effects of antibiotic and chemicals incorporated in the medium affecting the fungus morphology can be studied. It provides a rapid routine diagnostic aid on the morphologic study of the fungus. It is also possible to use infected hair or scales as inocula. For further details readers are referred to the original article.

As an alternative method materials of dermatophytes could be inoculated on agar drops. **Dr. Curry of Manchester Uviversity** demonstrated a simple method for this purpose (**Fig. 4.3**). A sterile Petri dish is inverted and on its lid, the melted media from a Sabouraud's agar-tube dipped in a boiling water bath are dropped rapidly from a Pasteur Pipette. Drops are put at a distance of about one inch or so in rows and columns. Agar drops solidify in no time. These drops are then inoculated with the infective material after a preliminary treatment in alcohol for about five minutes. About 20 such inocula may be put in one plate. Put a thin layer of sterile distilled water at the bottom of the Petri dish and put back the inoculated plate in position. Thus the agar drops do not dry until water is exhausted by evaporation. Water may be replenished if necessary from time to time. The lid of the Petri dish can be everted, placed on the stage of the microscope and examined for the reproductive organs.

The authors prefer to use **Dr Curry's method** in which details are available from primary culture or a simpler method by putting a drop of selected and melted agar (**Sabouraud's medium containing 0.5 per cent Difco agar**) with the help of a Pasteur pipette on a sterile coverslip which is inoculated with the fungus. The coverslip is inverted on a hollow ground slide (hanging drop culture) and kept sealed with a mixture of vaseline and paraffin (2 : 1) (**Fig. 4.3**).

GROWTH OF DERMATOPHYTES ON HAIR *in vitro*. The manner of invasion of hair *in vivo* is different from the way they grow on isolated hairs *in vitro*.

If sterilised human or animal hairs are placed in water and inoculated with arthrospores of an infected hair the fungus produces macroconidia and microconidia with aerial mycelia. Some species spread perforations in a wedge-shaped holes or tunnels borring the hair as a result of enzymic activity that dissolves the keratin.

Hair is perforated by all Microsporum species as well as by soil fungi like *Trichophyton (Keratinomyces) ajelloi* and other soil trichophytons (**Fig. 4.4**).

HAIR ON SOIL CULTURE

Hair baits of human or animal hair put on a Petri dish containing "moist soils" are inoculated for the **study of cleistothecia to determine the perfect state**. It is often positive for soil fungi like **Microsporum** and **Trichophyton show Arthroderma** or **Nannizzia states (Fig. 3.5)**.

CULTURE MEDIA ***for hair on soil culture*** (Young, 1968). She used (1) **a modification of Decroey's grain based medium**. Alternatively, (2) **Whiteman and Silva Hunter's modified alphacel medium** may be used (OV 8 medium)- Finely ground oat meal-10 g; $MgSO_4$, $7H_2O$-1g; $NaNo_3$-1g; v8 vegetable juice-10 ml, and distilled water-990 ml; pH adjusted to 5.6; autoclaved at 15 1bs. pressure for 15 minutes.

Plates are prepared by autoclaving garden soil, moistened with distilled water in glass Petri dishes together with horse hair scattered on the surface of the soil.

The plates are inoculated with spore suspension of the strains under investigation and moistened with sterile distilled water when required during culture period.

Hyphal fusion : This was originally devised by Davidson *et. al.* (1932). In this two inocula of each of the fungi to be studied are placed side by side in a **hanging drop culture in well-slide**, without touching each other. The fungus grows in the medium. When they are allowed to grow, mycelia of similar strains or same species only fuse with each other, but mycelia of different strains or species do not show any hyphal fusion.

BIOCHEMICAL REACTIONS

Biochemical reactions are studied after isolation of organism, like *Candida albicans* and other yeast like fungi. **Bacteriological sugar tubes** are used for this purpose. Fermentation results are fairly constant for these organisms and thus may help in the determination of species. Proteolytic activity of fungi is characterised by **liquefaction of coagulated serum, gelatin or peptonisation of milk**.

FILTERED ULTRAVIOLET (WOOD'S) light : This has been extensively used by American workers. By this, fluorescence is studied in lesions and in cultures, by using ultraviolet light passing through Wood's filter. This filter allows passage of **ultraviolet light of** wave length of about **3650 Å (365m μ)**. Certain fungi or infected materials show fluorescence in this light. The method has been utilised for the detection of infection caused by *Microsporum audouinii*, *M. gypseum*, and *M. canis* in the routine diagnosis of ringworm of the scalp. The changes detected under **Wood's filter** are mentioned in respective chapters.

HISTOPATHOLOGY

Fungi that invade deeper tissues or internal organs show formation of granulomas often with giant cell formation. The characteristic lesions are dealt with in respective diseases.

Some fungi may be visible with haematoxylin and eosin stain but many pathogenic fungi are not at all stained by this method although for the demonstration of tissue reaction this is the only stain to show the nature of infiltrating cells and may provide important clues for diagnosis. However, there may be a wide range of tissue reactions due to a single aetiological agent, and many aetiological agents may produce similar tissue reaction. It has also the advantage that eosinophilic deposit in phycomycosis is often seen with this stain without, however, demonstration of the fungus.

Special stains have therefore, been devised and those have stimulated pathologists to search for the aetiological agent and have immensely increased the possibility to specifically diagnose many fungal diseases from biopsy material (fixed tissue) without culture.

STAINS

Periodic Acid Schiff (PAS) stain with a **counterstain of light green** demonstrates many fungi in tissue **pink or rose red on a green background in** cases of **yeast forms, spores** or **hyphae**, but Gram positive filaments of Actinomyces and Nocardia are not satisfactorily shown by this technique. The actinomycotic granules stain well, but not the organism.

Gridley's stain for fungi **stains mucin and elastic tissue deep blue on a yellow background**. Morphologic **details of yeast forms and hyphae are easily visible; conidia stain deep rose red to purple**. But this method like the PAS stain does not stain the filaments of Actinomyces and Nocardia.

Gomori's methenamine silver stain (GMS) is useful in screening all forms of fungi which are filamentous like ***Actinomyces bovis*** and *Nocardia asteroides* or ***Myco. tuberculosis***. This stains the filaments blackish brown. For calcified foci of pulmonary nodules of ***Histoplasma capsulatum*** this method is useless, as chromic acid used in this technique dissolves calcium.

Leishmania donovani and *Toxoplasma gondii* are not stained by **PAS, Gridley or Gomori** stains.

Mayer's mucicarmine stain is **useful** for **differentiating cryptococcus from other fungi** of similar size. **Cryptococcus** polysaccharide **capsule** is beautifully stained by this method. *Rhinosporidium* (not considered to be a fungus at present) is also stained by this but the stain is not required for identification of this fungus. *Blastomyces* may be occasionally stained by mucicarmine but it is not sufficiently carminophilic to be mistaken for cryptococcus.

Gram stain (Brown & Brenn). For study of tissues harbouring bacteria or fungi this technique is preferred giving a purplish black appearance of **Gram positive micro-organisms** on yellowish background. Some fungi like **Actinomyces** and **Nocardia** are stained by this method.

Ziehl-Neelsen (Z - N) stain is indicated in granulomatous lesions for detection of tuberculous infection, but for **acid-fast Nocardia**

Kinyoun carbol fuchsin stain* should be used in which **1.0 per cent aqueous sulphuric acid instead of acid alcohol** is used to destain (decolorisation). **Counter-stain of methylene blue** is used **for 30 seconds, washed, blotted dry** and examined with oil immersion.

For screening tissues for aetiologic agents Gomori's silver methenamine stain, a Gram method and Z - N (Ziehl-Neelsen) method should be used. Two slides are taken for GMS method; one is kept in silver nitrate solution for 60 minutes and another for 90 minutes (see appendix for stains).

An experienced pathologist with the above stains can easily identify the following mycoses **in histopathological sections**, e.g., **Actinomycosis** and **nocardiosis, aspergillosis, North and South American blastomycosis, candidiasis, chromoblastomycosis, coccidioidomycosis, cryptococcosis, histoplasmosis, mucormycosis** and **rhinosporidiosis** (not considered a fungus) and so on.

ANIMAL INOCULATION

Inoculation of mice with infected material or culture is very important specially in cases of **blastomycosis, coccidioidomycosis, cryptococcosis and sporotrichosis**. This method should therefore, be resorted to along with usual examinations, namely **direct microscopic examination** and **culture**. The method is also useful to isolate the organism in pure form from contaminated materials. For inoculation, susceptible animals like mouse, rat, rabbit or monkey may be used. Again when an organism is isolated, animal inoculation is useful for finding out the pathogenicity; some of the saprophytes or contaminants can thus be excluded by adopting the method of animal inoculation.

In **ringworm infection**, when the fungus is isolated in culture, **animal inoculations** are undertaken as a routine measure for **pathogenicity test**. This is **done locally by shaving the skin on the back or abdomen** and the suspension of the culture is **applied** on the part **after scarification**. Fungi of animal sources usually produce lesions, whereas the lesions produced by human species are spontaneously cured in about 10 days.

*__Basic fuchsin__-4.0 g, **Phenol** crystal-8.0g, **Alcohol** (95%)-20 ml, **Distilled water**-100 ml. This is about 4 times the strength of Carbol fuchsin used for *Myco.tuberculosis*, The stain acts for 3 minutes at room temperature without heating.

REFERENCES

Brown, J. H. and Brenn, L. (1931). Methods for differential staining of Gram-positive and Gram-negative bacteria in tissue sections. *Bull. Johns Hopkins Hosp.* **48**, 69-73.

Dr. J. Curry-*Personal Communication*.

Davidson, A. M., Dowding, E. S. and Buller, A. H. R. (1932). Hyphal fusions in dermatophytes. *Canad. J. Res.* **6**, 1.

Georg, L. K., Ajello, L. and Papageorg, C. (1954). Use of cycloheximide in the selective isolation of fungi pathogenic to man. *J. lab. clin. Med.* **44**, 422-428.

Gridley, H. F. (1953). Technique used at the Armed Forces institute of Pathology. *Am. J. clin. Path.* **23**, 303-307.

Grocott, R. G. A stain for fungi in tissue sections and smears. *Am. J. clin. Path.* **25**, 975-979.

Malloary, F. B. Mayer's mucicarmine histological stain for Cryptococcus neoformans. Pathological Technique, Philadelphia, W. B. Saunders Co., page 130.

McDonogh, E. S., Georg, L. K., Ajello, L. and Brinkman, S. (1960a). Growth of dimorphic human pathogenic fungi on media containing cycloheximide and chloramphenicol. *Mycopathologia et Mycologia, Applicata* **13**, 113-120.

McDonough, E. S., Ajello, L., Georg, L. K. and Brinkman, S. (1960b). In vitro effects of antibiotics on yeast phase of *Blastomyces dermatitidis* and other fungi. *J. lab. clin. Med.* **55**, 116-119.

Riddel, R. W. (1950). Permanent stained mycological preparations obtained by slide culture. *Mycologia* **42**, 265-270.

Young, C. N. (1968). Pseudocleistothecia in *Trichophyton rubrum*. *Sabouraudia* **6**(2), 160-162.

5

Pathology of Subcutaneous and Systemic Mycoses

H. L. E. Grueber

Introduction, Close co-operation between **clinicians, pathologists, bacteriologists** and **mycologists** is most essential for a successful diagnosis of mycologic disease. Now, we do not grope (search) in the dark when confronted with the diagnosis of fungal diseases. Much progress has been made in this field recently. However, the subject is vast and includes not only histopathologic problems but also taxonomic-mycologic questions, a complex serology, biochemistry and often ill-defined clinico-pathological symptoms. This short chapter is therefore, not directed to the specialist, but it is hoped that the physicians may gain some insight into the pathology of mycologic disease and that the less experienced pathologists may find some practical diagnostic guidance.

THE DIAGNOSIS OF SYSTEMIC MYCOSIS RESTS ON SIX FUNDAMENTALS

(1) THE PRESENTING CLINICO-PATHOLOGIC AND RADIOLOGIC PICTURE

For instance, the combination of a"strange" radiologic lesion in the lung of a middle aged diabetic possibly in combination with "mental" symptoms, should trigger off the tentative diagnosis of an opportunistic, **Zygomycosis (phycomycosis mucormycosis)** in the mind of the suspecting physician. Quick laboratory confirmation of the suspicion is then often a matter of life and death question for the patient.

(2) DIRECT MICROSCOPIC EXAMINATION

Wound exudate, serous fluid, pleural, pericardial, and **peritoneal exudate, pus, necrotic tissue fragments, granules, CSF, teased tissue snippets**, all these may be **examined directly under the microscope**. Organisms can often be **identified in unstained material. Sincere interest matched with a little practice are almost more important than a very specialised knowledge**. Tissue material can be "cleared" by the addition of a few drops of 10 to 15% potassium hydroxide solution for some minutes prior to examination. By addition of a few drops of **lactophenol-cotton-blue under a cover slip preparation**, fungal elements may be stained and a better contrast may be obtained.*

***The authors get good result in a coverslip wet preparation with MCGUIRE STAIN which stains the fungus purple.**

(3) **HISTOPATHOLOGIC EXAMINATION OF FORMALIN-FIXED TISSUE**

Proper cultural identification of organisms should be the ultimate aim, but the histopathological method of diagnosis of fungal lesions is now so advanced, that most organisms can be accurately identified by an experienced pathologist. The method has the following advantages :

(a) Results can usually be obtained quicker than by cultural methods.

(b) Tissue reaction and invasion indicate to some extent aggressiveness and pathogenicity of the fungus as these informations cannot be obtained by culture.

(c) It is often the only evidence available either, because cultures are sterile or contaminated, or no unfixed material is available.

(d) Histopathologic diagnosis is sufficiently accurate for a proper assessment of optimal treatment.

(e) Spoilage of specimens through secondary bacterial growth in tropical climate is avoided because the tissue is sent in a fixed condition.

(4) **MYCOLOGIC CULTURE**

This yields best results in the hands of an experienced mycologist. If no such specialist is available, primary culture should be attempted in the local laboratory and cultures should then be sent to a reference mycologic laboratory for final identification.

(5) **IMMUNOLOGY AND SEROLOGY**

There are still vagaries and contradictions in the literature regarding the value of serologic tests in some fungal infections. The limitations of serologic tests are the following :

(a) Various tests, for instance complement fixation, gel precipitation, skin tests, etc., do not parallel each other, thus corresponding results in a given case cannot be expected.

(b) Each test reveals the presence of a different type of antibody, and each is of different significance for diagnosis as well as for the prognosis of the further clinical course.

(c) The immunology of different mycoses on the one hand, and the availability of proper antigens on the other, varies so greatly that serologic tests which might be useful in one infection are of no value in another.

(d) Smaller laboratories should use only approved methods and tried-antigens supplied by large reference laboratories. In India, seldom standardised antigens from Indian strains are available at present.

There has been such enormous progress in the field of diagnostic serology in mycotic disease during recent years, that details are beyond the scope of this short chapter. New methods include also the application of fluorescent antibody (FA) techniques. For more detailed information, the literature should be consulted (Seelinger, 1958, 1968; Wilson and Plunkett, 1965; Kaplan and Kraft, 1969 and chapter V of this book).

(6) **AUTOPSY**

The fungal aetiology of a disease can often only be established by a postmortem investigation. **No brain with a suspected inflammatory lesion should be fixed without normal tissue portions and exudate having been taken for mycological culture. The same applies to pulmonary lesions** and to ulcers of the oesophagus and of the gastro-intestinal tract. **It is surprising** how **many unsuspected fungal infections can be diagnosed in this way.**

Many other consideration may be helpful in the diagnosis of fungal disease. The ***geographic pathology*** of fungi should be kept in mind. The *ecology*, the *epidemiologic behaviour* of the organisms are often dependent on soil conditions, rainfall, and on the type of vegetation present.

Definition : Transitions between superficial and systemic mycoses are not so rare. Every superficial mycosis should therefore, be treated with respect. The organism may suddenly disseminate and the infection may end fatally.

I. SUPERFICIAL MYCOSES

See Chapter 10

II. CUTANEOUS MYCOSES

They are mostly The Ringworm infections. See Chapter 28 (Dermatophytis); it also includes **CANDIDIOSIS** as it affects the **SKIN** and **NAILS**.

III. SUBCUTANEOUS MYCOSES (INTERMEDIATE MYCOSES)

The infection is confined to the cutis and subcutis. This type is represented by :

1. **Chromoblastomycosis**
2. **Mycetoma**
3. **Sporotrichosis**
4. **Subcutaneous (phyco) Zygomycosis.**

Of these four, sporotrichosis has the greatest tendency to disseminate generally, though stray cases of dissemination have been reported of the other maladies.

IV. DEEP MYCOSES

Mycetoma is sometimes lumped together with subcutaneous mycoses, but during its natural course it is neither confined to the subcutaneous tissue as it penetrates invariably destructively through muscle, fascia and bone, like a malignant tumour, nor is a mycetoma causing generalised disease, though a local spread has occasionally been reported. The best term for a mycetoma is therefore, a ***deep mycosis***. In contrast to most other mycoses, mycetomas are caused by various types of fungi and also by higher bacteria. Mycetoma is therefore, not an aetiologically defined, but a clinical entity. Sometimes **Deep Mycoses refer to Systemic Mycoses.**

V. SYSTEMIC MYCOSIS

The term is often used **synonymously** with "**deep mycoses**". It connotes involvement of single or multiple internal organs or organ systems and it may finally end in a generalised and often fatal dissemination. The following systemic mycoses will be considered in the subsequent chapter :

(1) **Actinomycosis** (Also called **"subcutaneous)**

(2) **Aspergillosis* and other hyphomycoses**

(3) **Blastomycoses :**

  (a) **North American Blastomycosis**

  (b) **South American Blastomycosis and Lobo's disease.**

(4) **Candidiasis*** (Also called **"CUTANEOUS"** Mycosis)

(5) **Cladosporiosis* of the central nervous system.**

(6) **Coccidioidomycosis**

(7) **Cryptococcosis**

(8) **Geotrichosis**

(9) **Histoplasmosis**

(10) **Nocardiosis** (Also **cutaneous** and **subcutaneous)**

(11) **Zygomycosis (Phycomycosis), (mucormycosis)**

VI. OPPORTUNISTIC MYCOSES*

VII. MYCOTOXIC MYCOSES**

A small number of rather rare systemic mycoses cannot be considered here. The Bibliography under "reference books" should be consulted for further information.

***These fungi and/or fungal diseases along with other several fungi Constitute OPPORTUNISTIC MYCOSIS (see also end of Table 3.2).**

****See ch. 6, Table 6.1.**

BIBLIOGRAPHY

(Reference books and publications)

Ajello, L. and others ; Laboratory Manual for Medical Mycology, Public Health Service publication No. 994. U. S. Govt. Printing Office, Washington, D. C. 2nd ed., **1962**.

Bader, G. Die Visceralen Mycosen (in German). V. E. B. G. Fischer Verlag, Jena, DDR, **1965**.

Conant, N. F., Smith D. T., Baker, R. D., Callaway, J. L. and Martin, D. St. Manual of Clinical Mycology. W. B. Saunders Co., Philadelphia, Reprint, **1963**.

Emmons, Ch. W., Binford, C. H. and Utz, J. P. Medical Mycology. 2nd ed., Lea & Febiger, Philadelphia, **1970**.

Fejer, E., Olah, D., Szathmary, S, Szodoray, L. and Uri, J. Medizinische Mycologie und Pilzkrankheiten. Akadeiaikiado, Budapest, **1966** (in German).

Fegeler, F. Medizinische Mycologie in Praxis und Klinik. Springer, Berlin, Heidelberg and New York, **1966** (in German).

Fetter, B. F., Klintworth, G. K. and Hendry, W. S. *Mycoses of the central Nervous System.* The Williams and Wilkins Co., **1967**.

Fungus diseases in India, Calcutta School of Tropical Medicine, **1962**.

Kaplan, W. and Kraft, D. E. (1969). Demonstration of Pathogenic Fungi in Formalin-fixed Tissues by Immunofluorescence. *Am. J. cl. Path.* **52**, 420-432.

Laboratory procedures in clinical mycology, Department of the army technical manual : Washington D. C., **1964**.

Moss, E. S. and Mcqouwn, A. L. Atlas of Medical Mycology : Williams & Wilkins Co. Baltimore, **1960**.

Review of medical and veterinary mycology. Periodical Review. Commonwealth Mycological Institute : Kew, Surrey, England.

Seelinger, H. P. R. Mycologische Serodiagnostik Leipzig, J. and A. Barth Verlag. **1958**.

Seelinger, H. P. R. "Systemic Mycoses" edited by G. E. W. Wolstenholme and Ruth Porter, PP. 54-72. J. and A. Churchill, London. **1968**.

"The Pathologic Anatomy of Mycoses" in "Handbuch der speziellen pathologischen Anatomie und Histologie" vol. III, part 5. P. 1192, edited by R. D. Baker, Springer Verlag, Berlin, Heidelberg, New York, **1971**.

Wilson J. W. and Plunkett, O. A. The fungus diseases of man. University of California Press, Berkeley and Los Angeles, **1965**.

Winner, H. I. and Hurley, R. : Symposium on *Candida* infections. E & S. Livingstone Ltd., Edinburgh and London, **1966**.

6

Mycotoxins And Mycotoxicosis

TOXIC FUNGI

Toxic Mushrooms

Many of the mushrooms are toxic species and ingestion of erroneously identified as edible, or inadvertent ingestion causes acute discomfort, severe illness and rarely, death.

There is method to differentiate toxic from edible mushrooms.

Mushroom poisoning in adults usually is caused by ingestion of over-ripe specimens damaged by bacterial infection, and insect invasion, or to which the patient is allergic.

Mycetismus

Five types of recognized **mycetismus causes mushroom poisoning**. ***Mycetismus gastro-intestinalis*** designates a relatively mild illness characterized by nausea, vomiting and diarrhea. It follows ingestion of *Russula emetica, Boletus satanus, B. miniato-olivaceus, Lactarius torminosis, Entoloma lividum* and *Lepiota morgani. Mycetismus choleriformis* is a severe and sometimes fatal illness characterized by a latent period of 6 to 24 hours, followed by severe gastrointestinal symptoms. A temporary remission may be followed by a fatal relapse. The delay in appearance of symptoms contributes to the fatality rate. This type of poisoning following ingestion of *Amanita phalloides* and a few other species. ***Mycetismus nervosus*** involves the **parasympathetic nervous system** and causes **gastrointestinal distress**. It follows ingestion of ***Amanita muscaria, A. patherina, Inocybe infelix, Clitocybe illudens,*** and other species. ***Mycetismus sanguinareus*** is characterized by transient hemoglobinuria as well as jaundice due to ingestion of ***Helvella esculenta*** containing **heat labile hemolysins** and **thermo-stable toxins**. ***Mycetismus cerebralis*** is characterized by transient **hallucinogenic symptoms** as caused by ingestion of the species of Psilocybe, Panaeolus, and a few other genera as well.

FOUR TYPES OF MUSHROOM TOXINS

(a) The poisons are : ***Protoplasmic poisons*** of (**Amanita toxins and Helvella poisons**),

(b) Compounds exerting Neurologic effects (including **muscarine, "pilzatropine"** and **psilocybin**),

(c) Gastrointestinal irritants and

(d) Disulfide-like constituents. Most of the species of **toxic mushrooms** contain more than one toxin.

(A) PROTOPLASMIC POISONS

They are ***Amanita toxins*** the plants possible are ***Amanita phalloides***, *A. verna, A. virosa*, and *Galerina venenata* and *Lepiota cretacea*. The **toxins** are **cyclopeptides**, e.g., **phalloidin, phalloin** α-

amanitin, **β-amanitin,** and **γ-amanitin,** etc. They have an **LD_{50} for intraperitoneal injection** use white mouse of 1.4 mg of phalloin body weight and 0.1, mg of the **alfa amanitin**/kg.

Hepatic glucose mobilization with high blood sugar levels is the first toxic manifestation in mouse. Within an hour phalloidin begins to inhibit enzymes of carbohydrate, fat and protein metabolism. **Severe liver hemorrhage** occurs in rats and mice with **secondary fatty degeneration of the liver.**

In human, after ingestion of *A. phalloides*, delayed appearance of gastrointestinal distress and the severity of subsequent illness depends upon the α-amanitin content of the *A. phalloides* meal. There occur **cramps, drowsiness, dilatation of the pupils, albuminuria, hemoglobinuria, oliguria or anuria. Coma and death** may occur 3 to 5 days. Cause of death is **acute hepatic necrosis.** One third patients die of **jaundice,** with **deposition of fat in liver, kidneys** and **heart.**

TREATMENT of severe mycetismus is directed towards removal of the fungus by gastro-intestinal lavage, **emetics, cathartics,** and **colonic lavage.**

(B) COMPOUNDS WITH NEUROLOGIC EFFECTS

Small amounts *Muscarine* was isolated from ***Amanita muscaria,*** and ***A. patherina*** to 0.73 per cent of dry weight of the mushroom have been reported. It is also present in ***Boletus, Clitocybe, Lepiota, Hebeloma*** and ***Russula***. **Muscarine chloride is the quaternary ammonium salt** of 5-aminomethyl tetrahydro-3-hydroxy-2-methylsulfuran that exerts a specific peripheral **parasympathomimetic action** due to its effect on **postganglionic parasympathetic effector** sites. Its activity is enhanced by cholinesterase inhibition activity.

CLINICALLY symptoms appear 15 minutes after ingestion of Inocybe (mushrooms). **Perspiration, lachrimation** and **salivation** occur before **nausea, vomiting** and **diarrhoea**. Severe **vomiting** and continuous **diarrhoea** produce **weakness, dehydration** and **muscle cramps**. There is a **rapid, feeble pulse,** in an apathetic patient. Loss of body fluids (3 liters/day). **Blood pressure falls** for **decreased** content of **sugar** and **sodium chloride content of blood; blood urea nitrogen increases, potassium deficiency** and **leukocytosis occur**. Children and adolescents may succumb during this ***primary acute phase*** of poisoning. **Irregular pulse, constricted pupils,** and **asthmatic breathing** also occurs. **Death is due to cardiac or respiratory failure.**

TREATMENT consists of gastric lavage when vomiting is absent, subcutaneous injection of Atrophine sulfate to repeated every half hour when necessary in a dose of 0.5 to 1.0 mg, and fluid replacement.

(C) GASTROINTESTINAL IRRITANTS

They are resin-like objects present in species of ***Agaricus, Boletus, Cantharellus, Clavaria, Clitocybe, Lepiota, Paxillus*** and ***Russula***. **Mushrooms, *Lactarius torminosus, Rhodophyllus sinuatus*** and ***Tricholoma pardinum*** sometimes have caused severe or fatal illness in children. **SYMPTOMS** appear ½ hour after ingestion. **Nausea, vomiting and diarrhoea** may be severe and accompanied by severe **abdominal cramps** and pain. ***Rhodophyllus sinuatus*** and **other mushrooms cause violent gastroenteritis** and may be **hepatotoxic.**

TREATMENT includes gastric lavage and supportive treatment. Symptoms are over by a few hours.

(D) TETRAETHYLTHIURAM DISULFIDE (TETD)-LIKE CONSTITUENTS

Coprinus atramentarius mushroom, or up to 48 hours thereafter, produces symptoms (beginning within ½ to 2 hours after alcohol consumption) which resemble those of the **TETD-ethanol reaction** with alcohol consumption. **SYMPTOMS** are **flushing, palpitations, dyspnoea, hyperventilation** and **tachycardia,**

without vomiting and diarrhoea of short duration; if severe and prolonged, the **TREATMENT** is gastric lavage.

THE MYCOTOXICOSES AND MYCOTOXINS (AFLATOXINS)

Mycotoxicosis, as opposed to mycetismus, is poisoning by ingestion of toxins of fungal origin (*mycotoxins*) in foods which have been altered or damaged by the growth of toxin-producing molds. This **rye flour usually contained ergot and the gangrenous lesions** and other manifestations of chronic ergot poisoning were familiar to the physician and consumer as well. Ergotism might be designated a form of either mycetismus or mycotoxicosis because it is caused by ingestion of the sclerotium of ***Claviceps purpurea*** (the fungus has replaced most of the normal tissues of the rye seed). Ergotim might be designated a form of either mycetismus or mycotoxicosis. A heat labile toxins in maize meal overgrown by *Aspergillus contains a fumigatus* and ***A. flavescens mycotoxin*** etiologically being **related to pellagra**. Interest in toxins of ***A. fumigatus*** was enhanced by the recognized virulence of the fungus for birds and mammals, but some investigators failed in attempts to demonstrate toxin production by this pathogenic mold. Henrici (1939), isolated. An endotoxin from ***A. fumigatus Penicillium puberulum***, when grown on maize meal, produced **penicillic acid** ($C_8H_{10}O_4$), and a calculation stands that a subcutaneous injection of 0.3 g of penicillic acid killed the number of mice weighting 1 kg. Hodges *et. al.* (1964), demonstrated that this fungus produces also **aflatoxin**.

Hemorrhagic disease occurs following ingestion of moldy grain and forage (for horse and cattle) crops and causes "**Alimentary toxic aleukia**". It occurs in persons who and more so by eating cereal grain left in the fields during winter which is overgrown by ***Fusarium sporotrichioides***. Immediate effect is local **irritation of oral and pharyngeal mucosa, sweating, nausea, emesis and insomnia**. Ingestion of large amounts produces **gastritis, diarrhoea, fever** (40ºC) vertigo, cyanosis and convulsions occurs in large doses. Chronic consumption produces **leukopenia, malaise, vertigo, petechial rash, and ulcerative and gangrenous lesions of the pharynx** causing **aphonia and death** by strangulation. A milder disease is caused by ***Fusarium roseum***; a few more species also produce mycotic toxins.

Another type of mycotoxicosis, the (**Stachybotryotoxicosis**) is caused by ***Stachybotrys alternans***. The toxin was isolated toxin in crystalline form. There may be leukopenia, with death following 1 to 5 days in horses after severe agranulocytosis. **Human stachybotryo-toxicosis** occurs **by exposure to hay, by inhalation or absorption of the toxin through the skin. Skin rashs in the axillae, pharyngitis and mild leukopenia (2,000/mm^3) occurs**.

The **mycotoxicoses in domestic birds and animals occurs with hemorrhage. In chicken**, the disease occurs with ingestion of moldy food. Feeding of chickens with food contaminated with ***Penicillium purpurogenum*** and ***P. rubrum*** occurs. Study on mycotoxicoses by the occurrence of many deaths in turkey (bird) flocks by **aflatoxin** produced by ***Aspergillus flavus, A. parasiticus (A. flavus-oryzae* series)**, and ***Penicillium puberulum***. Some other fungi produce **aflatoxin**. **Aflatoxins B_1, B_2, B_{2a}, G_1, G_2, G_{2a}** have been isolated. **Administration of aflatoxin to rats produces hepatomas and renal damage. Chronic ingestion of aflatoxin** at low levels in peanuts or other moldy food **is a cause of human hepatic disease** in different parts of the world.

Aflatoxin are produced from molds (***A. flavus-oryzae*** series), and ***A. oryzae*** is an important agent in the production of **soy sauce**, and other foods and beverages but none of the strains of ***A. oryzae*** when tested produced **aflatoxin**.

MYCOSIS GROUP VII is MYCOTOXIC MYCOSIS (Referred to Classification of Mycoses, Ch. 3)

Table 6.1 : List of Fungi Producing Mycotoxins and Causing Myctoxicosis

GENUS/SPECIES	
Agaricus species	*Fusarium roselum*
Amanita muscaria	*Fusarium sporotrichioides*
Amanita patherina	*Galerina venenata*
Amanita phalloides	*Halvella esculenta*
Amanita verna	*Halvella gigas*
Amanita virosa	*Halvella underwoodii*
Aspergillus flavescens	*Hebeloma species*
Aspergillus flavus	*Inocybe infelix*
Aspergillus flavus-oryzae series	*Lactarius torminosus*
Aspergillus fumigatus	*Lepiota cretacea*
Aspergillus parasiticus	*Lepiota morgani*
Boletus mineato-olivaceus	*Mycetismus cerebralis*
Boletus satanus	*Mycetismus choleriformis*
Cantharellus species	*Mycetismus gastrointestinalis*
Claviceps purpurea	*Mycetismus nervosus*
Clavaria species	*Mycetismus sanguinareus*
Clitocybe illudens	*Penicillium puberulum*
Conocybe siligineoides	*Penicillium purpurogenum*
Coprinus atramentarius	*Penicillium rubrum*
Entoloma lividum	*Paxullus species*
	Pithomyces chartarum
	Psilocybe species
	Rhodophyllus sinuatus
	Russula emetica
	Stachybotrys alternans
	Tricholoma pardinum

REFERENCES

Ajello, L., Georg, L. K., Kaplan, W., and Kaufman, L. *Laboratory Manual for Medical Mycology*. Washington, Govt. Printing Office, **1966**.

Alsberg, C. L., and Black, O. F. : Contributions to the study of maize deterioration. *U. S. Dept. Agri. Bull.* **270**, p. 48, 1913.

Austwick, P. K. C, and Elphick, J. J. : The occurrence of toxin-producing isolates in the *Aspergillus flavus-oryzae* series. Proc. *10th Internat'l.* Bot. Congr., p. 69, **1964**.

Berry CL (1988) : The pathology of mycotoxins. *J Pathol.* **154**, 301.

Cann, H. M., and Verhulst, H. L. : Mushroom poisoning. *Amer. J. Dis. Children*, **101**, 128-131, 1961.

Ceni, C., and Besta, C. : Ueber die Toxine von *Aspergillus fumigatus* und *A. flavescens* und deren Beziehungen zur Pellagra. *Centr. Path. Anat.*, **13**, 930, 1902.

Dubos, R., and Hirsch, J. G. : *Bacterial and Mycotic Infections of Man*, 4th ed. Philadelphia, J. B. Lippincott Co., **1965**.

Ford, W. W. : A new classification of mycetismus (mushroom poisoning). *Trans. Assoc. Amer. Phys.*, **38**, 225-229, 1923.

Forgacs, J., and Carll, W. T. : Preliminary mycotoxic studies on hemorrhagic disease in poultry. *Vet. Med.* **50**, 172, 1955.

Forgacs, J., and Carll., W. T. : Mycotoxicoses, *In : Advances in veterinary Sci.*, **7**, 273-282, 1962.

Gajdusek, D. C. : Acute infectious hemorrhagic fevers and mycotoxicoses in the Union of Soviet Socialist Republics. Washington, Med. Sci. Publ. No. 2, Army Med. Ser. Grad. School, **1953**.

Henrici, A. T. : An endotoxin from *Aspergillus fumigatus*. *J. Immunol.* **36**, 319, 1939.

Herzberg, M., Ed. : Proc. of the U. J. N. R. Conference on Toxic Microorganisms, **1968**, Honolulu. Washington. U. S. Dept. Agr., **1969**.

Hesseltine, C. W., Shotwell, O. L., Ellis, J. J., and Stubblefield, R. D. : Aflatoxin formation by *Aspergillus flavus*. *Bact. Rev.*, **30**, 795-805, 1966.

Hodges, F. A., Zust, J. R., Smith, H. R., Nelson, A. A., Armbrecht, B. H., and Campbell, A. D. : Mycotoxins, aflatoxin isolated from *Penicillium puberulum*. *Science*, **145**, 1439, 1964.

Klein AS et al. (1989) : *Amanita poisoning:* Treatment and the role of liver transplantation. *Am J Med.* **86**, 187.

Louria, D. B. : Deep-seated mycotic infections, allergy to fungi and mycotoxins. *New England J. Med.*, **227**, 1065-1071, 1126-1134, 1967.

Riley RT, et al (1993) : Fungal toxins in foods: recent concerns. *Annu. Rev Nutr.* **13**, 167.

Smith, A. H. : *The Mushroom Hunter's Field Guide*. Ann Arbor, Univ. Michigan Press, **1958**.

Tyler, V. E. : Poisonous mushrooms. *Progr. Chem. Toxicol.*, **1**, 339-384, 1963.

Van der Werf, P. J. : *Mould Fungi and Bronchial Asthma*. Springfield, Ill., Charles C Thomas, 1958.

Wasson, R. G. : The hallucinogenic fungi of Mexico. *Bot. Mus. Leaflets, Harvard Univ.*, **19**, 137-162, 1961.

Wieland, T., and Wieland, O. : Chemistry and toxicology of the toxins of *Amanita phalloides*. *Pharmacol. Rev.*, **11**. 87-107, 1959.

Wogan, G. N., Ed. : *Mycotoxins in Foodstuffs*. Cambridge, Mass. Inst. Technol. Press, **1965**.

Wright, D. E. : Toxins produced by fungi. *Ann. Rev. Microbiol.*, **22**, 269-282, 1968.

7

Serodiagnosis of Fungal Diseases

Serodiagnosis includes **detection of antibodies in serum** and **detection of specific antigen in the organism** (fungus). They are detected by various **Antigen-Antibody reactions**, e.g., **Agglutination, Precipitation, Complement Fixation, (CF) ELISA, PCR, Microarray Tests,** etc.

Antibodies : These are useful indicators in determining the diagnosis and prognosis of many fungal infections. Paired sera should be collected, i.e., one at the height of the infection and another during convalescence or preferably several weeks after recovery. Four-fold or greater rise in titre is usually significant and acceptable diagnostic evidence of disease, although cross-reactions with heterologous antigens may appear early in illness causing confusion. Cross-reacting antibody titres, however, remain stable or move at a lower rate than the homologous titre with the progress of the disease.The tests that are useful are ; ***Complement fixation* test, (CFT)** *agglutination* and *precipitin* reactions and *fluorescent antibody* **technique (FAT)** *in vitro* or *in vivo* for the **identification of fungal antigens**. Fluorescent inhibition test is also useful in many cases. Detection of hypersensitivity is done by *skin test* with fungal antigen. The test sometimes may increase the circulating antibodies as in histoplasmosis.

Detection of Circulating Antibodies in cases of :

A. CANDIDIASIS

Agglutination and **CF test** have proved to be of **little value** because of positive responses by serum of healthy subjects or by individuals with superficial candidiasis without systemic involvement (Taschdjian *et. al.*, 1967). Recently, Stallybrass (1964) and Taschdjian *et. al.* (1964, 1967) demonstrated that **precipitins are formed only in response to systemic infection by candida species** and this reaction is not found in the sera of healthy persons. The antigen employed in the test by Taschdjian is one of the **cytoplasmic sonic extract of *C. albicans*,** designated as 'S' antigen. **Precipitins** to 'S' antigens were demonstrated **in 85%** of patients with **systemic candidiasis** but only in 8% of cases with superficial involvement. Again the **precipitin titre** may prove **useful as an index of recovery**.

B. CRYPTOCOCCOSIS

Immunological tests have only limited success in this disease but the most promising procedures are : ***indirect fluorescent antibody* (IFA) *technique*** (Vogel, 1966), ***agglutination test* for cryptococcal antibody** (Gordon *et. al.*, 1966). and *latex slide agglutination test* for cryptococcal antigen (Bloomfield *et. al.*, 1963). Vogel and

Padula (1958) used IFA method and detected *Cryptococcus neoformans* antibodies in sera of patients who were negative to conventional immunologic methods. Vogel *et. al.* (1961) by using **indirect fluorescent antibody (IFA) test detected anitbodies in six out of seven sera** from patients with culturally proved cryptococcosis. A reaction was also observed with 8% of 339 sera from persons without evidence of cryptococcosis. None of the latter titres, however, equalled those obtained in the sera from patients suffering from cryptococcosis. Vogel (1966) reported that 80% patients were positive by the IFA tests. He noted a **lack of response** in several patients with **central nervous system involvement**. Reactions with sera from non-infected cryptococcal individuals were observed but these were generally of low order of activity. No cross reactions were seen with sera from patients with blastomycosis, histoplasmosis or sporotrichosis. Although **indirect fluorescent antibody (IFA) test is not entirely specific**, it is a **sensitive method in the sero-diagnosis of cryptococcosis**. This is particularly evident in those cases of cryptococcosis which are negative for *C. neoformans* agglutinins and their antigens. ***The latex slide agglutination test*** has been successfully used by Bloomfield *et. al.* (1963) for specific detection of cryptococcal antigens, in sera and in cerebrospinal fluid of crude cases of cryptococcosis. The height of the titre appears to be co-related with the severity of the disease. A decline in the reactivity with an associated clinical response to chemotherapy, suggests that **the test has prognostic as well as diagnostic value**. The latex test basically involves the reaction between the latex particles sensitised with *anti-C. neoformans* globulin and the patient's body fluid suspected to contain cryptococcal antigen. Kaufman and Blumer (1968) observed that the maximal serodiagnosis of cryptococcosis may be obtained through concurrent use of 3 tests :

1. **Latex agglutination test with cryptococcal antigen.**
2. **Indirect fluorescent antibody (IFA) test.**
3. **Tube agglutination tests for** *C. neoformans.*

Concurrent use of these 3 tests showed a presumptive diagnosis in 61(92%) of 66 patients whose specimens were investigated.

C. HISTOPLASMOSIS

Serodiagnosis is the prime factor for histoplasmosis and the tests use are : CF, agar gel precipitin and latex agglutination tests, used either alone or better in combination.

Recent investigations indicate that **CF antibodies, precipitin and agglutinin to** *H. capsulatum* **increase significantly after an injection of** ***H. capsulatum*** **antigen in skin tests.** A comprehensive study by Kaufman *et. al.* (1967) confirmed the antibody inducing action of skin tests.

C. F. TEST performed properly either as a **tube or by microtitration method** can yield information **of diagnostic and prognostic values**. Two antigens are generally used in **NCDC** and **LBCF** tests. The use of one antigen fails to provide a diagnostic coverage as there is a chance of cross-reaction with sera of individuals suffering from other systemic mycosis, or these sera may react with sera of normal individuals or those suffering from non mycotic infections. The titre usually ranges from 1 : 8 and 1 : 16 and specially noted with yeast form of antigen. These titres with either of the antigens are considered to be presumptive evidences of histoplasmosis. **Titres of 1 : 32 are highly suggestive of** ***H. capsulatum*** **infection and of considerable diagnostic significance. The CF test is more frequently positive with** sera from **chronic active pulmonary histoplasmosis** than with primary histoplasmosis of the lung.

AGAR GEL PRECIPITIN TEST : This is a useful screening test, first described by Heiner in 1958 and considered it to be **an adjunct in the serological diagnosis of histoplasmosis.** He demonstrated **six precipitin bands with**

concentrated histoplasmin when sera from cases of histoplasmosis were used. These **two bands *viz.*; "h" and "m" had diagnostic value**, the former was detected in sera of cases of active histoplasmosis and was not affected by histoplasmin skin test, but the latter was found in acute and chronic histoplasmosis and was also affected by skin test sensitisation. **The 'm' antigen test has been considered to be presumptive evidence of infection with *H. capsulatum*** (Schubert *et. al.*, 1961). The results obtained by **agar gel precipitation test** was correlated in 82% of cases with those obtained by **CF test with histoplasmin antigen** (Gaufman *et. al*, 1962). The reaction with **'m' antigen only indicates active or inactive disease or sensitisation to skin testing** (Busey *et al.*, 1965; Wiggins *et. al.*, 1965).

LATEX FIXATION TESTS : It is widely used in the diagnosis of histoplasmosis. Commercially prepared antigen of histoplasmin-sensitised latex particles are available, and in the test a **two-fold dilutions of sera** ranging from **1 : 4 to 1 : 512** are prepared in tubes and optimal dilutions of antigen are added. The tubes are shaken, incubated for 2 hours in a water bath at 37ºC and refrigerated overnight before the tubes are examined for strong agglutination (Hill and Campbell, 1962; Bennet, 1966). Hill and Campbell consider a titre of 1 : 32 or higher to be a strong evidence for active or a recent infection.

D. BLASTOMYCOSIS

At present serological tests are inadequate and unsatisfactory as far as specificity and sensitivity of tests are concerned to establish unequivocal diagnosis of blastomycosis. CF reactions are positive with less than 50% of proved cases of blastomycosis. When the test is positive solely with *B. dermatitidis* antigens, one is inclined to diagnose blastomycosis. Out of 74 proved cases recently studied in **National Communicable Disease Centre (NCDC)** Laboratory 9 (12%) only reacted solely with homologous antigen, the titre ranging between **1 : 8 to 1 : 128** and other cases cross-reacted with both homologous **sera** and ***H. capsulatum* affected sera**. The yeast form of antigens of *B. dermatitidis* readily reacted with sera of patients who showed no evidence of blastomycosis. However, homologous antigens showing **titres of 1 : 8 or greater are considered to be presumptive evidence of blastomycosis.** The CF test for blastomycosis has limited diagnostic value but it is frequently of prognostic value in culturally proved cases.

E. COCCIDIOIDOMYCOSIS

The CF and tube precipitin tests are the common laboratory tests in the **diagnosis and prognosis of coccidioidomycosis** (Smith *et. al.*, 1950) with pooled soluble culture filtrate of *Coccidioides immitis*. The **antigens of CF and precipitin tests are different** and they appear to **be two different antigen-antibody systems.** The precipitin is effective in detecting **early primary infection** and *CF test* which appears later than the precipitin test, is most ***effective in determining the disseminated disease***. The **CF antigen is destroyed by heating at 60°C for 30 minutes** but not the on precipitinogen.

The CF titres of micro or macro tests tend to be parallel with the severity of infection (Smith *et al.*, 1950) rising as the disease progresses and declining as the patient improves. The specificity of the test was established by these authors and there is no cross-reaction with antigen of any kind. They found that by combination of these two tests, 90% of primary infection of coccidioidomycosis could be detected and a **titre of 1 : 2 and 1 : 4 indicated early meningeal or residual coccidioidomycosis** (Smith *et. al.*, 1950). Such a low titre obtained is conflicting with the diagnosis of coccidioidomycosis and the diagnosis must be confirmed by a second serological test and by clinical and mycological studies. **A negative, serological test does not exclude the diagnosis of coccidioidomycosis.**

LATEX FIXATION AGGLUTINATION TESTS (Hylana Laboratories, Los Angeles, Calif.) : In this, **latex particles are sensitised with coccidioidin heated at 60ºC for 30 minutes**. This is more sensitive than tube precipitation test, and yields a higher percentage of positive results in proved cases of coccidioidomycosis. Results are attained in 4 minutes (Hypperst *et. al.*, 1968).

IMMUNODIFFUSION TEST (Huppert *et. al.* 1965) : This test gives **results** that **correlate with those of CF test**. The antigen is heat labile toluene extract of mycelial mat. The test enables detection of 93% of specimens from diagnosed cases (Hyppert *el. at.*, 1968). The **results should be compared with CF and precipitin tests**.

SKIN TEST : **Coccidioidin skin test is valuable for detecting conversion of negative to a positive reaction**. It is **pathognomonic** and indicates the **earliest immunologic response to infection**. Smith *et. al.*, (1950) never observed positive result in patients with primary or with impending disseminated case without a positive skin test. Unlike histoplasmin skin test, **this does not produce a humoral antibody response** (Wallraff *et. al.*, 1967).

DETECTION OF HYPERSENSITIVITY

Most of the patients affected with **COCCIDIOIDOMYCOSIS** and **HISTOPLASMOSIS** develop hypersensitivity that can be readily and reliably detected with **coccidioidin and histoplasmin skin test** respectively. The **hypersensitivity can also be detected in** individuals suffering from **blastomycosis, candidiasis** and **cryptococcosis** but in these cases sufficient specificity to **warrant widespread use** of their respective skin tests **has not been established**.

The skin test is most useful in epidemiology and in defining endemic areas for a disease, but it has defeated the purpose for diagnosis as it does not distinguish present and past infections. **A positive reaction as indicated by erythema and induration is of diagnostic value**, if a negative result has been obtained prior to the clinical symptoms. Again except in infants, a **positive reaction with no past history has little diagnostic value. A negative reaction has greater significance** because it shows definite absence of the disease except in cases, when the patient is in the very early stage of the disease or at the terminal stage when the test becomes negative.

In **COCCIDIOIDOMYCOSIS** the skin test is a reliable guide in assessing the prognosis in different stages of the disease. A **reversion to negative state indicates ANERGY**.

Common antigens are shared by fungi. It is, therefore, suggested that **multiple skin tests are performed with optimally diluted extracts of fungi**, namely, **coccidioidin or histoplasmin**. Thus interpretation of tests is facilitated giving only one test positive or when one elicits a larger area than the others.

Skin test : 0.1 ml of the antigen injected intradermally with an appropriate dilution into the volar surface of the forearm. An area of induration and erythema appears at the site of injection in a sensitised person after 48 hours, when reading is taken. The **largest diameter of induration is recorded**. An **induration of 5 mm or more in diameter** is considered as **a positive reaction**.

FLUORESCENT ANTIBODY TECHNIQUE *IN VITRO* AND *IN VIVO* IDENTIFICATION OF FUNGI

Coons *et. al.*, (1942) discovered **fluorescent antibody technique (FAT)** for serological **identification of antigen** and demonstration of antibodies in sera. **This is** indeed an **immunological staining procedure** that allows **visualisation of an antigen antibody reaction** under the microscope through the use of **antibody coupled to a fluorochrome dye (RB200 or like fluorescein isothiocyanate). Fluorescein isothiocyanate gives an apple yellow green fluorescence** or **Rhodamine**

derivative giving orange fluorescence like the **colour of a flame.** The **fluorochrome is coupled to the antibody serum globulin and a labelled antiserum is produced.**

The **Fluorescent microscope** consists of a **standard microscope fitted with a cardioid dark field condenser,** a very bright light source and **selective filters for control of fluorescent excitation and emission wave lengths.**

FAT for : HISTOPLASMA CAPSULATUM

There are **3 ways of staining with labelled antibody** namely, direct, indirect, and fluorescent antibody inhibition.

1. ***Directly*** **(*DFAT*)**, by application of labelled antibody on smears of cultures or clinical material.
2. ***Indirectly*** **(*IFAT*), the infection can be detected rapidly when the antigen-antibody reaction is reinforced and which is visible by addition of a CONJUGATE directed against globulins of the animal producing the initial antibody.** The **indirect method** can be utilised **for detection of antibodies** or for the **identification of antigens.** Thus ***Histoplasma capsulatum*** **antibodies in human serum** may be **detected by first reacting the serum with smears** of ***H. capsulatum.*** **Then the preparation is treated with CONJUGATED ANTIHUMAN GLOBULINS.** In case the serum contains antibodies for *H. capsulatum,* the cells in smears of cultures or clinical materials combine. Thus smears are first reacted with known human or rabbit **anti-*H. capsulatum*** serum. The **preparation is then treated with fluorescent labelled antihuman or antirabbit globulins.** The fungus cells will thus fluoresce when the reaction is positive.
3. ***FA inhibition method.*** This involves the **blocking of the reaction between the antigen and conjugate by saturating antigen sites with homologous unlabelled antibody** as follows :

   (a) ***H. capsulatum*** antiserum plus ***H. capsulatum*** **cells.**

   (b) ***Add labelled H. capsulatum serum.*** **Inhibition of staining indicates** that the **test serum contained *H. capsulatum antibodies.*** **Thus the test can be carried out as one stage test as follows :** *H. capsulatum* plus equal parts of specific *H. capsulatum* antiserum and conjugate labelled serum are mixed. This is a **rapid method of one stage inhibition test,** which is preferred by many. The test is used as a control for the specificity of a **staining reaction given by labelled antiglobulin and inhibition of staining by proved specific antiserum** indicating specificity of the reaction.

The **fluorescent antibody technique (FAT)** for identification of other fungi :

CRYPTOCOCCUS NEOFORMANS

Eveland *et. al.,* (1957) employed FAT in the study of *Cryptococcus neoformans* or its polysaccharides in formalin fixed tissue. Kase and Marshall (1960) utilised the technique for the identification of cultures of *C. neoformans.* **They could stain 91 specimens out** of 92 **with labelled rabbit antiglobulins. All the conjugates used showed capsular staining.** These workers further reported that their conjugates stained 4 isolates of *C. neoformans* var. *innocuous* which is considered by many as a distinct species, *i.e.,* ***C. diffluens.*** No other cross-reaction was obtained.

CANDIDA ALBICANS AND OTHER CANDIDA SPECIES

Gordon (1958a) utilised 2 anti-rabbit ***C. albicans*** globulins one of which reacted not only with ***C. tropicalis, C. stellatoidea, C. pseudotropicalis*** and ***C. guilliermondii*** and also with 5 species of **heterologous yeasts** of genera other than candida. Adsorption of this antiserum with cells of *C. parakrusei* **eliminated all the above cross-reactions** except those with *C. tropicalis* and some strains of *C. stellatoidea*. Gordon's other conjugate stained both *C. albicans* and *C. tropicalis* with no cross-reaction with other candida species.

In the second report (Gordon, 1958b) **it was possible to differentiate *C. albicans* from *C. stellatoidea* with fluorescent antibody**. Kauffman and Kaplan (1961) applied the FA technique to the rapid detection of *C. albicans* globulins adsorbed with cells of *C. stellatoidea* and **revealed two distinct serological groups of *C. albicans***. They also observed that *C. tropicalis* and *C. stellatoidea* could be differentiated from *C. albicans* by the same technique.

SPOROTHRIX SCHENCKII

Kunz (1959) with labelled anti-sporothrix schenckii globulin successfully stained this fungus in culture, smears and in frozen sections. Kaplan and Ivens (1960) used the FA technique and demonstrated *S. schenckii* cell in cultures, smears and in the tissues of experimentally infected testes of mice and rats. Kaplan and Gonzalez-Ochoa (1963) rapidly identified *S. schenckii* cells in 16 out of 18 culturally confirmed human cases of sporotrichosis in smears from lesions.

HISTOPLASMA CAPSULATUM

Gordon (1959) **stained the YEAST PHASE CELLS** of ***Histoplasma capsulatum* by the FA technique**. Vogel (1966) **reported successful detection of 80 per cent patients of histoplasmosis by** indirect fluorescent antibody **(IFA) test** but he reported **lack of response in several patients with central nervous system involvement**. Kauffman and Kaplan (1961) stained ***anti-H. capsulatum*** antiglobulins by FA technique and in addition to the yeast phase cells of ***H. capsulatum*** the labelled globulins showed strong cross-reaction with ***Blastomyces dermatitidis*** *and other heterologous species.*

Gordon (1958a) stained by FA technique yeast cells of ***B. dermatitidis*** and ***C. albicans*** with *anti-B. dermatitidis* serum but this antiserum did not show cross-reaction with ***B. brasiliensis***, or ***H. capsulatum***.

Kaplan and Kauffman (1961) stained the yeast phase of ***H. capsulatum*** and ***B. dermatitidis*** with crude labelled anti-***B. dermatitids*** serum which also reacted strongly with many heterologous fungi. But adsorption of the conjugate yeast phase cells of ***H. capsulatum*** eliminated all the cross-reactions except that of ***Geotrichum candidum***. The conjugate was made specific by single adsorption with cells of ***G. candidum*** for yeast phase cells of ***B. dermatitidis*** in smears from lesions and exudates from experimentally infected mice.

DETECTION OF FUNGAL ANTIBODIES

Detection of fungal antibodies by Vogels and Padula (1958) made a preliminary study to determine the adaptibility of indirect FA method for the measurement of fungal antibody in human sera. The ***C. neoformans* antiserum complex was stained brightly from a case of cryptococcal meningitis. The FA inhibition test was used to detect antibodies against histoplasmin** and ***H. capsulatum* yeast cells** with sera from suspected and proved human cases of histoplasmosis (Kaufman, Schubert and Kaplan, 1961). **The test is simple and effective for rapid detection of antibody against whole yeast cells.**

REFERENCES

Bennett, D. E. (1966a). Histoplasmin latex agglutination test : Clinical evaluation and review of literature. *Am. J. med. Sci.* **251**, 175-183.

Benett, D. E. (1966b). Laboratory diagnosis of histoplasmosis. *A review. S. med. J.* **59**, 922-926.

Bloomfield, N., Gordon, M. A. and Elmendorf, D. F., Jr. (1963). Detections of Cryptococcus neoformans antigen in body fluids by latex particles agglutination. *Proc. Soc. exp. Biol. Med.* **114**, 64-67.

Busey, J. F. and Hinton, P. F. (1965). Precipitins in histoplasmosis. *Am. Rev. resp. Dis.* **92**, 637-639.

Coons, A. H., Creech, H. J., Jones, R. N. and Berliner, E. (1942). The demonstration of pneumococcal antigens in tissues by the use of fluorescent antibody. *J. Immunol:* **45**, 157-170.

Eveland, W. C., Marshall, J. D., Silverstein, A. M., Johnson, F. B., Iverson, L. and Winslow, D. J. (1957). Specific immunochemical staining of *Cryptococcus neoformans* and its polysaccharide in tissue. *Am. J. Path.* **33**, 616-617.

Gordon, M. A. (1958a). Differentiation of yeasts by means of fluorescent antibody. *Proc. Soc. exp. Biol. Med.* **97**, 694-698.

Gordon, M. A. (1958b). Rapid serological differentiation of *Candida albicans* from *Candida stellatoides. J. invest. Dermatol.* **31**, 123-125.

Gordon, M. A. (1959). Fluorescent staining of *Histoplasma capsulatum. J. Bact.* **77**, 678-681.

Gordon, M. A. and Vedder, D. K. (1966). Serological tests in the diagnosis and prognosis of cryptococcosis. *J. am. med. Assoc.* **197**, 961-967.

Heiner, D. C. (1958). Diagnosis of histoplasmosis using precipitin reaction in agar gel. *Pediatrics* **22**, 616-627.

Hill, G. B. and Campbell, C. C. (1962). Commercially available histoplasmin sensitized latex particles in an agglutination test for histoplasmosis. *Mycopathol. Mycol. Appl.* **18**, 169-176.

Huppert, M. and Bailey, J. W. (1965). The use of immunodiffusion tests in coccidioidomycosis. *Am. J. clin. Pathol.* **44**, 369-373.

Huppert, M., Peterson, E. T., Sun, S. H., Chitjian, P. and Derrevere, W. (1968). Evaluation of a latex particle agglutination test for coccidioidomycosis. *Am. J. clin. Pathol.* **49**, 96-102.

Kaplan, W. and Gonzalez-Ochoa, A. (1963). Application of the fluorescent antibody technique to the rapid diagnosis of sporotrichosis. *J. lab. clin. Med.* **62**, 835-841.

Kaplan, W. and Ivens, M. S. (1960). Fluorescent antibody staining of *Sporotrichum schenckii* in cultures and clinical materials. *J. invest. Dermatol.* **35**, 151-159.

Kase, A. and Marshall J. D. (1960). A study of *Cryptococcus neoformans* by means of the fluorescent antibody technic, *Am. J. clin. Path.* **34**, 52-56.

Kaufman, L. and Kaplan, W. (1961). Preparation of a fluorescent antibody specific for the yeast phase of *Histoplasma capsulatum. J. Bact.* **82**, 729-735.

Kaufman, L., Schubert, J. H. and Kaplan, W. (1962). A fluorescent antibody inhibition test for histoplasmosis. *J. lab. clin. Med.* **58**, 1033-1038.

Kaufman, L., Terry, R. T., Schubert, J. H. and MeLaughlin, D. (1967). Effects of a single histoplasmin skin test on the serological diagnosis of histoplasmosis. *J. Bact.* **94**, 798-803.

Kaufman, L. and Blumer, S. (1968). Development and use of a polyvalent conjugate to differentiate *Histoplasma capsulatum* and *Histoplasma duboisii* from other pathogens. *J. Bact.* **95**, 1243-1246.

Kunz, C. (1969). Fluorescenz-serologische untersuchungen an einem pathogenen pilstamm (Sporotrichum schenckii). *Arch. Klin. exp. Dermatol.* **209**, 200-205.

Schubert, J. H., Lynch, H. J. and Ajello, L. (1961). Evaluation of the agar-plate precipitin test for histoplasmosis. *Am. Rev. resp. Dis.* **84**, 845-849.

Smith, C. E., Saito, M. T., Beard, R. R., Kepp, R. McF., Clark, R. W. and Eddie, B. U. (1950). Serological tests in the diagnosis and prognosis of coccidioidomycosis. *Am. J. Hyg.* **52**, 1-21.

Stallybrass, F. C. (1964). Candida precipitins. *J. Path. Bact.* **87**, 89-97.

Taschdjian, C. L., Dobkin, G. B., Caroline., L. and Kozinn P. J. (1964). Immune studies relating to candidiasis, Paper II. *Sabouraudia* **3**, 129-139.

Taschdjian, C. L., Kozinn, P. J. and Caroline, L. (1964). Immune studies in Candidiasis, Paper III. *Sabouraudia* **3**, 312-320.

Taschdjian, C. L., Kozinn. P. J., Okas, A., Caroline, L. and Halle, M. A. (1967). Serodiagnosis of systemic candidiasis. *J. infec. Dis.* **117**, 180-187.

Vogel, R. A. (1966). The indirect fluorescent antibody test for the detection of antibody in human cryptococcal disease. *J. infec. Dis.* **116**, 573-580.

Vogel, R. A. and Padula, J. F. (1958). Indirect staining reaction with fluorescent antibody for detection of antibodies to pathogenic fungi. *Proc. Soc. exp. Biol. Med.* **98**, 135-139.

Vogel, R. A., Sellers, T. F. and Woodward, P. (1961). Fluorescent antibody techniques applied to the study of human cryptococcosis. *J. am. med. Assoc.* **178**, 921-923.

Wallraff, E. B., Van Liew, R. M. and Waite, S. (1967). Skin reactivity and serological response to coccidioidin skin tests. *J. invest. Dermatol.* **48**, 533-559.

Wiggins, G. L. and Schubert, J. H. (1965). Relationship of histoplasmin agar-gel bands and complement-fixation titres in histoplasmosis. *J. Bact.* **89**, 589-596.

8

Antibiotics And Chemotherapy For Mycoses

In view of the fact that there is substantial physiological differences between fungi and bacteria, the antibacterial drugs do not inhibit fungi; whereas, the antifungal agents are agains ineffective against bacteria. It is only in 1959 that some effective antifungal antibiotics have become available with the discovery of griseofulvin and polyene agents.

1. PENICILLIN AND SULPHONAMIDES

Actinomyces and Nocardia are sensitive to penicillin and sulphonamides; whereas true fungi are not affected by these agents.

2. GRISEOFULVIN (GRISOVIN; FIG. 8.1)

It is synthesised by several species of Penicillium and produces hyphal distortion in cultures of dermatophytes and other fungi. It is nowadays widely used in the treatment of dermatophytosis when administered orally for a given period of time; it is highly effective in cleaning many of these ringworm affections. It has action on some **anthropophilic dermatophytes** and not in **zoophilic species**. Reason for this discrepancy appears to be that the ingested drug accumulates in keratinous structures of cornified layer of epidermis, hair and nails, and as they are laid down, they render them resistant to infection. The keratinous structures already infected at the start of therapy

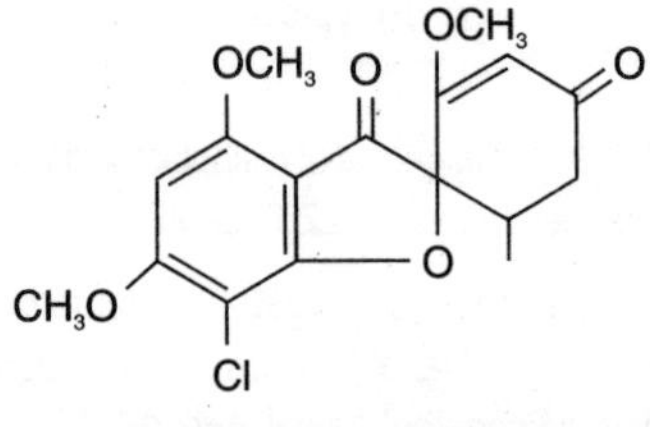

Fig. 8.1 : Griseofulvin

are not sterilised, as **the drug fails to attain a sufficiently high concentration in these compact fibrous materials**. Therefore, as the older infected structures are desquamated and replaced by younger keratinised tissue, its **concentration of griseofulvin subsequently increases and kills the organism**.

Because the **growth of epidermis and hair** is relatively **slow, the treatment must be prolonged** for weeks and months and this is specially applicable in case of infected nails. Fortunately, enough griseofulvin has only minimal toxicity in spite of prolonged administration.

Trichophyton rubrum often causing ringworm in the tropics and ***T. mentagrophytes*** infecting the skin and nails are best ameliorated by this antibiotic. The dose prescribed is 0.25 g; 4 times daily for 3 weeks and then 0.5 g daily till the condition is clinically and bacteriologically cured. There is no contraindication to local remedies as adjuvants while griseofulvin is being used as a *cure*. Although clinical response is favourable, 33 out of 71 patients (46.5%) were cured (Levechal, 1960).

This therapy is not effective in deep mycosis but may have some action on **subcutaneous mycosis**, like **sporotrichosis**.

SYSTEMIC MYCOSIS

For systemic and deep mycosis, **potassium iodide** was the mainstay of treatment but nowadays it has been largely abandoned except for sporotrichosis against which it is mostly effective.

3. SULPHONAMIDES

They are active against **(anaerobic) actino-mycosis** and **South American blastomycosis** cases, and ***dapsone* (diamino-diphenyl sulphone) is the drug of choice for nocardiasis.**The polyenes are fungicidal when added to growing cultures. They cause irreversible membrane damage as demonstrated by the leakage of small molecules of nucleotide, aminoacids and phosphorylated sugar from the cytoplasm. Moreover, polyene can be bound not only by intact fungi but also by membranes isolated from those fungi.

Polyenes exert their fungicidal effect by combining with sterols in the cytoplasmic membrane and **various sterols added to the medicine form complexes with polyene,** preventing their binding to the membranes and blocking their growth-inhibiting effect. If sterols are added shortly after the drug, they can reverse the drug effects; whereas, extraction of lipid from the isolated membrane prevents subsequent binding of the polyenes. The lost binding capacity can be largely restored to the lipid-free membranes by addition of lipid extracts or ergosterol or its esters, which are present in the fungal membrane.

The cell membranes of bacteria lack in sterols and, therefore, bacteria and not susceptible to polyenes. But some species of mycoplasma are exceptions to it. When these bacteria without cell-wall are grown in sterol-containing media, they incorporate sterols into the cytoplasmic membranes and become sensitive to polyene; whereas, when grown in sterol-free media they lack in sterols and become resistant to these drugs.

Toxic effects : Polyenes cause haemolysis *in vitro* owing to the fact that red blood cells contain sterols in their membrane. Therefore, in the **treatment of human mycotic infections with polyene, specially with amphotericin B(fungizone)**, one of the **toxic side effects is haemolytic anaemia**. The other **cytotoxic effect is damage of the kidney** but these effects are not severe or frequent probably due to the **binding of polyenes by serum albumin,** which reduces their free concentration in the body fluids.

4. NYSTATIN (MYCOSTATIN)

It is a TETRAENE, while **amphotericin B is a HEPTAENE**. Hazen and Brown (1950) extracted an antibiotic from a soil species of ***Streptomyces nauresi*** which possessed activity against some fungi like ***Candida albicans*** and ***Cryptococcus neoformans***. They called this antibiotic '**fungicidin**', but later it was named nystatin or mycostatin. Nystatin is a pale yellow, water soluble, almost odourless substance. It is inactive against bacteria but is most effective against the actively growing stage of yeast like fungi. It has no effect against actinomyces. **Nystatin is poorly absorbed from the gastrointestinal** tract, after a dose of 500,000 units 3 times a day. It shows its inhibitory effects against yeasts in the lumen of the intestine. *Toxicity*. It has a low degree of toxicity on oral administration, but very large doses may produce nausea, vomiting and diarrhoea. In many cases, no serious toxic or allergic reactions have been reported uptil now.

Therapeutic use : This antibiotic is used to prevent or treat **vaginal, intestinal and systemic candidiasis. The extension of gastrointestinal**

distress, diarrhoea, pruritus, vulvovaginitis and thrush is often encountered with therapy of broad spectrum antibiotics have been linked with the overgrowth of *C. albicans*. These fungal superinfections can be controlled by oral administration of tablets of **500,000** units of nystatin 2 of 3 times a day. The drug is also used in **systemic candidiasis, coccidioidomycosis, blastomycosis and histoplasmosis**. The drug is not likely to be effective unless it comes in contact with candida pathogenic to human. **The systemic antifungal antibiotics**, like **amphotericin B are more promising** in this respect.

5. CYCLOHEXIMIDE

This antibiotic inhibits protein synthesis of many yeasts, in animal cells, and in bacteria, but not in moulds. It is not a useful agent for treatment of fungal infections because of its toxicity on the host, but it is useful diagnostically as it **inhibits the growth of contaminants in media for isolation of fungi**. Its action is like that of chloramphenicol by inhibiting protein synthesis and by blocking the transfer to activated aminoacids on the ribosome from RNA to growing polypeptide chains.

TREATMENT OF DEEP MYCOSES : **Potassium iodide was the drug of choice in** almost every case of **deep mycosis** without much promising results but nowadays it has been practically abandoned in preference to antibiotics except in the treatment of sporotrichosis, against which it is most effective. At present **TRI-Azole**, like **CLOTRIMAZOLE** is exclusively used in candidiasis (and many ringworm fungi). Other drugs of choice which are gaining popularity are **fatty acids, ethyl vanilate, sulphonamides and sulphones, quinolines and acromatic diamidines**. Sulphonamides are effective against (anaerobic) **actinomycosis** although it cannot be claimed as an infallible remedy as only about 50 per cent cases may respond to this therapy. Penicillin in prolonged big doses are effective against **actinomycosis**. In **South American blastomycosis, diamino-diphenyl sulphone (DDS)** was the drug of choice; **KETOCONAZOLE.**

Antibiotics like antifungus polymer are effective against many deep mycoses. **Nystatin (Mycostatin) is active against candidiasis**. At present **amphotericin B (Fungizone)**, has been found to be effective in **cryptococcosis, coccidioidomycosis and histoplasmosis** and also in **North American and South American blastomycoses** and **disseminated candidiasis**. **Orally it is less active**, and gives better results when administered parenterally. The antibiotic is **dissolved in 5 per cent glucose (1 mg per 10 ml)** and the dose is **administered slowly intravenously covering a period of about 6 hours (by drip method)**. The **initial dose is 0.25 mg per kilogram** and then it is gradually increased to 1 mg per kilogram of body weight. All patients cannot tolerate the maximum dose but **good results are obtained in dose of 0.5 to 0.75 mg per kilogram**. The total length of treatment varies greatly and improvement is observed after 4-6 weeks when infusion may be given on alternate days. *Toxic effects* are seen in almost every case e.g., chill, fever specially seen in the beginning of treatment but prophylactic administration of antipyretics and antihistamines are useful. Nausea and vomiting are also other early side effects and dosage should be regularly checked.

Penicillin has been used successfully **in** many cases of **actinomycosis**. **Polyenes** : In contrast to griseofulvin polyene antibiotics are **highly effective in** the treatment of many **systemic mycosis** and ineffective in cutaneous and superficial mycosis.

6. AMPHOTERICIN B (FUNGIZONE; PHYTORAL/MYCORAL; FIG. 8.2)

It is a heptaene antibiotic derived from a species of streptomyces, has given encouraging results in cases of **coccidio-idomycosis, cryptococcosis, cryptococcal meningitis**

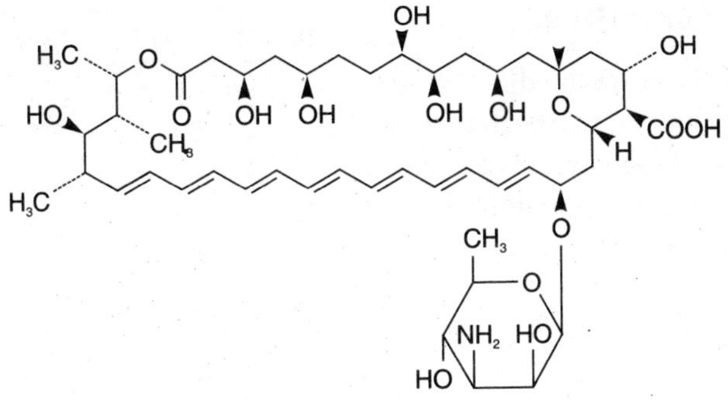

Fig. 8.2 : Amphotericin B

palmonary cryptococcosis acute primary and chronic cavitary form of histoplasmosis, in North and South American blastomycosis, disseminated aspergilloses, resistant sporotrichosis, paracoccidioidomycosis and also **in disseminated candidiasis** (Bindschadler and Bennett, 1969, Drutz *et. al.* 1968). It has some activities by oral route but the drug **gives best result parenterally**. The **calculated dose is dissolved in 5 per cent glucose solution (1 mg per 10 ml)** and **administered intravenously** for about 6 hours daily. The **initial dose is 0.25 mg per kg of body weight** and gradually **increased to 1 mg per kg of body weight**. Improvement is often seen in 4 to 8 weeks. Amphotericin B has been injected locally in **chromoblastomycosis** with good results (**Dey, 1970**). It has been given intramuscularly or even intrathecally in some cases and when locally administered it acts against **CHROMO-BLASTOMYCOSIS**.

Toxicity of "Amphotericin B"

Toxic effects occur in the beginning but subsides in time. **Headache, nausea and vomiting** are the common side effects, and the dosage is adjusted to avoid them. **Phlebitis** may occur. **The drug is highly nephrotoxic**. Blood urea and NPN should not be allowed to exceed 40 mg per 100 ml. **Renal tubular acidosis** and decreased clearance of urea, creatinine and inulin has been reported by Mc Curdy *et al.* (1968). In such cases, the treatment may be suspended, or **administered on alternate days along with I.V. hydrocortisone (25 to 50 mg) throughout the Amphotericin infusion**. Other toxic effects are **chills and fever, hypokalaemia, bone marrow depression with anaemia, profound muscle weakness and ECG changes**. To combat toxic effects, **orally chlorpromazine is given before amphotericin infusion**.

7. SARAMYCETIN (X 5079C, R02-7758)

It is another antibiotic, **polypeptide in nature** which is a polypeptid antibiotic administered subcutaneously has been reported to be of value in cases of **blastomycosis, histoplasmosis, sporotrichosis**, etc. This is an exceptionally well tolerated drug having **only one significant side effect** involving **changes in the hepatic function** which is **reversible**. The main liver function tests that go out of gear is the bromosulphthalein **(BSP) excretion** in association with a moderate **rise in conjugated and unconjugated serum bilirubin** values.

8. HAMYCIN

This polyene group of antibiotic was extensively tried in India and in the U. S. A., being administered orally in cases of **blastomycosis** and **disseminated** and/or **chronic cavitary histoplasmosis** (Ahn *et al.*, 1969). Side effects are nausea, vomiting and diarrhoea.

9. 2-HYDROXYSTILBAMIDINE

This is a number of the **aromatic diamidines** and is currently recommended **for** the treatment of **balstomycosis.** This drug is relatively well tolerated than the other drugs in blastomycosis. However, there is always a chance of relapse when this drug is administered; it has been a **failure in the cases of blastomycotic meningitis.** The daily dosage is 225 mg given intravenously until a total dose of 8 g. A second or even third course of treatment may be given for relapse or inadequate response (Lockwood, 1962).

10. FLUCYTOSINE (5-FLUOROCYTOSINE)

Flucytosine is the **fluorine analogue of cytosine.** It is used in the systemic treatment of some **deep-seated fungal infections,** e.g., **Candidiasis, Cryptococcosis,** and the drug of choice **in Chromomycosis.** Flucytosine-resistant strains emerge when the drug is used only once; so it is used with **Amphotericin B,** e.g., in **Cryptococcal Meningitis.**

Mechanism of Action :

I. Flucytosine enters the cell by ***cytosine permease.*** It is deaminated to 5-fluorouracil (5-FU) by *cytosine deaminase* and phsophorylated; then incorporated into RNA.

Another pathway is by formation of 5-FUdR monophosphate, a noncompetitive ***inhibitor of thymidylate systhetase.*** This interferes with DNA synthesis and causes defective cell division. Basically they cause :

(i) **Impaired DNA synthesis, and**

(ii) **Dysfunction of fungal RNA,** causing disturbed protein synthesis, which are the processes for antifungal activity.

Drug Resistance is attributed to many mechanisms. Resistant mutants possess deficient enzymes systems, causing.

(a) **Increased synthesis of** (competing) **Pyrimidines,** or

(b) **Compensating mechanisms for the abnormal RNA function.**

Table 8.1 CLASSIFICATION OF ANTIFUNGAL AGENTS

I. AGENTS FOR SUPERFICIAL FUNGAL INFECTIONS (INVOLVING SUPERFICIAL EPIDERMAL CELLS) :

1. PIEDRA :
   - (a) Lotio Hydrag. perchloride (1 : 2000)
   - (b) Lotio Resorcin
   - (c) Unguentum Hydrarg. ammon (3%)
   - (d) Broad spectrum Antibiotics to be locally applied, e.g.,

     IMIDAZOLES : (i) **CLOTRIMAZOLE (WITH GENTAMICIN)**, i.e. (Cloben G).

     (ii) Miconazole

     (iii) **Ketoconazole**

     TRIAZOLES

     Fluconazole

     Itraconazole

     Voriconazole

2. **TINEA NIGRA PALMARIS**

   After cleaning the lesion with soap water the keratolylic medicine (salicylic acid) and Azole compounds and basic medicines (for Piedra) would cure the condition.

3. **SUPERFICIAL STREPTOMYCOSIS :**
   A. HAJA (INTERTRIGO)

   (i) Mixture of 3 Antiseptic dyes. Lotio Gentian violet + 2% Lotio Acriflavin + Lotio Brilliant Green.
   B. CHALUNI (Pitted Soles)

   (ii) Lotio Gentian violet (2%) in 10% Ethyl Alcohol

   (iii) Lotio Formalin (10-20%) et Glycerol (50%)

   (iv) Clotrimazole (Lotion and ointment)

   (v) Miconazole (Lotion or ointment)

4. **TINEA VERSICOLOR**

   (i) Sulphur in any form

   (ii) Clotrimazole Lotion (recurrence occurs)

   (iii) Ketoconazole (recurrence occurs)

   The best and permanent remedy is :

   (iv) **Coal Tar preparation (ointment)** to apply for **15 nights** only; Not to APPLY during DIURNAL HOURS.

   For example THIOTAR OINTMENT (Pasteur Laboratories Pvt. Ltd.)

5A. **SEBORRHOEIC DERMATITIS**

   (i) Sulphur in any form, e.g., Sulphur precipitate Lotio (Thiosol - Pasteur Laboratories Pvt. Ltd.)

Contd.

(ii) Lotio Calamina with 10 gr. (0.6G) sulphur to one ounce of lotio calamira *i.e.*, sulphur precipitate, 0, 2g% sulphur)

(iii) Clotrimazole and other Antifungal Antibiotics.

5B. **For SEBORRHOEIC GRANULOMA**

(v) **Lotio Calamina et Sulphur precipitate (2g%)**

(vi) **Clotrimazole** may betried. If at all the cured that may be followed by a relapse.

5C. **For SEBORRHOIC BALDNESS**

(iv) **Lotio Resorcin et Hydrorg - perchloride (see Appendix)**.

II. **MEDICINES AND ANTIBIOTICS FOR CUTANEOUS MYCOSIS**

A. CRYPTOCOCCOSIS

(i) **CLOTRIMAZOLE**

(ii) **FLUCONAZOLE (particularly in CRYPTO-COCCAL MENINGITIS)**

(iii) ITRACONAZOLE (at times)

B. **CUTANEOUS (MUCO-CUTANEOUS AND BRONCHO-PULMONARY) CANDIDIASIS**

(i) NYSTATIN (Mycostatin - Topical)

(ii) MICONAZOLE (Topical)

(iii) **CLOTRIMAZOLE (Topical)**

(iv) GRISEO FULVIN - Prolonged treatment with oral administration

(v) **KETOCONAZOLE (IN CHRONIC MUCOCU-TANEOUS CANDIDIASIS)**

(vi) **FULCONAZOLE** (IN CASE OF OROPHA-RYNGEAL AND BRONCHO-PULMONARY CANDIDIASIS)

(vii) Lotio GENTIAN VIOLET (Topical)

C. **DERMATOPHYTES (RINGWORM) INFECTION**

(i) **CLOTRIMAZOLE** (antifungal) with GENTAMICIN (or some other broad spectrum, ANTIBACTERIAL COVERAGE

(ii) **KETOCONAZOLE (Topical)**

(iii) Salicylic Acid + Benzoic Acid ointment (Ephytol)

(iv) Ephytol + Steroid (**Dipsalic ointment and lotion**)

III. **REMEDY For Less Serious SYSTEMIC MYCOSIS ORAL AZOLE COMPOUNDS**

(i) **FLUCONAZOLE (Oral)**

(ii) **ITRACONAZOLE (Oral)**

IV. **CURE FOR SERIOUS SYSTEMIC INFECTIONS**

(i) **AMPHOTERICIN 'B' (FUNGIZONE Intravenous)**

This drug covers up all medically important fungi but it is highly toxic - As mentioned toxicity can be combated with Steroid and glucose in the same S. V. drip.

Oral Medicine can be administered by the name **Phytoral and Mycoral** which are also very toxic.

(ii) **ITRACONAZOLE**

(iii) **KETOCONAZOLE** - It offers best result with some systemic Mycoses, e.g., **Mucocutaneous candidiasis**, **Non-meningeal Blastomycosis**, (**North American Blastomycosis**) **Coccidioido-mycosis**, **Paracoccidioido-mycosis** (**South American Blastomycosis**), **Histoplasmosis**.

V. **ANTIBIOTIC FOR FUNGAL MENINGITIS**

(i) **FLUCONAZOLE** is the drug of choice as it offers maximum penetration through CNS tissue.

(ii) **AMPHOTERICIN 'B'** INFUSION along with its protective measures against Toxicity.

TABLE 8.2 :PHARMACOLOGY OF ANTIFUNGAL MEDICINES

ANTIBIOTIC CHEMOTHERAPEUTIC	THERAPEUTIC USE	MECHANISM OF ACTION	PHARMACOKINETICS	ADVERSE EFFECTS	COMMENT
AMPHOTERICIN-B (a polyene antibiotic)	Severe systemic fungal 8infections : 1. Systemic Candidiasis 2. Cryptococcal meningitis 3. Blastomycosis, and coccidio-idomycosis and histoplasmosis 4. Invasive sporotrichosis 5. Mucormycosis 6. Invasive aspergillosis	Punches holes in ergosterol; **increases membrane permeability** causing cell death	Oral absorption absent; IV Administration No crossing over blood-brain barrier; give intrathecally (CSF) Excreted via biliary tract and kidney (clearance not affected due to reversible kidney dysfunction)	1. Acute febrile reaction 2. Anaemia 3. Phlebitis IV (site) 4. Reversible Nephrotoxicity	1. **Monitor urea & creatinine** levels daily to follow kidney dysfunction 2. New lipid preparations are less nephrotoxic: ABLC (Anophotericin 'B' Lipid complex) & Amphocil (ABCD) and ambiosome (Liposomal Amphotericin B)
FLUCYTOSINE	1. Cryptococcal meningitis (in combination with amphotericin B) 2. Candidal endocarditis (Along with Amphotericin B)	Converted to 5-fluorouracil, which **inhibits fungal DNA & RNA synthesis**	Good oral absorption Excreted in urine Penetrates CSF as well	1. Nausea, vomiting and diarrhoea 2. Bone marrow suppression : A. Leukopenia B. Thrombocytopenia	Flucytosine inhibits DNA synthesis in rapidly dividing cells (bone marrow cells & GI epithelial cells)
KETOCONAZOLE	**Chronic mucocutaneous candidiasis** **Dermatophytosis**, Non-meningeal **Blastomycosis**, **PARACOCCIDIOIDOMYCOSIS** **Coccidioidomycosis,** **Histoplasmosis**,	Blocks ergosterol synthesis by inhibiting the cytochrome P - 450 enzymes. This causes depletion of ergosterol, resulting in **disruption of the permeability of the cell** membrane	Good oral absorption Better absorbed at low (Acid) pH; absorption hampered with antacids at and H_2 **blockers** Metabolised extensively in the liver	Nausea, vomiting and anorexia Hepatotoxicity Inhibits CY. - 450 system, responsible for androgen & testosterone synthesis : Results are : A. Gynecomastia B. Impotency C. Decreased sex drive D. Decreased sperm production E. Skin rash and pruritus	
Miconazole and **Clotrimazole** (Imidazoles)	A. Topical fungal infections 1. Tinea versicolor 2. Cutaneous candidiasis 3. Dermatophytosis B. Oral troches for thrush (oral candidiasis) C. Vaginal suppositories for candida vaginitis	**Blocks ergosterol synthesis** by inhibiting the cytochrome P - 450 enzymes	Available for Topical Application	Low toxicity when used topically	
Fluconazole (**triazole**)	Indicated in oral, vaginal and esophageal *Candid* infection **Alternative to amphotericin B for treatment of :** A. **Systemic candidiasis (amphotericin B)** B. **Cryptococcal meningitis** C. **Pulmonary & extrapulmonary coccidioidomycosis;** Treat coccidioidal and cryptococcal meningitis as a maintenance therapy.	**Blocks ergosterol synthesis** by inhibiting the cytochrome P - 450 enzymes	Good oral absorption Can also be administered by route IV	Less toxic than ketoconazole No interference with testosterone synthesis 1. Nausea 2. Skin rash 3. Headache	

Contd.

TABLE 8.2 :PHARMACOLOGY OF ANTIFUNGAL MEDICINES *(Contd.)*

ANTIBIOTIC CHEMOTHERAPEUTIC	THERAPEUTIC USE	MECHANISM OF ACTION	PHARMACOKINETICS	ADVERSE EFFECTS	COMMENT
AMPHOTERICIN-B (triazole)	Itraconazole (instead of amphotericin) 1. **Blastomycosis** 2. **Histoplasmosis** 3. **Coccidioidomycosis** 4. **Sporotrichosis** 5. **Chromomycosis** 6. **Invasive aspergillosis**	For treating systemic fungal infections **inhibiting the cytochrome** P - 450 **enzymes**	Blocks ergosterol synthesis by Metabolized and excreted products via liver Available also as **IV administration**	Good oral absorption No interference with testosterone synthesis 1. Nausea 2. Skin rash 3. Headache	**LESS TOXIC THAN KETOCONAZOLE**
VORICONAZOLE (Triazole)	Highly active against a broad range of fungi. May be useful in patients with **INVASIVE ASPERGILLOSIS** with therapy-limiting toxicity associated with other drugs	**Blocks ergosterol synthesis** by inhibiting the cytochrome P - 450 enzymes	Good oral absorption Also available for **IV Administration**	1. Liver enzyme increases 2. Photophobia 3. Skin rash	Drug has voracious appetitie for fungi
NYSTAIN (Mycostatin)	1. Oral, oesophageal or gastric candidiasis (oral administration) 2. Vaginal candidiasis (apply topically)	Punches holes in ergosterol : This **increases membrane** permeability, resulting in cell death	Not absorbed from GI tract. Oral therapy done for "**topical**" (**surface**) **treatment** on skin, Vaginal and G-1 mucosa	Highly toxic if given IV	
GRISEOFULVIN (Grixovin)	Dermatophytosis of the skin, hair and nails	**Inhibits cell mitosis** disrupting spindle formation	Good oral absorption; better with fatty foods Excreted unchanged in faeces	Headache, nausea, vomiting, photosensitivity, mental confusion and Bone marrow suppression	Works very slowly (6 months)
POTASSIUM IODIDE OR SODIUM IODIDE	Subcutaneous sporotrichosis & many others	3 to 5% aqueous solu. breaks down collagen & granuloma		Skin rash	

TABLE 8.3 :TYPE OF MYCOSES CAUSED BY PATHOGENIC FUNGI SHOWING THEIR CLINICAL/DIAGNOSTIC & TREATMENT FEATURES

MYCOSIS TYPE	FUNGI	SOURCE OF INFECTION	MORPHOLOGY	CLINICAL	DIAGNOSIS	TREATMENT	COMMENTS
SUPERFICIAL	***MALASSEZIA FURFUR Pityrosporum ovale***	Dandruff	(a) **MYCELIA** AND (b) **YEAST PHASE**	**Pityriasis versicolor** Hypo-pigmented patches on the skin; surrounding skin darkens with sunlight while the patches remain white	Sodium sulphide ("HYPO") solution Potassium hydroxide (KOH) prep: reveals short, curved, **unbranched hyphae** with **spherical yeast cells**	1. Dandruff shampoo (containing **selenium sulfide**) 2. **Topical imidazoles**	40% KOH Sol. or sodium Sulphide (Saturated) Soln.
SUPERFICIAL	*EXOPHIALA WERNECKII*	Dandruff	(a) Brown-pigmented, branched, septate **hyphae** (b) **budding yeast cells**	**Tinea nigra** Dark brown to black patches on the soles and palms	KOH prep: brown-pigmented, branched, septate hyphae and budding yeast cells	1. Dandruff shampoo (containing selenium sulfide) 2. **Topical imidazoles**	
CUTANEOUS	1. ***Microsporum species*** 2. ***Trichophyton species*** 3. ***Epidermophyton floccosum*** (only one pathogenic species)	Depending on the particular species: 1. Soil 2. Humans (by contact) 3. Animals (zoophilic)		**Dermatophytosis (Ringworm)** **Tinea corporis, cruris, pedis, capitis, unguium** etc.	1. KOH: branched hyphae 2. Wood's light: certain species of *Microsporum* will fluoresce under ultraviolet light	1. **Topical imidazole** 2. **Oral griseofulvin is** used for **tinea unguium and tinea capitis** 3. **Oral terbinafine**	Secretes ***kerratinase***, that digests keratin of cutaneous cells

Contd.

TABLE 8.3 :TYPE OF MYCOSES CAUSED BY PATHOGENIC FUNGI SHOWING THEIR CLINICAL/DIAGNOSTIC & TREATMENT FEATURES

MYCOSIS TYPE	FUNGI	SOURCE OF INFECTION	MORPHOLOGY	CLINICAL	DIAGNOSIS	TREATMENT	COMMENTS
SYSTEMIC CUTANEOUS OPPORTUNISTIC	***CRYPTOCOCCUS NEOFORMANS***	1. **FOUND IN PIGEON DROPPINGS** 2. **RESPIRATORY TRANSMISSION**	Polysaccharide capsule around Yeast body	**Cryptococcosis** 1. Subacute or chronic meningitis : A. Headache B. Fever C. Vomiting D. Neurologic or mental status changes 2. Pneumonia: usually self-limited 3. Skin lesions: look like acne	1. India-ink stain of cerebrospinal fluid CSF): Swollen capsulate around yeast 2. Cryptococcal antigen test of CSF: detects polysaccharide antigens 3. Fungal culture	1. **Amphotericin B** and **Flucytosine** (is superior to amphotericin B alone) 2. **Fluconazole**	Cases occur in AIDS/immuno-compromised people
CUTANEOUS/ SYSTEMIC/ OPPORTUNISTIC	***CANDIDA ALBICANS***	1. Normal inhabitant of the skin, mouth and gastrointestinal tract 2. Not found in blood	**DIMORMIC FUNGUS** **Pseudohyphae and yeast present**	**Candidiasis in a normal host** 1. Oral thrush, 2. Cutaneous, 3. Nails 4. Vulvovaginal candidiasis + Diaper rash, rash in the skin folds **Candidiasis in AIDS (Immunocompromised host)** Thrush, vaginitis and/or cutaneous lesions AND 1. Oesophageal Candidiasis; Fever, Dysphagia, Retrosternal pain 2. Disseminated candidiasis: in devitalised patients (in multi-organ system failure) 3. Chronic mucocutaneous candidiasis	1. KOH stain of specimen 2. Silver stain of specimen (GMS) 3. Blood culture: Growth occurs	It depends on the type of candidal infection and its severity: 1. **Thrush : Nystatin** and **imidazole candies (troches)** 2. **Cutaneous infection :** topical **Clotrimazole** 3. **AIDS** (thrush, esophageal, vaginal affections): **Fluconazole** 4. Systemic candidiasis : **Amphotericin B** 5. Chronic mucocutaneous candidiasis : **Ketoconazole** **Amphotericin B**	
BACTERIA-LIKE FUNGI / OR FUNGI-LIKE (HIGHER) BACTERIA							
Systemic/ Cutaneous (N. keratolytica) **(NAIL INFECTION)**	***NOCARDIA KERATOLYTICA*** ***NOCARDIA ASTEROIDES***	Respiratory transmission from atmosphere	1. Gram-positive rods 2. Partially **acid-fast:** (mycolic acids in the wall) 3. **Aerobic** 4. Grow as branching chains or beaded filaments	1. Nail involvement by ***Nocardia keratolytica*** 2. Formation of abscesses in the lungs, kidneys and CNS 3. Pneumonia	1. Gram stain 2. Modified acid fat stain : To Decolorize with 1% sulfuric acid : 3. **Aerobic culture**	**Trimethoprim (TM)** **Sulfamethoxazole** (S) **(TMS)**	*Nocardia* infection may occur in AIDs and immuno-compromised patients
Cutaneous/ Systemic	***BLASTOMYCES DERMATITIDIS***	North to the Great Lakes, Soil of Canada rotten wood Respiratory transmission	**Dimorphic** 1. Mycelial forms with spores (25°C) 2. **Yeast forms (37°C)**	**Blastomycosis** 1. Asymptomatic (uncommon) 2. Pneumonia: lesions rarely calcify 3. Disseminated (most common): present with weight loss, night sweats, lung involvement, and skin ulcers 4. Cutaneous: Skin ulcers	1. Biopsy Lung biopsy, skin biopsy, etc. A. Silver stain (GMS) specimen B. Culture on Sabouraud's agar 2. Serology 3. Skin test (for exposed persons)	**Amphotericin B** **Ketoconazole** **Itraconazole**	

Contd.

TABLE 8.3 :TYPE OF MYCOSES CAUSED BY PATHOGENIC FUNGI SHOWING THEIR CLINICAL/DIAGNOSTIC & TREATMENT FEATURES

MYCOSIS TYPE	FUNGI	SOURCE OF INFECTION	MORPHOLOGY	CLINICAL	DIAGNOSIS	TREATMENT	COMMENTS
CUTANEOUS/ SYSTEMIC	***COCCIDIOIDES IMMITIS***	1. Desert areas of southwestern USA, Northern 2. Respiratory transmission	**Dimorphic** 1. Mycelial forms with spores at 25°C 2. Yeast forms at 37°C	**Coccidioidomycosis** 1. Asymptomatic (in most persons) 2. Pneumonia 3. Disseminated: can affect the lungs, skins, bones, and meninges Note : A small percentage of individuals with his infection will develop painful erythematous nodular lesions called erythema nodosum	1. Biopsy: Lung biopsy, skin biopsy, etc. A. Silver stain (GMS) specimen B. Culture on Sabouraud's agar 2. Serology (Doubling dilution) 3. Skin test (tests for exposure only)	**Amphotericin B** **Fluconazole** **Itraconazole**	In AIDS Patients
SUBCUTANEOUS	1. ***PHIALOPHORA VERRUCOSA*** 2. ***CLADOSPORIUM CARRIONII*** 3. *Fonsecaea species*	These copper-colored saprophytes on **rotting wood** and in **soil**	Copper colored cells Sclerotic bodies	**Chromoblastomycosis** Following a puncture wound, a small, violet wart-like lesion develops. With time, clusters of these skin lesions can develop (resembling cauliflower)	Skin scrapings with unstained KOH prepn - copper-colored cells (sclerotic bodies)	Itraconazole Surgery (Local excision)	
SUBCUTANEOUS/ SYSTEMIC	***SPOROTHRIX SCHENCKII***	Rose thorns		**Sporotrichosis** 1. Subcutaneous nodule gradually appears at site of thorn prick 2. This nodule becomes necrotic and ulcerates 3. This ulcer heals, but new nodules pop up nearby along the lymphatic tracts	**Dimorphic** 1. Culturer at 25°C : Branching hyphae 2. Culture at 37°C : yeast cells	Oral potassium iodide Amphotericin B Itraconazole	
SUBCUTANEOUS/ SYSTEMIC	1. *Rhizopus* 2. *Rhizomucor* 3. *Mucor*	Molds (Saprophytic)	Broad, non-septate branching hyphae	**Mucormycosis** 1. Rhinocerebral (associated with diabetes): starts on nasal mucosa and invades the sinus and orbit 2. Pulmonary mucormycosis	1. Excisional Biopsy (if possible) 2. Black nasal discharge examination	Amphotericin B and Surgery	Thid disease is rapidly fatal
				THE FUNGI-LIKE BACTERIA (Higher Bacteria)			
SUBCUTANEOUS/ SYSTEMIC	***ACTINOMYCES ISRAELII*** ***A-BOVIS*** (in cows)	Normal flora of the mouth and gastrointestinal tract (GIT)	Gram-positive rods **Anaerobic bacteria** Grow as branching chains and beaded filaments	Eroding abscesses of the mouth, GIT and **lungs** **Types :** 1. Cervicofacial actinomycosis 2. Abdominal actinomycosis 3. Thoracic actinomycosis	1. Examine tissue or pus "for sulfur granules" 2. **Anaerobic culture**	Penicillin G Surgery	Yellow "sulfur granules" microcolonies of *Actinomyces* and cellular debris

Contd.

TABLE 8.3 :TYPE OF MYCOSES CAUSED BY PATHOGENIC FUNGI SHOWING THEIR CLINICAL/DIAGNOSTIC & TREATMENT FEATURES

MYCOSIS TYPE	FUNGI	SOURCE OF INFECTION	MORPHOLOGY	CLINICAL	DIAGNOSIS	TREATMENT	COMMENTS
SYSTEMIC	***HISTOPLASMA CAPSULATUM***	1. Universal 2. Present in bird and bat **droppings** 3. Respiratory transmission	**Dimorphic** 1. Mycelial forms with spores at 25°C 2. Yeast forms at 37°C Capsule absent	**Histoplasmosis** 1. Asymptomatic 2. Pneumonic lesions calcify, seen on chest X-ray (may look similar to tuberculosis like) 3. Disseminated: can occur in any organ, lung, liver, spleen, etc.	1. Lung biopsy A. Silver stain (GMS) B. Culture on Sabouraud's agar at 25°C - Hyphae and yeast at 37°C in Macrophages 2. Serology 3. Skin test (tests for exposure only)	Itraconazole Amphotericin B (in immunocompromised patients)	Can survive survive intracellularly with in macrophages
OPPORTUNISTIC SYSTEMIC	***Aspergillus fumigatus*** ***Aspergillus flavus*** ***Aspergillus niger***	Universal Aspiration of *Aspergillus spores = asthma*	Branching septated hyphae	**Aspergillosis** 1. Allergic bronchopulmonary aspergillosis (**IgE mediated**) : **asthmatic** reaction, high fever 2. Aspergilloma (fungus ball) : Haemoptysis 3. **Invasive aspergillosis :** Necrotizing pneumonitis Disseminate to other organs in immunocompromised patients 4. **Aflatoxin** consumption by ***flavus*** causing liver damage, cancer	A. Allergic bronchopulmonary aspergillosis : 1. High level of IgE and igG against aspergillus 2. Sputum culture 3. Wheezing patient and chest X-ray with fleeting infiltrates 4. Increased level of eosinophils 5. Skin test: immediate hypersensitivity reaction B. Aspergilloma: diagnosed with chest X-ray or CT scan C. Invasive aspergillosis: sputum examination and culture	1. Allergic bronchopulmonary aspergillosis: Corticosteroids & Ant; Aspergillus drugs 2. Aspergilloma: Surgery 3. Invasive aspergillosis: **Amphotericin B** and **others** (Rifampin, Itraconazole) Prognosis-Poor	Aflatoxins contaminate peanuts, grains rice, etc.

FLUCYTOSINE (Conldi)

Mechanism of Action of Flucytosine Contd.

II. Flucytosine is converted by the fungal enzyme ***cytosine deaminase to 5-fluorouracil*** and incorporated into 5-fluorodexoyuridylic acid monophosphate, that interferes with the activity of thymidylate synthetase and DNA synthesis. Mammalian cells lack ***CYTOSINE DEAMINASE*** protected from the toxic effects of fluorouracil. Resistant mutants emerge rapidly.

Range of Activity

Flucytosine (5-fluocytosine) is a fluorinated derivative of cytosie. It is an oral antifungal compound used in conjunction with amphotericin B to treat **cryptococcosis and candidiasis** effective against many **dematiaceous fungal infections** including **Chromoblastomycosis,** and penetrate all tissue hence effective in cryptococcal meningitis and prevents its relapse.

Side Effects

It has little toxicity and is well tolerated, conversion to fluorouracil results in a highly toxic compound that creates major side effects. Prolonged administration causes bone marrow suppression abnormal normal liver function, hair loss and colitis. Immunocompromised patients are more susceptible to bone marrow suppression.

AZOLE COMPOUNDS :

AZOLES are Classified into :

A. IMIDAZOLES

1. **CLOTRIMAZOLE**
2. **KETOCONAZOLE**
3. **MICONAZOLE**

B. TRIAZOLE

1. **FLUCONAZOLE**
2. **ITRACONAZOLE**
3. VORICONAZOLE

Mechanism of Action of AZOLES

The Azoles inhibit the **CYTOCHROME P-450 ENZYME SYSTEM** and thus block the **Cytochrome P450-dependent 14-alpha-demethylation of LANOSTEROL,** a precursor of ERGOSTEROL in fungal cells, while CHOLESTEROL in mammalian cells.

KETOCONAZOLE :

Ketoconazole (**Fig.** 8.3) is efficient in the treatment of **chronic mucocutaneous candidiasis, dermatophytosis, and nonmeningeal blastomycosis, paracoccidioidomycosis, coccidioidomycosis,** and **histoplasmosis.**

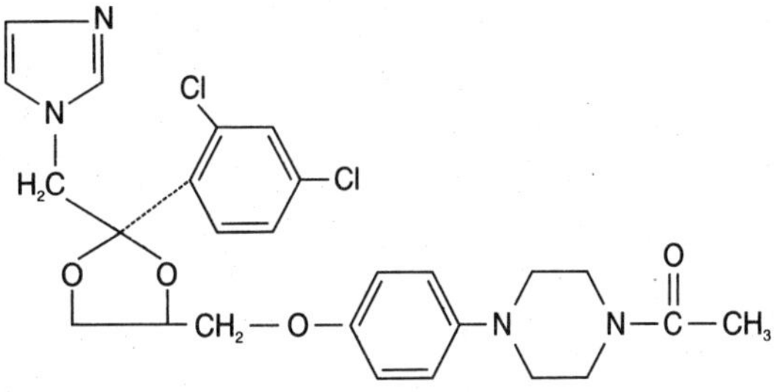

Fig. 8.3 : Ketoconazole

FLUCONAZOLE

Fluconazole (Fig. 8.4) is the best for penetration of the central nervous system (meningitis). It is used as **maintenance therapy for cryptococcal and coccidioidal meningitis. Oropharyngeal candidiasis and candidaemia in immunocompetent patients are treated with fluconazole.** Action of itraconazole overlap with those of fluconazole.

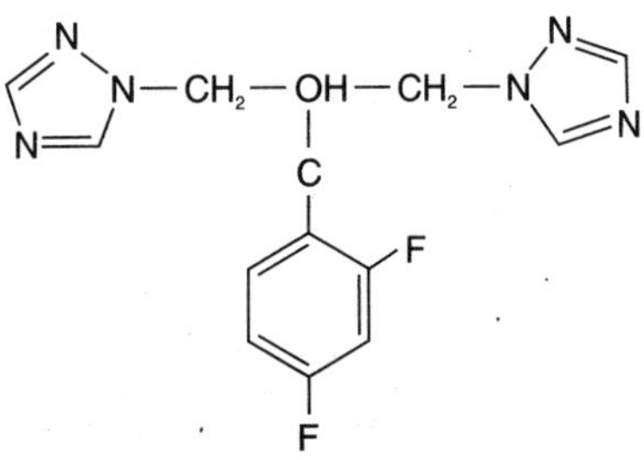

Fig. 8.4 : Fluconazole

ITRACONAZOLE (Fig. 8.5)

Itraconazole is also useful for the treatment of **INVASIVE ASPERGILLOSIS**, which generally responds poorly to amphotericin B. Itraconazole is efficient in the primary and maintenance therapy of histoplasmosis in immuno-compromised patients, chromoblastomycosis, meningeal coccidioidomycosis, sporotrichosis, balstomycosis, aspergillosis, and dermatophy nail infections.

TOPICAL ANTIFUNGAL AGENTS

Nystatin

Nystatin is a polyene antibiotic, structurally related to amphotericin B and having a similar mode of action. It cures **local candidal infections** of the **mouth and vagina**. Nystain suppress **oesophageal candidiasis and gastrointestinal overgrowth of candida**. No systemic absorption occurs (see before also).

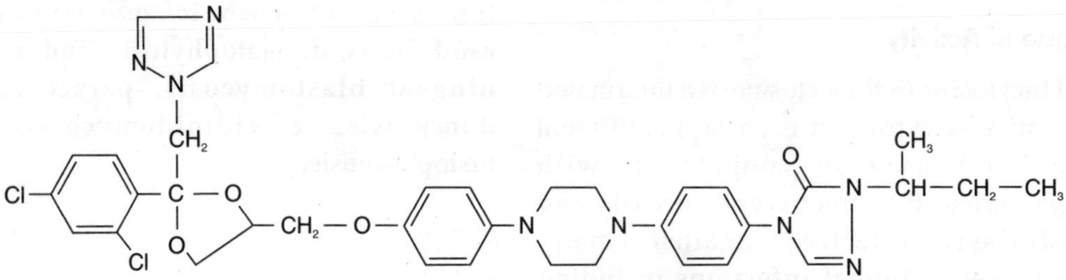

Fig. 8.5 : Itraconazole

Side Effects

The adverse effects of the azoles are primarily related to their ability to **inhibit mammalian cytochrome P450 enzymes**. Ketoconazole is the most toxic, and therapeutic doses may inhibit the synthesis of testosterone and cortisol, which may cause a variety of reversible effects such as gynecomastia, decreased libido, impotence, menstrual irregularity, and occasionally adrenal insufficiency. Fluconazole and itraconazone at recommended therapeutic doses do not cause significant impairment of mammalian steroidogenesis. All the systems i.e. antifungal azoles can raise the values of liver function tests. Since antifungal azoles inferct with P450 enzymes that are also responsible for drug metabolism, some important drugs interactions can occur. Increased antifungal azole concentrations occur with isoniazid, phenytoin, or rifampin therapy. Azole therapy can also lead to higher than expected serum levels of cyclosporine, phenytoin, oral hypoglycemics, anticoagulants, digoxin, etc.

CLOTRIMAZOLE (Fig. 8.7), MICONAZOLE (Fig. 8.6), & OTHER AZOLES

A variety of antifungal azoles too toxic for systemic use are available for **topical administration**. Clotrimazole and miconazole are available in several formulations. butoconazole, econazole, terconazole tioconazole, and are also available. All of these compounds seem to have comparable efficacy.

Topical azoles have a broad spectrum of activity almost in all dermatophytes and tinea versi-colour and cutaneous candidiasis by local application of creams or powders. Vulvovaginal candidiasis is treated by vaginal cream and suppositories. Clotrimazole oral torches are indicated in oral and oesophageal thrush.

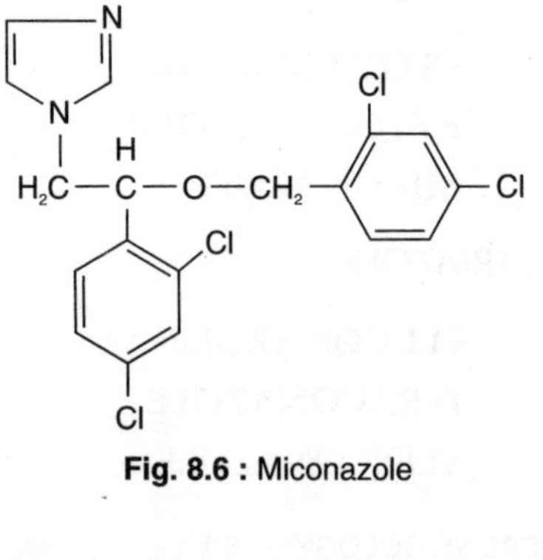

Fig. 8.6 : Miconazole

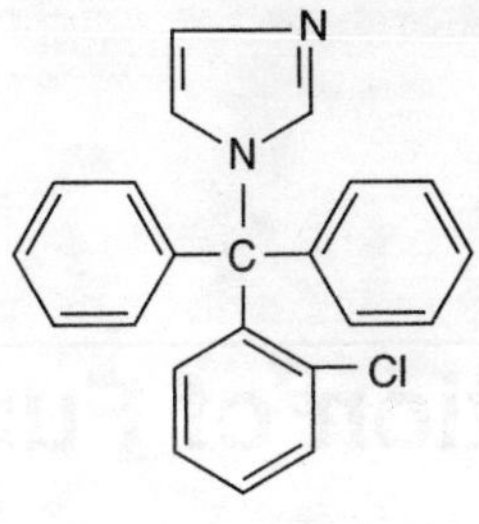

Fig. 8.7 : Clotrimazole

TERBINAFINE

Terbinafine is an allylamine drug; it blocks ergosterol synthesis by inhibiting squalene eposidase. Terbinafine is an oral drug to treat **DERMATOPHYTE NAIL INFECTIONS** and other dermatophyte infections. Side effects are gastrointestinal distress, headaches, skin reactions, and loss of taste sensation. For the long-term treatment of tinea unguium, **terbinafine, itraconazole and fluconazole are given intermittently.**

Table 8.4 : LIST OF ANTIFUNGAL AGENTS

1. **ANTIFUNGAL ANTIBIOTICS :**
   (a) Used *systemically* : **Amphotericin B, Griseofulvin.**
   (b) ***Locally*** : **Nystain, Hamycin, Natamycin, Candicidin.** Imidazoles and Triazoles.
2. **FATTY ACIDS :**
   Undecylenic acid, Zinc undecylenate, Sod. caprylate, Sod. propionate (used locally in ringworm infections).
3. **AROMATIC ACIDS :**
   Benzoic acid, Salicylic acid : Whitfield's ointment in Ringworm infections.
4. **AROMATIC DIAMIDINES :**
   (a) ***Hydroxy-stilbamine*** **(in Blastomycosis).**
   (b) 8-Hydroxyquinoline (locally in ***C. albicans*** infections).
5. **IMIDAZOLES :**
   Clotrimazole, Ketoconazole, Miconazole.
6. **TRIAZOLES :**
   Fluconazole, Iraconazole, Voriconazole.
7. **MISCELLANEOUS :**
   Diamthazole, Brilliant green, Ocrosol, Tolnaftate, Flucytosin, Thiobendazole.

Tolnaftate

It is useful in **Microsporum, Epidermophyton**, and **Trichophyton infections**, and **used locally.**

Hamycin

Antifungal antibiotic; active against **Candida, Aspergillus.**

Use :

It is also used locally in **vaginal candidiasis.**

Natamycin

It is used as aerosol, or locally applied in Aspergillosis. It *increases permeability of fungal cell membrane.*

Candicidin is an antifungal antibiotic useful in ***candida albicans*** infection; used as *ointment* and *vaginal tablet.*

AZOLE compounds (see 5 and 6 above).

REFERENCES

Ahn. C., *et al.* (1969). The therapy of cavitary pulmonary histoplasmosis. *J. Tboracic Surg.* **57**, 42-51.

Bindschadler, D. D. and Bennett, J. E. (1969). A pharmacologic guide to the clinical use of amphotericin B. *J. infect. Dis.* **120**, 427-436.

Dey, N. C. (1970). *Unpublished record.*

Drutz, D. J., *et al.* (1968). Treatment of disseminated mycotic infections. A new approach to amphotericin B therapy. *Amer. J. Med.* **45**, 406-418.

Hazen, E. L. and Brown, R. (1950). *Science* **112**, 423.

Levechal (1960). *Personal communication.*

Lockwood, W. R., *et al.* (1962). Experiences in the treatment of North American blastomycosis with 2-hydroxystibamidine. *Ann. int. Med.* **57**, 553-562.

McCurdy, D. K., *et al.* (1968). Renal tubular acidosis due to amphotericin B. *New England J. Med..* **278**, 124-131.

9

Taxonomy and Classification of Fungi

There is hardly any subject where classification (of fungi) requires constant review as well as revision. Hence, there is also some chance of overlapping and repetition of the genera of various **phyla and class**. To start with the classification, Fungus itself is assumed to be the **KINDOM : FUNGUS.**

Table 9.1 : KINGDOM : FUNGI

SUBKINGDOM : I. SCHIZOMYCOPHYTA, II. EUMYCOPHYTA	
I. SCHIZOMYCOPHYTA	**II. EUMYCOPHYTA**
PHYLUM I : **ACTINOMYCOTA** **CLASS** : ACTINOMYCETES **Order :** Actinomycetales **Family :** Actinomycetaceae **Subfamily :** Actinomycetoma **Genera :** 1. Actinomyces 2. Nocardia 3. Streptomyces	**PHYLA :** 1. **MYXOMYCOTA** 2. **ASCOMYCOTA** 3. **BASIDIOMYCOTA** 4. **DEUTEROMYCOTA** (Fungi imperfecti, i.e., sexual stage not known) or **HYPHOMYCOTA** 5. **ZYGOMYCOTA**

Table 9.2

PHYLUM II : Myxomycota
Class : Myxomycetes

MYXOMYCETES includes **slime and molds** with vegetative form comprising "aseptate protoplasmic mass" — the "**plasmodium**". Example : "**Dictylostelium**" mold.

Table 9.3

PHYLUM III : Ascomycota
Class : Ascomycetes

Reproductive process of Class : Ascomycetes

1. **Ascomycetes** occurs in the "ASCUS", an elongated sac-like structure that contains Ascospores which are produced after **KARYOGAMY** (nuclear union) and meiotic division.

Many imperfect fungi, *i.e.*, whose sexual stage was not known before has been discovered at present.

Medically important fungi belong to one order called "**ORDER - EUROTIALES**" "*Piedraia hortae*" (the organism for **black piedra** is included in it). Two **families** are well known :

1. **FAMILY : GYMNOASCACEAE**

It includes

(a) ***Arthroderma (uncinata)*** (Perfect stage)
(b) ***Nannizzia*** (Perfect stage)
(c) ***Ajellomyces*** or ***keratinomyces ajelloi*** (Imperfect stage)
(d) **Blastomyces** (Perfect stage)
(g) **Histoplasma** (Perfect stage)
(h) Perfect Stage of **Emmonsiella**.

The **CLEISTOTHECIA** of Gymnoascaceae have unorganised **peridial hyphae**.

2. **FAMILY : EUROTIACEAE**

The **Cleistothecia** is surrounded by compact **peridial hyphae.**

(a) Sexual stage of **Mycetoma fungus** the ***PSEUDOALLESCHERIA BOYDII***.
(b) Teleomorphic state of many **ASPERGILLUS** Fungi and many **PENICILLIUM** Fungi (Taleromyces and carpenteles).
(c) Others are Eurotium, Emericella, etc.

Contd.

Class 2. Archiascomycetes

This class includes the organism ***Pneumocystis carinii*** which was formerly thought to be a protozoa of undetermined classification. It is associated with AIDS or in other immuno-compromised individuals causing **"PLASMA CELL PNEUMONIA"**.

Table 9.4

PHYLUM IV : BASIDIOMYCOTA
CLASS : BASIDIOMYCETES

In the class Basidiomycetes, basidiospores are formed on the surface of a specialised structure called **Basidium**.

This phyllum includes a large complex group of fungi of plant, and animal (both mammalian and insect) hosts, e.g., Smuts, Rusts, Puffballs, Mushrooms and Bracket fungi, the **genus Filobasidiella, a non pathogenic Cryptococcus**.

The class includes the most important human pathogenic fungus family **Cryptococcaceae** that contains the teleomorphic (sexual) state of 1. **many Cryptococcus species**.

The other fungi are 1. **Mushrooms**, 2. **The Amanita**, 3. **Agaricus**, and 4. **Filobasidiella**.

Table 9.5

PHYLUM V : DEUTEROMYCOTA or HYPHOMYCOTA

This phylum includes those fungi whose **sexual (Perfect/ Teleomorphic) stage is not known** or could not be discovered as yet.

All of them have been put together into 2 class.

CLASS I : BLASTOMYCETES

The members include the **IMPERFECT** or Hyphomycotic or Deuteromycotic or **ASEXUAL YEASTS**, *i.e.*, organisms showing morphologically yeast phase including the **DIMORPHIC FUNGI** with their **YEAST PHASE** and **MYCELIAL PHASE**, *e.g.*,

(a) **Candida** species, the **Dimorphic Fungus**,

(b) ***Cryptococcus** species the **Unimorphic Yeast Phase Fungus**, and

(c) **Trichosporon species**, the **Unimorphic Mycelial Phase with arthrospores and terminal elongated budding cells**.

CLASS II : HYPHOMYCETES

They mostly include the following fungi whose sexual stage is not known as yet.

A. Most of the Ringworm fungi :

(a) **Microsporum**

(b) **Trichophyton**

(c) **Epidermophyton**

B. Gastrointestinal infection-causing Fungus : GEOTRICHUM

C. Systemic infections-causing Fungi :

(i) **Histoplasma**

(ii) **Cladosporium**

D. Dermatomycotic Fungi

(i) **Phialophora**

(ii) **Exophiala**

E. Other fungi :

(i) Aspergillus

(ii) **Penicillium** (some species)

(iii) **Alternaria**

*Many cryptococcal species whose **sexual stage** has been discovered belong to **BASIDIOMYCOTA**.

Table 9.6

PHYLUM VI : ZYGOMYCOTA
CLASS : ZYGOMYCETES

Zygomycetes fungi comprise unseptate mycelia containing flagella and showing motility. The compatible gametes mate together by fusion to form a **ZYGOSPORE** and hence they reproduce sexually. They can also reproduce **ASEXUALLY** by formation of **conidiospores** (borne on conidia or condiophore) or **sporangiospores** borne on sporangiophores.

ORDER I : MUCORALES : Members of this order are

(i) **Absidia**

(ii) **Mucor**

(iii) **Mortierella**

(iv) **Rhizopus**

(v) **Pilobolus**

(vi) **Cunninghamella**

(vii) **Saksenaea**

(viii) **Phycomyces**.

They asexually reproduce by **sporangiospores**.

ORDER II. ENTOMOPHTHORALES

The members of this order reproduce **ASEXUALLY BY** condiospores (forming conidia) and/or **Zygospores**.

They belong to the Genera :

(i) Basidiobolus

(ii) Canidiobolus

(iii) Entomophthora.

10

Superficial Mycosis PHYLLUM : DEUTEROMYCOTA (Hyphomycota)

CLASS : I. Blastomycetes II. Hyphomycetes

In superficial mycosis the fungus affects the outer most keratinised layer of the skin or grows in the secretions on the surface of the mucous membrane. The pathological process does not extend deep like those in **cutaneous mycosis (Dermatophytosis)** and **candidiasis.**

The diseases in this group are

Erythyrasma

Trichomycosis Axillaris

1. **Piedra** — (a) Black; (b) White
2. **Tinea nigra**
3. **Superficial streptomycosis** causing chaluni with Haja (intertrigo) and Phata (cracks).

Skin lesions of **pityriasis versicolor** and **pityriasis capitis** caused by *Malassezia furfur* and *Pityrosporum ovale* respectively, belong to this group. Dermatitis caused by *Candida albicans* is also included in this group (for details see chapter on Candidiasis).

THE BACTERIAL INFECTIONS

1. ERYTHRASMA

Definition : ERYTHRASMA IS A SUPERFICIAL BACTERIAL INFECTION OF THE SKIN of the moist parts of the body, *viz.*, the **axilla (Fig. 10.1), groin, fold of the breast and other intertriginous regions,** caused by the bacterium *C. minutissima*. The causative organism is ***CORYNEBACTERIUM MINUTISSIMA*** (von Barensprung) n. comb.

Since *Erythrasma* as well as *Trichomycosis axillaris* are caused not be Nocardia but by Corynebacterium minutissima and c tenuis respectively, they are considered is Author "Medical Bacteriology". The causative organism was formerly through to be Nocardia minutissima. (Fig. 10.1, 2, 3)

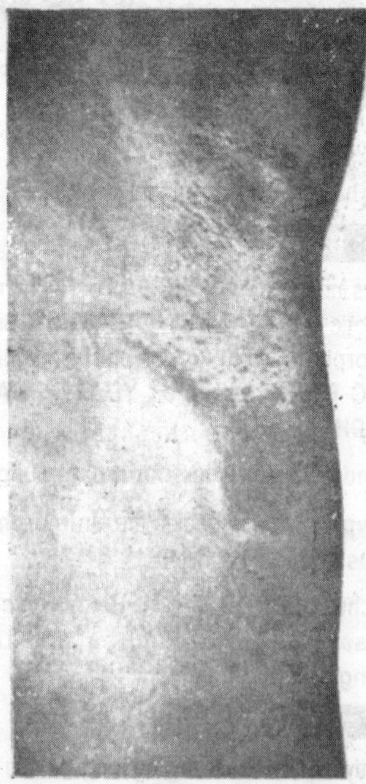

Fig. 10.1 : Erythrasma on the axilla caused by ***C. minutissima***

A depreciation of C. *minitissimum* (Von Barensprung) n.comb. as a causative agent of erythrasma previously attributed to *Nocardia minutissima* as stated above.

2. TRICHOMYCOSIS AXILLARIS

(Trichomycosis nodosa)

DEFINITION : IT IS NOT A FUNGAL BUT A SUPERFICIAL BACTERIAL INFECTION

Organism : According to Castellani, trichomycosis axillaris was previously known to be caused by *Nocardia tenuis*. Montea *et. al.* (1963) studied in Argentina infected hairs from a 45 years old man with red variety of Trichomycosis axillaris. The hair showed concretions consisting almost entirely of diphtheroids which were also present in the horny layer of the hair, supporting the observation of Crissey *et. al.* (1952) that ***CORYNEBACTERIUM TENUIS*** caused yellow form of infection in 28 out of 100 patients in Philadelphia area.

So, this disease is discussed in Authors' Bacteriology book.

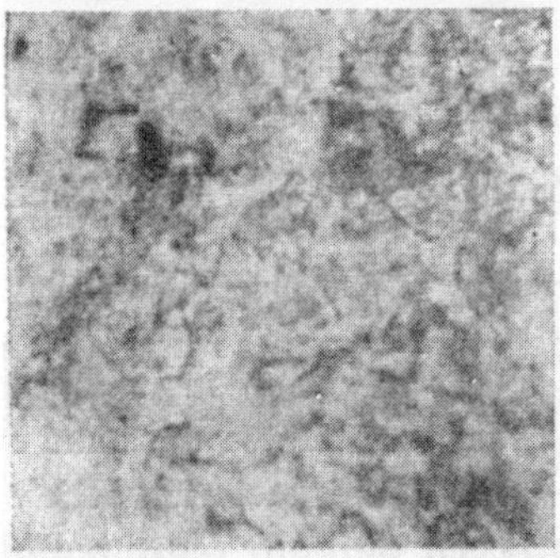

Fig. 10.2 : *Corynebacterium minutissima* in scales (McGuire's stain)

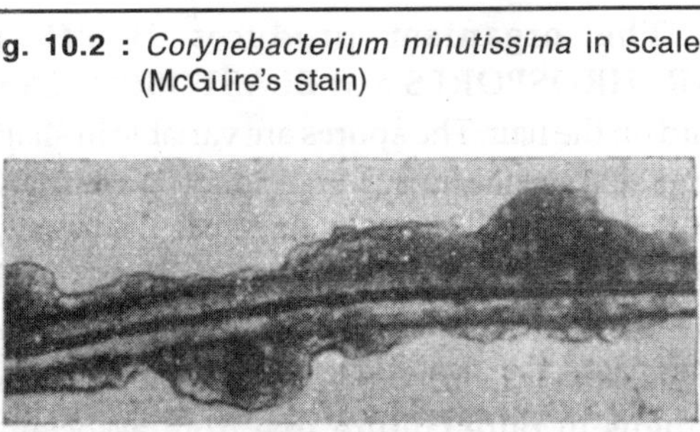

Fig. 10.3 : Trichomycosis nodosa (axillary hair), i.e., T. Axillaris, bacterial infection

THE FUNGAL INFECTIONS

PIEDRA

Definition

Piedra is a fungus infection of the hair of the scalp, moustache or other hairs of the body caused by :

(a) ***TRICHOSPORON BEIGELII*** *(cutaneum)* producing white, nodular excrescences, **WHITE PIEDRA**, they are easily detachable from the **long hair of the scalp**, and

(b) ***PIEDRAIA HORTAE*** forming **BLACK PIEDRA**, which produces **hard nodules** firmly adherent **to the stiff hair of moustache** or **beard**. The condition is contagious and caused by ***EXOPHIALA (Cladosporiun) WERNECKII.***

History

The disease was first described by Beigel (1865) in Germany, forming nodules on the hair of wigs. Behrend (1890) described *Trichosporon ovoides* recovered from the hair of the scalp in Germany. Vuillemin (1902) found light coloured nodules on the moustache and called these **TRICHOSPORON BEIGELII**.

Black piedra was originally described by Dessene (1878). Horta (1911) studied black piedra of the hair in Brazil, and Fonseca and Leao (1928) called it ***Piedraia hortae***. According to the work of Macinnon (1942) **Black Piedra** in Brazil is due to ***P. hortae***; white piedra in South America is due to *T. beigelii*; other species, namely, *T. giganteum* Unna, 1895, *T. ovoides*, Behrend, 1890, *T. glycophila*, Dubois, 1910, etc. are synonymous with *T. beigelii* (Rabenhorst) Vuillemin, 1902. However, the currently used synonym is ***T. beigelii (cutaneum)***. ***T. beigelii*** *(cutaneum)* is used in this book.

Geographical distribution

The disease occurs in the tropical and subtropical areas of Africa and Asia including India and Latin America. The African infection has been confined to primates but human infection is yet unknown in that continent. Black Piedra has been reported from Columbia and other countries of South America, East Indies, Malayasia and India. **Dey and Kakoti (1954)** reported a case of white piedra in Assam (India).

Epidemiology.

P. hortae infection is seen in epidemic form, more often in the moustache (seen by the author in India) than in the long hair of women, in whom infection with *Trichosporon (beigelii) cutaneum* is more common.

Types : **There two types of Piedra** :

Aetiology :

1. **Black Piedra caused *by Piedraia hortae*.**
2. **White Piedra,** caused by ***Trichosporon beigelii (cutaneum).***

Pathogenicity of PIEDRA

T. beigelli (cutaneum) causes sporadic infection of the long hair of the scalp and is much less contagious than *P. hortae*; whereas, *P. hortae* affects the coarser hair of the beard and moustache, rather than the scalp hair.

BLACK PIEDRA (*P. hortae*)

BLACK PIEDRA causes an infection of the course or stiff hair caused by ***Piedraia hortae*** which belongs to the class ***Ascomycetes***. It is characterised by the development of *hard brown or black nodule along the shaft of scalp hairs* or *moustache.*

Symptoms

P. hortae usually affects stiff hair of beard and moustache *penetrating* the *shaft of the hair*. The *fungus proliferates* and then *breaks* and *forms a dark, hard gritty, nodular mass* around the hair. In the white variety caused by *T. beigelii (cutaneum)* the ***nodosities in long hairs*** remain more or less *loose* and can be *removed easily* from the hair.

Mycology (*P. hortae*)

CLASS : ASCOMYCETES

BLACK PIEDRA : In ***DIRECT EXAMINATION*** under a coverslip preparation with *10 per cent KOH* or *sodium sulphide* solution, the *nodosities* of the infected hair contain round or *polygonal cells*. They form *dark, hard, adherent mass* by *broad, closely septate hyphae, branching dichotomously*. Numerous ***oval asci*** are seen, *30-50 μ long* containing *2-8 curved* or *fusiform ascospores*, and *10-30* μ long having simple terminal cilia like appendages.

Culture

P. hortae grows readily but slowly on agar. *Smooth black* colonies are formed which are raised and smooth. The colony is sclerotia-like and formed by dark mycelia and **chlamydospores.** Asci are rarely produced in culture in the thicker part of the colony.

WHITE PIEDRA [*T. beigelii (cutaneum)*]

In a coverslip preparation **with 10% KOH,** *DIRECT EXAMINATION* shows a mass of **gelatinous, mycelial spores** composed of round or **oval hyaline cells**. The fungus **reproduces by arthrospores** and also by budding and both the elements are held together in a **gelatinous substance,** (**Pl. 10.I a, b**). The **arthrospores** are hyaline **3 to 10** μ in diameter.

Mycology [*T. beigelli (cutaneum)*]

The organism produces **HYPHAE, ARTHROSPORES** and **BLASTOSPORES** in and on the hair. The **spores** are variable in shape and size measuring from **2 to 4μ in diameter** but many attain a size of **8μ** in its **longest dimension**. A group of cells or individual **cells** are **surrounded by a thick gelatinous capsule,** which creates difficulty for identification of the fungus in pure culture free from associated bacteria and for biochemical and nutritional studies.

Culture

On **antibiotic-free Sabouraud's medium, *T. beigelli (cutaneum)* grows fast, forms soft brittle, cream coloured, wrinkled colonies** (Pl. IId, e). They show thick **double walled ARTHROSPORES** often **forming chains** (Pl. 10.I e).

Treatment

Topical and systemic applications of Broad spectrum antifungal antibiotics are the treatment of choice (preferably topical imidazoles); (see Ch. 8)

The part should be kept dried by and antiseptics and Clotrimazole powder should be applied.

The hair should be cropped close to the crown or shaved and ***lotio Hydrarg. perchloride*** (1 : 2000) or ***Ung. Hydrarg. ammon.*** (3 per cent) should be applied on the scalp. **Lotio resorcin**

PLATE 10.I

TRICHOSPORUM (BEIGELII) CUTANEUM

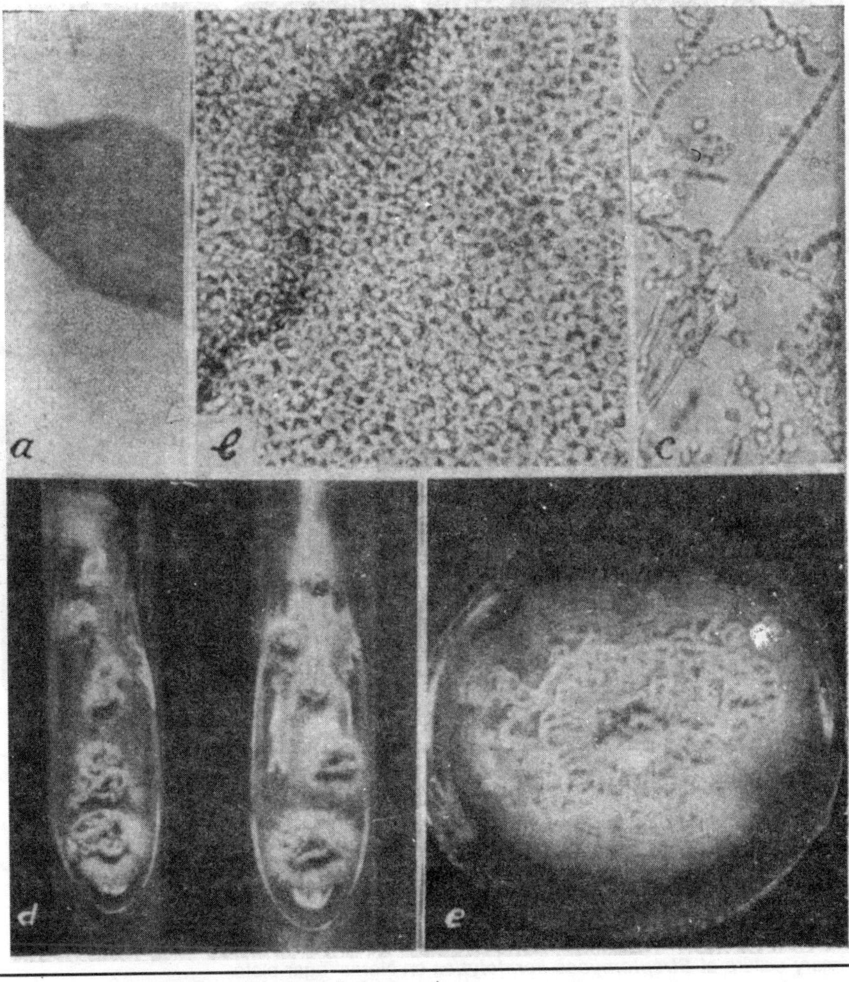

(a) Nodosity of the hair (low power);

(b) Same under high power mag. × 215;

(c) Hanging drop culture showing arthrospores in chain;

(d) Growth of *T. beigelii* on Sabouraud's medium-three weeks old;

(e) Giant colony, six weeks old.

PLATE 10.II
COMMON PATHOGENIC NOCARDIA and Corynebacteria

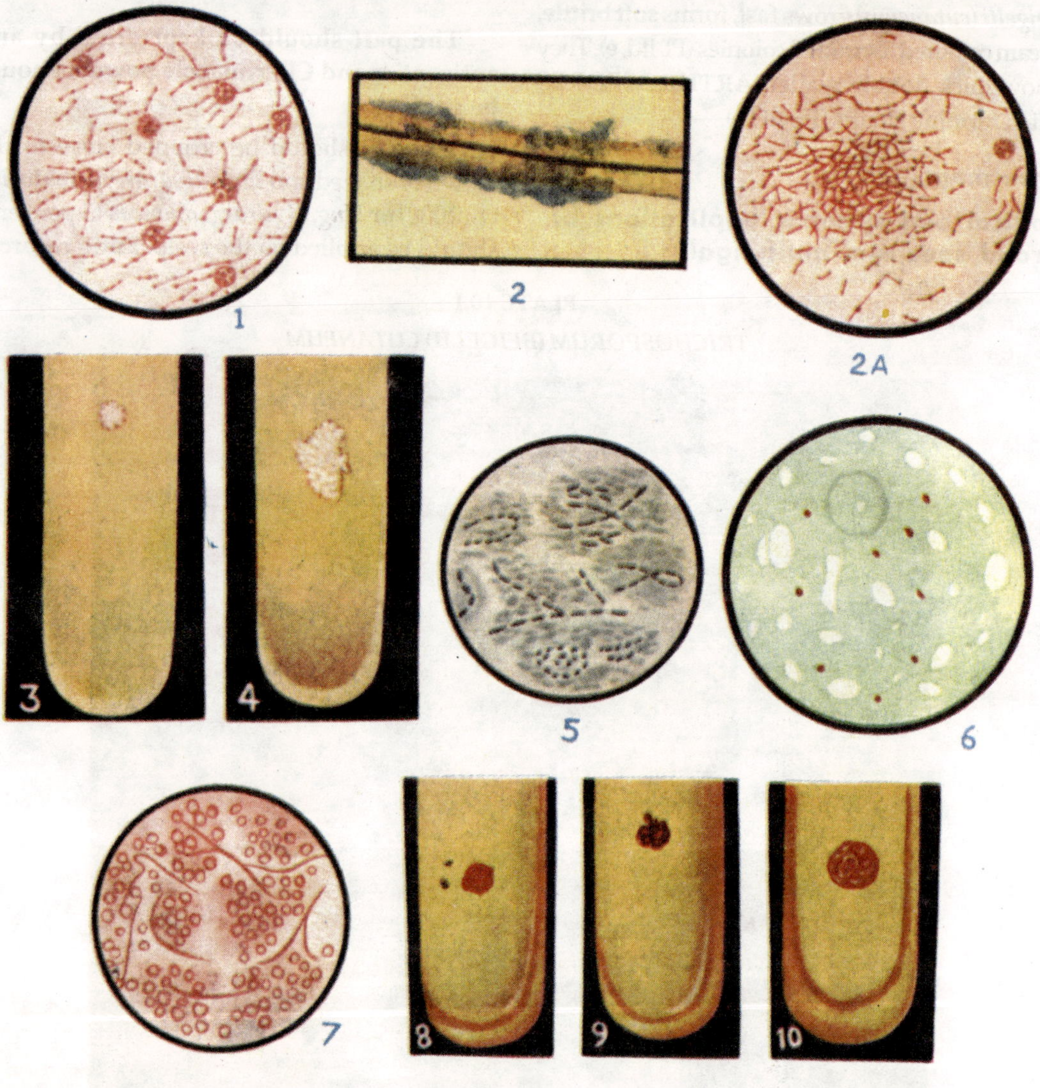

Fig. 1 : ***Corynebacterium minutissima*** is scales (McGuire's stain)

Fig. 2 : Hair showing ***Corynebacterium tenuis*** affection

Fig. 2A : ***Corynebacterium tenuis*** showing metachromatic granules and branching in places (McGuire's stain)

Fig. 3 and 4 : *Corynebacterium tenuis* in pure culture

Fig. 5 : Skin scraping showing ***Streptomyces keratolytica*** bacillary and coccal forms

Fig. 6 : Primary culture of ***Streptomyces keratolytica*** on Norris' agar plate

Fig. 7 : Mycelia and spores with terminal clubs in culture of ***Streptomyces keratolytica***

Fig. 8 : Early colonies of ***Streptomyces keratolytica*** (Sabouraud's glucose agar)

Fig. 9 : Old colonies of ***Streptomyces keratolytica*** showing black colouration at the periphery

Fig. 10 : Growth of ***Streptomyces madurae*** (Sabouraud's glucose agar)

is equally efficacious in this condition (**see appendix B**).

TINEA NIGRA (PALMARIS)

PHYLUM : DEUTEROMYCOTA

CLASS : HYPHOMYCETES, UNIMORPHIC (MYCELIAL) FUNGUS

(The causative organism ***EXOPHIALA*** *(Cladosporium)* ***WERNECKII*, Hortae 1921**

SYNONYMS : *Microsporon mansoni* Castellani, 1908; *Cladosporium mansonii* (Castellani) Pinoy, 1912; *Dematium wernecki* Dodge, 1935, **EXOPHIALA.**

Definition : Tinea nigra palmaris is essentially a disease of the tropical countries, however, occasional cases are seen in subtropical areas. The disease appears as *dark blotches on the palms of the hands* or more rarely on other parts of the body. ***EXOPHILIA*** *(Cladosporium)* ***WERNECKII*** **Horta (1921)** is the aetiological agent.

History

In 1905 and subsequently Castellani published superficial dermatophytoses referring Manson's observations in 1872. It is probable that **TINEA NIGRA PLAMARIS** refers to dirty, stained lesion of **PITYRIASIS VERSICOLOR**. In 1957, he quoted Manson when a lesion of *tinea nigra palmaris is cleaned with soap and water* when it *becomes* ***TINEA ALBA***. In 1919, he called the fungus of true tinea nigra palmaris lesions caused by Exophiala formerly *(Cladosporium)* ***mansoni*** and differentiated it from pityriasis versicolor in 1923. *Exophiala (C.) mansoni* is characterised in the epidermis by hyphal fragments which are bent or **banana shaped and clusters of large globular cells**. The fungus could be easily cultivated from the scraping on maltose agar with greenish hyphae. Horta (1921) described the disease and the fungus, and named the latter ***Exophiala*** *(C.)* ***werneckii***. Neves and Crista in Brazil (1947) have clarified the clinical aspects of the disease and characteristics of the fungus.

Taxonomy

The fungus is Dematium Person 1794; Dodge put it in the genus *Dematium werneckii*, Dodge 1935. The genus Dematium (Link, 1816) belongs to the Family Cladosporiaceae Dematiaceae Saccardo (1886) and Class Fungi Imperfecti. Cooke (1959), however, in a review of nomenclature accepted *Aureobasidium pullulans* a saprophytic fungus, and ***Exophiala*** *(Cladosporium) werneckii* described as different species.

Morphology

In the epidermal scraping, ***E. werneckii*** appears in the form of olivaceous, sinuous, branched septate hyphae 2-3μ in diameter as hyphal fragments, or **arthrospores** with close septa and elongated budding cells.

DIRECT EXAMINATION

They are seen readily in 10 per cent potassium hydroxide mounts or by ***McGuire's stain.***

Culture

Cultures of ***Exophiala (C.) werneckii*** (formerly Cladosporium) are obtained on *cycloheximide agar*. It *grows slowly* and produces a *moist, greenish-black raised colony*. With age, *blackish aerial mycelium* covers the growth. The dark mycelium consists of of *one-or two-celled budding spores* 2 to *4μ* by *4 to 8μ in size.*

SAPROPHYTIC LIFE

Fungi like *E. werneckii* are almost ubiquitous and found in soil, compost, sewage or dirts. They grow on wood and painted surfaces in humid climates. It is not yet clear, whether *Exophiala werneckii* is different from *Aureobasidium pullulans* (Cooke, 1959).

SUPERFICIAL STREPTOMYCOSIS OF THE SKIN

Caused by ***STREPTOMYCES*** **(Acton and McGuire) 1957**

Phyllum : Actinomycota

Family : streptomycetaceae unimorphic fungsus (Mycelial structure)

Definition

This is a keratolytic condition of the hands and feet caused by *Streptomyces keratolytica,* characterised by keratolysis of the epidermis of the palms and soles, and other lesions variously described as **pitted sole, mangoe toe, cracked heel, paronychia, onychomycosis and vesicular eruptions.**

Aetiology

The disease is caused by *S. keratolytica,* (enormously present on the damp and manured soil containing organic matters). The disease is common in cultivators and gardeners who often walk bare-footed. It is also common amongst housewives and maidservants, who are in the habit of handling mud with their ***hands kept moist for a long time***. The disease is generally seen during the monsoon months in India. (It has also been found to occur amongst European sailors on coastal vessels (ship) of India, as they have acquired the habit of walking bare-footed on the deck).

Mycology

DIRECT EXAMINATION :

Scraping from the margin of the pit, sulci or top of the vesicle is soaked in normal saline to soften it, teased and placed on a slide with a drop of **McGuire's stain** which is allowed to act for 2 to 5 minutes and then mopped off with a piece of filter paper. It is then mounted with a drop of neutral glycerine under a coverslip. A little pressure on the coverslip will flatten out the epidermis. It is then examined with oil immersion objective and × 10 ocular. The hyphae are seen to consist of fine, segmented branching filaments. The oldest hyphae are very closely segmented and give a beaded appearance like the chain of streptococci **(Pl. 10.II, fig. 5)**. Conidia may be seen at the end of some hyphae.

INDIRECT EXAMINATION

1. Culture

The diseased area is well washed with soap and water to remove all dirts and then swabbed with 75 per cent alcohol. The tissue from the margin of the lesion is obtained and soaked for about 2 hours in normal saline to soften it. The material is then teased with a pair of needles into small strips and treated with absolute alcohol for one minute and these strips are planted directly into the Norris' medium at pH 7.4.

NORRIS' MEDIUM

Soluble starch	2 g
Dipotassium phosphate	0.5 g
Hydrated magnesium sulphate	0.2 g
Calcium chloride	0.05 g
Ferric chloride	0.01 g
Sodium nitrate	0.05 g
Asparagin	0.05 g
Agar	20.00 g
Water	1000.00 ml

Plates prepared from this medium are inoculated with small bits of infected tissue and incubated at a temperature of 25° to 35°C or at room temperature. This synthetic medium inhibits the growth of most of the secondary organisms. In **four days**, the primary culture shows small, raised, pink colonies **(Pl. 10.II, fig. 6)**. The colonies become larger and deeper in colour with the age of the culture and still later develop into flat, moist colonies with areas of darker or black tint. In Sabouraud's medium, subcultures show the same character of growth **(Pl. 10.II, figs. 8 and 9). On blood agar the culture shows a zone of haemolysis**.

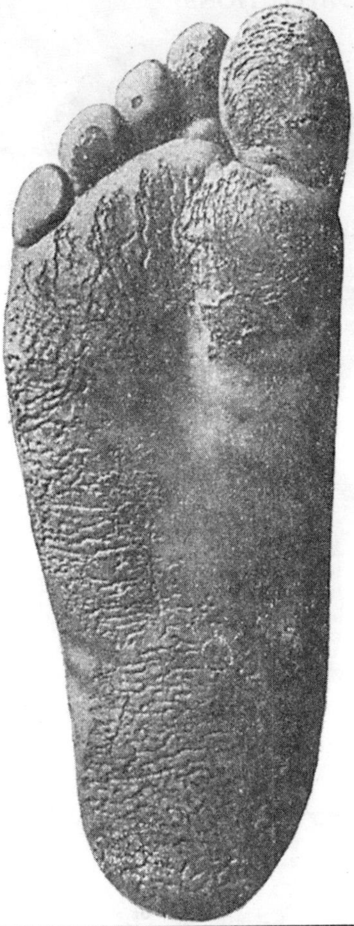

Fig. 10.4 : Sieve-like keratolysis of the sole of the foot (Chaluni) by ***STREPTOMYCES KERATOLYTICA***

2. Biochemical reaction

Sugar or glycerol is not fermented. Milk becomes slightly acid and clotted.

Morphology

The hyphae are about 0.8μ in diameter; they are segmented, branched and slow conidia (1 to 1.5μ in diameter) which may be single or in cluster; some mycelia may have clubbed-ends due to swelling of the end of the hyphae. There are often intercalary chlamydospores with thickened walls.

Sings and Symptoms of STREPTOMYCOTIC affection

The disease is characterised by pitted appearance of the affected parts. But as the disease persists, cracking and exfoliation takes place. According to the type of the lesion, several clinical types have been described by Acton and McGuire (1930, 1931).

CHALUNI or sieve-like pitted sole **(Fig. 10.4)** shows **small, multiple, pitted areas with moth-eaten appearance (Pl. 10.III, Fig. 1)**.

They may show similar **pitting on the palms** of the **hands (Pl. 10.III, Fig. 7)**.

PHATA

This is a condition of **cracked-heel** and **fissures** appearing at the margin of the sole **(Pl. 10.III, figs. 2 and 3)**. **Hyperkeratosis** is often associated with it **(Pl. 10.III, fig. 3)**.

HAJA OR INTERTRIGO (Pl. 10.III, figs. 4, 5 and 6) is an interdigital affection of the webs of the toes and fingers caused by *S.keratolytica*. The condition somewhat resembles "**mango-toe**" **caused by** ring worm fungi **(Pl. 10.III, fig. 4)**. The infection starts as an intertriginous lesion on the webs of fingers and toes and extends to the palmar or the plantar surfaces as gyrate areas of keratolysis **(Pl. 10.III, figs. 5 and 6)**. The ulceration generally progresses to a considerable depth, producing in some cases spontaneous amputation of the toe.

PARONYCHIA

The Streptomyces fungus has been found to produce inflammation around the soft tissues of the nails of the hands and feet and sometimes the base of the nail is involved showing ulceration **(Pl. 10.III, fig. 4)**. The nail plate, when affected, may give a moth-eaten or pitted appearance with darkening of the nail.

POMPHOLYX (Streptomyces infection)

Vesicles may **appear on the palm**, instep or sole. They leave **pitted scars on** the affected areas of **palms (Pl. 10.III, fig. 7)** and **soles**.

PLATE 10.III : STREPTOMYCOTIC AFFECTIONS

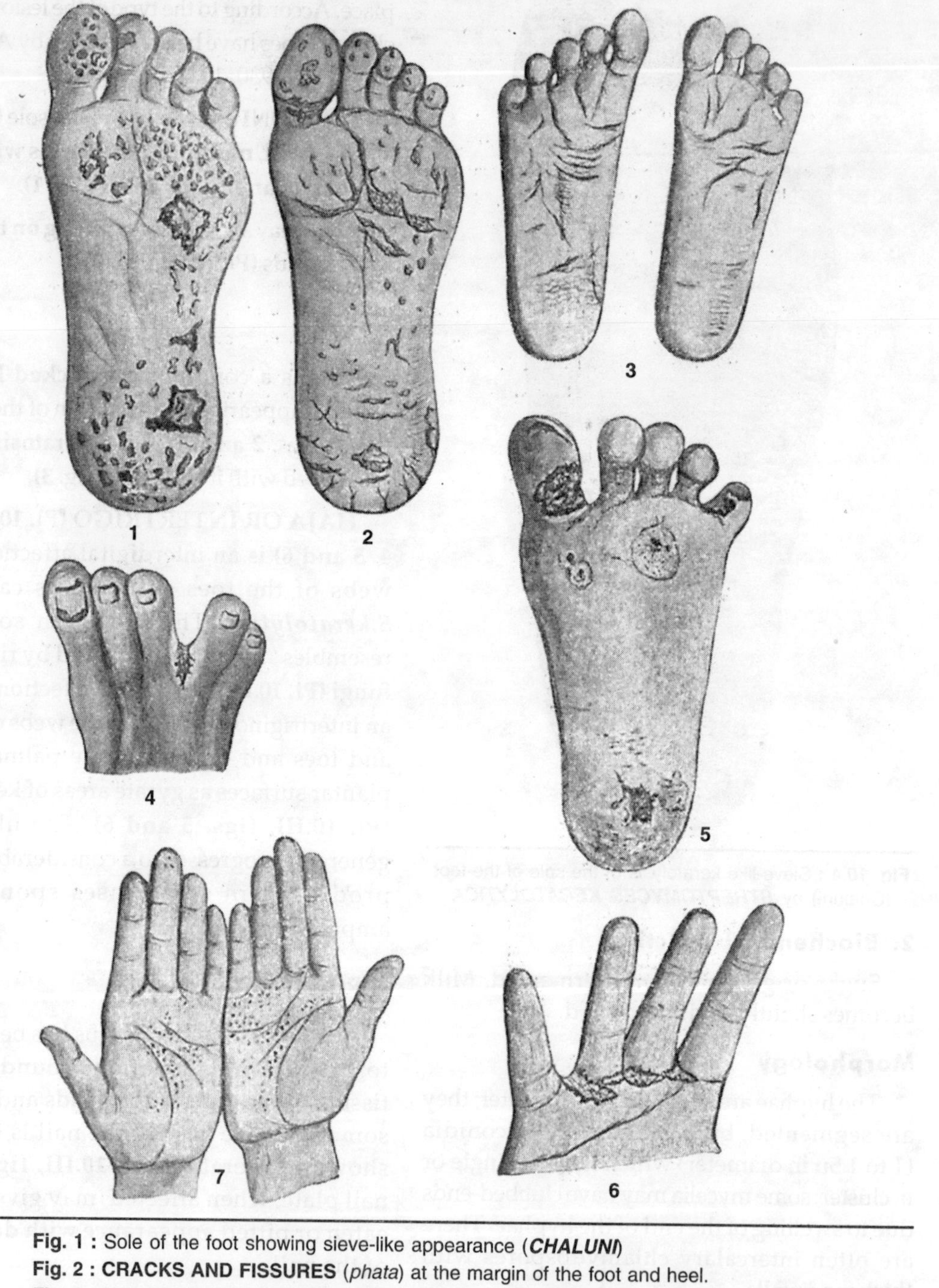

Fig. 1 : Sole of the foot showing sieve-like appearance (***CHALUNI***)
Fig. 2 : CRACKS AND FISSURES (*phata*) at the margin of the foot and heel.
Fig. 3 : Hypertrophic lesions on the sole of the foot with cracks and fissures
Fig. 4 : Sodden condition in between the toes (*Haja*) and ulcer at the base of the nail
Fig. 5 : Depp ulcers on the great and little toes as a result of extension of (*Haja*). A deep area of keratolysis is seen at the heel
Fig. 6 : Ulcer on the palm of the hand as a result of extension of intertrigo
Fig. 7 : Minute pitting of the plams of the hand due to keratolysis (**POMPHOLYX**)

Differential Diagnosis

Phata or fissures at the margin of the foot and heel, and *Chaluni* or **pitted sole** are easily diagnosed by the typical appearance of the lesion.

HAJA (intertrigo) or interdigital erosion should be differentiated from those produced by the ringworm fungi. In the latter, **microscopic examination of scrapings should show numerous ringworm fungi** under the high power objective. When the examinations of scale is negative for ringworm fungi, one should consider investigation for *S. KERATOLYTICA*. Interdigital affections of the webs of the fingers show a sodden, worm-eaten appearance, which spreads on the palm in the contiguous parts. The deep fissures of mango-toe are more characteristic of *S. keratolytica* infection than those of ringworm; this is due to the marked keratolytic action of the fungus.

HYPERTROPHIC LESIONS of the palm and sole must be differentiated from **tylosis, keratodermia of palms and sole caused by syphilis and yaws**, and hyperkeratosis caused by ringworm. Localised keratolysed areas caused by *S. KERATOLYTICA* often simulate perforating ulcers of leprosy and it should be differentiated from the latter condition.

PARONYCHIA caused by *S. KERATOLYTICA* should be differentiated from similar lesions produced by *Candida albicans*; in the former condition the base of the nail shows keratolysis and pitting of the nails is often associated with interdigital affections; but in the latter, the margin of the nail shows **bolster-like swelling**.

NAILS affected by *S. keratolytica* are like those affected by ringworm fungi, but in ringworm of the nail a positive finding of the ringworm fungus clinches the diagnosis; whereas, in the affection caused *S. keratolytica* fine mycelia can be demonstrated with **McGuire's stain under oil immersion objective**. Pitting of the nail is found in this as well as in psoriasis; therefore, psoriasis must be excluded in these cases by exclusion of the lesion in other parts of the body.

Prophylaxis

The practice of walking barefooted should be avoided. In the case of cultivators who are inevitably exposed to the infection, the prophylactic application of *lotio formalin et glycerine* is advantageous.

Treatment

For chaluni, the application of *lotio formalin et glycerine* is useful (see Appendix). For other conditions, a 2% solution of gentian violet in 10% alcohol is quite satisfactory. Lotio triple dye has also been used. It is quite as effective as in mango toe and other interdigital affections. In paronychia, the lotion should be applied well inside the margin of the nail.

MALASSEZIA FURFUR

Phyllum : Deuteromycota

Class : Blastomycetes (Dimorphic Fungi)

SYNONYMS : *Microsporon furfur* Robin, 1853, *Sporotrichum furfur* Saccardo, 1880, *Oidium furfur* Zopf, 1890; *Monilia furfur* Vuillemin, 1931.

Historical

This is a yeast-like fungus and was detected and described by Eichsted in 1846 and Sluyter in 1847. Robin gave an excellent description of the fungus and called in *Microsporum furfur* in 1853. However, his adaptation of the genus Microsporum Gruby 1843 for the fungus of of pityriasis versicolor was erroneous. Baillon in 1889 created the generic name Malassezia and the fungus is now called *Malassezia furfur*. Gordon (1951) isolated a yeast-like fungus from tinea versicolor and called it *Pityrosporum orbiculare*. He suggested that it might be the aetiological agent for pityriasis versicolor. Gordon's culture shows a growth of yeast like organism without formation of mycelia. It is likely to be a growth of *Pityrosporum ovale* which

does not show growth of mycelia like those of *Malassenia furfur*. The organism was cultivated by Panja (1927), and Mcleod and Dowling (1928) in lipophilic media and credit goes to these pioneer workers.

TINEA VERSICOLOR

(The causative organism is *Malassezid furfur*, **Robin, Baillon, 1889)**

SYNONYM : **PITYRIASIS VERSICOLOR**, TINEA FLAVA, TINEA NIGRA, ETC.

Definition

The disease is characterised by scaly lesions of the face and the body, and associated with hypopigmentation and discolouration of the skin of the affected parts.

Aetiology

The disease is common in young adults of both sexes, but it is more common in males than in females. It is rare in children. **In the winter** season, the disease may remain latent but with the advent of the **summer, there is a rapid spread** of the disease. This can be explained owing to excessive perspiration, as seen in **hyperidrosis which often predisposes to it**.

Mycology

DIRECT EXAMINATION

Malassezia furfur is the causative organism of tinea versicolor. *Direct examination*. The fungus can be demonstrated in the **scales under a coverslip preparation** with **caustic potash**, but **better results** are obtained **with McGuire's stain** which stains the **spores and mycelia purple** (Pl. V). When properly prepared, the affected **scales** can be **identified under the low power**, but under the **high power characteristic morphology of *M. furfur*** is seen, namely;

(a) **Groups of spores** with double contour;

(b) **mycelia** which are short, stout, sinuous, often **septate with blunt ends** and **occasionally show branching**.

INDIRECT EXAMINATION

Culture

There is a conscentious of opinion whether the organism can be cultivated but the author is of opinion that the organism can be **cultivated** in the same **way as that of *P. ovale*** (Method of **Panja**). From the primary growth subcultures may be obtained in **glucose or glycerol agar** containing sterile **olive oil spread on the surface of the slope. Budding forms are frequently** seen in the culture. The organism has great similarity with *P. ovale* and their difference is shown in Table 10.1

Table 10.1 :

Particulars	M. furfur	P. ovale
SMEAR :		
Microscopical-(McGuire's stain) :		
Mycelia	Always present	Rare
Budding	May be present	Always present
Spores (Double-walled)	Plenty	Absent
CULTURE :		
Growth	Difficult to grow on ordinary media	
Petroff's medium	Heaped up, chalky in appearance	
Sabouraud's glucose agar with and without olive oil	Heaped up, glabrous growth in both	

Signs and symptoms

The disease starts as **tiny, hypopigmented, scaly macules** on the **face, chest, neck, back or abdomen.** The macules coalesce and form hypopigmented areas which exhibit a wide diversity of shape and size. On close observation the **macules**, often **gyrate due to coalescence**, are seen to be **covered with shiny scales** and **show yellowish brown dark colour** in contrast to the healthy skin. According to the colour, they have been variously called tinea flava, tinea alba or tinea nigra. The **condition is non-inflammatory** and does not produce any subjective symptoms except a **little irritation** or **discomfort specially** when **exposed to the sun**. The disease causes **mottled disfigurement of the neck, face** and **trunk (Pl. 10.VI a, b, c); Fig. 10.5** and interests the patient mainly from the **aesthetic point of view**. The affected parts show **fluorescence under the Wood's light**.

Diagnosis

This can be done **directly** by examination of the **scale scraped from** the lesion. When **McGuire's method is used**, the **morphology of the stained fungus** can be very well studied; namely **groups of spores and mycelia**. **Indirectly** the condition can be diagnosed by **demonstration of fluorescence under** the **ultraviolet light passed through Wood's** filter and the colour may vary from **light to dark**

PLATE 10.IV
MALASSEZIA FURFUR

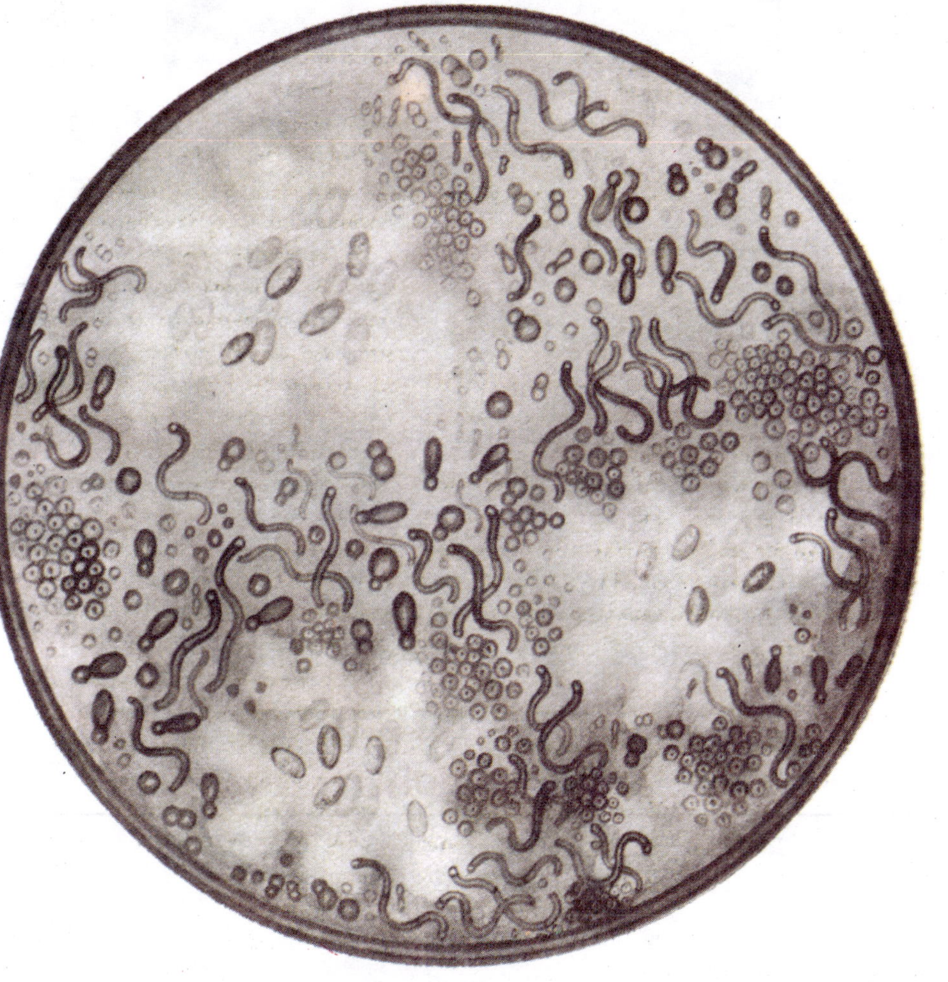

Scales of Tinea versicolor (McGuire's stain) showing spores in cluster and mycelia scattered irregularly; budding forms are also seen. **Some of the cells are nucleated (Parakeratosis)** with small **pyknotic nuclei indicating necrosis**

PLATE 10.V

TINEA VERSICOLOR

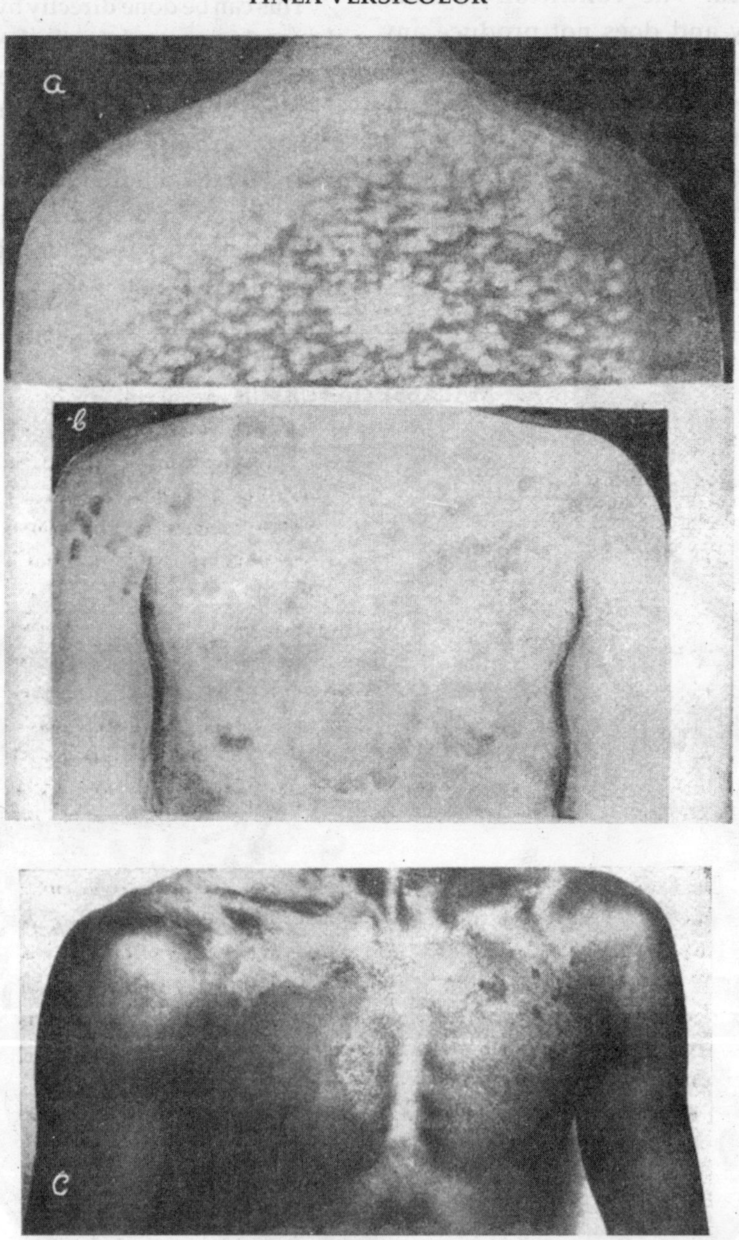

Fig. (a) : Back; **(b)** and **(c)** : Upper part of the chest

brown according to the degree of discolouration.

Differential diagnosis

Hypopigmented macules are often confused with : (1) **Dermal leishmanoid,** (2) **Neuromacular leprosy.**

(1) Dermal leishmanoid

Hypopigmented macules of dermal leishmanoid are erythematous rather than scaly. Scrapings do not show presence of *M. furfur*. Characteristic feature of dermal leishmanoid is that the granuloma more often affects the mask area of the face with an erythematous flush. In well-advanced cases small nodules may be present specially on the chin. In case of dermal leishmanoid, the patient always gives a history of kalaazar. Smears from the erythematous, hypopigmented lesions do not usually show *Leishmania donovani*.

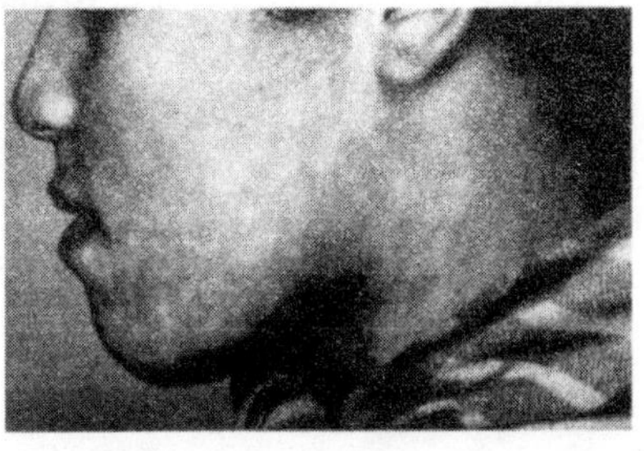

Fig. 10.5 : Tinea versicolor of the face and neck

(2) Neuromacular leprosy

In this condition there is always partial loss of pigment as in a case of tinea versicolor. The patch is hypopigmented and anaesthetic but rarely there is hyperaesthesia or paraesthesia. Deep analgesia is often present. The scraping from the lesion does not show any fungus.

Treatment

Referred to Ch. 7

Sulphur in any form will effect a cure (for details see seborrhoea). To avoid recurrence, personal cleanliness is essential and underwears of garments coming in direct contact with the skin should be changed and boiled daily to avoid reinfection. *Pulv Sulphur et Camphor* is very useful in cases of profuse perspiration.

Coaltar preparations are the treatment of choice by local application for fifteen nights only.

PITYROSPORUM OVALE

Kile and Engman, 1938

Phyllum : Deuteromycota

Class : Blastomycetes

UNIMORPHIC YEAST FUNGUS

Synomyms : **Flask-bacillus of Unna;** *Malassezia ovalis* (Bizzozero, 1884) Panja, 1927; Pityrosporon of Malassez; *P. orbiculare.*

Mycology

DIRECT EXAMINATION

Smear : Its an yeast-like organ is first described by Malassez in 1874. The organism is also called **bottle bacillus of Unna,** found in the dandruff or scraping from the scalp. It is a **commensal of normal skin,** and according to some authors, is the causative organism of dandruff and seborrhoeic dermatitis. The fungus comprises ovoid or spherical cells with buddding on a broad base and often appear to be flask or bottle-shaped (**Fig. 10.6**). The cells vary from 2 to 10μ in diameter.

INDIRECT EXAMINATION

A. Culture

Culture of ***P. OVALE*** is one of the problems to the dermatologists as it is difficult to cultivate them. Cultures do not give uniform results and failure is often a rule than an exception. Failures are often due to contaminations. Successful cultures have been obtained by Panja (1927) in cent per cent of cases; by MacLeod and Dowling (1928), Ota and Huang (1933), Benham (1935) claimed to have cultivated ***P. ovale*** in fort agar (Difco) and after three or four days, subcultures were obtained in other media from the primary growth. He reported success in only about 10 per cent of cases.

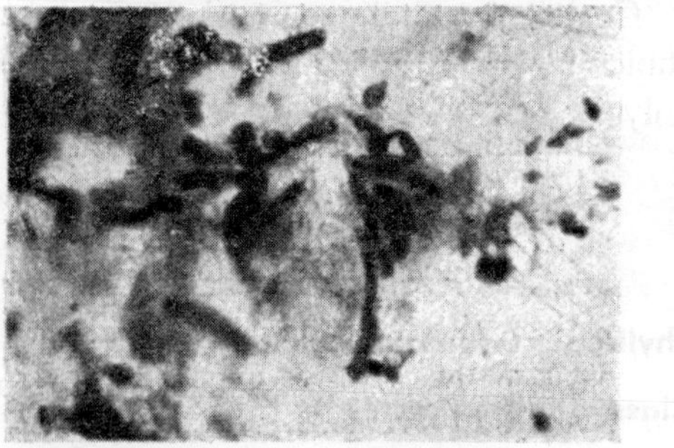

Fig. 10.6 : Bottle bacilli in scales : × 970

(a) *Method of Panja*

This gives uniformly good results in all the cases provided suitable media and proper precautions are employed. **Scales** are obtained from the scalp by scraping with the help of a scalpel and the material is collected **between a pair of sterile slides**. The scales are picked up with a needle, **put on a watch glass** containing sterile normal **saline** and **washed for some time**. Small **bits** of scales are then picked up **with** the help of a **platinum loop** and transferred **to the upper**, drier **part of Petroff's medium with 0.004 per cent gentian violet** (see appendix). Four or **five inocula** may be **made in each tube** and 3 or **4 tubes** should be **inoculated** at a time. The tubes are **incubated at 37°C**. The **growth** is visible **in two or three days'** time. **Two types of growths** are seen in the primary culture :

(1) **Chalky growth** specially seen on the dry part of the above medium near the tail end of the slope. The chalky surface gradually extends at the periphery by fine concentric rings. The centre of the growth often forms a pyramidal mass with a ring at the top, resembling a truncated cone.

(2) On the moist part of the medium **bead-like masses** develop, often **mammillated**, with a **pinkish colour** which may develop further into **orange colour after a few days.**

Subcultures** on glucose or **Sabouraud's glucose agar. The growth is appreciable in 2 or **3 days time** as a crenated mass with **irregular, spike-like projections on the surface** or a **star-like growth with distinct, radial striations**. They become **mammillated with radial striations in different directions**. The **striations** are best observed **in transmitted light**. Growths on glucose or glycerine agar are scanty but **if sterile olive oil is spread on the surface** of the medium, the **colonies** assume **large, thick, creamy characters in 24 hours**. *Ghee*, butter, oleic acid or lanoline may also be used **instead of olive oil**. The **oil** is **autoclaved in a test tube**. A few drops of the sterile oil are **put into the media-tubes** and spread on the surface **of the slope** by tilting the tube. The excess of the oil is allowed to **settle at the bottom**. The medium is **then inoculated** with the culture of ***P. OVALE***. The colonies appear as **opaque, creamy, large colonies like** those of **yeasts**. The growth is more profuse near the bottom of the slope. To maintain profuse growth, the oil is to be spread on the surface of the medium by tilting the tube once a day. *Microscopically*, they show budding forms, the majority being 5 to 10μ in diameter (**Fig. 10.7**).

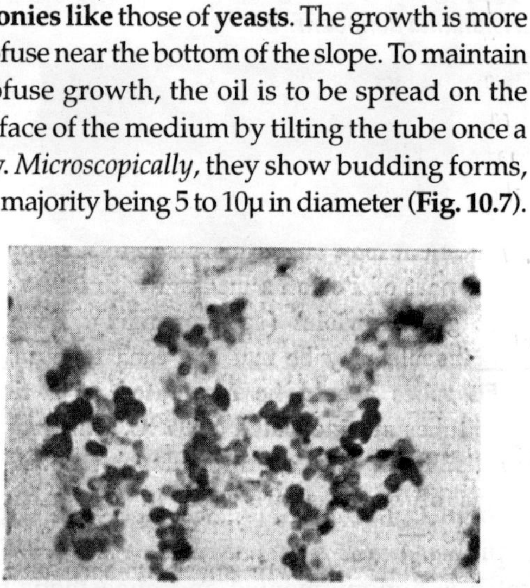

Fig. 10.7 : Bottle bacillus in culture : × 950

(b) ***Moore's method***

Wort agar medium (Bacto) is used for the culture of *P. ovale*. Growth appears on the fourth day after inoculation. The growth is slow and colonies are compact, smooth and glistening. For subculture maltose, dextrose or peptone agar is used (see appendix). Wheat-germ oil or butter is added to this medium, as it has been found that the organism requires a fatty environment. According to Moore, colonies of *P. ovale* become 2 cm in diameter after forty days. These are cushion-shaped with radiating ridges at the periphery and the surface is rough with small excrescences. The colour is dull and varies from light ochraccoussalmon to pinkish buff. *Microscopically*, cell are 3 to 15μ in diameter, the majority being 4 to 5μ. Budding cells are present resembling the shape of a flask or gourd. The larger, old cells are encapsulated with thick gelatinous coating.

Growth factors of *P. ovale* have been worked out by Rhoda W. Benham. She has shown that the fungus is lipophilic, and *thiamine* and *oxaloacetic acid* act as growth factors. Thiamine causes increase in the growth, when added to a basal medium with *asparagin* or **ammonium chloride** as nitrogen source. The growth is much better with **asparagin** than with ammonium chloride. ***Ethyl oxaloacetate*** can replace thiamine. Again ethyl oxaloacetate and ammonium chloride can be substituted for asparagin and thiamine in the cultivation of *P. ovale*.

B. Animal inoculation

Inoculation of animals and human subjects with *P.ovale* has been reported to be successful by Moore *et. al.* (1926) as it produced clinical and pathological picture (dermatitis showing erythema or brown scaliness) resembling seborrhoeic dermatitis. But it has been frequently noticed that dermatitis of similar type may be produced by application of yeasts and yeast-like organisms on the abraded skin.

SEBORRHOEIC DERMATITIS

Synonyms : Pityriasis capitis, Seborrhoea (Hebra), **Hyperidrosis oleosa** (Unna) Flux, Sebacea (Rayer), etc.

Definition

The condition is supposed to be caused by infection of the scalp by ***P. ovale*** which grows on the surface layers of the epidermis and produces a furfuraceous desquamation. The infection of the epidermis is predisposed by **excessive oiliness of the skin** as a result of **overactivity of the sebaceous glands.**

Aetiology

During the first thre years of life the soil is suitable for implantation of ***P. ovale***. According to Acton, the cause of seborrhoea is as follows ;

Primitively the embryonic skin consists of two layers :

(a) **Basal layer of columnar cells** and

(b) **Rauber's layer consisting of swollen cells which do not keratinise till birth.** This is due to the formation of some fatty substances or *vernix caseosa* which acts as a waterproof coating, preventing maceration of the skin by liquor amnii. It is this superficial layer of the skin, where *P. ovale* first infects during the first three years of life. After birth, presence of remnants of **vernix caseosa can also favour the growth of this organism.** The **infection primarily affects the scalp** and then **spreads on to face and extremities.** In the adult life the **infection occurs in relation to sebaceous glands**. To understand the sites of infection, distribution of sebaceous glands of the skin is to be understood. Sebaceous glands are of two main types;

(A) SEBACEOUS GLANDS

Which are not connected with hair as an appendage. They are large or racemose in character and

(i) Connected with sexual organs as in the **scrotum, labia, prepuce** and **axilla.** They produce volatile fatty oils imparting a distinct odour to the individual or species.

(ii) Sebaceous glands connected with lubrication of various orifices of the body, e.g., **ceruminous glands of the ear, Meibomian glands of the eyelids.** In these areas ***P. ovale* may cause boils of the external auditory meatus, styes, blepharitis, retention cysts, etc.**

(B) PILOSEBACEOUS GLANDS

Occur as appendages at the root of the hair follicles and found in places where hair is present. These are further differentiated into three varieties.

(i) **Sebaceous glands that are at the root of long hairs**. These are single, saccular or bilobed, i.e., long hairs have small glands.

(ii) **Sebaceous glands at the root of lanugo hairs are large and racemose in character,**

(iii) **Multilobed, large, racemose glands** are present in areas **where glands have lost** the **relationship** to the **hairs**. They are mostly seen in the **forehead, naso-labial folds, back of the neck, mid-thoracic** and **inter-scapular regions**. There is a minimum amount of subcutaneous fat in these regions. These are the **sites where seborrhoeic dermatitis is seen**.

PREDISPOSING CAUSES

Physiological overactivity of the sebaceous gland occurs during the first three months of life and at puberty. Oily condition of the skin favours the growth of *P. ovale*. According to Sabouraud, greasy skin is not essential for seborrhoea but the moist condition, which predisposes seborrhoea, is essential and is due to "serum and not sebum". At puberty along with sex glands sebaceous glands become more active.

Heredity plays an important part in determining the **number and size of sebaceous glands**. Certain individuals are predisposed to this condition by virtue of inheritance of the skin containing numerous sebaceous glands.

Sex : Baldness is comparatively rare in women and this might be due to the influence of different gonadal hormones in males and females.

Endocrine dysfunction : The condition of seborrhoea is common at the advent of puberty, when there is an imbalance of endocrine glands, specially the gonads. At that time the sebaceous glands become hypertrophic and more active, producing a condition of seborrhoeic state. The male gonad or, androgen is particularly responsible for this condition and helps in the formation of a suitable soil for the growth of *P. ovale*.

Mode of life ; A sedentary, city-life with **lack of exercise predisposes to seborrhoea** and **exposure to the sun** often **increases** the condition.

Food : Diet rich in carbohydrate or fat producing lack of digestion predisposes to seborrhoea. It has been observed that deficiency of vitamin B-complex, which makes the patient sensitive to sunlight, sets up keratosis and exfoliation of the naso-labial folds.

Relative acidosis : **Urine is highly acid** in seborrhoeic patients and the alkali-tolerance of the patient is high (as large doses of alkalies are necessary to make the urine alkaline).

Age : Seborrhoea commences in early life shortly after birth due to extraneous infection of *P. ovale*, which is the determining factor. A few days after birth, **vernix caseosa is shed off and the superficial layer of the skin begins to keratinise. Scalp, with the presence of vernix caseosa, forms a good soil for the growth of *P. ovale*.** Once the infection occurs, the **disease progresses upto the age of three years**, after which the **seborrhoeic manifestations disappear** temporarily. At **puberty** it reappears and continues for the rest of life. At **puberty** the activity of the gonads increases giving rise to a peculiar **condition** of the skin that predisposes to **pityriasis capitis and acne**. Darier has described the condition as kerosis; there is a modification of the keratinisation of the epidermis with a **lowering of the resistance of the epidermic cells to bacterial infection**, in which the **complexion** of the patient is **muddy** and the **skin is coarse, greasy** and **thickened, mouths of the sebaceous glands** are **large and patent**, but many of them are blocked with dried up, inspissated sebum forming **black-heads (comedones)**. The condition often disappears spontaneously at the age of 25 years or so but it may persist throughout the whole life.

EXCITING FACTOR

Organisms : The role of the organisms responsible for seborrhoeic condition is still disputed. It is a fact that ***P. ovale, Staphylococcus albus*** and ***Corynebacterium acne*** form the ***seborrhoeic triad***. *P. ovale* is found in **flask-shaped from (bottle-bacillus of Unna)**, the coccal form (**morococcus of Sabouraud**), or rarely in the rudimentary mycelial form (**Fig. 10.6**). Coccal forms are seen when the disease is spreading rapidly.

C. acne is specially found in acne vulgaris. Experiments of MacLeod and Dowling have proved that **seborrhoeic infection is a pure fungus infection** with *P. ovale*. The **organism grows well in subjects with moist and greasy skin**.

Sings and Symptoms

Seborrhoeic lesions are best described under **three age groups**. The types of the lesion and their symptoms vary according to the age and the area affected.

(a) INFANCY

Milk crust

This condition develops within a few days after birth, if vernix caseosa is not thoroughly removed. The child becomes infected with *P. ovale* from the mother, or the nurse. In this case, vernix caseosa acts as an ideal culture medium for its growth. The scalp becomes inflamed and very irritable resulting in serous exudation and crust formation. The scalp is covered with a cap of hair matted with exudate mixed up with dirts and contains the infecting fungus. Impetigo is superimposed on it in many cases due to infection of the skin caused by streptococci.

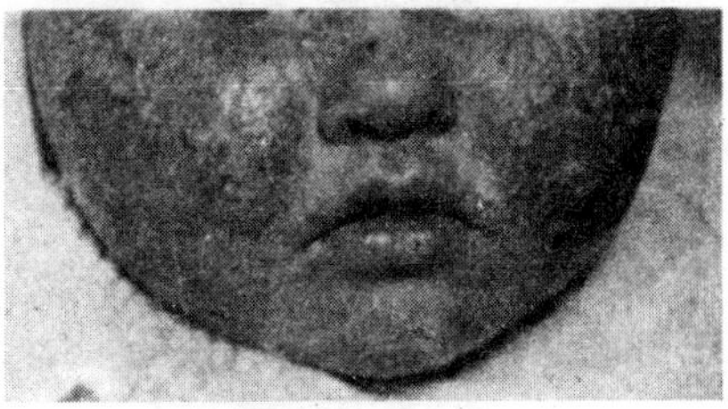

Figs. 10.8 : Infantile eczema. Typical distribution on face

Infantile eczema

There are various causes of infantile eczema but in certain cases, it is due to extension of the lesion from the scalp to the skin of the forehead, cheek, behind the ears, flexures of the body, outer side of the legs (**Fig. 10.8**). It is characterised by serious exudate due to secondary streptococcal infection, producing yellow crusts like impetigo. There is an intense itching associated with it. Food allergy may maintain the condition or aggravate it. These children are often subjects of prickly heat, impetigo and boils in the later life.

Liver spots

These are localised patches of seborrhoeic dermatitis with superficial exfoliation and development of hypopigmented patches on the face and back, popularly known as live spots.

(b) PUBERTY UP TO THE AGE OF 25 YEARS

Seborrhoea capitis (Pityriasis sicca of Hebra-dandruff).

It is a dry, scaly condition of the scalp which is very common during adolescence. Brank-like scales accumulate on the scalp causing exfoliation of the outer horny layer and mild irritation of the skin.

PLATE 10.VI

(Annexure)

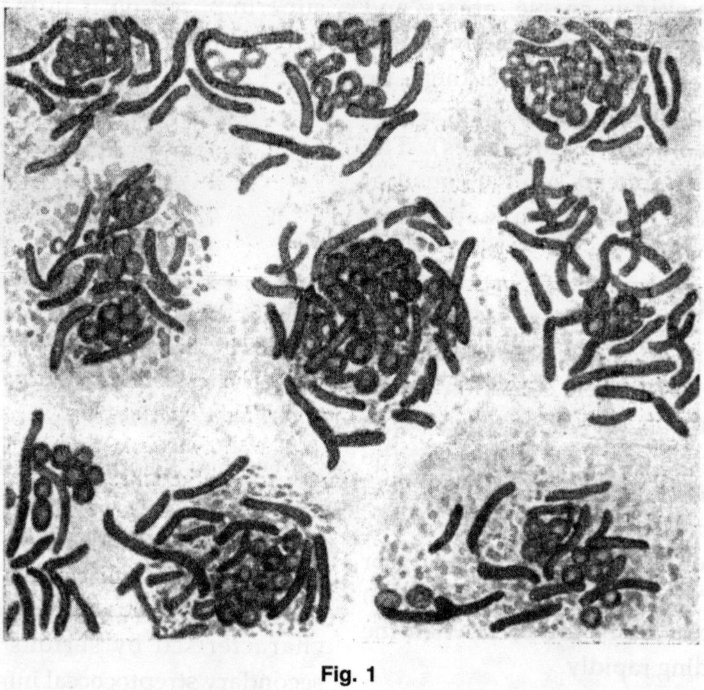

Fig. 1

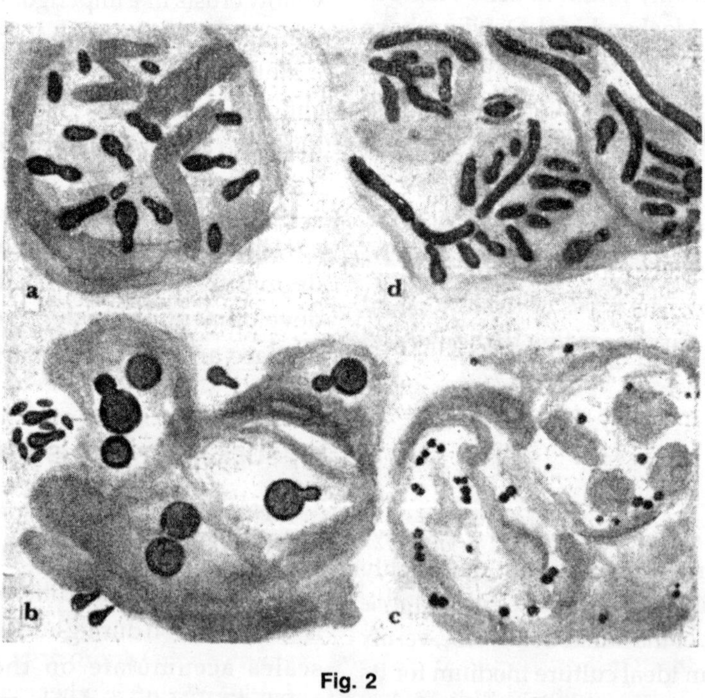

Fig. 2

Figs. 1& 2 : Bottle bacilli in scales; **a** and **b** : flask forms; **c** : coccal form; **d** : mycelial form

Seborrhoea oleosa (**Hebra**)

This usually occurs at puberty and is characterised by the greasy condition of the skin, containing *P. ovale*. They are the subjects of acne vulgaris and seborrhoeic alopecia in later life.

Seborrhoea corporis

The condition is also called **flannel-rash** owing to the old conception that it was due to wearing flannel garments next to the skin. The infection spreads from the scalp to the front of the chest and the trunk. There are irregular, hypopigmented, circinate patches of different sizes, varying from a 25 paise coin bit to that of a rupee coin. Lesions are seen on the face, neck, sternal and interscapular regions but they may spread to other parts of the body, viz.; arms, thighs, **genito-crural region** and perineum.

Seborrhoeic eczema

This is a dry, scaly condition of the neck, face and eye-lids, but later they may show a weeping, eczematous condition on the flexural aspects of the axillae, elbow, or knee joints.

Seborrhoeic lichen

This occurs as a result of irritation or itching and is generally seen on the nape of the neck and flexures of the elbow and other regions, where there is a condition of seborrhoeic eczema. This starts as a group of popular rash in patches due to hyperkeratosis of the hair follicles. It is associated with thickening of the skin on the nape of the neck with raised margin, scaliness, and pigmentation in seborrhoeic subjects.

(c) FROM 30 TO 45 YEARS

Seborrhoeic baldness : During this period the infection is restricted to the scalp and involves the true pilosebaceous glands. The infection destroys the glands as well as the hair follicles, producing seborrhoeic alopecia or baldness. The **baldness may start from the crown and extend towards the forehead or it may start from the forehead and spread towards the crown of the head**. Baldness is diffuse and complete within a few years. The whole of the **forehead and vertex of the scalp is denuded of the hair**, the **skin becomes shiny** and only a little area on the side and back of the head shows thin growth of hair (**Fig. 10.9**).

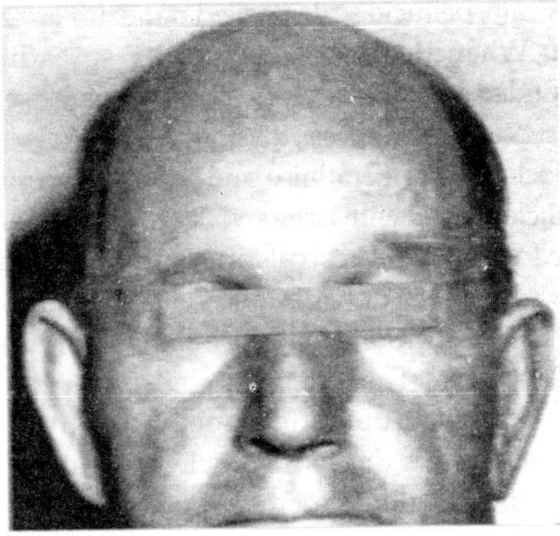

Fig. 10.9 : Seborrhoeic baldness

PREDISPOSITIONS : Prickly heat is one of the **seborrhoeic predispositions** and the condition is characterised by the infection of sweatpore by staphylococci. The epidermis at the top of the vesicle shows both staphylococci and *P. ovale*. It is usually seen in persons suffering from seborrhoea of the scalp.

At the age of puberty, the condition of keratosis is commonly seen. The pilo-sebaceous glands, which undergo hypertrophy, are exposed to infection by staphylococci and ***P. ovale. C. acne*** also invades the sebaceous glands, when **mouths of the glands are obstructed and partial anaerobiosis is produced; this is favourable for the growth of** ***C. acne***. The glands that are usually attacked are those of naso-labial folds, forehead, sternal and interscapular regions. In deep seated lesions, they produce scars which may subsequently form keloids. The mouths of the glands are blocked and form **black heads (comedones)** due to drying up of **sebum mixed with dirts**, but in other cases **retention cysts** are formed.

PATHOLOGY

P. ovale grows on the superficial layers of the epidermis and appears as yeast-like budding forms. During multiplication it causes alternation in the superficial cells, preventing them from perfect keratinisation. In the affected areas, tiny, delicate, pellucid scales, called dandruff are cast off. When the disease is rapidly spreading, rounded coccal or rarely mycelial forms are seen. In seborrhoeic alopecia, the horny layer in the hair follicle fails to keratinise and the mouth of the follicle is filled with large swollen cells. Deeper in the follicle minute coccal forms and yeast forms are seen as far down as pilo-sebaceous glands, which are surrounded by an inflammatory zone of leucocytes. The hair is distorted in shape; later on the hair root undergoes atrophy and subsequently the hair follicle is obliterated resulting in baldness of the scalp.

TREATMENT

In treating cases of seborrhoea it should be remembered that the primary site of infection is the scalp and *sulphur* in various forms is the drug of choice. Secondary infections are responsible for the persistence of eczematous lesions in infants.

(a) *MILK CRUST*

In this condition, prophylactic measure is very important and thorough removal of the vernix caseosa should be strictly adhered to. This can be done by rubbing **olive oil** on the scalp **followed** by gentle rubbing with **soap and water to remove every trace of oily matter**. Should any scaliness persist, ***lotio Resorcin*** **diluted with equal parts of water** should be applied on the scalp. Application of 0.5 per cent *Hydrarg. oleata in olive oil* is also useful. In established cases of infection. ***sulphur-salicylic ointment*** (15 gr of each in an ointment base) should be applied.

(b) *INFANTILE ECZEMA*

Treatment in these cases should be local and general. Weeping, eczematous lesions on the face (eczema rubra) are best treated with *lotio Calamine* in the weeping stage, till the part dries up. When the part is dry and weeping has stopped, *liniment Calamine with Sulphur* (10 gr. to an ounce or 2g%) should be applied thrice daily followed by application of 3 per cent *Resorcin ointment* at night. In this, primary infection of the scalp should be remembered and *lotio Resorcin* (diluted should be used as mentioned before. The general treatment consists of relieving constipation by administration of *grey powder* in doses of 10 to 15 mg according to age, twice daily. Treat the coexisting food allergy by changing the milk or the diet as the case may be.

Mometasone (Elocon) and other steroids are beneficial.

SEBORRHOEA CAPITIS

In this and other seborrhoeic conditions in adults, remove the scales and promote the free flow of sebum by keeping the gland orifices clean with application of spirit-soap, or **ether-soap shampoo** (see appendix). The shampoo may be rubbed twice weekly. If it is rubbed more frequently, the affected hairs which become brittle may break. As an antiseptic and exfoliative, *lotio Resorcin et Hydrarg. perchloride* (see Appendix) is an effective application on the scalp. For seborrhoea oleosa 1 per cent *oleate of mercury in olive oil* is effective for cleaning the scalp. For seborrhoea of the face and body, *lotio Sulphuris alba* and *Pulv. Sulphur et Camphor* should be applied alternately, each twice daily. The former is a time honoured remedy for seborrhoea of the body and the latter is used in the summer when one perspires profusely. *Lotio sulphuris* (see Appendix) *flava* is almost as efficacious as lotio sulphuris alba. On flexural eczema and other irritable conditions ***liniment Calamine with sulphur*** **(600 mg/30 ml)** is indicated along with other remedies. In seborrhoeic lichen, the above treatment should be combined with application of unguentum acid salicylic on the affected part at night as an exfoliative. The patient responds very well with this treatment.

ACNE VULGARIS

In this condition along with the local applications, general treatment is also indicated. Constipation, dyspepsia, indigestion should be treated. The quantity of carbohydrate and fat in the diet is reduced. Intake of alcohol and highly spiced food are restricted. Administration of vitamin B-complex, yeast, marmite and liver extract are useful. All attempts should be made to improve the general health of the patient and iron should be administered as a tonic.

The final treatment of Acne vulgaris and Allied conditions are treated with special Prickly heat powder for 15 days application twice a day. Face is cleaned by soap three times a day.

Treatment of

TINEA VERSICOLOR

Imidazole - topical - BID × 2 wks.

Zinc pyrithionone (2%)

Sulfur salicylic acid (2%); shampoo base

Selenium sulfides 2.5% shampoo form apply to all area at bed time except scrotum for 3-4 days, washed off in the morning.

Short term oral **Itraconazole** (200 mg/day for 7 days) is effective (well tolerated).

REFERENCES

Erythrasma

Sarkany. J., Taplin, D., Blank, F. (1961). Erythrasma. Common bacterial infection of the skin. *J. am. med. Assoc.* **177**, 2, 130-132.

Sarkany, J., Taplin, D., Blank, F. (1962). Organism causing erythrasma. *Lancet*, **1**, 304-305.

Trichomycosis axillaris

Crissey, J. T., Rebell, G. C. and Laskas, J. J. (1952). Studies on the causative organism of Trichomycosis axillaris. *J. Inv. Dermatol.* **19**, 187-197.

Tinea nigra palmaris

Area Leao, A. E., Cury, A. and Ferreira Filho, J. M. (1945). "Tinea nigra" ("keratomycosis nigricans plamaris") observacao e estudo de uf caso. *Rev. Brasil. Biol.* **5**, 165-177.

Castellani, A. and Chalmers, A. J. *Manual of tropical medicine.* 3, London, Balliere, Tindal & Cox, 1919.

Cooke, W. B. (1959). An ecological life history of Aureob silium pullulans (de Bary) Arnaud. *Mycopatb. Mycol. Appl.* **12**, 1-45.

Piedra

Behrend, G. (1890). Ueber Trichomycosis nodosa. (Juhel Renoy); *Piedra (Osorio)* **27**, 464.

Dey, N. C., Kakoti, L. M. (1955). White Piedra in India. *Ind. Jour. med. Sciences*, **9**, 544.

Fonseca, O. Da, Jr. and Area Leao, A. E. (1938). de, Sobre as cogamelos da piedra brasilara. *Mem. Inst. Oswaldo Cruz. (suppl.)* **4**, 124.

Horta, P. (1911). Sobre uma nova forma de piedra. *Mem. Inst., Oswaldo Cruz*, **3**, 87.

Mackinnon, J. E. and Schouten, G. B. (1942). vestigaciones shobre las enfermedades de los calbe denomina das "piedra". *Arch. de la Soc. De Biol. Montevideo*, **10**, 227-266.

Pityrosporum ovale and Seborrhoeic dermatitis

Acton, H. W. and Panja, G. (1927). Seborrhoeic dermatitis or pityriasis capitis : Lesion caused by Malassezia ovale. *Ind. Med. Gaz.* **62**, 603.

Benham, R. W. (1939). The cultural characteristics of Pityrosporum ovale — lipophylic fungus. *Invest. Dermatol.* **2**, 187-203.

Emmons, C. W. (1940). The isolation and pathogenicity of Pityrosporum ovale. *Publ. Health Rep.*, **55**, 1306-1312.

MacLeod, J. M. H. and Dowling G. B. (1928). Experimental study of Pityrosporon of Malassex : Morphology, cultivation and pathogenicity. *Brit. J.* Dermatol. **40**, 139.

Malassez, L. (1974). Note sur le champignon de la pelade. *Arch. de Physiol. Norm. et Path.* **1**, 203.

Moore, M. (1935). Cultivation and study of pityrosporum ovale, the so-called bottle bacillum of Unna. *Arch. Dermatol. & Syph.* **31**, 661.

Moore, M., Kile, R. L ., Engman, M. E., Jr. and Engman, M. F. L. (1926). Pityrosporum ovale (Bottle bacillus of Unna, spore of Mallassez) : Cultivation and possible role in seborrhoeic dermatitis. *Arch. Dermatol. and Syph.* **33**, 457.

Ota, M. and Huang, P. (1933). Sur les champignons du giare pityrosporum Sabouraud. *Ann. de parasitol.* **11**, 49.

Panja, G. (1927). The Malassezia of the skin, their cultivation, morphology and species. *Trans. Far Eastern Assoc. trop. med.* **1**, 422-456.

11

Actinomycetes(Schizomycotic Actinomycetoma)

PHYLLUM : SCHIZOMYCOTA; CLASS : Schizomycetes Order : Actinomycetales; Family : Actinomycetaceae (Subcutaneous and Systemic Mycosis)

SCHIZOMYCETES

Classification (Modified from Waksman and Henrici, 1943)

Medical mycologists are mainly concerned with the **order Actinomycetales**.

ORDER : Actinomycetales - True mycelium present

(i) **FAMILY : ACTINOMYCETACEAE**

(a) **ANAEROBIC** or microaerophilic, non-acid fast **GENUS : Actinomyces**; species : ***A. israeli, A. bovis, A. neaslundi*** causing actinomycotic infection.

(b) **AEROBIC** — Acid fast or weakly acid fast **GENUS : Nocardia**, causing systemic nocardiasis e.g., ***N. asteroides; N. brasiliensis*** and ***N. caviae*** causing **ACTINOMYCOTIC MYCETOMA**.

Vegetative mycelia do not fragment into bacillary or coccoid elements; they are nonacid-fast; bear spores in some species.

(ii) **FAMILY : STREPTOMYCETACEAE**

Spores in chains - **GENUS : Streptomyces**; they are :

S. madurae, S. pelletieri, S. paraguayensis, S. somaliensis.

S. keratolytica causes keratolysis of the epidermis of the sole of the foot, being contracted from soil contamination.

ACTINOMYCETES

GENERAL CHARACTERS

Actinomycetes belong to the **order Actinomycetales that includes mycobacteria and actinomycetes (ANAEROBIC AND AEROBIC)**.

Actinomycetes are Gram positive rod shaped, filamentous organisms which tend to grow slowly as branching filaments. In some genera, filaments show segmentation with their growths, and yield pleomorphic club-shaped cells resembling those of Corynebacteria or Mycobacteria. Some Actinomycetes are **weakly acid-fast (Nocardia)**. Their colonies are "felt-like" and show chronic subcutaneous granulomatous lesions forming abscesses. The name **"ray-fungus"** (*Gr.* **Actino = ray**) is derived from the characteristic radial arrangement of club shaped elements, formed in infected tissues by the microcolonies formed by the major human pathogens of this group.

The actinomycetes **resemble bacteria** in the following ways :

1. The **cells are prokaryotic**, lacking a nuclear membrane.
2. The filaments are 1μ or less in diameter and in the more fragile form, they often segment into **bacillary and coccoid forms** with typical dimensions of bacteria.

3. Their cells wall contains muramic acid and **diaminopimelic acid (DAP; lysine** in some species) and these are characteristics of bacterial cells. They **lack chitins and glucans** characteristic of cell walls of yeasts and molds.
4. Their **growths are inhibited by sulphonamides, penicillin, tetracyclines** and **other antibacterial drugs**. They are not affected by polyenes and other effective antifungal agents.
5. Some actinomyces are affected by viruses (**actinophages**) like those of bacteria.

Common members of the Order Actinomycetales (Based on the classification of Waksman and Henrici, 1943) :

ORDER : ACTINOMYCETALES

FAMILY : Mycobacteriaceae — Branching absent or rudimentary.

GENUS : Mycobacterium, *e.g., Myco. tuberculosis, Myco. leprae, Myco. phlei, etc.*

FAMILY : Actinomycetaceae — Grow as fragile branching filaments that form bacillary or twig-like elements.

GENUS : ***Actinomyces*** **— ANAEROBIC** or micro-aerophilic, non acid-fast, e.g., ***A. neaslundi***.

GENUS : ***Nocardia*** — AEROBIC; some species are weakly acid-fast, e.g. ***N. asteroides***, *N. brasiliensis, N. caviae.*

FAMILY : ***Streptomycetaceae*** — Grow as stable branching filaments, form aerial mycelia with spores; filaments not readily fragmented; non-acid-fast, except for spores in some species.

GENUS : STREPTOMYCES — spores form in chains; ***S. madurae, S. pelletieri, S. paraguayensis, S. somaliensis. S. keratolytica.***

ACTINOMYCOTIC MYCETOMA

*ACTINOMYCOTIC MYCETOMA** differs from **MADUROMYCOTIC MYCETOMA** in that the granules consist of mycelial filaments each being one micron or less in diameter. The granules may or may not show clubbing.

ANAEROBIC ACTINOMYCOTIC species, namely, ***A. israeli*** and ***A. bovis*** may cause localised lesion and is called actinomycotic mycetoma. These lesions often develop on the face or any part of the body with draining sinuses from which **'sulphur yellow' granules** are excreted with the pus. *AEROBIC ACTINOMYCOTIC* species belong to the family **Actinomycetaceae comprises the genera NOCARDIA and STREPTOMYCES.**

Family : Actinomycetaceae

GENUS : **Actinomyces (Anaerobic or microaerophilic)** causing actinomycosis.

Nocardia and Streptomyces (Aerobic) causing Nocardiosis.

STUDY OF ACTINOMYCES (ANAEROBIC) ACTINOMYCETES : Two species of anaerobic pathogenic Actinomycetes are recognised, e.g., *A. israeli* and *A. bovis*; the former is responsible for diseases in man and the latter for diseases in cattle causing "lumpy jaw". These are anaerobic or microaerophilic and require enriched media containing blood or brain heart infusion. The growth is stimulated by CO_2 and is poor at a temperature below 37°C. The pathogenic species are often confused with *A. neaslundi* occurring as saprophytes of the oral cavity, or anaerobic diphtheroids.

Aetiological agent

A. israeli and ***A. bovis*** are the aetiological agents of actinomycosis, the former is usually isolated from human actinomycosis; whereas, the latter causes "**lumpy jaw**" in cattle. However, there is one authenticated case of *A. israeli* isolated from a lumpy jaw in cattle (Pine *et.al.*, 1960). The earlier workers did not distinguish between *A. israeli* and *A. bovis* and anaerobic actinomyces isolated either from cattle or human cases were called *A. bovis*. Nowadays *A. israeli* usually isolated from human cases is distinguished from *A. bovis* causing actinomycosis in cattle.

Clinical materials

These are obtained from pus from draining sinuses, subcutaneous abscesses or mycetoma. Pus may be obtained from an unbroken abscess with a sterile syringe and put in a sterile Petri dish. When sinuses of mycetoma are washed they show small granules, **"sulphur granules", 2 to 4μ** in size. The wall of the sinus may be scraped for obtaining material and similarly washed with sterile saline in a Petri dish. In pulmonary infection sputum may be received in a Petri dish and washed with saline to obtain "**sulphur granules**".

Washed granules either from pus, curretted material or sputum should be received in a Pasteur pipette and studied as follows :

LABORATORY DIAGNOSIS

A. **DIRECT EVIDENCE** : Examination of a granule is made as fresh mount from pus or sputum by pressing or crushing it under a cover slip. For this, put a drop of **McGuire stain*** on a slide and put the lobulated granule in the drop. Place a coverslip over it and press gently to spread the granule. This stains the mycelium nicely. The preparation examined under low power reveals Actinomyces as a "**ray fungus**".

2. Smears stained by Gram and Ziehl Neelsen method reveal that they are **Gram positive and non-acid fast**.

B. INDIRECT EVIDENCES

(i) CULTIVATION of Anaerobic Actinomyces

Media — For primary isolation anaerobic Actinomyces are cultivated on **Brain Heart Infusion (BHI) agar**; some strains, however, may require blood, i.e., **BHI blood agar**. Lentze (1938) recommended the use of the beef infusion agar containing 30 per cent ascitic fluid. For maintenance of the culture, liquid maintenance medium for isolation. For a solid medium, add 20 g agar per litre (2%).

After the colonies have been isolated from a **solid BHI agar plate** the suspected colonies are picked up and **subcultured in** freshly **boiled thioglycollate broth**. The boiling removes the dissolved oxygen. The growth thus obtained is reinoculated and restreaked on BHI plates in order to determine whether a pure culture has been recovered.

The pure single colonies are inoculated again in :

1. **Brain Heart infusion broth (sugar free)** enriched medium.
2. **Thioglycollate (Anaerobic) broth.**
3. **Maintenance medium for Actinomyces** (see later).

Further morphological and biochemical reactions are studied from pure culture of thioglycollate broth. For sugar fermentation the inoculum should be taken from a sugar free BHI agar medium.

ANAEROBIOSIS

For primary culture a strict anaerobiosis is to be observed in an anaerobic jar on BHI agar or **BHI blood agar** but subsequent cultures may be done on slants in tubes as anaerobic condition is maintained with **pyrogallol-carbonate seal.**

***McGuire stain.** It contains : **Toluidin blue** — 1 g., **Alcohol** — 2 ml., **Acetic acid** — 4 ml. and **Water** to 100 ml, A better substitute for this is a stain containing : Thionin —1 g., Phenol crystal — 5 g., Alcohol — 30 ml., Water add to 100 ml.

The granules are put on a slide; a few drops of procedure the stain are added and kept for a few minutes. A coverslip preparation is examined under oil immersion objective. The method is simple and stains mycelia well.

(ii) Pyrogallol-carbonate sealing

This method is suitable for individual tubes and slants only. *Procedure*

(a) Clip off the portion of the non-absorbent cotton plug sticking out at the end of the tube.

(b) Heat the mouth end of the tube and push the plug inside leaving a spaces of about ½ inch at the top of the tube.

(c) Insert a small pad of absorbent cotton into the space and soak it be adding 5 drops of pyrogallol solution (100 gm of pyrogallic acid in 150 ml H_2O and 10% Na_2CO_3 solution.

(d) Plug with a rubber cap or stopper and seal with melted paraffin.

Isolation : If actinomycosis caused by *A. israeli* is suspected by smear preparation, granules should be cultivated. For *A. israeli* and *A. bovis* an anaerobic culture is put up in enriched media with a pH of 6.8 to 7.4 for primary isolation.

Enriched agar

(a) **Brain heart infusion (BHI) agar with 0.2% dextrose** is satisfactory for the culture of *A. bovis*. The washed inoculum is streaked in a series of BHI agar plates and incubated in anaerobic jar at 37ºC. CO_2 stimulates the growth in an anaerobic environment and this is why anaerobic jar is filled with 5 per cent CO_2 and 95 per cent N_2.

(b) It is likely that BHI agar may not give growth under the above conditions and Garrod (1952) encountered a few strains which grew well on BHI blood agar in primary isolation. A number of workers have found addition of starch to be useful and used casitone starch maintenance (see maintenance medium) medium to be a useful isolation medium. **A duplicate set of BHI plates should be incubated at 37ºC anaerobically and also in a candle jar** in order to determine the oxygen requirement of the isolate. Cultures are examined for appearance of growth 48 hours after inoculation in order to determine the morphology of young colonies. After examination the plates are returned back to the jar to produce the anaerobic condition again and held there from 7-10 days to study the "mature colony".

For subculture and further study thioglycollate broth is recommended. It is not used for primary isolation as broth is not suitable for it. Pick up suspected colonies of *A. israeli* or *A. bovis* from streaked BHI or BHI blood agar plate and inoculate in thioglycollate broth to study the branched form *in situ* separately, as the organism develops abundantly in this medium. A pure culture is necessary for biochemical reactions.

Colony characters

On BHI agar two types of colonies are seen in *A. israeli* namely "Rough" (R) and "Smooth" (S) types, but an intermediate type is sometimes encountered. It is convenient to study the plates from anaerobic jar once after 48 hours and again after 10 days when the colonies are mature on BHI agar plates. **Rough or "R" type of growths on BHI agar are characteristic** of most strains of *A. israeli*.

(iii) Growths of 'R' type

On 48 hours' old BHI agar plates the colonies are usually invisible to the naked eye. Under low power (10 × objective they appear as spider-like colonies. Seven to ten days old culture, however, show 'mature' colonies of *A. israeli*.

MATURE 'R' COLONIES are dull, white, heaped up with irregular surfaces. Many of them show depressed centres and irregularly raised surfaces often with **mulberry**

appearance. They are dry and crumbly when felt with a loop. They stick to the medium with a depression around the colonies on the agar surface.

"R" type of growth IN LIQUID THIOGLYCOLLATE MEDIUM : The colonies are discrete, lobulated bread-crumb-like with small discrete granules. The growth occurs about 1 cm below the surface of the broth (microaerophilic).

Direct examination of a loopful of this "R" type growth shows branched filamentous colonies under a coverslip preparation.

(iv) Gram stain

Smears from agar plates or broth cultures show Gram-positive branching elements but longer branched elements are observed in broth cultures. They often break is smears into bacillary and coccoid forms. Bacillary forms are long and irregular in width and often show knobbed or thickened ends. It is difficult to find this in most strains.

SMOOTH OR "S" COLONIES : 2 days' old *A. israeli* are pin-head in size as seen by the naked eye. Under low power (10 ×) objective of the microscope they appear as flat, finely ganular.

MATURE 'S' COLONIES (on BHIagar) : After 7 to 10 days the mature colonies appear fairly large, convex and circular with smooth edges. The surface is smooth and glistening. They are soft to touch with loop and easily broken. Some of the 'S' colonies are larger, more dense and scalloped in outline. These have a pimply or lumpy surface and they are firm in consistency.

(v) In liquid thioglycollate medium

Several types of growth are common : (1) **Large loculated bread-crumb-like colonies** which break easily on shaking, (2) **Soft diffuse growth** which stream down to the bottom of the tube, (3) **Flaky type of growth**, (4) **Mucoid** growths which form spirals like ropes when the tubes are rotated.

Intermediate type : *A. israeli* and *A. vovis* show colonies that are intermediate between 'R' and 'S' forms. Even a single strain changes in colony characters in different media or upon prolonged cultivation.

Morphological characters of *A. israeli* and *A. bovis* are compared as follows :

The morphological characters of both the species are more or less similar, e.g., **Rough (R)**

Table 10.1 : Differences Between 'R' Type & 'S' Type Colonies

Colonies	*BHI agar 7-10 days*	*Liquid thioglycollate medium*	*Gram stain of smears*
'R' type	Dull, white heaped up often with depressed centre and irregularly raised border. The colonies are dry and crumbly and easily moved over the surface of the agar. A depression on the surface is often observed around the colonies	Large, discrete, lobulated "bread crumb" colonies 1 cm or more or small, discrete, below the surface of the broth. Direct examination shows branched, filamentous colonies clearly	Gram stain of smears from agar or broth show Gram positive branching elements and coccoid forms. Bacillary forms irregularly thickened with thicker ends. True branching is difficult to demonstrate
'S' type	Smooth, large, convex, circular, glistening with entire edge. They are soft and easily broken when touched with the inoculating loop. Some colonies are scalloped in outline	(1) Large lobulated, **bread crumb colonies** soft and easily broken (2) Soft diffuse growths at the bottom of the tube (3) Flaky growth (4) Mucoid growth forming rope	From solid & liquid media — **Gram positive diphtheroid forms**. In some, branching is observed. In general highly irregular, thick and thin, or with knobbed ends. Branching **X & V forms are seen as in coryne bacterium**

and **Smooth (S) colonies**. The chief difference between the two species is the predominance of rough (R) forms in the former and the Smooth (S) form in the latter. There may be some colonies which are of intermediate type. Most of the strains of *A. israeli* produce R colonies and rarely S type; whereas, *A. bovis* produce mostly S type of colonies (Pine *et. al.*, 1960). The characters of different colonies on BHI agar and thioglycollate media are tabulated below :

General Physiological Characters

(*A. ISRAELI* AND *A. BOVIS*) :

1. Both these actinomyces grow on primary isolation in anaerobic or micro-aerophilic conditions.
2. They require enriched media like brain heart infusion (**BHI) agar or BHI blood agar or thioglycollate broth**.
3. pH required between 6.8 and 7.4.
4. Incubation at 37ºC.
5. In most culture media they die due to depletion of growth factor and development of acid pH. This is why a maintenance medium is necessary for stock strains in the laboratory.

BIOCHEMICAL TESTS FOR IDENTIFICATION OF ACTINOMYCES

1. FERMENTATION TUBES (ENRICHED) *for Actinomyces*

Heart infusion broth (Difco) — 25 g, casitone — 4g, yeast extract — 5 gm, Bromocresol purple (0.04%) — 15 ml, and distilled water — 1000 ml.

Dissolve together and set pH at 7, distribute 5 ml in each (15 × 100 mm) tube with inverted Durham tubes and plugged with cotton wool. Autoclave at 15 1bs for 15 minutes.

Sterile sugar solutions are added to give a final concentration of 0.5%.

The following sugars are useful — Glucose, xylose, mannitol, raffinose, starch and glycerol. *Test method*. Fermentation test is prepared by growing culture in a sugar free medium like Heart infusion broth (Difco). When a good growth in thioglycollate broth is obtained in 3-8 days time the inoculum is taken with a Pasteur pipette of wide bore after a uniform suspension and then removing some broth depending on the density of the growth. It is then inoculated into the fermentation tubes. The tubes are sealed with **pyrogallol carbonate seal** (see before) and **incubated at 37ºC for 4 weeks**. The tubes are read for acid and gas at weekly intervals. Presence of good growth should occur before the results are recorded.

2. NITRATE REDUCTION TEST

The cultures of be tested are inoculated into liquid nitrate medium. These are **incubated at 37ºC in an anaerobic jar or pyrogallol carbonate seal.**

Tests for nitrate are made after 4-7 days' inculation by transferring a small portion of medium in which the culture has grown to a clean test tube and several drops of each of solutions (1) and (2) (see below) are added. The development of a red or brown colour indicates presence of nitrite due to reduction of nitrate.

If after 14 days the nitric test is still negative, add a little pulverised zinc to the aliquot part being tested. This reduces any remaining nitrate to nitrite. Now retest as above. If the test with zinc is positive (i.e., nitrate is still present) it is clear that the organism has not attacked nitrate at all. If both the tests are negative, the organism has reduced all nitrates to nitrite and then to N_2 or NH_3.

Liquid nitrate medium

Heart infusion broth — 25 g, **Casitone** (Difco) — 4g, **yeast extract** — 5 g, **KNO_3** — 1 g, distilled **water** — 1000 ml. Dissolve the ingredients and adjust pH at 7. Distribute 4 to 5 ml in 15 × 100 mm test tubes plugged with cotton wool and autoclave at 15 1bs pressure for 15 minutes.

Solution 1 : *Sulphanilic acid solution.* Glacial acetic acid — 100 ml, Water 250 ml and 2.8 g Sulphanilic acid.

Dissolve sulphanilic acid in glacial acetic acid with slight warming if necessary, then add water.

Solution 2 : ***Dimethyl alpha-naphthylamine solution.*** Glacial acetic acid 100 ml, Dimethyl alpha-naphthylamine 2.1 ml, and Water 250 ml. Both the solutions remain stable in a glass stoppered bottle.

3. STARCH HYDROLYSIS TEST

For this, duplicate sets of slants or plates of casitone starch medium (see below) is streaked with culture. Place them in an anaerobic jar or test tubes with **pyrogallol carbonate seal** and incubate at 37ºC. Test one set after 3 to 4 days incubation pouring Gram's iodine on the surface of the culture. Hydrolysis of starch is indicated by clear area around the colonies, the rest of the medium-surface being blue to purple. If all the starch in the medium is hydrolysed as in *A. bovis*, no blue colour will appear at all.

Casitone starch medium : Heart infusion broth (Difco) — 25 g, Casitone (Difco) — 4 g, Yeast extract — 5 g, **Soluble starch** — 5 g, **Agar** — 15 g and distilled water — 1000 ml. Dissolve together and set pH at 7.0. Distribute 8 ml in each tube (of 15 × 150 mm), cotton plugged and autoclaved at 15 1bs pressure for 15 minutes and prepare slants.

4. CATALASE TEST

Actinomyces species are catalase negative but **anaerobic diphtheroids are catalase positive which breaks down H_2O_2 when added with the formation of gas bubble.**

Procedure : The test is performed by taking a small amount of the growth on solid medium, preferably from a single colony and placing it on a microscopic slide. A drop of fresh H_2O_2 is added and the preparation is placed under the low power. Observe for appearance of bubbles.

5. LITMUS MILK TEST

The test culture is inoculated into reinforced litmus milk (see below) tubes and sealed with pyrogallol carbonate seals and then incubated at 37°C for 4 weeks. Changes in the litmus milk observed, namely :

(a) Production of acid (red),

(b) Development of acid curd,

(c) Peptonisation, or

(d) Reduction, are noted at weekly intervals.

Reinforced litmus milk : **Skimmed milk** powder — 100 g, **Yeast extract** — 5 g, **Dextrose** — 3 g, **Litmus powder** — 750 mg and Distilled **water** — 1000 ml. Dissolve together, adjust pH at 7.0. Distribute in (15 × 100 mm) tubes, cotton plugged. Autoclave at 10 1bs pressure for 10 minutes.

6. GELATIN LIQUEFACTION TEST

Inoculate culture to be tested into tubes containing gelatin medium (see below). Apply **pyrogallol carbonate seal** and incubate **at 37ºC for 4 weeks**. Test for liquefaction weekly by removing the tubes from the incubator to the refrigerator or iced water to test the ability of gelatin to harden. If the tubes harden, they are again reincubated and tested at weekly intervals.

Gelatin medium : Heart infusion broth (Difco) — 25 g, Casitone (Difco) — 4 g, Dextrose — 5 g, Yeast extract — 5 g, Gelatin — 150 g per litre of media. Dissolve together, adjust pH at 7.0. Distribute 2/3 rows of tube (15 × 100 mm); autoclave at 15 1bs for 15 minutes or do a **fractional sterilisation**.

Maintenance media for Actinomyces species : It is difficult to maintain cultures of Actinomyces species and anaerobic diphtheroids. The following methods have been suggested for **maintaining stock cultures**.

Table 11.2 : Biochemical Characters of Actinomyces and *Corynebacterium acne*

Characteristic	***A. israeli***	***A. bovis***	***A. neaslundi***	***C. acne***
Maximum growth	7-10 days	2-3 days	1-3 days	3-4 days
O_2 requirement	Anaerobic & then microaerophilic after isolation	Anaerobic & then microaerophilic after isolation	Facultative in presence of CO_2	Anaerobic or microaerophilic
Air with 10% CO_2 **(Candle jar)**	0 or 1+	0 or 1+	4+	0 or 1+
Anaerobic with N_2 + CO_2 (5%)	4+	4+	4+	4+
Catalase production	0	0	0	+
Starch hydrolysis	± or 0	4+	± or 0	0
Nitrate reduction	80%+	0	90%+	+
Indole formation	0	0	0	+
Gelatin liquefaction	0	0	0	+ may take 1-3 weeks
Litmus milk reactions	0 or 1 + Acid Slight reduction No coagulation, No peptonisation.	0 or 1 + Acid Slight reduction No coagulation or peptonisation	0 or 1 + Acid Slight reduction or Acid clot No peptonisation	Acid followed by peptonisation in 1-3 weeks
Sugar fermentation : Acid only				
Glucose	+	+	+	+
Starch	20%+	+	20%+	0
Mannitol	80%+	0	0	+
Raffinose	Variable	0	80%+	Variable
Xylose	80%+	0	0	0
Glycerol	0	Occasionally +	0	+

1. *Lyophilisation*

This is the most effective method for maintaining Actinomyces cultures and when properly done it may survive or 5 to 10 years. Cultures are inoculated into Brain Heart Infusion broth in centrifuge tubes and incubated at 37°C for one week. The growth is then spun down and the supernatant broth withdrawn. **Sterile milk** is then added and the culture is well-mixed with the milk. With the help of a capillary pipette 0.2 ml of the suspension is transferred to several **lyophilisation tubes** which are then lyophilised in the usual manner.

2. *Semisolid maintenance media for Actinomyces*

This is a modification of medium recommended by Pine and Watson (1959).

(a) ***Salt solution*** : K_2HPO_4—60 g, $(NH_4)_2SO_4$ — 4 g; $MgSO_4$, $7H_2O$ — 0.8 g, $CaCl_2$ — 0.08 g; Distilled water — 500 ml.

(b) ***Heart infusion broth*** : **Heart infusion** broth powder — 25 g; **Dextrose** — 5 g; **Cysteine hydrochloride** — 1 g; **Casitone** (Difco) — 4g; **Yeast extract** — 5 g; Soluble **starch** — 1 g; **Agar** — 7 g; **Distilled water** — 750 ml.

Final media : Add salt solution (a) — 250 ml to 750 ml of (b) Heart infusion to make 1 litre. Adjust pH to 7.2. Distribute 8 ml in each tube (18 × 150 mm), cotton plugged. Autoclave at 15 lbs (120°C) for 10 minutes. Final pH should be 6.8-7.

3. ***Liquid Actinomyces maintenance mdium*** : Prepared as No. 2 above without agar.

4. ***Solid Actinomyces maintenance medium*** is prepared as No. 2 above but 20 gm (1.5%).

Epidemiology : Once it was thought that infection might originate from grains, since fragments of the latter in the mouth were occasionally seen in relation to primary lesion.

A. israeli is a strict parasite and incapable of saprophytic existence. Hence it is considered to be of endogenous origin due to fall in the local defensive mechanism of the host. Its commensalism is proved by its existence in the teeth or oral cavity.

There are some puzzling facts in the incidence of the disease; males are more frequently affected than females (3 : 1) and although cases occur at all ages, most of the cases fall in the age group of 20-30 years and the second age group falls in the age group of 10-19 years. These two decades account for more than half the incidence.

Although there is no relationship of human with bovine actinomycosis, **male agricultural workers are more liable to suffer** from infection with *A. israeli* than men in other occupations (Porter, 1953). The alleged greater infection among farmers is not readily reconcilable with the prevailing view that the disease arises from invasion of tissue by an indigenous organism in the mouth.

Trauma caused by grain probably incites infection in the buccal mucosa and it has been recognised that dental extraction may precipitate infection. Similarly, appendicectomy is associated with abdominal infection. Trauma, either accidental or intentional has been noted to precede the onset of the disease.

PATHOGENESIS

A. israeli occurs as a commensal in human oral cavity and can be inoculated from carious teeth and tonsils in 5 per cent of otherwise healthy individuals (Sullivan and Goldsworthy, 1940).

In man actinomycosis is esentially an endogenous infection of **cervicofacial region** (65 per cent), **abdominal infection** (19 per cent) is the next common one. **Other sites** involved are the **thorax and skin**. They show suppurative tendency due to secondary pyogenic infection with cocci. The **commonest oral** infection is through the mucosa of the **mouth and throat related to carious tooth** or **tooth extraction**. The initial stage is probably a low grade tissue invasion and trauma may be an exciting factor in the causation of the disease. Metastatic lesions may occur by blood stream to the **kidney, lung, brain and liver** from a primary **intestinal lesion from the appendix**.

Signs and symptoms : The incubation period is not definitely known but it varies from a few days to a few weeks. The **lesions** are generally **painless**. The disease starts as a firm, purplish **nodule** which gradually enlarges and **softens** to form an **abscess**. The abscess **bursts, forms a sinus** from which **"sulphur granules" are discharged**. The disease spreads by continuity and the surrounding areas become studded with nodules. The affected area has board like resistance on palpation and shows multiple sinuses, discharging granules and sero-purulent fluid. These multiple openings are due to formation of intractable sinuses. The underlying bone may be involved in the process. The spread of the disease does not occur through the lymphatics and this is how it differs from other granulomas. **Metastasis** may take place in the **liver, brain, kidney and lungs by haematogenous route**. The breast or ribs may be involved as a process of extension from the lungs.

It is customary to classify actinomycosis into **cervicofacial, thoracic, adbominal**, and **cutaneous types** according to the site of involvement.

Cervicofacial actinomycosis is the commonest type (**Pl. 11.I, Fig. 5**). It accounts for 50 per cent cases. The route of infection is the mucous membrane of the mouth and frequently it starts from a carious tooth or tonsil. From the site of "primary lesion" it spreads to the deeper tissues and ultimately to the skin of the face and neck or to the tongue and involve the neighbouring structures. Maxillary bone, vertebrae or orbit may be secondarily involved.

Thoracic actinomycosis accounts for 15 percent of cases. It originates from the mouth or throat by aspiration of the infective material in the lungs or as an extension from the abdomen or as metastasis. The lesion is mainly found in the lungs where small abscess cavities are formed. Extensive lesions may cause dissemination through the bronchogenous route. From the **lung** the lesion may **spread to pleura, pericardium, heart, or chest** wall by a process of direct extension.

Abdominal actinomycosis accounts for 20 per cent of cases. It is commonly seen in the caecum and appendix. The abdominal infection is probably derived **through the intestinal mucous membrane** which may not be involved. Actinomyces are frequently found in the intestine of normal persons. **Appendix as a "locus minoris resistance"** is often the **portal of entry of this type of actinomycosis**.

Suppurating foci may be adherent to the abdominal wall by a process of fibrosis involving the skin. **From the appendix the liver and gallbladder are commonly involved** by a process of metastasis or extension **(Pl. 11.I Fig. 6)** and extension of the lesion to the genital tract like ovaries and kidney is relatively frequent **(Pl. 11.I, Fig. 7)**.

Skin actinomycosis accounts for 15 per cent of cases. It is difficult to say whether skin is primarily affected in every case as actinomycosis of the skin. It is rather commonly secondary to lesions of the underlying tissue or organ.

Other locations of lesion ; Actinomycosis affect other locations as a result of haematogenous dissemination or as a result of continuity of infection. In untreated cases generalisation is seen affecting **skin, bone, internal organs including abscess of the brain with formation of giant cells and proliferation of fibroblasts**. Fat-laden macrophages are present at the peripheral portion of the abscess and this may give yellow coloration of the lesion, visible even to the naked eyes.

Mycological Diagnosis

The clinical diagnosis should always be confirmed by microscopical examination of the granule. Granules are crushed and examined under coverslip preparation with a drop of **MCGUIRE'S STAIN**. They show **'ray-fungus', i.e., filaments with ray-like arrangement**. The isolation and identification of the organism can only be made by culture, which should be done both aerobically and anaerobically. For this, granules are washed repeatedly in saline, treated with alcohol for 5 minutes and crushed into small bits. The tiny bits are then inoculated in 1 per cent glucose agar shake culture tubes. Several tubes should be inoculated and incubated at 37°C anaerobically for about a week. **Mulberry-like colonies** about 2 or 3 μ in diameter appear in the depth of the agar. A dense zone of growth is seen about 2 cm below the surface. Bacto brain-heart infusion (BHI) agar with **5 per cent CO_2** may be used **for culture under anaerobic condition**. Once the anaerobic growth has appeared, subcultures may be maintained aerobically in a suitable medium. Serological and allergic cutaneous tests have proved to be unsucessful.

X-ray (Radiological Diagnosis)

In the early stage of the disease, the skiagram of the jaw does not show any involvement of the bone but later there may be periosteitis or true osteomyelitis with destruction of the bone.

Animal pathogenicity

Attempts to experimentally infect laboratory animals like guinea pigs or rabbits with an aerobic actinomyces have failed (Rosebury *et. al.*, 1944). The organism was successfully inoculated by Meyer and Verges (1950), Gale and Waldron (1955) who recommended the use of gastric mucin or heart infusion broth as adjuvant in producing infection in mice.

Hazen and Little (1958) recommended 3 to 4 weeks old golden hamsters for experimental lesion to prove the pathogenicity of actinomyces. The suspension of cultures are prepared from broth cultures in duplicate sets which have been incubated for 7-10 days. The cultures are centrifuged and washed 3 to 4 times with 0.85 per cent saline solution. The tough strains contain large aggregates which should be uniformly broken with a syringe by drawing in and expelling until it can be injected through 22 gauge needle. One ml of the uniform suspension is injected intraperitoneally into 2 or more hamsters and the animals are studied for 4 weeks for palpable lesions and then sacrificed and examined in an autopsy for gross infection. Pus from abscesses is examined microscopically for actinomycotic granules showing branching filaments. Culture should be taken as well. ***A. israeli*** was injected intraperitoneally, which formed ***clubbed granules in*** 13 out of 15 (81.3%) **golden hamsters**; whereas Howel *et al.*, (1959) and Pine *et. al.* (1960) could produce actinomycotic lesions in smaller number of cases and think, the experiment of limited value.

LABORATORY DIAGNOSIS

Clinical materials

This may be pus, sputum or aspirated material from early pulmonary cases, granules in the pus or draining sinuses. These are received in a sterile Petri dish and examined for "sulphur granules". These granules are yellowish white or tan in colour and range in visible specks of 2 to 5 mm in diameter.

A. DIRECT EXAMINATION

Examine in (a) **fresh mounts,** (b) **mount after MCGUIRE STAIN** under a cover slip, and fixed preparations of smears are stained and then cultures are made.

After direct examination, a granules are crushed, smears prepared. The dried smear is stained with Gram stain and acid-fast stain. Under high power or oil immersion, the granules are seen to be composed of interwined branched filaments, 1 micron or less in diameter.

Gram stain : The filaments in granules of pus or sputum are Gram positive but some short fragmented Gram negative, rod shaped bacteria will also be noted. Secondary organisms are often seen in a draining sinus.

B. INDIRECT METHOD TO IDENTIFY FUNGUS

1. **Culture** : The granules are washed repeatedly with sterile saline, crushed and inoculated by streaks in plates of enriched **BHI agar media at pH 7.2 to 7.5** and incubated at 37°C under **anaerobic conditions**. Brain heart infusion (BHI) agar with 0.2 per cent dextrose is satisfactory. *Anaerobic cultures* are put in

   (i) **Anaerobic jar** with 95 per cent N_2 and 5 per cent CO_2. In the *primary cultures* blood may be added to support the culture (Garrod, 1952). The **casitone starch maintenance medium** is also a useful isolation medium.

(ii) **A duplicate set of plates** are incubated at 37°C **anaerobically in candle jars**. The colonies are examined after 24 to 48 hours to determine the characteristic morphology of young colonies, and again after 7-10 days in order to examine mature colonies.

(iii) ***The thioglycollate broth is not recommended for primary isolation***; characteristic growth may be examined in this medium for the study of the organism as branched forms develop abundantly in this medium. A second plating may be necessary to ensure pure culture before biochemical tests are performed.

2. **Animal inoculation** : Inoculations in guinea pigs and rabbits have failed (Rosebury *et. al.*, 1944) for pathogenicity. Hazen and Little (1958) have recommended **golden hamsters for animal inoculations** and 81.3 per cent of hamsters developed clubbed granules inoculated with *A. israeli*. Howel (1959) and Pine *et. al.*, (1960) could, however, produce lesions in a small percentage of animals.

3. **Histopathology** : Abscesses form which intercommunicate or coalesce. The **mycelium and the granule is formed in the centre of an abscess** and is, therefore, seen surrounded by pus cells is sections of the granule with a characteristic ray fungus appearance. **Giant cells** are frequently found **close to the granules**. On the outer edge of the abscess there is granulation tissue of varying amounts depending on the healing which has occurred.

H & E method does not stain the fungus and for demonstration of fungi **Gomori's methenamine silver (GMS) stain should be carried out.**

Since different species of **Eumycetes** may produce similar changes, the nature of the organism should be determined to identify the aetiological organism by anaerobic* and aerobic methods in different enriched media.

TREATMENT

In the early stage, when there is no involvement of the deeper tissue, excision of the nodule may be possible; in the later stage of the disease, all the sinuses should be opened up, scaped and drained. **Liq. iodi mitis** can be used locally as an antiseptic.

Penicillin should be administered intramuscularly in doses of 400,000 units every three hours to a total dose of 6,000,000 units. It should be noted that, different strains vary in their sensitivity to penicillin *in vitro*.

Sulphonamides have been administered with good results. Sulphadiazine is administered in doses of **4 to 6 g per day** and the dose is reduced to 2 g per day after clinical improvement of the patient. The results are, however, inconsistent with this treatment. **X-ray, penicillin and sulphonamide** treatment may, however, be **combined** with advantage.

ACTINOMYCES BOVIS

A. bovis was formerly considered synonymous with ***A. israeli*** but is **serologically different** from the other.

*(1) A simple method for **ANAEROBIC CULTURE** described by Mathews (1961) may be utilised by incubating single plate culture by polythene inserting BHI agar plate with blood in 6"-7" in wide bags. **(2)** A fine porosity seitz filter pad 2½" diameter is placed on a slightly larger disc of aluminium foil folded over the pad leaving 1¼" of the pad exposed on the top. **(3)** The absorbent disc is impregnated with 2 ml each of pyrogallol solution and 10 per cent Na_2Co_3 and inserted in the bag with the foil side next to the Petri dish. **(4)** The open end of the bag is then folded and sealed lightly with masking tape. **(5)** The bags containing the plates are incubated at 37°C. The process is satisfactory for actinomyces as strict anaerobiosis is not necessary in this case.

The anaerobic condition can be maintained using pyrogallol-carbonate seals in tubes or putting plates in anaerobic jars (see before).

Morphology : It is similar to *A. israeli* but the colonies are smoother and softer in consistency and less sticky to the medium. The growth is less tolerant to oxygen (comparatively more anaerobic).

Pathogenicity : A bovis is a commensal of the mouth of the healthy cattle but it may produce lesions locally, producing lumpy jaw. Like *A. israeli* it may spread from the primary site by blood stream.

ACTINOMYCES BAUDETII

A. baudetti was also considered synonymous with ***A. israeli****. Morphologically*, it differs from *A. israeli* in that, it forms clubs, absorbs basic stains and not the acid stains. The organism is microaerophilic and grows slowly, producing dull, whitish colonies of about the size of a pin head after incubation of 4 to 5 days.

Pathogencity : It produces lesions in cats or dogs but not pathogenic to man or cattle.

PATHOLOGY OF ACTINOMYCOSIS

Introduction : For an intelligent understanding of the pathology of actinomycosis a few peculiarities of this infection should be understood.

1. ***Pathologically,*** actinomycosis is a condition of mycetoma, which is defined as a chronic suppurative, granulomatous infection running a relentless course if not treated, which destroys all types of contiguous tissues including muscles, tendons and bones almost like a slow growing malignant tumour. Several authors have, therefore, classified actinomycosis with other **schizomycetomas** caused by **pseudomycetes** and especially by actinomycetales (Winslow and Steen, 1964; Zaias *et. al.*, 1969). If we consider actinomycosis separately here, it is not only for historical reasons, but because actinomycosis differs in various respects from other mycetomas.
2. ***Source of infection*** : There appears to be no exogenous source of infection. It is really an endogenous infection occurring from the gums, tonsils and oral mucous membrane and carious tooth.

   Actinomyces israeli responsible for human infections has never been recovered from outside the human body. Very rare instances of primary cutaneous actinomycosis have been recorded, showing a rather doubtful aetiologic connection with a preceding injury (Briney, 1965).
3. Many such **records are inaccurate** because the authors ignore two basic facts.

   One : The pathogenic *A. israeli* is strictly anaerobic. This means that the pathogenicity of a doubtful organism must be proved by anaerobic culture.

   Two : There is no reliable histopathological differentiation of an *Actinomyces israeli* granule from a Nocardia granule.

   Both may be of the same size, both may or may not have a peripheral fringe of **eosinophilic "clubs"**, and in both infections actinomycosis as well as in *Nocardia* mycetoma, the tissue reactions are almost identical. All records based on the histopathological appearance only, as for instance the one of Punvamurthy and Shenoy (1968) of a **perianal "actinomycosis"**, have to be viewed rather critically. Only bacteriological demonstration of an anaerobic actinomycetes is the only method for a definite confirmation of an actinomycosis.

   For all practical purposes, however, one can fairly safely assume that **cervico-facial, ileo-caecal and thoracic lesions are almost always actinomycotic**. Similar lesions of the foot or from other areas of the skin and of the subcutaneous tissue showing granules, represent most frequently a ***Nocardial*-mycetoma**. It has to be remembered, that an ileo-caecal actinomycosis may involve the abdominal wall and break through the skin from within in the lower right quadrant of the abdomen or in the right groin.

4. Though *A. israeli* may habitually lodge in crypts of tonsils, at the root of carious teeth and on the oral mucosa, it must somehow gain entrance into the soft tissue. **Tooth extractions**, injuries of the mucosa by fish-bones etc., are considered to be mechanisms for paving a way for the infection. In addition, however, pathogenic *Actinomyces* is almost invariably associated with a variety of bacteria in the tissue, particularly with ***Bacterium actionomycetem comitans***. The role of these bacteria is not clear and some authors believe that *A. israeli* alone is unable to cause the disease if not accompanied by various bacteria.

5. Lastly, **an actino "mycosis" is not a mycosis, but a bacterial infection with clinical manifestations rather similar to some true mycoses**. It has, therefore, traditionally been discussed together with the mycoses in the textbooks. The fact that ***Actinomyces***, which are related to mycobacteria, can produce branching filaments and a mycelium like network, has contributed to the erroneous name of a **ray**- "**fungus**". It would be superfluous to mention this in a chapter on pathology if the possibility of effective treatment of actinomycosis with **penicillin** or **sulphonamides** would not alter the course, prognosis and pathology of this disease profoundly.

Gross Pathology : Lesions begin first with micro-abscess formation, and considerable swelling of the soft tissues. From the primary abscess, **tentacle-like extensions burrow** their way through the tissue, carrying with them the organisms, breaking through the skin, or forming communicating sinuses into body cavities. Soon, **granulation tissue, cicatrication and scar formation sets in** and may transform the tissue involved in a **firm to very hard irregular lumpy area**.

Larger cavities filled with pus and necrotic material may be present and contain small yellow granules, barely visible to the naked eye. The infection spreads preferably along fascias and tendon sheaths, but finally does not respect any boundaries or organs. *CERVICO-FACIAL ACTINOMYCOSIS* may spread to the spinal column and cause fatal osteomyelitis and spinal meningitis. *ILEOCOECAL ACTINOMYCOSIS* may destroy the liver **(Pl. 11.I, Fig. 6)**. *Thoracic actinomycosis* may involve the lungs, the ribs and the wall of the thorax with a maze of inter-communicating sinus tracks and fistulae. The case had been regarded as tuberculosis for many years. Sinus tracks may be demonstrated radiologically by injection of contrast medium and may be of surprising length and complexity.

Histopathology

The clue to the diagnosis of actinomycosis is the **granule**. Its **diameter** is usually about 100-**400 microns** and it is found in the centre of an abscess. It is round, oval or slightly irregular in shape. There is a central tangle of branching filaments, which are not readily recognisable in H & E sections **(Pl. 11.II, Fig. 1 and 2)**. They are **gram positive, non-acid-fast** and not very adequately stained with PAS, but can be well **demonstrated with silver methenamine stain**. The granule looks pale blue in the centre in H & E sections, and more pinkish at the periphery. Often, a peripheral fringe of eosinophilic radiating "**clubs**" **of 10-20 microns length** is present. The "**clubs**" probably consist of an immune precipitate similar to the **eosinophilic perihyphal substance in subcutaneous** phycomycosis (Zygomycosis) and to that of

PLATE 11.I

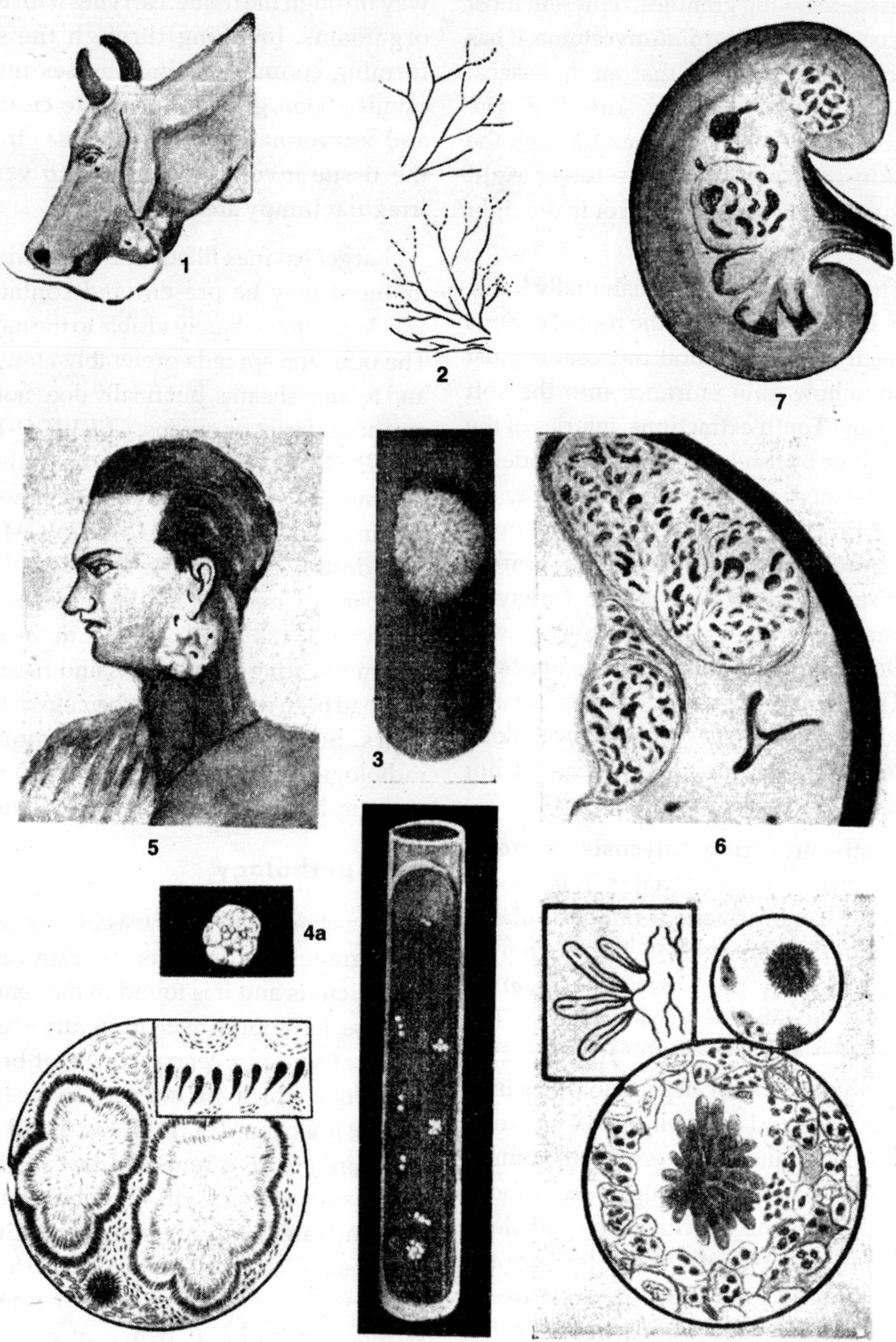

Figs. **(1)** Lumpy jaw in cattle; **(2)** branching of mycelia of *A. bovis*; **(3)** shake culture of *A. bovis* in glucose agar; **(4, 4a)** anaerobic culture of *A. bovis* on glucose agar showing mulberry character; **(5)** cervicofacial actinomycosis; **(6)** actinomycosis of liver; **(7)** actinomycosis of kidney; **(8)** ray fungus (inset-high power; showing clubs); **(9)** a granule with club formation surrounded by polymorphonuclear neutrophils.

PLATE 11.II

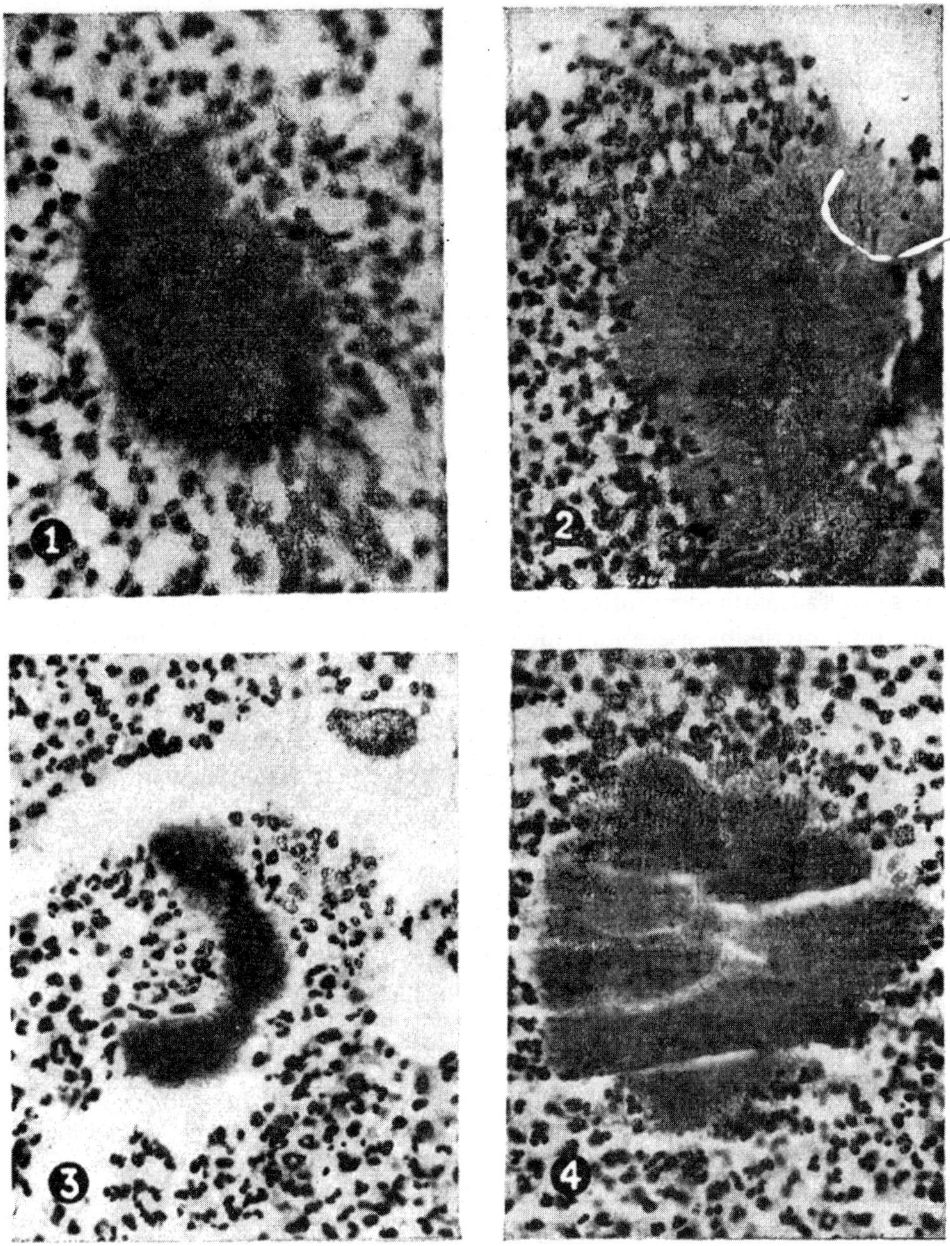

Figs. 1-2 : Actinomycosis. **Fig. 1 :** *Actinomyces israeli* granule in tissue section : *without "clubs".* **Fig. 2 :** *Actinomyces* granule *with clubs.* **Figs. 3-4 :** Nocardia mycetoma. **fig. 3 :** For comparison : *Nocardia* granule in tissue *with clubs.* The appearance is slightly more powder, than that of the actinomyces granule. **Fig. 4 :** *Nocardia* granule *with clubs.* Note that no definite differentiation of *Nocardia* and *actinomyces* is possible. All figs. H & E; × 560 other fungi and parasites. The matter has been discussed by Bader and Grueber (1970), by Winslow and Steen (1964), and by others. Pus cells are usually entangled in the "clubs". The "clubs" are *not* pathognomonic of ***Actinomyces* granules** and may be present around *Nocardia* **(Pl. 11.II, Fig. 4)**, and ***Streptomyces madurae*-granules** as well, both being agents of mycetoma. However, **clubs may be completely absent in actinomycosis (Pl. 11.II, Fig. 1)**. **A few giant cells** may be present particularly in long standing actinomycosis. The periphery is formed by an abscess wall with macrophages, showing transitions in more chronic fibroblastic granulation tissue. **Epithelioid reaction is not a typical feature** of actinomycosis. Sections of skin biopsies may show epithelium which is lifted off by serosanguinous and purulent exudate. Small steaks of pus may lead into the depth of the tissue, where **multiple micro-abscesses** and **focal collections of plasma cells, polymorphs and lymphocytes may be found**. The histological picture is nonspecific, and if there is clinically a suggestion of actinomycosis it is essential to examine numerous step sections until granules in the tissue are found or excluded (For **differential diagnosis of granules**, see under "Mycetoma").

Autopsy

Since adequate **penicillin treatment results in complete cure; actinomycosis is rarely seen on the autopsy table**. Lesions have been well described by Conant *et. al.*, (1971). **Cervico-facial actinomycosis** may kill the patient by extension upward along the spine into the cranial cavity, or by downward progression to the thorax. Destruction of the spine and the skull, the ribs, and the meninges are possible. From the ileocaecal area, ramifying abscesses may spread around the kidney **(Pl. 11.I, Fig. 7)** and along the psoas muscle, with sinuses draining the flank or the groin. The whole **ileocoecal area may be transformed into an inflammatory tumour** but **with innumerable small abscesses included in the fibrous tissue**. The liver is frequently involved showing multilobular abscesses **(Pl. 11.I, Fig. 6)**. From there, the process may spread to the diaphragm, the pleura and the lungs.

Clinical pathology : Pus from draining sinuses must be **examined for "sulphur granules"**. They consist of colonies of the organisms with entangled pus cells and may be up to one mm in diameter but they are mostly smaller **(Pl. 11.1, Fig. 9)**. If recognisation is difficult pus should be vigorously shaken in a tube of water. The granules, not emulsified, will sink to the bottom and may be removed with a pipette (Bronner and Bronner, 1969). **They are easily crushed between two microscopic slides and the radiating pattern of filaments can be recognised in unstained sections (Pl. 11.1, Fig. 8)**. Subsequently, add a drop of water, make two smears, fix over flame, stain one with Gram's stain and the other with Ziehl Neelsen method. **Actinomyces filaments are** about 1 micron wide, **Gram positive** but **not acid fast**. Actinomyces filaments may be indistinguishable from those of *Nocardia sp.* in smears, but ***Nocardia* strains are acid-fast**, in contrast to *Actinomyces*. *Nocardia* **filaments tend also to be more fragmented** than *Actinomyces*.

Important Points

For direct culture of ***A. israeli*** from clinical specimens, granules should be isolated and washed several times in saline or in thioglycollate broth containing 0.05 per cent cystein hydrochloride, as described by Emmons *et. al.* (1970). This is necessary in order to free the culture from concomittant bacteria. A diluted suspension of filaments may then be anaerobically incubated in brain heart infusion **agar, or in thioglycollate broth at 30ºC to 37ºC**. When colonies appear after 2-5 days, they should be transferred to fresh media and again be incubated anaerobically. Cultures are often very small, and hardly visible with the naked eye. They consist of a white spidery radiating system of branching filaments and later of white subspherical colonies with an angular or lobulated surface. **Biochemically** *A. israeli* produces acid without gas in glucose, galactose, lactose, levulose, maltose, raffinose, saccharose and xylose. Starch is slightly hydrolysed but not the cellulose.

Summary

Suppuration, sinus and scar formation are characteristics of actinomycosis along with burrowing of the abscesses. The reacting cells are the polymorphonuclear neutrophils, but giant cells are also found in contact with the sulfur granules occasionally. Macrophages containing sufficient fat (to give a yellow colour to the lesion) may be present in large number at the periphery of the abscess.

Tissue removed by surgical biopsy is taken from sinuses, bone, and lymph nodes and tongue, or isolated lesions.

The HISTOLOGIC there is pyogenic infection with infiltration by polymor-phonuclear cells; there may not anything significant except the presence of chronic inflammatory cells. Sometimes there is formation of granulation tissue dense scar tissue formation. Granules may be absent in the abscess wall.

Histological diagnosis is confirmatory by finding of the actinomycotic granules. The presence of separate mycelial segments also suggests the diagnosis. In H & E sections the central portion of the granule tends to stain heavily with hematoxylin, and the periphery is tained with eosin; variations may occur due to fixation process.

Granules are better found in sections stained with H & E and Gram's method. Branching mycelia are Gram positive.

In infections caused by **Monosporium apiospermum** (see **MADUROMYCOSIS**), spores forms are identified near the periphery of the granule. It is desirable, also, to stain the section to determine if the organism is acid-fast.

In biopsy material, actinomycosis may be confused with tumor or with chronic suppurative processes due to other causes. Tissue from lesions of the neck, for example, may resemble sarcoma. Actinomycotic lesions of the breast may histologically simulate either tumor of the breast or bacterial abscesses. Actinomycotic biopsy for histologial examinations may resemble a sarcoma.

Autopsy

Sometimes the diagnosis of actinomycosis is not confirmed except at the autopsy (examinations). The appearances at autopsy will depend largely upon the portal of entrance of the fungus. Actinomycosis of the NECK and FACE, the commonest form of actinomycosis, is rarely encountered at autopsy, but death may result from extension of the process from the angle of the jaw upward along the cervical spine and into the cranial cavity or downward to the superior portion of the thoracic cage.

Occasionally in Head and neck affections of actinomycosis, and extensive osteomyelitis of the spine and of the skull an epidural abscess with thickening of the skull occurs.

The involvement of the orbital cavity may cause unilateral exophthalmos.

Ramifying abscesses of the right lower quadrant and other portions of the abdominal cavity may occur. Rather characteristic is extension from the retrocecal region along the iliopsoas muscle and perirenal region causes sinuses draining to the skin in the inguinal region.The **LIVER** is affected by extension to the portal vein. Actinomycotic **pylephlebitis and multilocular liver abscesses** are not uncommon. Extension from the liver to the subdiaphragmatic region, pleural cavity and the lungs does occur. Rarely extension by the hematogenous route from the intestine may cause infection of the lungs, brain, meninges or other tissues.

Primary actinomycosis of the **LUNGS** may result in the formation of chronic abscesses, bronchiectasis or dense fibrosis. Extension to the pleura and chest wall or to the pericardium and heart may also occur.

IMMUNOLOGY & SEROLOGY

Although opsonins, agglutinins, precipitins and complement fixing antibodies occur in the sera of actinomycotic patients results are not sufficiently confirmatory due to technical difficulties encountered in the preparation of smooth homogenous antigenic suspensions. Cross-reactions do occur with unrelated fungal antigens.

Hypersensitivity

Reactions patients of actinomycosis sensitisation. Intradermal injections of vaccines and or culture filtrates have resulted in local reactions with erythema, edema and pain.

Drake and Henrici were able to produce specific hypersensitivity in guinea pigs and rabbits by the intratesticular infection of oily suspensions of the acid-fast N. asteroides. Intracutaneous injection of heat-killed vaccines produced no reactions in these animals, but injection of suspensions of living Nocardia produced non-specific toxic reactions in the infected and control animals. Intradermal injection of a crude extract prepared from powdered organisms, (rendered lipoid-free by acetone extraction) resulted in a **specific allergic reaction of the tuberculin type**. The polysaccharide fraction produced a maximal reaction in 24 hours which disappeared after 4 days.

REFERENCES

Bader and Grueber, H. L. E. (1970). *Personal communication.*

Bollinger, O. (1877). Ueber eine neue Pilzkrankheit beim Rinde. *Zentralbl. f. med. Wissensch.* **15**, 481-485.

Bostroem, E. (1891). Untersuchungen Uber die Aktinomycose des Menschen. *Beitr. z. path. Anat.* **9**, 1-240.

Bronner, M. and Bronner, M. *Actinomycosis*. John Wright and Sons Ltd., 1st Ed., Briston, 1969.

Burden P (1989) Actinomycosis. J Infect. 1995.

Chatterjee, P. K. (1964). Ray-Fungus Disease, Editorial *J. Ind. med. Assoc.* **42**, 85-86.

Conant, N. F., Smith, D. T., Baker, R. D. and Callaway, J. L. *Manual of Clinical Mycology*. W. B. Saunders Co., 3rd Ed., Philadelphia, 1971.

Emmons, C. W., Binford, C. H. and Utz, J. P. *Medical Mycology*. Lea & Febiger, 2nd Ed., Philadelphia, 1970.

Gale, D. and Waldron, C. A. (1955). Experimental actinomycosis with *Actinomyces israelii. J. infect. Dis.* **97**, 251-261.

Garrod, L. P. (1952). Actinomycosis of the lung. Aetiology, diagnosis and chemotherapy. *Tubercle.* **33**, 258-266.

Harz. C. O. (1877). In Bollinger ; Ueber eine neue Pilzkrankheit beim Rinde, *Zentralbl. f. med. Wehnscbr.* **15**, 481-485.

Harz. C. O. *Actinomyces bovis*, ein neuer Schimmel in den Geweben des Rindes, Jahresb. d.k. *Zentralbl. - Tierarznei-Schule in Munchen*, p. 125, 1879.

Hazen, E. L. and Little, G. N. (1958). *Actinomyces bovis* and "anaerobic diphtheroids" : Pathogenicity for hamsters and some other differentiating characteristics. *J. lab clin. Med.* **51**, 968-976.

Howell, A., Murphy, W. C., Paul, F. and Stephan, R. M. (1959). Oral strains of actinomyces. *J. Bact.* **78**, 82-95.

Israel. J. (1878). Neue Beobachtungen auf dem Gebiete der Mykosen des Menschen. *Virchow's Arch. f. path. Anat.* **74** 15-53.

Lentze, F. A. (1938). Die mikrobiologische Diagnostic der Actinomykose. *Muenchener med. Wehscbr.* November, 1826-1829.

Meyer. E. and Verges, P. (1950). Mouse pathogenicity as a diagnostic aid in the identification of *Actinomyces bovis. J. lab. clin. Med.* **36**, 667-674.

Mikelsen. O. A. (1965). Actinomycosis simulating malignant tumours. *Nord. Med.* **73**, 286-287.

Pine. L., Howell. A. and Watson, S. O. (1960). Studies on the morphological, physiological and biochemical characters of *Actinomyces bovis. J. gen. Microbiol*. **23**, 403-424.

Punyamurthy, M. and Shenoy, B. V. (1968). Multiple perianal fistulae due to actinomycosis. *J. ind. med. Assoc.* **51**, 396-397.

Rosebury. T., Epps, L. J. and Clark, A. R. (1944). A study of the isolation, cultivation and pathogenicity of *Actinomyces israelii* recovered from the human mouth and from actinomycosis in amn *J. infect. Dis.* **74**. 131-149.

Schaal KP, Lee H-J (1992) : Actinomycete infections in humans: A review *Gene* **115**, 201.

Scheibe, G. (1965). Metastasization in actinomycosis. *Bruns. Beitr. clin. Chir.* **210**. 60-79.

Waksman, S. A. and Henrici, A. T. (1943). The nomenclature and pathogenicity of actinomycetes. *J. Bact.* **46**, 337-341.

Weese WC, Smith IM (1975) : Study opf 57 cases of actinomycosis over a 36-year period. *Arch intern Med.* **135**, 1562.

Winslow, D. J. and Steen, F. G. (1964). Consideration in the histologic diagnosis of mycetoma. *Am. J. clin. Path.* **42**, 164-169.

Wolff, M. and Israel, J. (1891). Ueber Reincultur des actinomyces and seine Uebertragbarkeit auf Their. *Virchow's Arch. f. path. Anat.* **126**, 11-59.

Zaias, N. Taplin. D. and Revel, G. (1969). Mycetoma. *Arch. Derm.* **99**, 215-225.

12

Aerobic Actinomycetes: NOCARDIA

Schizomycotic Actinomycetoma (NOCARDIA AND STREPTOMYCES)

Streptomyces species are characterised by stability of their filaments and by the formation of spores on the aerial mycelia. They resemble Nocardia in their growth and appearance and the mutants that lose the ability to form aerial mycelia are difficult to be distinguished from Nocardia. Antigenically all actinomycetes have common antigen and cross react amongst them as a rule. **These are *AEROBIC*** and unlike Nocardia they are **NON-ACIDFAST**.

A number of actinomycotic abscesses (mycetomas) are caused by streptomycetes which were formerly classed as Nocardia (see Table).

BIOLOGY: Streptomycetes are ubiquitous in soil and the infection is attributed to contamination of scratches of penetrating wounds by thorn prick.

Innumerable isolates from soil samples from all parts of the world have systematically scrutinised by Waksman & others and various species (about 500 in number or more) have been found to produce antibiotics which are mainly produced from cultures of streptomycetes. Thus these species are medically important for their antibiotic properties rather than their pathogenic effects. Streptomyces and actinomycetes also act as widespread scavengers in soil by breaking down proteins, cellulose and other organic matters.

STUDY OF NOCARDIA

On the basis of preliminary studies the following methods and tests should be undertaken for the identification of Nocardia (Georg *et. al.*, 1961).

1. **AEROBIC GROWTH**
2. **ACID FASTNESS**
3. **Slide culture technique for branching mycelia**. Nocardia and streptomyces show branched mycelia.
4. **Pellicle formation in thioglycollate broth**.
5. **Casein hydrolysis**.
6. **Tyrosine utilisation**.
7. **Paraffin utilisation**.
8. **Growth in dilute** (0.4 per cent) **gelatin**.
9. **Sugar utilisation and acidification**.
10. **Ability to grow at various temperatures**.
11. **Animal pathogenicity test**.

1. Acidfastness

A modified Kinyoun cold staining method may be utilised. In this 1 per cent aqueous H_2SO_4 is used in place of acid alcohol as a decolourising agent as recommended by Drake and Henrici (1943). Because decolourisation is not complete by this technique, partially acid fast organisms remain stained instead of being decolourised.

2. Slide culture technique for demonstration of branched mycelium

Mycelia in many Nocardia species fragment readily in stained smears made from growth on solid medium; frequently they reveal only bacillary or coccoid forms. **Riddell's agar block culture technique** described in Chapter 4 is adopted here. The organism adheres to the slide or coverslip after the agar block is removed may be mounted separately with or without mounting fluid.

Excellent illustrations on the early growth of *Nocardia* in microcultures are to be found in McClung's publication on Nocardia (1963).

3. Pellicle formation in thioglycollate broth (Carter and Buhler, 1957)

It is of great value in the differentiation of unclassifed (yellow) mycobacter from Nocardia species. The unclassified chromogenic acid fast mycobacteria (when grown on thioglycollate broth) form clumps of growth just beneath the surface of the broth. In contrast the Nocardia and Streptomyces produce a surface pellicle composed of a fine network of branches. *Myco. tuberculosis* does not develop in this medium.

4. Casein hydrolysis

Gordon and Smith (1955) recommended this method as a means of differentiation of *N.asteroides* from *N. brasiliensis* when casein plates were used to demonstrate in contrast for hydrolysis. **Streptomyces hydrolysed casein plates completely**, and *N. brasiliensis* did it slowly after incubation for 2 weeks; whereas, *N. asteroides* did not hydrolyse at all. Casein medium was prepared as follows :

Skimmed milk (dehydrated) or instant non-fat milk is dissolved in distilled water (10 g per 100 ml). An equal volume of 2 per cent agar in water is prepared as well. The two solutions are then autoclaved separately and cooled to 45°C, mixed and poured into sterile Petri dishes.

The Petri dishes are inoculated and incubated at 28ºC. If the test organism grows poorly, the plates are incubated at 37ºC. The plates are examined and the readings are taken over on the 7th day and another on 15 days of inoculation (**Fig. 12.1**).

According to Georg *et. al.* (1961), this is the most reliable test for differentiation of *N. asteroides* from *N. brasiliensis* and Streptomyces, when grown on milk plates, though all of them grew well in this medium.

5. Tyrosine utilisation (Gordon and Smith, 1955)

Cultures were streaked in a medium containing : Peptone — 5 g; Beef extract- 3 g; Agar — 15 g; Tyrosine — 5 g; Distilled water — 1000 ml at pH 7.0. Care should be taken to distribute insoluble tyrosine uniformly throughout agar. Plates were incubated at 37ºC to 28ºC and examined after 14 to 21 days of growth for the disappearance of crystals.

6. Paraffin utilisation (Schneidau and Shaffer, 1957)

Glass rods coated with paraffin were placed in tubes containing 5 ml of **Czapek's fluid** without carbohydrate so that 1 to 5 cm of the rod remained above the surface of the fluid. The test cultures were grown on beef extract agar (sugar free) medium and a small pellicle of growth was placed in Czapek's fluid. The tubes were placed in the incubator at 37ºC and the cultures observed for evidence of growth on the paraffin over 4 week period.

7. Growth in dilute (0.4 per cent) gelatine

The utilisation of gelatin as sole source of nutrition was shown by Bojalil and Cerbon (1959) in the differentiation of ***N. brasiliensis*** from *N. asteroides* and various Streptomyces species. The medium is produced by dissolving 4 g of gelatin in 1000 ml of distilled water and

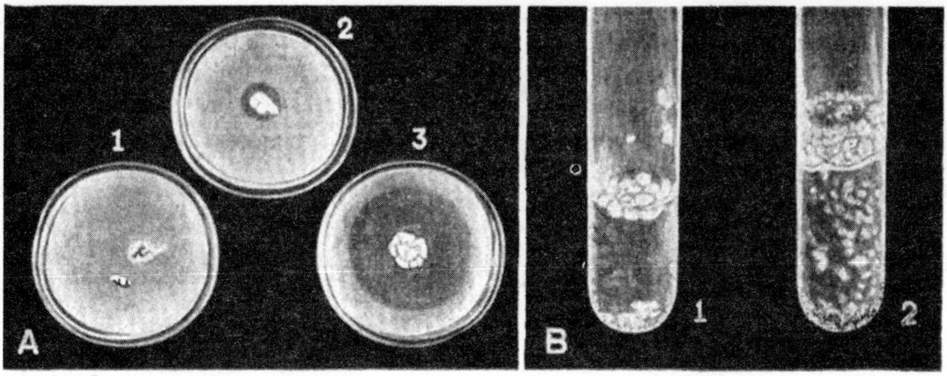

Fig. 12.1 : A — Casein plate for demonstration of casein hydrolysis (1) *N. asteroides;* (2) *N. brasiliensis* showing hydrolysis and (3) **Streptomyces showing rapid hydrolysis**. B(1 & 2). Growth of two strains of *N. brasiliensis.*

the pH is adjusted at 7.0 and sterilised by autoclaving. A small fragment of growth was inoculated into a tube containing 5 ml of medium. The tubes were inoculated at 30ºC and examined at intervals over a 25 days period. The amount and type of growth were recorded.

8. Sugar utilisation and acidification

The technique employed is a modification of the method employed by Mackinon and Artagaveytia-Allende (1956) and by George *et. al.* (1961). The medium used was one of Pridham and Gottlieba (1948) which contained the following.

$(NH_4)_2$ SO_4-2.6 g; KH_2PO_2-2.38 g; K_2HPO_4-5.65 g; trace mineral solution ($CUSO_4$, $MgSO_4$, $FeSO_4$, $MnCl_2$, $ZnSO_4$) — 1 ml; and 15 g of agar in 1000 ml of distilled water. In this modification by George *et. al.* (1961), 0.1 per cent **glutathione** was added to improve the rate of growth. **Bromothymol blue** and **Bromothymol purple** (0.1 g each in 20 ml of 95 per cent alcohol) were used as pH indicators. The ingredients were added to the medium before autoclaving. The test sugars were autoclaved separately in 10 per cent aqueous solutions. These were added aseptically to each tube of melted cooled agar to give one per cent concentration. The tubes were inoculated with fragments of growth (sugar free medium).

Tubes are read by noting the growth and acidification after 25 days, and this long time was necessary, since some strains required this length of time to acidify glucose tubes. Galactose and inositol may also be prepared for the test. ***N. brasiliensis*** is much more active in the production of acid than *N. asteroides* (Georg *et. al.*, 1961).

9. Ability to grow at different temperatures

A pin head sized inoculum is used to inoculate a set of 3 dextrose agar tubes. The growth was recorded after 10 days' incubation on a scale of 0 to 4 plus. ***N. asteroides*** strains develop profusely at 46°C; whereas, many strains of ***N. brasiliensis*** and saprophytic strains failed to do so (Georg *et al.*, 1961).

10. Animal pathogenicity test

For this 4 Sabouraud dextrose agar slants were heavily inoculated and incubated either at 28°Cor 37ºC depending on the temperature favourable for the growth. When heavy growth covered at least one half of each slant in five days to two weeks (depending on the strain) the growth is scraped off into a mortar with an equal volume of 5 per cent gastric mucin and the mixture was ground into a fine suspension.

One millilitre of this mixture was injected into each of two guinea pigs intraperitoneally. The animals that died were autopsied and examined for lesion for the presence of acid-fast mycelium. If neither of the guinea pigs died after 2 weeks, one is sacrificed and the second was held for four weeks before being sacrificed and examined. The test strain is considered pathogenic when lesions and acid-fast organisms in the lesion are demonstrated in the guinea pigs. The results suggest that ***N. asteroides*** is more virulent for guinea pigs than *N. brasiliensis*.

SYSTEMIC NOCARDIOSIS

Nocardia and streptomyces are free living aerobic organisms 1μ or less in diameter and grow readily on simple media, e.g., Sabouraud's glucose agar, over a wide range of temperature. An aerobic **Actinomyces are obligatory commensals in animals**; whereas, **Nocardia and streptomyces are inhabitants of soil.**

History : The first aerobic pathogenic actinomycetes was discovered by Nocard in 1889 in *farcin du boeuf* and later named *Nocardia farcinica* by Trevisan in 1889. However Eppinger in 1880 described the first pathogenic aerobic actinomycete found in human infection, pseudotuberculosis with brain abscesses and meningitis and the organism was called *Cladothrix asteroides* but later named as *Nocardia asteroids* by Blanchard in 1896. Recently Gordon and Mihm (1962b) reported that *N. farcinica* and ***N. asteroides*** are probably identical and the name of human species ***N. asteroides*** has been retained because of its importance in medicine.

In 1894 Vincent described ***Streptothrix madurae*** as the aetiological agent of actinomycetoma in cases of Madura foot in India. Since then only few species of *Nocardia* and *Streptomyces* have been reported in about 480 cases of mycetoma from different parts of the world (Abott, 1956; Mariat, 1963; Orio *et al.*, 1963).

Distribution : ***Nocardia asteroides*** which is free living in nature, has been isolated repeatedly from soil (Henrici and Gardner, 1921, Gordon and Hagan, 1936; Emmons, 1954, Ajello, 1956) and also as laboratory contaminant. *N. brasiliensis* was isolated from the soil in Mexico (Gonzalez-Ochoa and Sandoval, 1955). They probably exist as saprophytes in nature. From soil they are introduced into traumatised tissues to initiate localised chronic infection. Hence mycetoma is more prevalent in tropical and subtropical regions where prople are exposed to trauma and infections from the soil (Abbott, 1956; Mariat, 1963). However, systemic nocardiasis has a world-wide distribution and is found as a **rare *primary pulmonary infection*** which follows inhalation of *N. asteroides* (McQuown, 1955).

Epidemiology

Pathogenic aerobic acid-fast actinomycetes (Nocardia) are free living in nature and cause air borne infection from extraneous sources or by introduction into tissues through trauma. This view is supported by the study of systemic nocardiasis in which pulmonary infection is the rule and also in cases of trauma with subcutaneous infections which develop localised abscesses. The extent of subclinical nocardiosis is not yet known. However, dermal tests with the antigens mentioned before in large population group may reveal the occurrence of this self limited condition.

Many animals may develop **spontaneous nocardiosis, e.g., rabbit, cat, dog, goat, horse and cattle** are naturally affected (Ajello *et al.*, 1961). There is, however, no report of animal to man or man to man transmission.

Morphology

Nocardia are aerobic, weakly acid-fast actinomycetes whose vegetative mycelium readily fragments into bacillary and coccoid elements; on the other hand *Streptomyces* are aerobic nonacid-fast actinomycetes whose vegetative organism does not fragment but form conidia in chains from aerial hyphae. Both are Gram-positive.

Examination of several isolates by Gonzalez-Ochoa and Sandoval (1955) showed conidia formation in *N. madurae, N. somaliensis* and *N. pelletierii*. Gordon and Smith (1955) have demonstrated that physiological characteristics could be used for differentiation of either *Nocardia* or *Streptomyces*. Therefore on morphologic and physiologic criteria of the pathogenic aerobic actinomycetes are put into two groups, namely (1) ***Nocardia asteroids, N. brasiliensis***, *N. caviae* and ***S. madurae***, (2) ***Streptomyces somaliensis, S. pelletierii***, and *S. paraguayensis*.

Aetiology

N. asteroides is the recognised aetiological agent of systemic nocardiasis; whereas, the remaining ones, viz.; other **Nocardia and Streptomyces are aetiological agents of actinomycotic mycetoma**. In **systemic nocardiasis** the organisms are found in **pulmonary and cerebrospinal infections** and **subcutaneous abscesses**, and are **demonstrated in sputum, empyemic fluid, cerebrospinal fluid** and **in pus from subcutaneous abscesses as loose, scattered, branching, Gram-positive and\and acid-fast bacillary bodies** often **mistaken for tubercle bacilli**.

Other lesions : In actinomycotic mycetoma, however, the organism appears in tissue or exudate from draining sinuses in the form of yellowish-white or red, granules with or without clubs at the periphery.

Culture

Aerobic actinomycetes grow readily at 37°C and at room temperature in simple **solid media** like **beef infusion, glucose agar, Sabouraud's glucose agar, Czapek's agar**, etc. **They grow** slowly and should be incubated **for 3-4 weeks** till a typical appearance is obtained **to show giant colonies**. They show wide variations in gross appearance, texture and pigmentation in different media. Although most strains of *N. asteroides* and *brasiliensis* grow at room temperature, some strains of ***N. asteroides*** develop poorly or not at all at this temperature **(Fig. 12.2)**.

On solid media the growths are glabrous, wrinkled, granular resembling tubercle bacillus. Occasionally aerial mycelia grow, giving a chalky powdery appearance of the growth and on subculture this appearance may be lost. ***Pigmentation* is best produced in Czapek's agar and varies from cream to yellow or orange, ochraceous and pink to coral or brick red**. Sometime some sectors of growth are pigmented. The consistency may be soft or hard or moist and mucilagenous or dry and granular. Antibacterial antibiotics like chloramphenicol cannot be used, as ***N. asteroides*** and ***N. brasiliensis*** are sensitive when incorporated in isolation media.

On liquid media, they develop as wrinkled **surface pellicle** with the **medium clear**.

Fresh clinical material as well as concentrates may be cultured, although some of the routine digestion and concentration procedures used in tubercle bacillus to destroy pathogenic fungi, but in many instances *N. asteroides* survives.

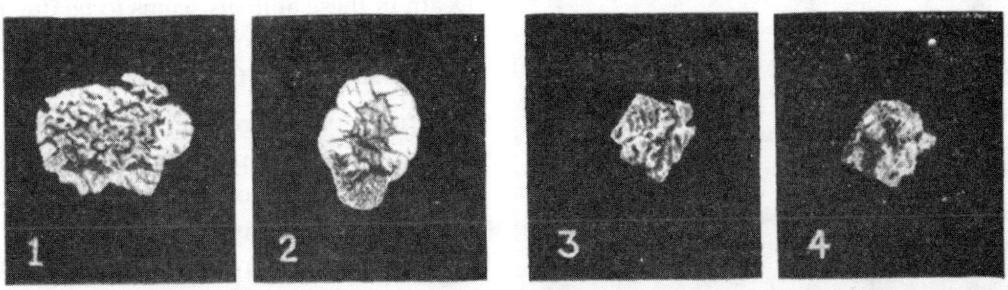

Fig. 12.2 : Growth on Sabouraud's glucose agar — 2 weeks old culture 1. *N. asteroides*. 2. *N. brasiliensis*. 3. *Str. madurae*. 4. *S. peilletierii*.

It is important to know that these species grow well in usual isolation media of tubercle bacilli. The colonies appear earlier (within 1 to 2 weeks) than *Myco. tuberculosis* and when incubated at 37ºC appear as moist glabrous growth in screw capped vials. They appear as "atypical mycobacteria". They should be studied by acid fast staining particularly by slide culture technique. **Nocardia are weakly acid-fast** and **show branching mycelia**. Therefore a searching study for colonies which develop early on tuberculosis media may be rewarding revealing pathogenic fungi which have survived the concentration method.

Biochemical reactions

Nocardia and streptomyces are so similar morphologically that it is necessary to compare **biochemical reactions** not only **for separation of genera Nocardia and Streptomyces,** but also **for separation of species** (Gordon and Smith, 1955). The biochemical reactions of aerobic actinomycetes remain constant when several strains of the same species are tested. According to Gordon and Smith (1955), genera streptomyces and Nocardia could be separated as follows :

Table 12.1 : Characteristic Biochemical Reactions in Nocardia and Streptomyces

	Nocardia asteroides	*N. Brasiliensis*	*Steptomyces*
Hydrolysis of casein	0	0	+
Decomposition or tyrosine	0	0	+
Growth in gelatin	0	+	+
Fermentation (acid production) :			
Lactose	0	0	+
Maltose	0	0	+
Xylose	0	0	+
Mannose	0	0	+

Animal pathogenicity

Nocardia asteroides is pathogenic to rabbit, guineapig and mice as described by *Drake and Henrici (1943). However, isolates of this species vary in their ability to produce infection (Waksman, 1943; Georg *et al.*, 1961). It is difficult to establish infection in laboratory animals which gives variable results with different isolates but suspension of the organism in hog gastric mucin apparently enhances the pathogenicity in guineapigs. When the organism is inoculated with such suspension of ***N. asteroides*** the animal develops abscesses to which it occasionally succumbs.

Rabbits : Injected **intravenously** with a large inoculum develop a generalised infection with miliary abscesses throughout the entire body. **Intramuscular and subcutaneous injections** produce localised abscesses.

Guineapigs : When injected intraperitoneally, they show diffuse peritonitis with formation of abscesses on the peritoneal surface. **Intraperitoneal injection of guineapigs** with large doses of virulent strains of *N. asteroides* may result in acute fatal peritonitis. The lesions are like those of bacterial peritonitis except that there is a tendency to formation of walled-off abscesses in the omentum and on the peritoneal surface of abdominal organs. The **pancreas frequently shows extensive abscesses. Lungs are usually congested and may show abscesses from extension through diaphragm**. But if the animal survives for a week or more the lesion appears to heal with recession of the inflammatory reaction. Death in these animals seems to be due to the toxic effects of the inoculated material than from massive invasion of the fungus.

Smears of pus from abscesses or from the exudate from the peritoneal surface reveal partially acid-fast organism with branching mycelia, or rods or cocci. From acid-fast mycelial form cultures are to be taken to obtain

---

*Pathogenic strains can be separated by additional tests (Mariat, 1962).

characteristic nocardia cultures. Strauss and Kligman (1951) demonstrated that addition of gastric mucin to the inoculum potentiates development of lesions.

Mice : **Injected intraperitoneally with *N. asteroides* mixed with 5 per cent hog gastric mucin** die with sufficient regularity and this may be used for testing chemotherapeutic agents (Strauss and Kligman, 1951). A rapid method of testing *N. asteroides* and *N. brasiliensis* is to inoculate mice **intravenously** with small inoculum followed by **intraperitoneal injection** with **0.5 ml of 5 per cent suspension of gastric mucin with the organism** (Mohapatra and Pine, 1963).

Occasional isolates of *N. brasiliensis* are also pathogenic for guinea pigs (Ajello *et al.*, 1961). They appear as Gram stained in smears of pus from the lesion. **Suspended in an adjuvant of paraffin oil or killed tubercle bacilli**, cultures of ***N. asteroides*** and ***N. brasiliensis*** produce granules (Destombes *et al.*, 1961).

Pathogenesis in man

Diseases caused by Nocardia are referred to as nocardiosis. The organism is often isolated from soil and hence there are two common modes of infection.

Systemic nocardiasis is caused by *N. asteroides* and chiefly is primarily of pulmonary origin (Murry *et al.*, 1961, Saltzman *et al.*, 1962 and Freese *et al.*, 1963).

The symptoms are headache, nausea and vomiting suggesting either a brain tumour or abscess or tuberculous meningitis, and with sputum, night sweats and loss of weight suggesting pulmonary infection; X-ray shows signs of infiltration leading to multiple cavity formation. Haematogenous transmission results in **METASTATIC LESIONS**.

1. **Pulmonary nocardiosis** arises from inhalation of the organism, *E.g., N. asteroides.*
2. **Chronic destructive *subcutaneous abscess*** called *mycetomas* form from contamination of skin with soil. The latter often affects feet and hands of labourers and cultivators.

Mycetoma : Different species are responsible to cause actinomycotic mycetoma in different parts of the world, *e.g.*, ***N. brasiliensis*** in Mexico. They form chronic subcutaneous abscesses clinically similar to those due to Streptomyces. *Actinomycotic mycetoma* is produced often due to introduction of the organism directly on wounds to cause localised mycetoma and subcutaneous tissues of the body may also be infected. The characteristic lesion is a swelling with pain followed by sinus formation and marked deformity of the part may develop after chronic infection for months or years. Infection spreads by extension to adjacent tissues with multiple abscesses and have destruction with no systemic reaction till secondary bacterial invasion is established.

N. brasiliensis and some Streptomyces cause mycetoma as mentioned in the Table 12.1.

Importance of *N. asteroides* isolated from sputum : Murry *et al.*, (1960) in a review of the problem of nocardiosis are of opinion that nocardiosis is frequently associated with underlying disorders of immune mechanism of the host with adreno-cortical steroid therapy. They suggest that the ability of ***N. asteroides*** to produce disease in a host with reduced resistance appears to be similar to those of ***Candida albicans, Aspergillus*** species and the antibiotic resistant bacteria, which take advantage of alteration of host's resistance and tissue.

C. albicans and *A. bovis* **are normal** components of the **microflora of the human body** and they are recovered from normal individuals with great frequency. It may be that *N. asteroides* normally exists in the saprophytic or commensal state but this is considered to be remote since in the series of Hosty *et al.* (1961) only 0.2 per cent of sputum specimens yielded *N. asteroides*. A further cultural survey of sputum, bronchial and gastric washing from individuals free from pulmonary diseases should be carried out ot determine whether *N. asteroides* is normally present in the human body.

Antigen : **A purified polysaccharide** obtained from ***N. brasiliensis*** has been **shown to be specific in an agar gel diffusion precipitin test with sera from patients** infected with this organism. No gross reaction occurred with sera from tuberculous patients, patients with mycetoma caused by ***S. madurae*** or with sera from rabbits infected with *N. asteroides* (Bojalil and Zamora, 1963).

Purified proteins obtained from *N. brasiliensis* showed delayed type of skin reaction in patients with mycetoma caused by the organism (Bojalil and Magnusson, 1963).

Diagnosis

A. *DIRECT EXAMINATION* of clinical materials like **sputum, cerebro-spinal fluid** and **pus** from subcutaneous abscesses are stained to demonstrate **Gram positive acid-fast, branching mycelium.**

B. *INDIRECTLY : Culture of sputum* : Blood agar plates and Sabouraud's glucose agar tubes are inoculated and incubated at 37ºC and also at room temperature. Although most strain of *N. asteroides* and *N. brasiliensis* grow well at room temperature; some strains of *N. asteroides* develop poorly or not at all at this temperature. So both 37ºC and room temperature and preferred to incubation.

Cerebrospinal fluid should be used stained by Gram's method. **Gram positive, acid-fast hyphae are seen.** For culture of sputum or pus antibiotic should not be added to media as the aerobic actinomycetes are sensitive and may fail to grow.

Tuberculosis media can be inoculated, where the **growth is obtained earlier (1-2 weeks) in case of** ***N. asteroides*** **and** ***N. brasiliensis.***

Treatment

Response to sulphonamide therapy is dramatic in systemic nocardiosis and a high concentration (15 to 20 mg per cent) should be maintained in the blood for 4 to 6 weeks during hospitalisation of the patient and a lower blood concentration should be maintained at least for months after the patient is discharged. Surgical drainage of cutaneous and other abscesses as well as empyema are often necessary for effective during therapy (Freese *et al.*, 1963).

In actinomycotic mycetoma the extent of infection determines the effectiveness of the therapy. In most cases excision of the affected part is an adjunct to chemotherapy. After administration of commonly used sulphonamides for a period of months. DDS and long acting sulphonamides like sulphadimethoxine or sulphamethoxy pyridiazine have been successfully administered (Mariat and Satre, 1961) in nocardiosis.

THE PATHOGLOGY OF SYSTEMIC NOCARDIOSIS

Introduction : *Nocardia* species cause frequently mycetomas, which have been discussed under **"Actinomycetomas"** (see Ch. 11). However, *Nocardia*, particularly ***N. asteroides*** may be responsible also for systemic infection. In contrast to mycetomas, there are usually **no bacterial colonies (granules) found in systemic nocardiosis,** but the organisms spread diffusely in the tissue, causing suppurative lesions. Systemic nocardiosis remains confined to the lungs in about half of the cases, whilst evidence of generalisation is found in the remainder. Involvement of the central **nervous system** is not a rare event. **Spleen, kidney, liver, heart, bone, joints, lymph nodes, adrenals, pancreas, thyroid, eye, ear, spinal cord** and **bladder** have all been found involved in the disease.

Frequently, there is **association with** other diseases like **leukaemia, Hodgkin's disease, lupus erythematosus, pemphigus, carcinoma,**

PLATE 12.I

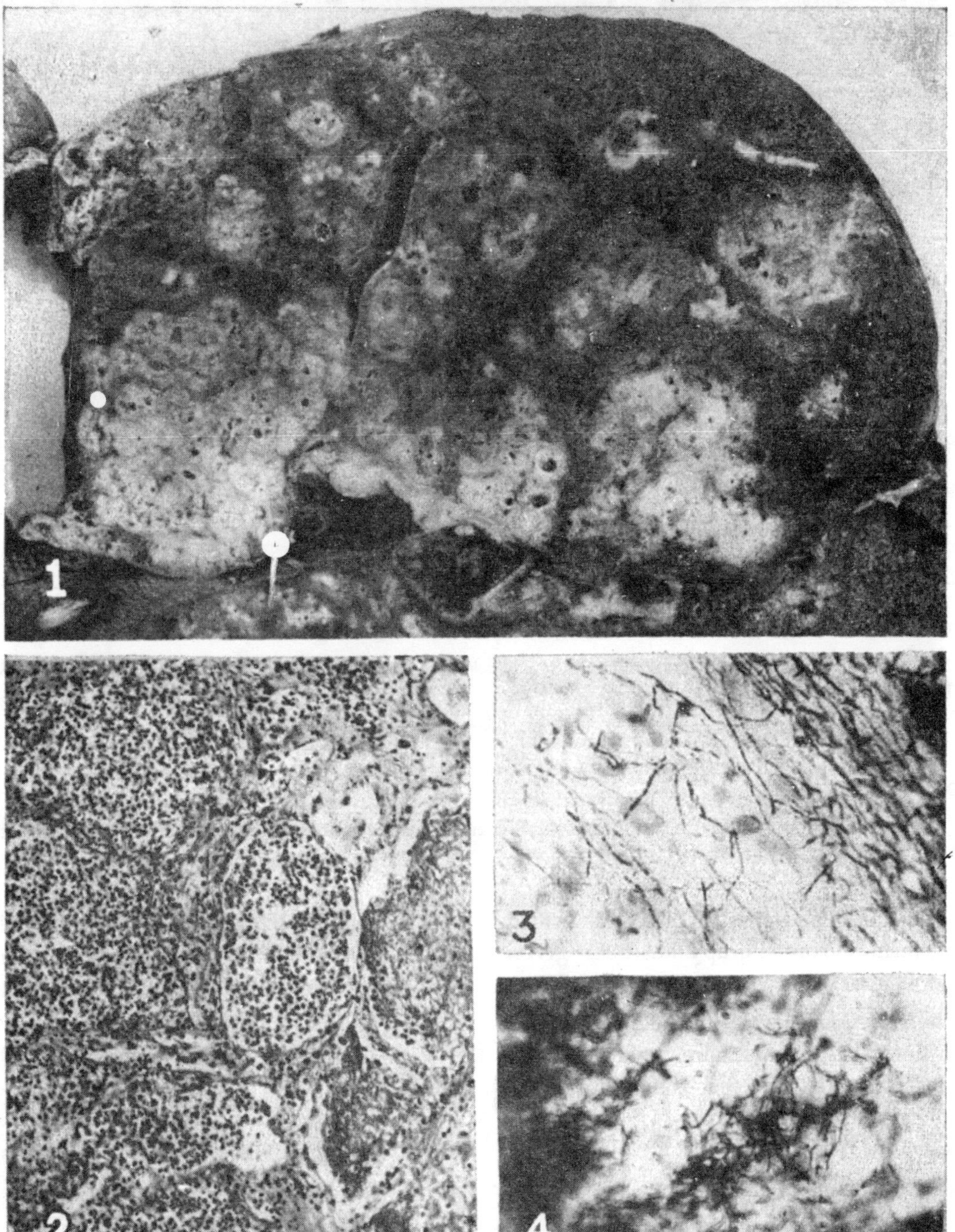

For legend see next page

PLATE 12.I (*contd....*)

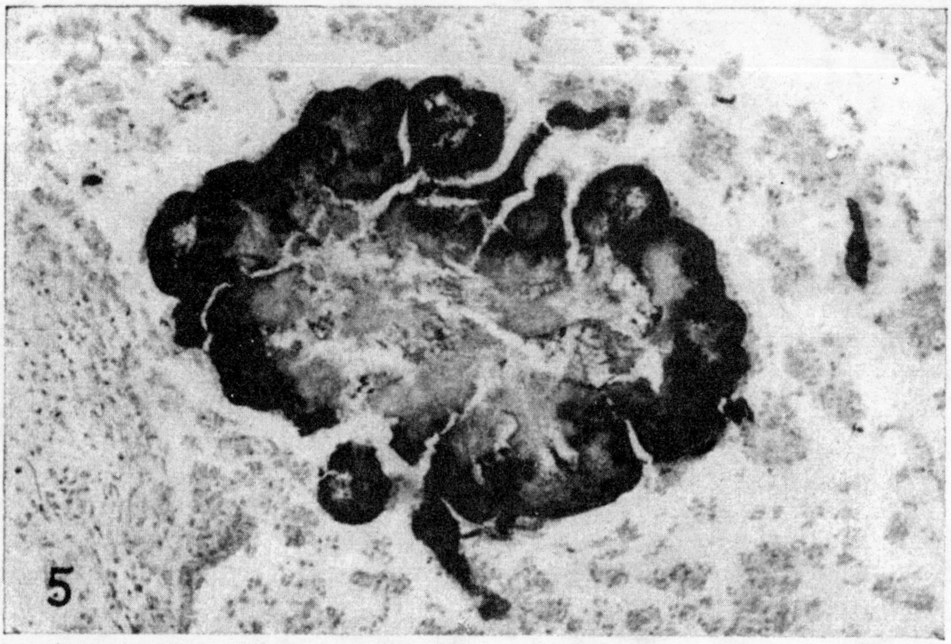

Fig. 1 : Pulmonary nocardiosis in a 60 year old patient. Showing greyish acinar confluent necrotising lesions with central abscess and cavity formation.

Fig. 2 : Areas of acute Pneumonia with polymorphonuclear and fibrinous exudate in pulmonary nocardiosis. H & E, 130 ×

Fig. 3 : Short branching filaments of Nocardia in histological section of pulmonary nocardiosis. Gram, 1000 ×

Fig. 4 : Fragmented Nocardia filaments in direct smear from sputum of a case of pulmonary nocardiasis. Gram, 1000 ×

Fig. 5 : Giant colony of *Nocardia* from a case of pulmonary Nocardiosis. Colony formation in systemic Nocardiosis is a rare occurrence. Gram. 130 ×

bronchiectasis and chronic disease in which long term steroid therapy has been given (Murray *et al.*, 1961).

Pulmonary Nocardiosis : *Nocardia asteroides* infection of the lungs is characterised by acute necrosis with abscess formation and widespread pneumonic or broncho-pneumonic lesions. **Pl. 12.I Fig. 1** shows the typical appearance of a lung with lesions due to *Nocardia* species.

The organ was from a 60 year old patient who had been treated for "**suppurative pulmonary disease**" for about a fortnight in the hospital. The lung showed considerable **adhesions of the pleura** including pockets filled with pus. The right lung showed **patchy nodular confluent consolidation with numerous small abscesses and areas of necrosis**. The bronchi were dilated, and contained mucopurulent material. The left lung in the periphery of the upper lobe showed a ragged cavity with thick fibrous wall, which was communicating with a localised pleural empyema. The remainder of the lung tissue again was studded with acinar, confluent, greyish areas of consolidation including small central abscesses. In other, more chronic cases, pulmonary lesions resemble clinically and pathologically tuberculosis.

Chronic cavities containing purulent necrotic material may be present. It has been variedly estimated that between 2 and 5 per cent cases in tuberculosis sanatoria, suffer actually from Nocardiosis. They are often those cases resistant to antituberculous therapy.

Histopathology

In the acute form, the histological picture may resemble lobar pneumonia, in areas. However, there is patchy distribution of necrosis and abscess formation **(Pl. 12.I, Fig. 1)**. Large fields with almost pure fibrin exudation may alternate with areas of more purulent exudate. The patchiness of the lesions is characteristic. Tuberculoid granulomas with Langhans' giant cells are rarely present in acute cases. In chronic infection, however, cavities may be surrounded by epithelioid granulation tissue, indistinguishable from tuberculosis.

Nocardia **organisms** are not recognisable in Haematoxylin and eosin sections. Gram stain may show gram positive branching filaments, which have a great tendency to break up into coccoid and bacilloid fragments **(Pl. 12.I Figs. 3 and 4)**. Fite Farace stain is most helpful because it often shows acid-fast filamentous branching organisms diffusely distributed throughout the wall of a cavity or in purulent intra-alveolar exudate. Though it was believed that large colonies are nerver found in pulmonary Nocardiosis, the **formation of giant colonies** (by some authors called "**fungus balls**") in abscesses, bronchiectases and cavities have been seen in rare cases (Kurup *et al.*, 1968; **Pl. 12.I, Fig. 5**).

Disseminated Nocardiosis

Contiguous spread of the infection from the lungs to the pleura, pericardium and to the wall of the thorax is not infrequently seen. Lesions are more diffusely spread as in Actinomycosis. If haematogenous dissemination takes place, numerous necrotic nodules and abscesses may be found in the internal organs, and in the subcutaneous tissue.

An intense **pyogenic reaction** usually accompanies pyogenic spread. Numerous nodules and abscesses in the subcutaneous tissue and in the internal organs may be formed. On the one hand the organisms cannot be demonstrated in tissue sections with Haematoxylin and Eosin stain while on the other hand, **not all strains are acid-fast** and even if they are, they may be mistaken for tubercle bacilli by the inexperienced workers because of their tendency to break up into **cocco-bacillary forms**. Proper cultural identification is, therefore, most desirable in doubtful cases.

Nocardiosis of the Central Nervous System : Lesions consist usually in solitary or multiple abscesses, not different in appearance from other abscesses caused by bacteria. In cases of **nocardial septicaemia, a leptomeningitis with thick creamy yellowish purulent exudate** may be present. Taking of culture material before fixation of the brain is most important.

Direct examination : Autopsy material from cavities, abscesses and sputum show short branching **filaments 1 micron in diameter**. Filaments are **Gram positive**.

REFERENCES

Abraham, S. Randhawa, H. S. Kurup, P. V., Damodaran, V. N., Basu, B. and Mazumdar, P. R. (1967). Pulmonary Nocardiosis. *Ind. J. chest. Dis.* **9**, 165-169.

Abbott, P. (1966). Mycetoma in the Sudan. *Trans. roy. Soc. trop. Med. Hyg.* **50**, 11-30.

Ajello, L. (1956). Soil as a natural reservoir for human pathogenic fungi. *Science* **123**, 876-879.

Ajello, L., Walker, W. W., Dungworth, D. L. and Brumfield, G. L. (1961). Isolation of *Nocardia brasiliensis* from a cat. *J. am. Vet. med. Assoc.* **138**, 370-376.

Beaman BL, Beaman L (1994) : *Nocardia* species : Host-parasite relationships. *Clin. Microbiol Rev.* **7**, 213.

Blanchard, R. (1896). Parasites vegetaux a l'exclusion des bacteries. *Traite de path. gen.* **2**, 811-926.

Bojalil, L. F. and Cerbon, J. (1959). Schema for the Differentiation of *Nocardia asteroides* and *Nocardia brasiliensis. J. Bact.* **78**, 852-857.

Bojalil, L. F. and Magnusson, M. (1963). Specificity of skin reactions of humans to *Nocardia sensitins. Am. Rev. resp. Dis.* **88**, 409.

Bojalil, L. F., Trujillo, A. and Cerbon, J. (1959). Differentiation bioquimica de algunas especies de actinomycetes pathogenos. *Mycopath. Mycol. Appl.* **11**, 287.

Bojalil, L. F. and Zamora, A. (1963). A precipitin and skin tests in the diagnosis of mycetoma due to *Nocardia brasiliensis. Proc. Soc. exp. Biol. Med.* **113**, 40.

Destombes, P., Mariat, F., Nazimoff, O. and Satre, J. (1961). A propos des mycetomes a Nocardia. *Sabouraudia* **1**, 161-172.

Drake, C. H. and Henrici, A. T. (1943). *Nocardia asteroid*. Its pathogenicity and allergic properties. *Am. Rev. Tuberc.* **48**, 184-198.

Emmons, C. W. (1954). The significance of saprophytism in the epidemiology of the mycoses. *Trans. N. Y. Acad. Sci.* **40**, 157-166.

Eppinger, H. L. (1890). Ueber eine Neue Pathogene Cladothrix and eine durch sie hercorgerufene Pseudotuberculosis. *Beitr. Z. path. Anat.* **9**, 287-328.

Freese, J. W., Young, W. G., Sealy, W. C. and Conant, N. F. (1963). Pulmonary infection by *Nocardia asteroides*. Findings in eleven clinical cases. *J. thorac. card. Surg.* **46**, 537.

Georg, L. K., Ajello, L., McDurmont, C. and Hosty, T. S. (1961). The Identification of *Nocardia asteroides* and *Nocardia brasiliensis. Am. Rev. resp. Dis.* **48**, 337.

Gonzales-Ochoa, A. and de los Angeles Sandoval, M. (1955). Caracteristicas de los Actinomicetes Patogenos mas comunes. *Rev. Inst. Salub. y Enferm. Trop.* **15**, 149.

Gordon, R. E. and Hagan, W. A. (1936). A study of some acid-fast Actinomycetes from soil with special reference to pathogenicity for animals. *J. infect. Dis.* **59**, 200-206.

Gordon R. E. and Mihm, J. M. (1962 a). Identification of *Nocardia caviae* (Erikson) nov. comb. *Ann. N. York Acad. Sci.*, **98 Art. 3**, 628-636.

Gordon R. E. and Mihm, J. M. (1962 b). The type species of the genus *Nocardia. J. gen. Microbiol.* **27**, 1.

Gordon, R. E. and Smith, M. M. (1955). Proposed group of characters for the separation of *Streptomyces* and *Nocardia. J. Bact.* **69**, 147-150.

Henrici, A. T. and Gardner, E. L. (1921). The acid-fast actinomycetes. *J. infect. Dis.* **28**, 232.

Hosty, T. S., Mcdurmont, C., Ajello, L., Georg, L. K., Brumfield, G. L. and Calix, A. A. (1961). Prevalence of *Nocardia asteroides* in sputum examined by a tuberculosis diagnostic laboratory. *J. lab. clin. Med.* **58**, 107-114.

Kurup, P. V., Sharma, V. N., Viswanathan, R., Sandhu, R. S., Randhawa, H. S. and Damodaran, V. N. (1968). Pulmonary fungus ball due to a *Nocardia* species. *Scand. J. resp. Dis.* **49**, 9-14.

Lerner PI (1996) : Nocardiosis. *Clin Infect Dis.* **22**, 891.

Mackinnon, J. E. and Artagavetia-Allende, R. C. (1956). The main species of pathogenic aerobic actinomycetes causing mycetomas. *Trans. roy, Soc. trop. Med. Hyg.* **50**, 31-40.

Mamelak AN et al. (1994) : Nocardial brain abscess: Treatment strategies and factors influencing outcome. Neuro-surgery **35**, 622.

Mariat, F. (1962). Criteres de determination des principales especes d'Actinomycetes aerobies pathogenes *Ann. Soc. belge med. Trop.* **4**, 651.

Mariat, F. (1963) Sur la Distribution Geographique *et la* Repartition des Agents de Mycetomes. *Bull. Soc. path. Exot.* **56**, 35.

McClung, N. M. (1960). Isolation of *Nocardia asteroides* from soils. *Mycologia* **52**, 154-156.

McClung, N. M. (1963). Nocardia and nocardiosis. *Acta Tubercul. Japon.* **13**, 1-13.

McNell MM, Brown JM (1994) : The medically important aerobic actinomycetes: Epidemiology and microbiology. *Clin. Microbiol Rev.* **7**, 357.

Mcquown, A. L. (1955). Actinomycosis and nocardiosis. *Am. J. cl. Path.* **25**, 2.

Mohapatra L. N. and Pine. L. (1963). Studies on the pathogenicity of aerobic actinomycetes Inoculated into Mice Intravenously *Sabouraudia* **2**, 176.

Murray, J. F., Finegold, S. M., Froman, S. and Will, D. W. (1961). The changing spectrum of nocardiosis. A review and presentation of Nine Cases. *Am. Rev. resp. Dis.* **83**, 315-330.

Murry, I. G., Spooner, E. T. C. and Walker, J. (1960). Experimental infection of Mice with Madurella mycetomii *Trans. roy. Soc. trop. Med. Hyg.* **54**, 335.

Nocard, M. E. (1888). Note aur la Maladie des Boeufs de la Guadeloupe Connue sous le Nom de Farcin. *Ann. Inst. Pasteur* **2**, 293-302.

Orio, J., Destombes, P., Mariat, F. and Segretain, G. (1963). Les mycetomes en cote francaise des Somalis. Revue de 50 cas observes entre 1954 et 1962. *Bull. Soc. path. Exot.* **56**, 161.

Pier, A. C., Grady, D. M. and Fassatti, M. J. (1958). *Nocardia asteroides* - a newly recognized pathogen of the mastitis complex. *Am. J. vet. Res.* **19**, 319-331.

Runyon, E. H. (1951). *Nocardia asteroides* : Studies of its pathogenicity and drug sensitivities. *J. lab. clin. Med.* **37**, 713-720.

Saltzman, H. A. Chick, E. W. and Conant, N. F. (1962). Nocardiosis as a complication of other diseases *Lab. Invest.* 11 : **part 2**, 1110.

Schneidau, J. D. and Shaffer, M. F. (1957). Studies of nocardia and other actinomycetales. I. Cultural studies. *Am. Rev. Tuberc.* **76**, 770.

Strauss, R. E., Kligman, A. M. and Pillsbury, D. M. (1951). The chemotherapy of actinomycosis and nocardiosis. *Am. Rev. Tuberc.* **63**, 441-448.

Trevisan, V. **I** genere e le species delle bacteriaceae, Milan, 1889.

Waksman, S. A. and Henrici, A. T. (1943). The nomenclature and classification of the actinomycetes. *J. Bact.* **46**, 337.

13

Study of Blastomycetes

(Yeast and Yeast-Like Fungi : Cutaneous and Systemic Mycosis)

General consideration : Yeast-like fungi in this book refer to a group of Imperfect fungi having morphological and physiological characters of true yeasts of **Saccharomyces that show sexual cycle** or **perfect stage**. All the fungi in this group show creamy, large, soft, colonies. **Morphologically**, they are unicellular organisms with budding but some of them may produce mycelia in addition to the budding form. The imperfect stage of the fungus is concerned with pathogenesis in medical mycology and these are **Cryptococcus, Pityrosporum, Trichosporon, Candida** and **Geotrichum** and the **Dermatophytes**.

GENUS : *CRYTPOCOCCUS*

Blastopores are **unicellular, oval or round yeast-like organisms** which **reproduce by budding (blastospores)**. The buddings are pinched off from the "mother" cell. **Unlike Saccharomycetes**, they **do not produce ascospores** and hence **belong to IMPERFECT FUNGI**. They **do not form** any other **developmental form** like **PSEUDOMYCELIUM. WHERE AS IN CANDIDA, ELONGATED BUDDING YEAST CELLS LOOK LIKE HYPHAE, BUT THEY ARE NOT DETACHED.**

GENUS : *PITYROSPORUM*

Pityrosporum species **reproduce by blastospores**, which unlike the budding of cryptococcus are **cut off** from the mother cell by the **development of cross wall or septa**. The subject has been dealt in **Ch. 10, "Superficial Mycoses"**.

GENUS : *CANDIDA*

Candida species reproduce by **yeast-like budding cells** but they also show formation of **pseudomycelium**. These pseudo-mycelia are chains of elongated cells formed from buds and the **buds elongate without breaking of the mother cell**. They are very fragile and separate easily. **Mycelia also form** by the **elongation of the germ tube** produced by a mother cell.

GENUS : *Trichosporon* : These fungi are characterised by formation of true mycelia which break into small fragments called *arthrospores*. Members of this group are better known to show formation of true mycelia forming arthrospores which are rectangular at first and they eventually round up and **look like blastospores** without showing buds.

CLASSIFICATION (BLASTOSPORINAE)

CLASS : **Deuteromycetes** (Hyphomycetes) (Fungi Imperfecti)

ORDER : Pseudosaccharomycetales.

FAMILY : **Cryptococcaceae.**

A. SUBFAMILY

Cryptococcoideae — show unicellular budding cells.

I. **GENUS : CRYPTOCOCCUS : Reproduce by blastospores** which are pinched off from the mother cells without any separating wall.

II. GENUS : **PITYROSPORUM : Reproduce by blastospores** that are **cut off** from the mother cell **by** formation of a **cross wall**.

B. SUBFAMILY

Candidiodeae — Unicellular budding forms that may form pseudomycelium or true mycelium.

I. **GENUS : CANDIDA** : They reproduce by budding ***blastospores*, PSEUDOMYCELIUM*** or true mycelium and show **chlamydospore formation**.

II. **GENUS : TRICHOSPORON : Reproduce by blastospores and arthrospores**. They may form pseudo-mycelium and a true mycelium.

III. **GENUS : GEOTRICHUM : Reproduce, by arthrospores** only. They may form **true mycelium** that **readily fragments into arthrospores**.

MORPHOLOGY OF CANDIDA AND OTHER YEASTS (PL. X)

1. *C. krusei*	**Crossed stick** appearance of **mycelia**.
2. *C. parapsilosis*	**Blastospores** often at **septate ends of cells**.
3. *C. tropicalis*	Well developed **mycelia** with numerous **blastospores**. No chlamydospores.
4. *C. guillermondi*	**Mycelium well developed with** small clusters of **blastospores** at septa.
5. *C. pseudotropicalis*	**Mycelium poorly developed**; chlamydospores absent.
6. *C. stellatoidea*	**Ball-like mycelia** with **blastospores**.
7. *C. albicans*	**Tree-like mycelia** with **blastospores** and **chlamydospores**, which do not develop at 37ºC. **GERM TUBE FORMATION WITH BLASTOCONIDIAL GERMINATION OCCURS AT 37ºC FOR 3 HOURS WHEN THE CULTURE IS MIXED WITH HUMAN SERUM.**
8. *Geotrichum candidum*	True **mycelia and oidia** formation.
9. *Trichosporon cutaneum.*	**Mycelia, blastospores** and **arthrospores**.
10. *C. viswanathii*	

*Fungi with yeast cells grow partly as yeasts and partly as elongated budding cells (see Plate 13.1 Fig. 3; 13.3A) which are joined end to end to form **pseudomycelium**

METHODS OF STUDY FOR IDENTIFICATIONS OF YEAST-LIKE FUNGI

(Cryptococcus, Candida, Geotrichum and Trichosporum)

DIRECT FINDING

Wet cover slip preparation : A coverslip mount is prepared with saline, lactophenol cotton blue or McGuire stain. The mount is examined under high power objective for the presence of unicellular budding cells, pseudomycelium or true mycelium. If mycelium is not present, determine by India ink preparation if the unicellular yeasts have capsules.

Crypt. neoformans shows a **broad capsule.** It will grow at 37°C and pathogenic to mice.

If mycelium is present, determine whether it is accompanied by **blastospores (Candida)** or **arthrospores (Geotrichum)** or **both in the case Trichosporon.**

INDIRECT METHODS

Culture : Isolation of pure cultures without contamination are necessary for proper biochemical tests.

A. **ISOLATION OF YEASTS IN PURE CULTURE.**

1. **Inoculate** a tube of Sabouraud's glucose broth **(SDA)** and incubate at 37°C over night.
2. **Shake and inoculate a blood agar** plate by streaking.
3. **Fish out single colonies** of different types and transfer to Sabouraud's dextrose agar slants for further study. Incubate at 37ºC over night.

B. **Isolation of pure yeast colonies** without bacterial contamination.

The above method may fail and the following may be followed : subculture original isolate in a Sabouraud's glucose agar plate containing 0.05 mg/ml of chloramphenicol and incubate at 37ºC over night. A pure culture of yeasts may occur.

STUDY OF PURE CULTURE

Make a microscopic study in a cover slip preparation and observe :

(a) **If no mycelium** — it may be cryptococcus, candida or a true yeast and follow the following steps.

(b) **If** mycelium is present determine if **blastospores and/or arthrospores** are produced.

(c) If both **blastospores and arthrospores are produced** the fungus may belong to **Trichosporon**, but when arthrospores alone are produced, the fungus is probably ***Geotrichum candidum.***

IDENTIFICATION OF CANDIDA ALBICANS

STUDY OF CHLAMYDOSPORES FOR CANDIDA ALBICANS. Pour 25 ml of chlamydospore agar medium in a 10 cm plate or corn meal agar with 1 per cent Tween 80. Inoculate by plating out with a thick nicrome wire loop (gauge 22) and cut the agar at some places at 45° angle. The cut must extend across the plate and care must be taken not to have too thick an agar layer.

A *control* cut must be made an each plate with a known *Candida albicans* organism to ensure that the medium is working properly and correct incubation temperature is utilised.

Four or five test cultures can be inoculated (cut) to conserve plates. Incubate the plates 25°C or room temperature for 2-3 days. When the control steaks show formation of chlamydospores, slide mount of the submerged growth are made from the edge of the inoculated streaks.

In case **chlamydospores fail to develop** (as morphologically characteristic of *C. albicans*), **carry out fermentation and assimilation studies**. Report the unknown as *C. albicans*, if it shows the morphology and chlamydospore formation.

If no **mycelium forms** and it is **not *Crypt. neoformans*** report it as **"nonpathogenic yeast"**.

If the **morphology is that of candida species and it is not *C. albicans, i.e.,* do not show chlamydospore formation in chlamydospore agar**, report as **candida species.**

FOR IDENTIFICATION OF *Crypt. neoformans*, a combination of the following tests is required.

1. Ability to grow at 37°C,
2. Demonstration of a capsule,
3. Urease test positive,
4. Nitrate assimilation test negative,
5. Animal pathogenicity test positive.

PLATE 13.I

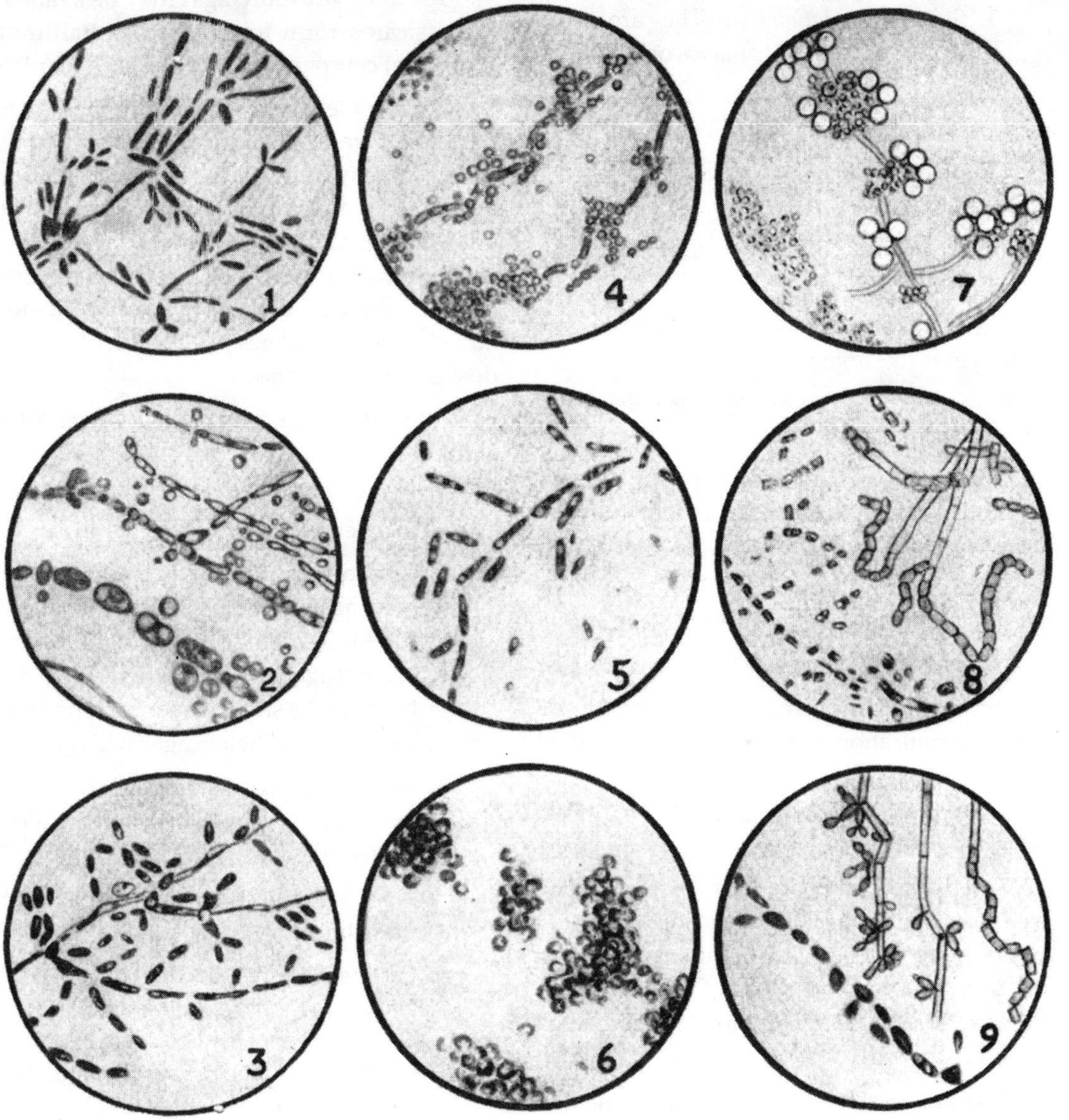

1. C. *krusei*
2. *C. parapsilosis*
3. *C. tropicalis*
4. *C. guillermondii*
5. *C. pseudotropicalis*
6. *C. stellatoidea*
7. *C. albicans*
8. *G. candidum*
9. *Trichosporon*
10. *C. viswanathii*

The other candida species are *C. lipolytica, C. lusitaniae* and *C. rugosa.*

TECHNIQUES FOR IDENTIFICATIONS OF YEAST-LIKE FUNGI

1. *Sugar fermentation tests* : They are useful in identification of species of candida.
2. *Sugar assimilation tests* : They are useful for C. *albicans* of 25°C.
3. *Nitrate assimilation test. Crypt. neoformans* does not assimilate nitrate.
4. *Urease test* : It is positive for *Crypt. neoformans.*

Except the above, study for the morphology of the yeast-like organism.

Growth at 37°C must occur specially in case of *Crypt. neoformans.*

Growth at 25ºC is necessary for demonstration of chlamydospores in Chlamydospore agar in case of *C. albicans.*

Sugar assimilation test : It is useful for identification of yeasts and yeast-like fungi.

Basal medium : Difco yeast nitrogen base.

According to the direction on the bottle, use 10 × concentration (6.7 g/100 ml of distilled water) — sterilise by filtration. Store in vaccine vials and keep in the refrigerator to be used as needed.

Test plates are prepared as follows : Prepare a dilute suspension of the organism to be tested in sterile saline. Place 2 ml of the suspension in a 10 cm plate and add 1.5 ml of yeast nitrogen base and 13.5 ml of **2 per cent melted agar** cooled to 50°C and swirl.

Test proper : Place sterile, flame-warmed antibiotic assay cups in different areas of the plate at well separated distance (5 or 6 may be placed in a 10 cm Petri dish). Fill cups with 20 percent solution of test sugar (carbohydrate) which are to be tested. Label on the back of the plate. Incubate at 25°C or room temperature. Observe the growth around the cup of various sources. Usually **dextrose, galactose, lactose, sucrose, melibiose** are usually put up.

A pathogenic *C. albicans* shows glucose, galactose, melibiose and sucrose assimilation as shown by the growth around the cup.

(3) **Chlamydospore agar** may be available in dessicated form from **Difco** or **Baltimore biological company** and may be used for plate.

(2) **Corn meal agar** can be combined with 1 **per cent Tween 80**. Corn meal agar without Tween 80 may be available in dessicated form and 1% Tween 80 may be added before autoclaving. **Corn meal agar is an excellent medium** for the study of the morphology of candida. The addition of 1% Tween 80 (before autoclaving) stimulates production of chlamydospores of *C. albicans.*

Slide culture for morphology of yeast-like cultures

1. Place a slide on a **'U' shaped glass rod** in a Petri dish; cover and sterilise.
2. Take a sterilised tube containing corn meal agar preferably with 1% Tween 80. Melt it in a water bath and pour a thin layer of agar on the slide in the Petri dish holding the slide in a slanting position. Allow to cool, when agar will set.
3. **Yeast culture is inoculated** by a nichrome wire by two cut streaks along the length of the slide.
4. Place a sterile cover slip over a part of these lines.
5. Let the slide rest on the U shaped rod and a thin layer of sterile water is poured at the bottom of the Petri dish to prevent drying of the agar. Cover the Petri dish; *the culture is **incubated at 25°C**.*
6. **After 2-5 days**, study the slide under the microscope after removing it from the Petri dish. The back portion of the slide may be cleared with a sterile clean gauze. Parts of inoculation lines both under and beside the coverslip are studied. If the culture has not developed sufficiently, replace it in the moist chamber.

Nitrate assimilation test

This is necessary for the identification of **different species of Cryptococcus**. Whereas, ***Cr. neoformans*** and some other nonpathogenic sprecies **do not assimilate nitrate, *Cr. albidus*** and some other species assimilate nitrate.

TECHNIQUE

Basal medium

Yeast carbon base (Difeo) Make 10 × concentration of the base according to the direction of the makes (16.7 g/ 100 ml distilled water). Sterilise by filtration, Store in vaccine bottle in the refrigerator for use when needed.

Test plate

Make a dilute suspension of the organism to be tested in sterile saline. Place 2 ml of the suspension on the plate and then add 1.5 ml of yeast carbon base and **13.5 ml of melted agar at 45° to 50°C and swirl.**

Test method

Place flamed antibiotic assay cups in the centre of the plate pressing slightly on the agar and **sealing the bottom with agar**. Fill up the cup with **1 per cent solution of KNO_3**.

Incubate at 25°C. Observe if **growth** has occurred around the cup **after 24 or 48 hours**.

Cr. neoformans, Cr. laurentii and *Cr. lutedus* do not assimilate nitrate; whereas, *Cr. albidus, Cr. diffluens* and *Cr. terreus* assimilate nitrate. A few strains of *C. terreus* may not assimilate nitrogen.

UREASE TEST

This test is utilised for differentiating cryptococcus from other yeast-like organisms, the former is able to split urea.

For urease test **CHRISTENSEN'S MEDIUM** is used. This medium consists of an agar base to which **sterile urea solution (sterilised by filtration)** is added.

Agar base

Peptone — 6g, sodium chloride — 5 g, dipotassium hydrogen phosphate (K_2HOP_4) — 2 g, phenol red-0.012 g (1 : 500 aqueous solution — 6 ml), agar — 20 g, distilled water — 1000 ml, pH adjusted to 6.8-6.9. Sterilise the basal medium by autoclaving **at 121°C for 15 minutes**. Cool at 50°C and add sterile solution of glucose to give a final dilution of 0.1 per cent and 100 ml of a 20 per cent solution of urea, previously sterilised by filtration is added. The **medium is tubed as deep slopes**.

Difco urea agar base may also be used as dehydrated powder. For this 29 g of urea agar is dissolved in 100 ml of distilled water and sterilised by filtration.

Dissolve 15-20 g of agar in 900 ml of boiling distilled water sterilise by autoclaving for 10 minutes at 120°C. Cool at 50°C and add 100 ml of filter-sterilised urea agar base. Mix and tube as deep slopes.

Urease test

Streak the Christensen's urea agar slope with the unknown yeast-like organism. Incubate at room **temperature of 25°C. A positive result** is indicated by the appearance of **deep red (on majenda colour) colour spreading throughout the medium**. The tubes are rejected after 48 hrs as negative if no red colour develops.

Cryptococcus always gives a positive urease test and is, therefore, useful in screening and identifying cryptococci (***J. Bact.* 72**, 121-136, 1956).

Sabouraud's dextrose agar

This is meant for **isolation of pathogenic fungi**.

Composition :		
	Dextrose	40 g
	Peptone	10 g
	Agar	20 g
	Distilled water	1000 ml

Adjust **(acid)** pH to 5.8 to 6; may be obtained commercially in a dehydrated form.

Preparation

it is obligatory that the medium should not be overheated or remelted. Distribute media in tubes. Best way of sterilising in tubes (pH 5.8-6) is to raise the pressure to 121ºC and then reduce the gas supply so as to stabilise the temperature to 110ºC, when it is sterilised for 10 minutes. **Overheating or remelting** makes the **agar surface soft and useless for culture.**

Selective media

Sabouraud's cycloheximide chloramphenicol media or similar dehydrated media are available as Mycobiotic agar of **Difo or Mycosel agar of Baltimore Biological Co**. for selective isolation of pathogenic fungi inhibiting contaminants. The medium is used often for routine isolation of pathogenic fungi.

Composition

(1) Sabouraud's dextrose agar with cyclohexamide. **Acti-dione (Upjohn company, Kalamazod, Michigan)** 0.5 mg/ml and chloramphenicol (Chloromycetin, Parke Davis & Co., Detroit, Michigan) 0.05 mg/ml.

For preparation of 1 litre

1 litre of Sabouraud's agar medium is put in a boiling bath before autoclaving. Add 50 mg of chloramphenicol suspended in 10 ml of 95 per cent alcohol to the boiling medium.

Add cycloheximide solution (500 mg in 10 ml acetone). Mix well and distribute in tubes. Autoclave as for Sabouraud's medium. Stand and allow to harden.

Pathogenicity test for Crypt. neoformans: Obtain a saline suspension of culture from a 4-5 days old culture **in Sabouraud's dextrose agar** approximately in a suspension of 1 : 100.

Inoculate into 2 or 4 mice intracerebrally with 0.2 ml of the suspension using a ¼ inch 26 gauge needle. Mice often die in 2-3 days. The skull of the inoculated mouse may be swollen. **Autopsy** is undertaken when the mouse dies or sacrifice mice at the end of two weeks in case death does not occur in this time. **Remove the top of the skull** and examine **Gelatinous material from brain**, which is **PREPARED IN A DROP OF INDIAN INK FOR THE PRESENCE OF ENCAPSULATED YEAST-LIKE CELLS**.

14

Cryptococcosis

Blastomycetes (Systemic Mycosis)

Synonyms : Torulosis, European blastomycosis, Busse-Buschke's disease.

Introduction

CRYPTOCOCCUS NEOFORMANS causes a variety of lesions in human beings. As a rule, the lungs are primarily affected. The infection is highly dangerous, because of the tendency for **dissemination**, particularly **to the brain** and meninges, and also **other organs**.

Definition

It is an infectious disease of human and animals caused by **Cryptococcus neoformans**, which is the **only pathogenic member of the genus**. The most dangerous and the common lesion is **cryptococcal meningitis**. Other sites of infection are **respiratory tract** and less commonly **skin**, bone, lymph node, etc.

History

Although Senker (1864) described an yeast like organism which could be this fungus, the first authentic isolation of the fungus was made by Busse (1894); Buschke (1895) described the organism in details. Benham (1934) confirmed from further study of the original culture of Busse that the organism was *Cryptococcus neoformas*. Stoddard and Cutler (1916) made detailed study of pathology and clinical characteristics of the fungal infection. Emons (1955) noted the close association of the fungus with **PIGEON DROPPINGS**.

Geographical distribution

The infection is **world-wide** in distribution. In India, cryptococcosis appears to be much more common than was generally assumed. More than one hundred fifty cases of cryptococcosis were reported from this country. Mohapatra (1969) reviewed the publications from India. Further cases were recorded by Bhaktaviziam *et al.*, 1969; Klokke, 1967; and Ichaporia *et al.*, 1970. Since the introduction of **ROUTINE INDIA INK PREPARATION OF CEREBRO-SPINAL FLUID** by Profesor **Grueber**, two **(unpublished)** cases of cryptococcal meningitis were discovered and **mycologically confirmed** by Martin and Mohana Kumar. Geographically the climative factor is probably negligible. Basu Malik *et al.* (1967) in an epidemiological study showed that the symptoms occurred mostly during the rainy-warm season. Although cryptococcosis usually affects persons between 40 and 60 years of age, out of 20 patients in Calcutta, there were eight cases only of 10 years age or below. About **half of the patients had contact with pigeons and other birds**. The disease was reported from different parts of India (Khan *et al.*, 1959; Basu Malik *et al.*, 1967).

Epidemiology : The organism is ubiquitous in distribution in nature. ***Cryptococcus neoformans*** has been isolated from soil (Senfelice, 1894), soil (Emmons, 1951), pigeon droppings (Emmons, 1954). Source of infection in many cases has been traced to pigeon or other bird droppings, though in most cases such corelation is absent. It has been shown that the rich creatinine content of the bird excreta is responsible for the growth of the fungus (Staib, 1964).

General characteristics of cryptococcosis and CRYPTOCOCCUS

1. Cryptococcal lesions may be entirely localised, mostly **affecting lungs** and also the skin and mucosae. If localised lesions appear to be tumour-like, they are called a *CRYPTOCOCCOMA*.
2. Any localised lesion is a potential focus of dissemination which is extremely dangerous, even if symptoms of spread may not be evident for several years. The majority of cryptococcal infections terminate in cryptococcal meningo-encephalitis which may linger for years without causing more symptoms than a headache.
3. Any seemingly localised cryptococcal focus other than in the skin or in the lung indicates in most cases that dissemination has already taken place even if there are no general symptoms to indicate this clinically.
4. The tissue reaction to *Cr. neoformans* is often extremely poor. Pus formation is usually absent. Foci have often a gelatinous appearance due to abundant production of capsular substance. Localised cryptococcal granulomas are not infrequently mistaken for myxomas, chondromas and other tumours. These circumscribed crypto-coccomas may exist for years without spread.
5. Many patients affected by cryptococcosis suffer from mother debilitating diseases like Hodgkin's disease, myeloprolife-rative or lymphoproliferative disorders, tuberculosis, or sarcoidosis.

ORGANISM : STAIN and Reproduction

Cryptococci are Gram positive; stain well with **P A S** (Periodic Acid Schiff's) stain, and **silver methenamine** (GMS — Gomori's Methenamine Silver) stain. They reproduce entirely by a small, narrow neck.

Aetiology : The common portal of entry is through respiratory tract. Cutaneous infection may occur from direct contact. The disease, has been reported at all ages from infancy to old age. Men are more frequently affected that females, probably because of the greater chance of exposure. People from lower socio-economic strata suffer from this infection more frequently.

Although in many cases no predisposing factor is detectable, certain diseases of the lymph node like Hodgkin's disease, lymphosarcoma or sometimes prolonged steroid therapy may predispose to this infection. Normal human serum contains some substance with inhibts the growth of *Cr. neoformans*. It is lacking in the serum of the patient.

Pathology : In cryptococcal meningitis, the surface of the brain is covered with a thin mucoid opalescent exudate. One may find small gelatinous cysts in the cortex just beneath the surface. Pulmonary lesion may be solitary granulomatous mass usually in the lower lobe. Less often the lungs may be studied with multiple small discrete nodules or cavities lined with mucoid material. Large cavitation is uncommon. Calcification never occurs in pulmonary lesions.

In histological examination of the sections from different organs, the fungi can be easily demonstrated even by routine haematoxylin and eosin preparation. Fungi are more clearly stained by PAS or mucicarmine stain. One interesting feature of the cryptococcal lesion is the absence of any inflammatory tissue reaction round the organism in many cases. Occasionally granulomatous lesion with foreign body type of giant cells and histiocytes are detected.

CLINICAL FEATURE

Clinical manifestations depend on the site of infection. Patients with intracranial lesions show signs of **CHRONIC MENINGITIS**, like tuberculous meningitis with which condition it is often confused. **Low grade pyrexia, headache, vomiting** are the common symptoms that develop insidiously. **Drowsiness** and even **coma** are late features. Another group of patients present signs and **symptoms of space occupying lesion**.

Cerebrospinal fluid is under tension. Lumbar puncture often relieves the headache which, however, recurs within a few days as more fluid accumulates.

CHEMICAL EXAMINATION of C S F shows **slight increase of protein**, and **reduction of sugar and chloride**.

CELL COUNT is **increased** moderately with **relative** of lymphocytosis.

India ink preparation of C S F shows the fungus with its capsule is present as a halo (**Fig. 14.1a**).

Patients with **PULMONARY CRYPTOCOCCOSIS** may be asymptomatic, and diagnosed from sputum examination.

The patients complain of **cough, low grade fever, pain chest, expectoration of mucoid sputum**, etc.

Chest Skiagram shows varying pictures depending on the type of pulmonary lesion (**Pl. 14.I, Fig. 5**). The commonest appearance is a rounded shadow usually in the lower lobe without any enlarged hilar shadow. Less common radiological findings are **diffuse miliary shadows**, or **areas of consolidation**. Unlike histoplasmosis, **calcification** of the healed lesion is **uncommon**. Repeated sputum examination will give the clue to the diagnosis.

Lesions elsewhere in the body shows symptoms and signs related to of those sites. **CUTANEOUS LESION** has the signs of a chronic ulcer.

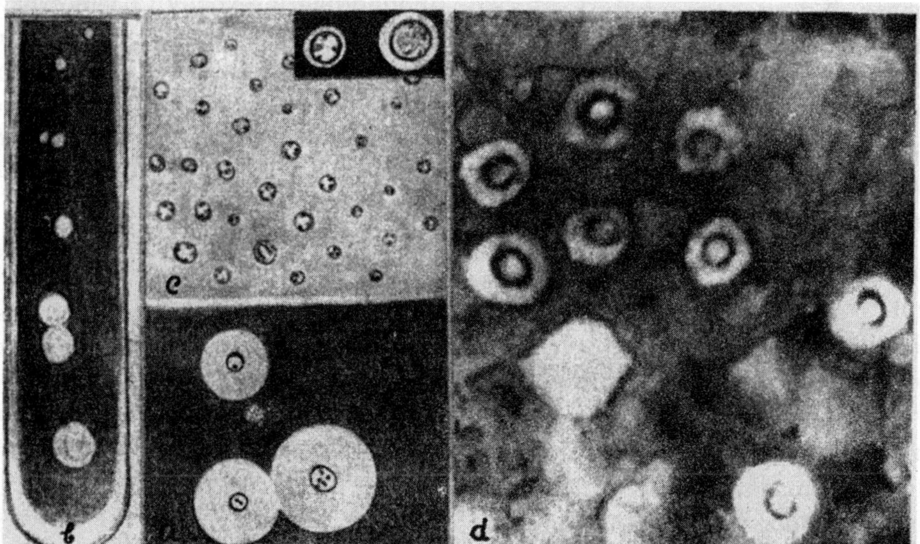

Fig. 14.1(a) : Cerebrospinal fluid showing **capsules** of ***Cr. neoformans in Indian ink preparation.*** (b) Culture on glucose agar. (c) Yeast form in culture (saline preparation). Inset same in Indian ink preparation. (d) Budding forms as seen in section of the brain

PLATE 14.I

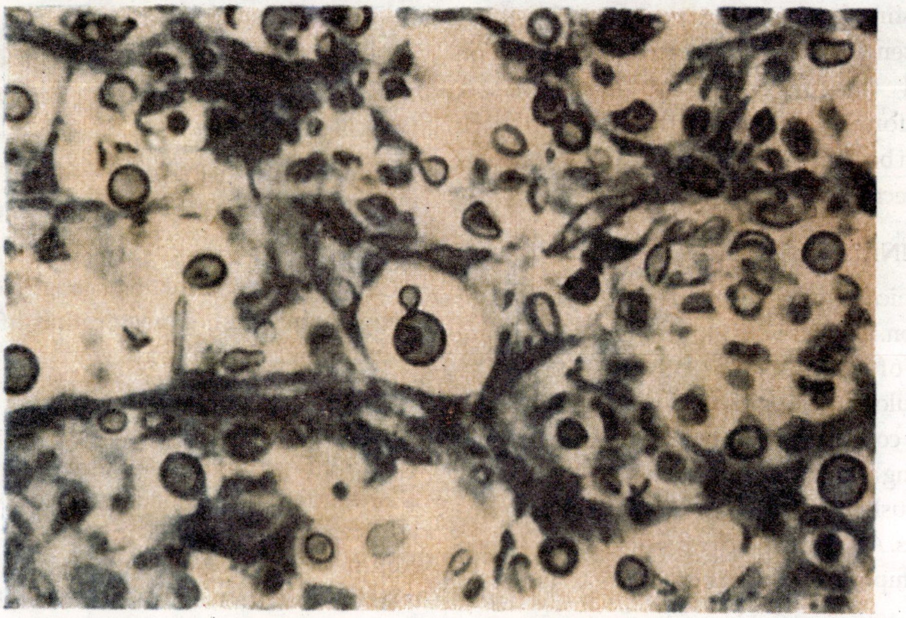

Histological appearance of cryptococcosis of the mouse spleen, Budding *Cr. neoformans* is seen in the centre of the plate. Silver methenamine stain; × 1000

PULMONARY CRYPTOCOCCOSIS

Gross and histopathological lesions : In some cases the infection is silent, leaving no trace of the disease and spontaneously healing. Localised ***PULMONARY*** *cryptococcoma* consists of **circumscribed subpleural, pulmonary** or **mediastinal tumour-like lesions** composed of **WELL DEFINED GELATINOUS MASSES. The size may be up to 10 cms** and lesions are sometimes ***multiple***.

Histologically, the centre consists of masses of irregularly spherical **fungus cells called. Spherules that stain pale** blue in the centre with haematoxylin and eosin and are surrounded each by an "**empty**" **halo**, caused by **a large gelatinous capsule**. During histological procedure, the remnants of the capsule shrink and are converted into small strands which are sometime radiating from the fungus cell. **Capsular substance remains unstained** and fungus cells stain well with mucicarmine. This stain proves to be most useful when there are only few fungus cells present in tissue.

Diffuse pulmonary lesions often consist of [budding (asexually)] haemorrhagic and oedematous exudative patchy pneumonic lesions (**Pl. 14.II, Figs. 5 and 6**), which may be uni-or variety. **Cavitation, miliary nodules** occur and mediastinal lymph nodes may be involved. Lesions tend to be localised in the base of the lungs.

Thus, in the X-ray appearance, cryptococcosis resembles a great variety of chronic pulmonary diseases, for instance, pulmonary abscess, tuberculosis, sarcoidosis or neoplasm. A primary pulmonary form cannot be differentiated from disseminated cryptococcosis. **Any lung lesion may disseminate**, even without apparent symptoms in the initial stages of spread.

In **haemorrhagic and pneumonic pulmonary lesions** cryptococci are often found only in small groups scattered in the tissue or in the

PLATE 14.II

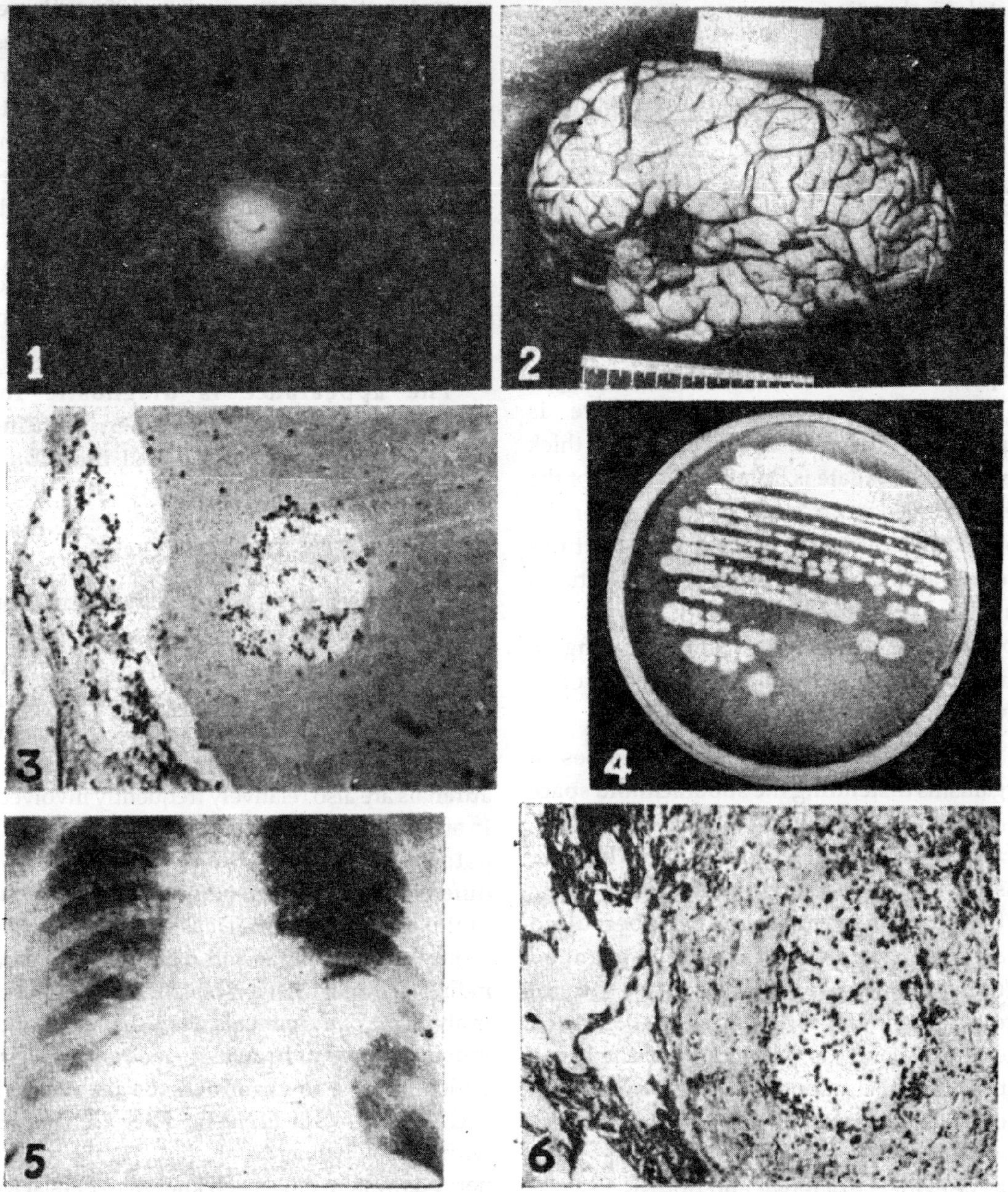

Fig. 1 : India ink preparation of *Cryptococcus neoformans* from CSF of a patient

Fig. 2 : Brain from cryptococcal meningitis one month after infection

Fig. 3 : Cryptococcal meningitis; histological section showing cryptococci accumulated around a small meningeal artery. Note absence of inflammatory reaction. H & E; × 560

Fig. 4 : 7 day old culture of *Cryptococcus neoformans* from CSF of a patient

Fig. 5 : Radiological appearance of cryptococcal pneumonia

Fig. 6 : Cryptococcal pneumonia; Cryptococci are scanty in this case. They are recognisable with PAS and Mucicarmine stain. PAS; × 560

exudate, so that diligent search is necessary in histological sections.

CRYPTOCOCCOSIS OF THE CENTRAL NERVOUS SYSTEM

GROSS AND HISTOPATHOLOGICAL LESIONS

Approximately 9 in 10 cases of pulmonary cryptococcosis disseminate to the central nervous system. Brain and meninges may appear to be almost normal on naked eye inspection, because the organisms do not cause a violent reaction. Usually, there is **leptomeningitis** present. In some cases, thick **gelatinous exudate** is found at the base of the cerebrum or in circumscribed areas of the convexity. Occasionally, **tubercle like structures** are seen **along the vessels**. On coronal sections, small "**soap bubble**" like cystic spaces may appear in the grey matter, in the **basal ganglia, mid brain and pons. Spinal meninges** are often involved, but not the cord.

Histologically, *there* may be masses of organisms **distending the subarachnoid space** almost like a pure culture. In such cases, practically no tissue reaction is seen. In other instances, there are chronic lymphocytic infiltrates, or granulomatous tuberculoid reactions. Cryptococci tend to be **grouped around meningeal vessels** if they are not present in abundance (**Pl. 14.II, Fig. 3**). Though the clinical course is usually rapid, one has to remember the duration that varies **from a few months to 20 years** which explains the variety of histological lesions and the fact that there may be evidence of repair and fibrosis.

Laboratory Investigation of Cerebrospinal Fluid

Cryptococcal meningitis often resembles a mild tuberculous meningitis. The **pressure of the CSF may be increased**. It shows sometimes normal, but usually there is **REDUCED CHLORIDE** and sugar content, but **increased protein**. The cell count is often low or even normal, but may be as high as **1000 cells per cubic millimeter**. Cells are mostly **Lymphocytes. A pellicle forms** instead of cob-web coagulum in Tb. meningitis on standing. It is essential to search for cryptococci which is best done by the ***INDIA INK METHOD*** : **One drop** of centrifuged **CSF** is mixed with a **drop of India ink** on a slide and **covered with a coverslip (Pl. 14.II, Fig. 1)**.

On immediate examination, cryptococci appear as **thick walled spherules** varying from **10 to 20 microns in diameter.**

The appearance is diagnostic for *Cryptococcus*, particularly when budding forms can be found. Not in all individual cryptococci produce capsules.

DISSEMINATED CRYPTOCOCCOSIS

In the majority of all cases, **dissemination** to the **central nervous system takes place**. In addition, about 20% of the patients show affection of the **lymph nodes, spleen and liver,** sometimes **simulating Hodgkin's disease** or malignant lymphoproliferative disorders. The **adrenals** are also relatively frequently involved. In about 10% of the patients, there are multiple **osteolytic bone lesions**, which must be differentiated from sarcoidosis, carcinomatous metastases; in America from coccidio-idomycosis. **Pustular and ulcerated disseminating lesions of SKIN AND MUCOSAE** occur in about 5% of the patients. They may discharge **mucinous grey or brownish-reddish material** which should be **examined for cryptococci. PRIMARY AND FINAL SKIN LESIONS,** without involvement of the lungs or the central nervous system have been known to persist for years, but **spontaneous healing may occur.**

Heart, blood vessels, kidneys, testes, prostate and other organs **may be involved** in the disease process.

BIOPSY

Lesions from the skin, mucosae, liver, prostate, etc., may show masses of cryptococcal spherules dispersed in "empty" gelatinous material. **Inflammatory reaction** in the periphery of the foci may be **minimal** or absent. Particularly in **frozen sections**, the picture has been mistaken, on occasion, for a **chondrosarcoma or for a myxoma. Cryptococcomas** may also **grossly resemble myxomas**. In other instances, cryptococci in tissue sections have been mistaken for *Candida sp.* Lesions in lymph nodes and spleen have been confused with Hodgkin's disease, particularly if only a few organisms were present. Application of **Mucarmine stain is most helpful** in such cases. Scrapings from fresh and even from fixed biopsy material may be examined successfully **by the India ink method**.

Mycology

The organism is a spherical or oval body, 4 to 20μ in diameter. It has a **mucoid gelatinous capsule** which is well-developed in animal tissues. **REPRODUCTION IS BY BUDDING.** The organism grows in Sabouraud's medium, blood agar, etc. ***Primary culture* from animal** source in Sabouraud's medium **may be difficult** for which **enriched media** may be necessary. Several such media are there (creatinine **media** of Shields and Ajello, 1966; **Potato-carrot medium** of Kapica and Shaw; 1969).

The fungus grows in Sabouraud's medium at room temperature as well as at 37°C. The latter property distinguishes this species from other nonpathogenic species of the genus Cryptococcus, the **saprophytic species fail to grow at 37°C**. The colonies in solid medium are white opaque and mucoid. The colour deepens to creamy yellow later on, which may subsequently turn to a **orange** hue.

Various biochemical reactions are shown in Table (by courtesy of Dr. L. Ajello).

Mouse is the animal of choice for animal pathogencity test. *Cr. neoformans* is the only pathogenic species of this genus for mice. After intraperitoneal injection, the animal usually dies within 4 to 6 weeks. The organism can be demonstrated in the peritoneal exudate and at a later date in smear form the brain.

Immunology It is difficult, to detect antibodies in the sera of the patients suffering from systemic cryptococcosis. This was presumed to be due to the *poor antigenicity of the organism*. In recent years employment of the more sophisticated immunological techniques like **haemagglutination**, demonstration of **fluorescent antibody** and **complement fixation tests** has clearly indicated that there is **appearance of antibodies in the sera** of these patients. It is now believed that the **difficulty of detection** of antibodies in the sera of these patients is due to the excess of circulating antigen which **interferes with the detection of antibody**.

Based on the **assay of polysaccharide factor**, the fungus has been typed into three **groups A, B and C**. Apart from the demonstration of circulating antibodies, for the diagnosis of this infection, a skin hypersensitivity test has been introduced. However, the test is as yet not so reliable.

LABORATORY DIAGNOSIS

It depends on the isolation and identification of the fungus from the site of lesion.

PULMONARY CRYPTOCOCCOSIS

Diagnosis is made from the examination of sputum.

DEMONSTRATION OF Cr. neoformans

Sputum is to be treated with 10% NoOH, then mixed with a little amount of India ink on a glass slide, covered with a cover slip and examined first under the low power, then under the high power with diminished light as they are highly refractile.

ISOLATION

Sputum is treated **with 10% NaOH, centrifuged** at 300 rpm for 10 minutes and then **cultured both at 37°C** and at **room temperature**.

MEDIA

Sabouraud's dextrose agar with 0.05 mg/ml **chloramphenicol or creatinine** medium. Growth occurs in **48 to 72 hours**. Brown colonies are picked up for identification on the basis of

Table 14.1 : Differentiation and Identification of Cryptococcus Neoformans from Saprophytic Cryptococci (Courtesy, Dr. L. Ajello)

Cryptococcus Species	*Sabouraud*		*Cyclohexamide Media*	*Urease Production*	Sugar Assimilation (Yeast-Nitrogen Base)							*Nitrogen Assimilation (yeast carbon base)*		*Formentation*	*Pathogenicity for mice*
	25°C	37°C	25°C	25°C	Dextrose	Sucrose	Lactose	Cellobiose	Melobiose	Dulcitol	Trehalose	KNO_3	Urea	Dextrose	
Cr. albidus	+	–	–	+	+	+	+	+	–*	+	+	+	+	–	–
Cr. diffluens	+	– (few +)	–	+	+	+	–	+	+*	–	+	+	+	–	–
Cr. laurentii	+	– (few +)	+	+	+	+	+	+	+*	+	+	–	+	–	–
Cr. luteolus	+	– (few +)	–	+	+	+	–	+	+	+	+	–	+	–	–
Cr. neoformans	+	+	–	+	+	+	–	+	–	+	+	–	+	–	–
Cr. terreus	+	–	–	+	+	–	+	+	–	+	+	+	+	–	–
Cr. uniguttulatus	+	–	–	+	+	+	–	–	–	–	+	–	+	–	–

* Some isolates of these three species may give opposite reactions

(a) morphology, (b) ability to hydrolyse urea (urea agar slant) at 20°C, (c) ability to grow at 37°C, (d) mouse pathogenicity (3 million cells in 0,03 ml).

ANIMAL PATHOGENICTY

The material is injected **intraperitoneally** or **intracerebrally in 2 white swiss mice** and animals are observed for 3 to 6 weeks. **Autopsy shows gelatinous mass in the peritoneum or brain**. From these lesisons ***Cr. neoformans*** can be **demonstrated** and **cultivated**.

CRYPTOCOCCAL MENINGITIS

India ink preparation of the cerebrospinal fluid when examined under microscope will show the fungus with the characteristic **CAPSULAR HALO**. It should be cultured in Sabouraud's medium at **room temperature** as well as at **37°C**. Animal pathoginecity test in mice should be performed.

PROGNOSIS

Cryptococcal meningitis is almost always **fatal**. Pulmonary cryptococcosis has a better prognosis unless in association with intracranial lesion. With the introduction of **Amphotericin B therapy** alongwith **FLUCYTOSINE FLUCONA ZOLE** the outlook of meningitis cases has improved to a great extent. Maintenance dose with Fluconazole prevents relapse.

TREATMENT

Amphotericin B is the drug of choice. The drug should be administered intravenously in doses of **1 mg per Kg body weight daily** or on **alternate days**. The drug is very toxic and should be used with caution specially in patients with renal damage. For **pulmonary granuloma, surgical excision** under the cover of Amphotericin B therapy has been advocated. Since thiamine is a growth stimulant of the fungus, it has been advised to **withhold thiamine** during treatment (Littman, 1956). Please also see **FLUCYTOSINE & FLUCONAZOLE** under **PROGNOSIS**.

Summary

Members of this genus are encapsulated and possess urease. *C.neoformans* differs from nonpathogenic species of cryptococcus by its ability to **grow at 37°C** with **production of phenol oxidase**. The identification is confirmed by demonstrating the presence of **phenol oxidase** or a specific pattern of **carbohydrate assimilations**.

Adsorbed antisera show five serotypes : A, B, C, D and AD. There are two varieties of *C.neoformans - var.neoformans* and *var.gattii* - which differ in their ecology and some biochemical reactions. Serotypes **A, D, and Ad are found** on ***C.neoformans var.neoformans,*** is universally **distributed *c.neoformans var.gattii* is associated with eucalyptus trees** in the tropical and is represented by **serotypes B and C.**

Sexual reproduction can be demonstrated experimentally; mating results in the production of mycelia and basidiospores; **the corresponding sexual stages of two varieties are** *Filobasidiella neoformans var.neoformans* and ***Filo*basidiella neoformans var.basillispora.**

THE ANTIGENIC STRUCTURE

The capsular polysaccharides are long, unbranched polymers comprising an **alpha-1, 3-linked polymannose backbone** with **beta-linked monomeric branches of xylose and glucuronic acid.** In vivo during infection the capsular polysaccharide is dissolved in CSF, serum, and urine and can be detected by **agglutination of latex particles coated with antibody to the polysaccharide**. This is a confirmatory diagnostic test of cryptococcosis. Patient antibodies to the capsule can also be measured.

PATHOGENESIS

Infection follows inhalation of the yeast cells, which are dry, minimally encapsulated in nature, and easily aerosolized. **The primary pulmonary infection** may be asymptomatic or may mimic an influenza-like respiratory infection, often resolving spontaneously. In patients who are compromised, the yeasts may multiply and disseminate to other parts of the body but preferentially to the central nervous system, causing cryptococcal meningo-encephalitis. Other common sites of dissemination include the eye, and prostate gland and skin. The inflammatory reaction is granulomatous usually.

CLINICAL FINDINGS

A chronic meningitis with spontaneous remissions and exacerbations is an important clinical entity. The meningitis may resemble a brain tumor, brain abscess, degenerative CNS disease, and tuberculosis or any other fungal meningitis.

Patients complain of headache, neck rigidity, and mental disorientation. There may be lesions in skin, lungs, and other organs. CSF pressure, and protein may be increased cell count is a little elevated. CSF sugar is normal or low.

The course of cryptococcal meningitis may fluctuate over long periods, but ultimately all untreated cases are fatal. About 5-8% of patients with AIDS develop cryptococcal meningitis. The infection is not transmitted from person to person.

A. Specimens

Specimens include spinal fluid, tissue, exudates, sputum, blood, urine, and serum. Spinal fluid is centrifuged aseptically before microscopic examination and culture.

B. Microscopic Examination

Specimens are examined in wet mounts, both directly and after mixing with India ink, which delineates the capsule.

C. Culture

Colonies develop within a few days on most media at room temperature or 37ºC. Media with cycloheximide inhibit C.neoformans and should be avoided. Cultures can be identified by **growth at 37ºC and detection of urease.** Alternatively, on an appropriate diphenolic substrate, the phenol oxidase of C.neoformans produces melanin in the cell walls and colonies develop a brown pigment.

D. SEROLOGY

Tests for capsular antigen can be performed on cerebrospinal fluid and serum which is not confirmatory. The **latex slide agglutination test** for cryptococcal antigen is positive in 90% of patients with cryptococcal meningitis. With effective treatment, the antigen titer drops - except in AIDS patients.

REFERENCES

Basu Mallik, K. C., Chatterjee, B. D., Banerji, P. L. and Dutta, M. (1967). An epidemiological study of systemic cryptococosis in Calcutta area *Ind. J. med. Res.* **55**, 529.

Benham, R. W. (1935). Cryptococci, their identification by morphology and serology. *J. inf. Dis.* **57**, 255.

Bhaktaviziam, C., Mathai, R. and Mammen, A. (1969). Unusual deep mycoses. *Ind. J. Derm. Ven.* **35**, 139-142.

Buschke, A. (1895). *Deutsch. med. Woch.* **21**, 14.

Busse, O. (1894). Uber parasitars Zellinschlusse und ihre Suchtung. *Zentralbl. f. Bakt.* and Parasit. **16**, 176.

Emmons, C. W. (1951). Isolation of *Cryptococcus neoformans* from soil. *J. Bact.* **62**, 685.

Emmons, C. W. (1955). Saprophytic sources of *Cryptococcus neofarmans* associated with pigeon. *Amer. J. Hyg.* **62**, 227.

Evans, E. E. and Mehl, J. W. (1951). Qualitative analysis of capsular polysaccharide from *Cryptococcus neoformans* by filter paper chromatography. *Science* **114**, 10.

Grueber, H. L. E. (1970). Unpublished study.

Ichaporia, R. N., Grant, K. B. and Waria, R. S. (1970). Cryptococcal meningitis. *J. ind. med. Assoc.* **54**, 425-427.

Khan, M. J., Meyers, R. and Koshi, G. (1959). Pulmonary Cryptococcosis - A case report and experimental study. *Dis. Chest* **36**, 656.

Klokke, A. H. (1967). Generaliseerde cryptococcosis met huidulceratie bijeen Indiase vrouw. *T. Geneeskund*, **111**, 1748-1754.

Kopica, L. and Shaw, C. E. (1969). Improvement in laboratory diagnosis of pulmonary cryptococcosis. *J. canad. med. Assoc.* **101**, 47.

Littman, M. L. and Zimmerman, L. E. *"Cryptococcosis, Torulosis or European Blastomycosis"*, Gruene & Stratton, New York, 1956.

Mohapatra, L. N. (1969). Systemic Mycoses in India. *Bull. A.I.I.M.S.* **3**, 7-19.

Senfelice F. Quoted from *"An introduction to Medical Mycology"*, Yearbook Publishers, Chicago, 1894.

Shields, A. B. and Ajello, L. (1966). Medium for selective isolation of *Cryptococcus neoformans. Science* **151**, 208.

Staib, F. (1961). Vorkommen von *Cryptococcus neoformans* in Vogelmist. *Zentralbl. Bakteriol. Parasitol. Abt.* **182**. 562.

Stoddard, J. L. and Cutler, E. C. (1916). Torula infection in man. *Rockefeller Inst. med. Res.* 25, 1.

Zenker, F. A. (1861). *Jaershc. Gesell. f. nat. Heilk* in *Dresd.* **62**, 51.

15

Candidiasis And *Candida Albicans*

(Robin) Berkhout, 1923; Genus : Candida

(Cutaneous Mycosis) and Systemic Mycosis

Introduction

Candida albicans is present in sputum and cases of about a third of all normal persons. In Britain, it was found in hospital specimens with the following frequency: Sputum 35%; Faeces 30%; vaginal swabs (pregnant) 25%, Vaginal swabs (non pregnant) 16% (Winner, 1966). Similar figures have been reported from all over the world. **Superficial or "intermediate" infections** with *Candida sp.* are very common and consist in epidermophytoses, **paronychia and oncholysis, intertrigo, oral candidiasis, perleche, perianal and vulvovaginal infections, *candida* infections of the eye and the external auditory meatus**. The immense amount of literature which exists about these infections is in inverse proportion to the seriousness of the disease. However, it is now recognised that the **"harmless commensal"** *candida* **can turn into a deadly foe which may cause disseminated, fatal candidiasis**. Superficial as well as systemic candidiasis are most often found in persons suffering from other traumatic conditions or deep seated diseases. In pregnancy and in infancy *candida* infections get a much easier foothold. Disorders of the endocrine system like diabetes mellitus, hypoparathyroidism, Addison's disease; or malnutrition, antibiotic and steroid therapy, leukaemia, agranulocytosis, other malignant disease and post operative states all favour the systemic form of *candida* infection. Wilson (1962) wrote: "...... ***C. albicans*** is a better clinician and can discover abnormalities in persons much earlier in the course of the development of such abnormalities than we can with our chemical tests." **Septicaemic candidiasis,** occurs without any recognised predisposing factor (Symmers, 1963). Because so little is known about the factors which contribute to dissemination of a *candida* infection, experimental work is important and has recently shed new light on the mechanism of *candida* infection. Stanley and Hurley (1969) have shown, that although *candida* are actively phagocytosed in macrophage cultures, the most virulent species of *candida* were rapidly producing filaments that ruptured the cells and killed them. From these experiments it was concluded that the cells of the reticuloendothelial system do not protect the host in candidiasis but aid the proliferation and dissemination of the organisms by acting as vectors.

History

Legenback (1893) demonstrated a yeast-like fungus in the lesion of thrush which was named by Robin (1923) as *Oidium albicans*. The organism was called by Zopf (1890) as *Monilia albicans*. The generic name Candida was suggested by Berkhout to include fungi which reproduced by budding and developed **pseudomycelia**. A number of species was included in genus since then. An excellent review of the genus Candida was published by Skinner and Fletcher (1960).

PSEUDOMYCELIUM

Yeasts are single cells, usually spherical to ellipsoid in shape and varying in diameter from 3 to 15 um. Most yeasts reproduce by budding. Some species produce buds that characteristically fail to detach and become elongated; continuation of the budding process then produces a chain of **elongated yeast cells** called **pseudohyphae** or **pseudomycelium**.

General Characters

The genus Candida is characterised by **yeast-like fungi** which reproduce by budding and form **septate pseudomycelia or mycelia** (Ajello *et. al.*, 1963).

Species : Common species of candida other than *C. albicans* are 1. **C. *tropicalis*.** 2. ***C. pseudotropicalis*.** 3. ***C. krusei*.** 4. ***C. parakrusei*.** 5. ***C. stellatoidea*.** Of all *Candida* species, ***C. albicans*** is **the commonest pathogenic species**. Occasionally other Candida species may be pathogenic.

Morphology

C. albicans is a small budding yeast-like fungus 2.5 μ by 6 to 12 μ. In culture medium, pseudomycelial forms are seen along with clusters of blastospores (**Fig. 13 1a**).

The **characteristic feature** of *C. albicans* is **formation of chlamydospore**, which is not formed by other species of Candida. These spores are thick walled and bigger in size than blastospores. **Chlamydospores are best cultured in corn-meal agar or rice-infusion agar**. Addition of 1% Tween 80 in corn-meal or rice infusion agar appears to stimulate production of chlamydospores of ***Candida albicans*** (Ajello *et. al.*, 1963) (**Fig. 13.1b**).

Cultural Characters

C. albicans grows easily in Sabouraud's dextrose agar at room temperature or at 37°C (Incubator temperature) in about 48 hours.

The colonies have yeast-like odour. Giant colonies (**Fig. 15.2**), when old, show honey-comb appearance in the centre and may show radial furrows (**Pl. 15.II, Fig. 9**). **For chlamydospore** production, the **temperature should be 25°C**.

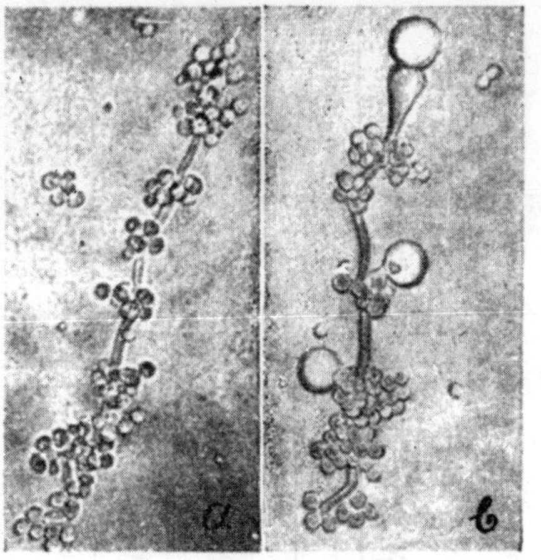

Fig. 15.1 : (a) Blastospores in culture;
(b) Chlamydospores and blastopores in crop-meal agar

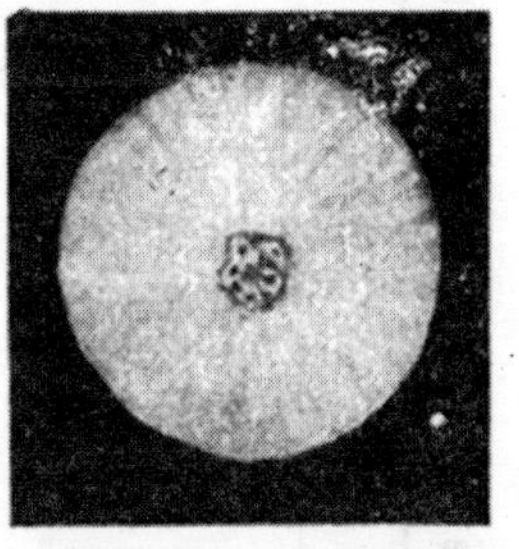

Fig. 15.2 : Giant colony of *C. albicans*

C. albicans is not inhibited by antibacterial antibiotics or cyclohexamide (which inhibits growth of saprophytic fungi). The use of these antibiotics in media is helpful in isolation of *C. albicans* from contaminated clinical materials.

Biochemical Reactions : There are **two** peculiar **reactions** : sugar **fermentation** and sugar **assimilation**. These reactions are utilised as diagnostic tests for identification of ***C. albicans. C. albicans* ferments dextrose** and maltose with **acid and gas production**. It does not ferment lactose but **ferments sucrose** with production of acid only.

***C. albicans* assimilates dextrose, galactose, maltose and sucrose** but does not assimilate lactose, raffinose and cellobiose (Table , Ajello *et al.*, 1963).

Table 15.1 : Diagnostic Biochemical Reactions of the Candida Species of Medical Importance (Courtesy, Dr. L. Ajello)

Species	Sugar Fermentation Reactions				Starch Synthesis Reaction	Sugar Assimilation Reactions									Nitrate Utilisation KNO_3	Urease Reaction
	D	L	M	S		Dextrose	Cellobiose	Galactose	Inositol	Lactose	Maltose	Raffinose	Sucrose	Trehalose		
C. albicans	AG	O	AG	A or O	O	+	O	+	O	O	+	O	+	+	O	O
C. guilliermondii	O or AG	O	O	O or AG	O	+	+	+	O	O	+	+	+	+	O	O
C. krusei	AG	O	O	O	O	+	O	O	O	O	O	O	O	O	O	O
C. parapsilosis	AG or A	O	O	O or A	O	+	O	+	O	O	+	O	+	+	O	O
C. pseudotropicalis	AG	AG	O	AG	O	+	+	+	O	+	O	+	+	O	O	O
C. stellatoidea	AG	O	AG	A or O	O	+	O	+	O	O	+	O	O	+	O	O
C. tropicalis	AG	O	AG	AG	O	+	+	+	O	O	+	O	+	+	O	O
C. viswanathii	AG	O	AG	Weak	O	+	+	+	O	O	+	O	+	+	O	O

Pathogenicity

Rabbit injected intravenously with 1 ml of a saline (1 per cent) suspension of ***C. albicans*** die in 4-5 days with abscess formation in the kidneys. Rabbits having high agglutinin titre after previous immunisation are not protected against a subsequent injection of a lethal dose of the organism.

Meyer and Ordal (1946) produced lesions on the chorio-allantoic membrane with *death* of the embryo after inoculation of *C. albicans*, *C. stellatoides, C. tropicalis* causing mild lesions, while *C. krusei* and *C. parakrusei* produced no lesion.

Enhancement of infection : Gastric mucin, tetracycline and or cortisone enhances infection (Seligmann, 1952; Winter and Foley, 1956).

Mice : Adriano and Schwazz (1955) reported the pathological changes, distribution and nature of lesion of experimental candidiasis in mice. Hill and Gebhardt (1956) have suggested the possible relationship of rapid conversion of yeast cells to mycelial forms with *C. albicans* in animal tissues.

Immunology

Demonstration of agglutinin in blood of patients has little diagnostic significance as normal individuals may have agglutinin in blood. Immunised animals with high agglutinin titre in blood are not protected from challenging doses of ***C. albicans***. Peck *et. al.*, (1955) discussed the **limitation of serodiagnosis** in candidiasis.

Antigen

Recognised species of candida are related antigenically by serological methods (Benham). Negrone (1936) showed that agglutinogens were present in the cell wall of *C. albicans* and complement fixing and precipitating antigens in the polysaccharide capsule. He also showed that sensitisation to give a positive skin test to oidiomycin is due to a capsular polysaccharide. Almon and Stoval (1934), Stone and Garrod (1931), and Martin *el at.* (1937) showed by various serological tests like agglutination, precipitation or complement fixation tests that cross reaction between the species was pronounced. Although **by absorbed sera some of them are identical, *C. albicans*** and ***C. tropicalis*** are antigenically identical. Tsuchiya and others (1954, 1955, 1955a, 1956, 1956a, 1961) have demonstrated the presence of several kinds of thermolabile and thermostable antigens in different species of candida, some of the antigens being shared by many of the species.

Isolation and Identification of *C. albicans* from Clinical Materials.

Clinical materials such as skin scraping or nail scrapings or part of nails should be examined in 10% KOH solution. Scrapings from mucous membrane should be directly mounted in **McGuire's stain or Amman's lactophenol cotton-blue or stained by Gram's method**. Spinal fluid, bronchial washings, sputum and stools are examined directly and also in gram-stained smears. Both mycelial forms and budding cells may be found in the specimens.

These clinical specimens should also be cultured in the Sabouraud's medium both at room temperature and also at 37ºC. These cultures should be made repeatedly from consecutive repeat clinical specimens (Ajello *et al.*, 1963). Tsuchiya *et al.* (1961) advocated serological identification of Candida species, which is not clinically helpful.

GROWTH OF CANDIDA IN TISSUE

Candida sp. usually grows only on the surface of skin or mucosae (thrush) or in damaged or dead tissue. However, if the *tolerance threshhold* of the tissue is lowered, invasion and dissemination may result (Rieth, 1964). **Two factors are important as precursors of invasion** :

(1) **Abundant growth of *candida*,** even if only superficial, often precedes invasive growth. Treatment with antibiotics may eliminate saprophytic bacteria and thus tip the balance in favour of *candida* proliferation.

(2) **Debilitating disease** may lower the tolerance threshold considerably. When invading

tissue, the oval yeast cells of *candida* stretch and form *pseudohyphae* consisting of elongated yeast cells, resembling hyphae, but propagating still by budding. Under certain conditions, true hyphal growth with apical stretching may be found. If *oxygen supply is insufficient*, yeast cells are produced. Most people agree that invasive growth of *candida* is often accompanied by formation of pseudo-hyphae. **Chlamydospores and blastospores may also be found in tissue sections.**

The identification of Candida sp. in tissue rests mainly on three criteria :

1. **Slender pseudohyphae** are usually smaller than real hyphae of fungi and measure about 2-4 microns in diameter.
2. **Oval yeast forms and pseudohyphae** together are most suggestive of *Candida sp.* which stain **GRAM POSITIVE**.
3. In disseminated or invasive candida infection in humans there is usually relatively **little inflammatory tissue reaction.**

Staining of *candida* with **haematoxylin and eosin** is variable. Pale blue staining **pseudohyphae may be recognised (Pl. 15.I, Fig. 3)** but sometimes they remain unstained. With **PAS and silver stains**, *Candida* can be demonstrated excellently. Gram stain is not **always successful in tissue sections.** Often, only the interior of hyphae stains in a lumpy fashion. A conclusive histological differentiation of *candida* from other fungi producing hyphae is not always possible though most invasive *candida* foci are sufficiently characteristic for a diagnosis.

CANDIDIASIS

This is also known as Moniliasis as also Candidosis. This is caused by *C. albicans* or other opportunistic fungs like Aspergillus, Rhizopus and Mucor.

Definition

Candidiasis is an infection caused generally by the fungus ***Candida albicans***, occasionally by other Candida species and may cause lesions in the mucous membrane of the lesions in the mucous membrane of the **mouth (thrush), skin (dermatitis), nails (paronychia and onychia)**, vagina (vulvo-vaginitis) and bronchopulmonary tracts and rarely endocarditis, meningitis, pyelitis and septicaemia (mycaemia).

Systemic Candidiasis in India : Mahapatra (1969) reviewed the reports on systemic candida infections in India. Besides, a considerable amount of experimental work has been done in India by Chitale and Bhende, 1965; Sandhu *et al.* 1966; Chakravarty and Damodaran, 1967; Andleigh and Sharma, 1968; Chugh *et al.*, 1968.

Aetiology

Candidiasis is reported universally and occurs at all ages and both sexes. ***Candida albicans* is an opportunist fungus** and is often found in healthy subjects. It is present in 19.0 per cent of normal throat (Chakravarty and Sandhu, 1963). It is also present in **skin, vagina and intestine of healthy subjects.** In most cases the infections spreads endogenously, when resistance of host becomes low. **Oral thrush is seen in infants and in old age** as also in debilitated persons from wasting diseases like **tuberculosis, cancer and diabetes mellitus. Angular cheilitis** is associated with candidiasis in patients **having vitamin B-complex deficiency (Pl. 15.III, d, h).**

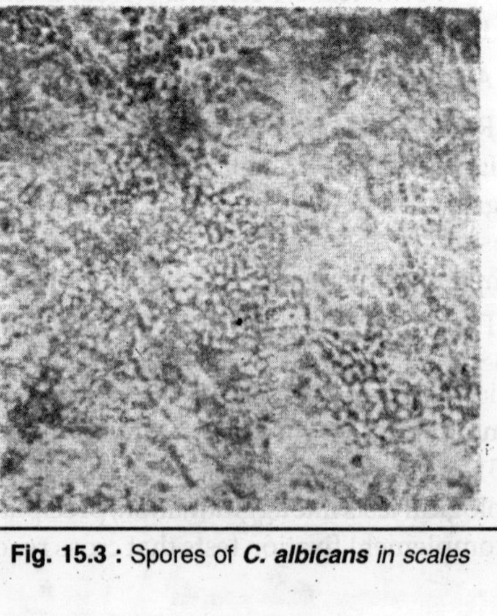

Fig. 15.3 : Spores of ***C. albicans*** *in scales*

PLATE 15.I

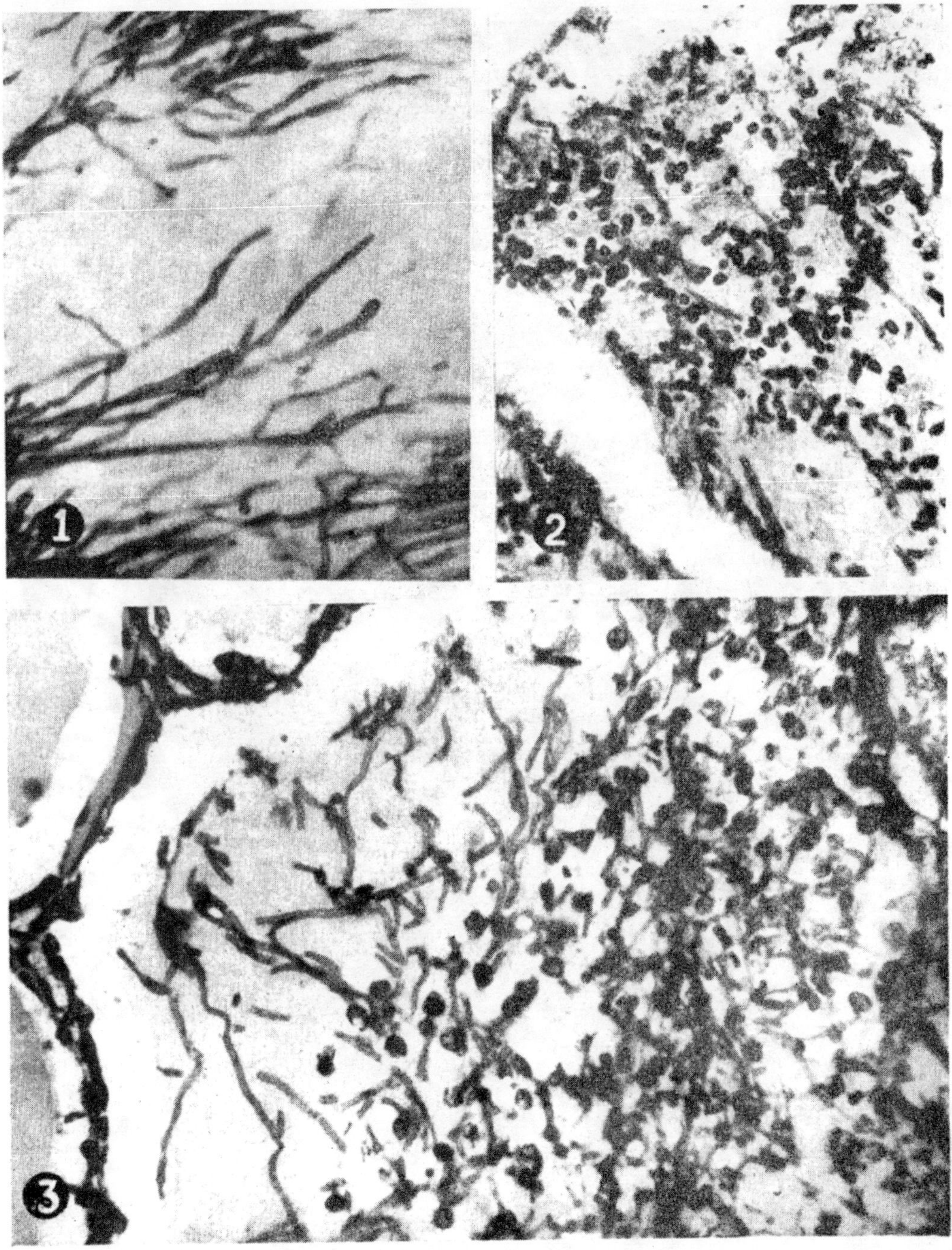

Fig. 1 : *Candida albicans*, pseudohyphae in ulcer of oesophagus. **Silver-methenamine**, 560 X.

Fig. 2 : *Candida sp.* growing in necrotic tissue from bronchial biopsy. Small globular structures are transverse sections of pseudohyphae, not Chlamydospores. Silver-methenamine, 560 X

Fig. 3 : Generalised candidiasis. Focus in thyroid. Pseudohyphae are only faintly stained by H & E.

Weigerts Iron H & E, 560 X

PLATE 15.I (continued)

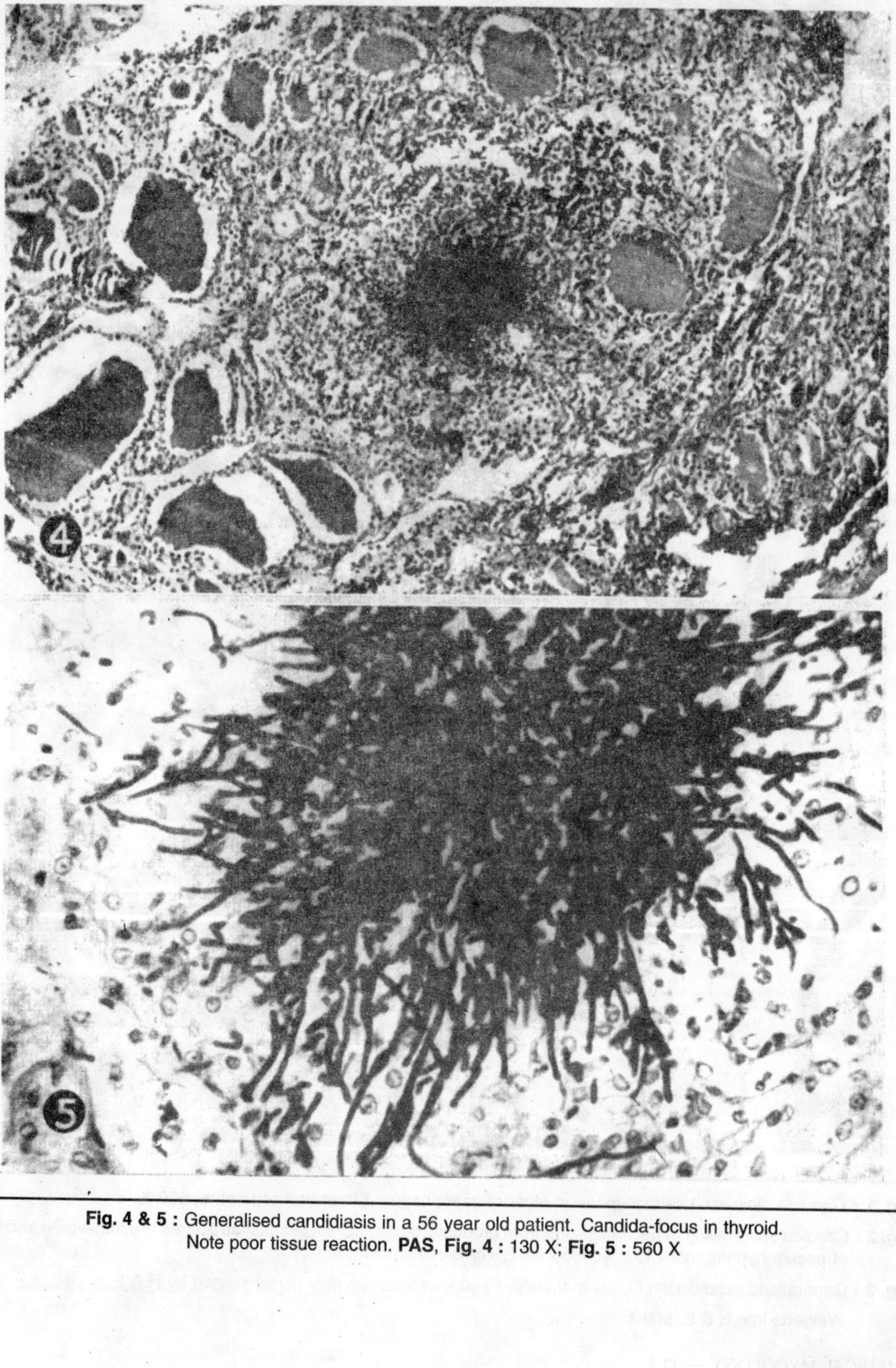

Fig. 4 & 5 : Generalised candidiasis in a 56 year old patient. Candida-focus in thyroid. Note poor tissue reaction. **PAS**, **Fig. 4 :** 130 X; **Fig. 5 :** 560 X

EROSIO INTERDIGITALIS BLASTOMYCETICA (INTERTRIGO) : It is a kind of dermatitis caused by *C. albicans*. The interdigital skin is sodden and macerated in the affected areas with occasional vesicle formation in the margin. This usually affects the web of fingers and toes, but axilla, folds of breast and groin and intergluteal regions may also be affected.

Generalised dermatitis may be associated with oral infection of ***C. albicans*** in children.

Candidid (also known as "**Monilid**" or "levurid"). Some individuals develop sterile vesicular lesion of skin due to "Id" or allergic reaction from the haematogenous spread of ***C. albicans*** from a primary focus. The allergic lesion clears up after healing of primary lesion. Fungus is not present in this "**Candidid**" lesion. **Candidid lesions are commonly found in hands and fingers; but may be found anywhere in the body. Similar sterile lesions occur in the course of dermatophyte infection (Pl. 15.III).**

Candida Granuloma : This condition, variedly called "Generalised cutaneous candidiasis", "monilial granuloma", or "muco-cutaneous candidiasis" is not a truly disseminated process, but a chronic been used to indicate superficial secondary *candida* infection of cavities, persistent presence of candida in bronchial secretion, a chronic granulomatous infection of the bronchial mucosa, etc. (Chakravarty and Sandhu, 1963; Chakravarty, 1967) **(Pl. 15.I, Fig. 2)**. *Lung abscesses* due to *candida* infection are nearly always the result of *candida septicaemia. Candida pneumonia* is probably rarely a primary disease and the diagnosis of previously published cases was often based on clinical observation only which is not acceptable. However, candida may invade the lungs extensively in the hours and days before death in severely ill patients. *Histologically*, candida infections of lungs are characterised by areas of tissue necrosis, nuclear fragmentation, some polymorphonuclear reaction and stellate centrifugal growth of pseudohyphae from the point of view of infection.

In *pulmonary type*—cough is present with mucoid expectoration, rise of temperature and occasional pleural effusion. There might be broncho-pneumonia or pneumonia due to candidiasis of the lung. It may also be associated with haemoptysis. Pulmonary candidiasis is a serious disease. If the patient survives, fibrosis along with minute abscesses are found in lungs. Pulmonary candidiasis is difficult to differentiate from pulmonary tuberculosis.

In Candida endocarditis*, bulky, soft, greyish vegetations**, not much different from bacterial endocarditis, are seen. ***Microscopically, the fibrinous vegetations are infiltrated by a dense network of **pseudomycelia of candida**.

Systemic Candidiasis and Candida Septis : In systemic candidiasis, the infection tends to localise preferably in the kidneys, lungs and endocardium. However, the spleen, urinary bladder, brain and other organs may be affected. **Pl. 15.I, Figs. 3-5** show the typical appearance of such disseminate *candida* lesions. If there are haematogenous *candida* lesions in one organ, one can assume that there are *microscopic foci* in other organs too, but they are easily missed in routine sections of post mortem material (Taschdjian *et al.*, 1969).

Septicaemic candidiasis may be the result of a severe mucosal infection or of faulty asepsis during blood transfusion or surgery. Symmers (1966) has pointed out, that septicaemic candidiasis and comparable infection "art not rare curiosities of clinical practice but frequent enough to make it necessary for every practising doctor to be familiar both with the circumstances under which they may appear with their manifestations."

Infection of the kidneys is so frequent because of the "filter effect" of the gomerular capillaries. Early lesions are therefore, found in the cortex as small whitish or yellowish abscesses and granulomas. *Histologically*, yeast cells may be

PLATE 15.II

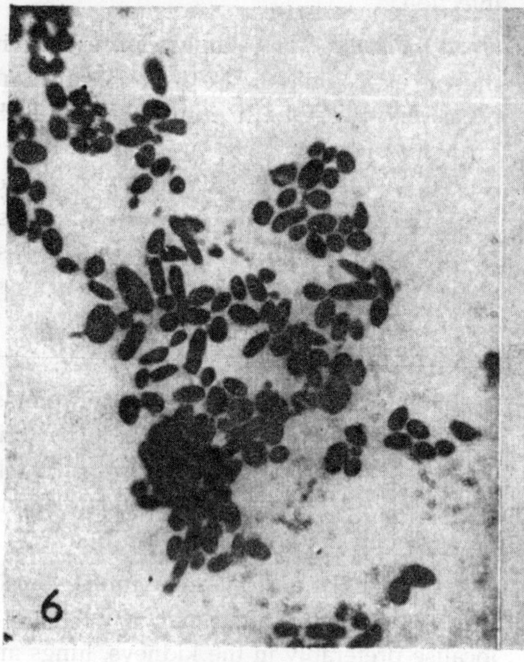
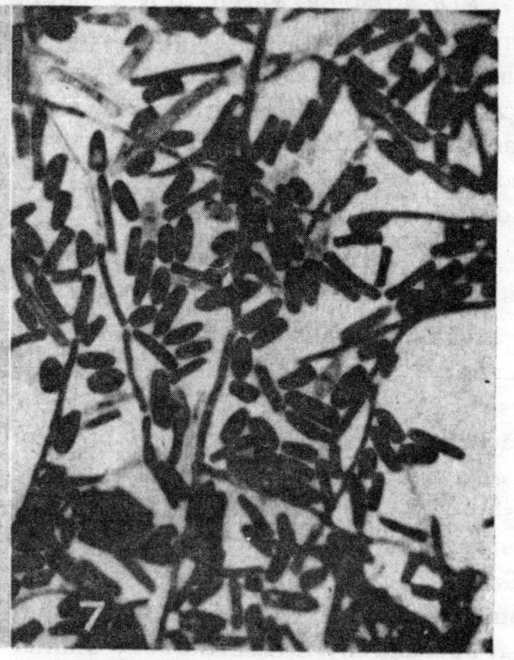
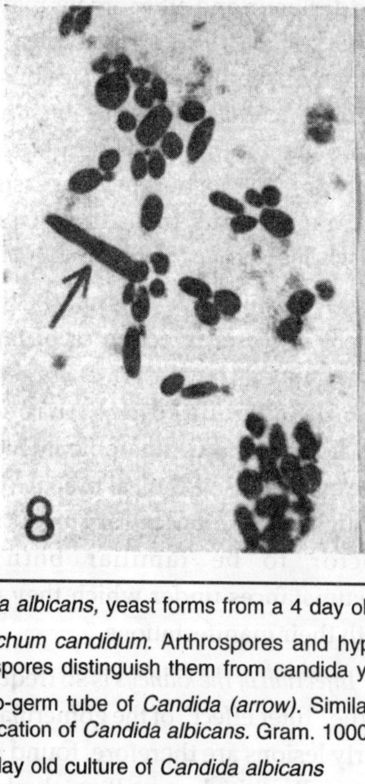
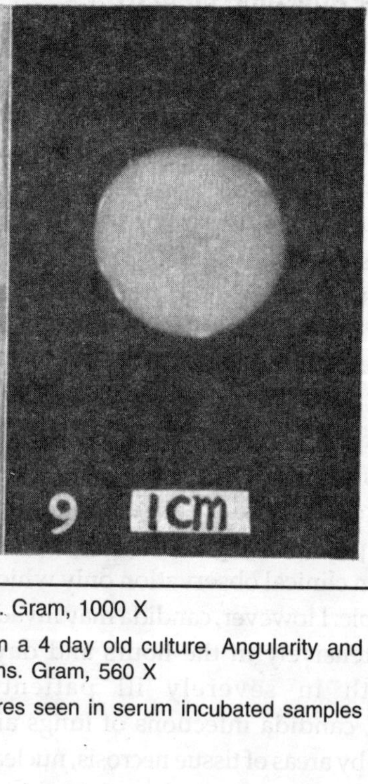

Fig. 6 : *Candida albicans,* yeast forms from a 4 day old culture. Gram, 1000 X

Fig. 7 : *Geotrichum candidum.* Arthrospores and hyphae from a 4 day old culture. Angularity and larger sizes of arthrospores distinguish them from candida yeast forms. Gram, 560 X

Fig. 8 : Pseudo-germ tube of *Candida (arrow).* Similar structures seen in serum incubated samples serve for rapid identification of *Candida albicans.* Gram. 1000 X

Fig. 9 : Three day old culture of *Candida albicans*

PLATE 15.III

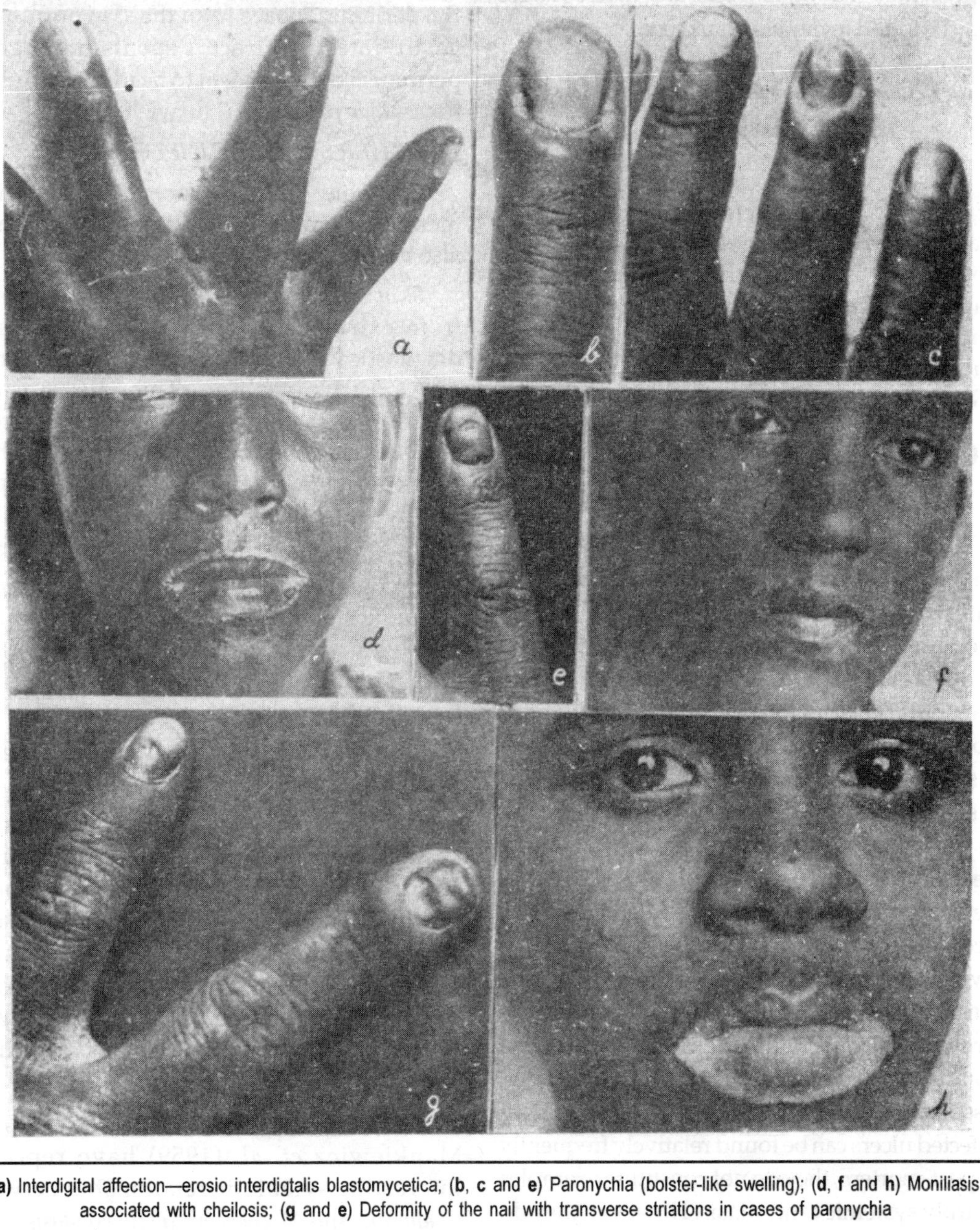

(a) Interdigital affection—erosio interdigtalis blastomycetica; (**b**, **c** and **e**) Paronychia (bolster-like swelling); (**d**, **f** and **h**) Moniliasis associated with cheilosis; (**g** and **e**) Deformity of the nail with transverse striations in cases of paronychia found in or around glomeruli, and later in the centre of small abscesses, or foreign body granulomas. From these, foci of pseudohyphal growth may develop. The perifocal tissue reaction is poor, though few giant cells and macrophages may be found in more chronic cases.

CONGENITAL SYSTEMIC CANDIDIASIS

This is a very rare systemic *candida* infection of the new born affecting multiple organs leading rapidly to death. The route of infection appears to be ascending infection from the vagina, penetration of the intact, membranes with infection of the amniotic fluid and subsequently of the infant by aspiration (Gherardi, 1965; Dworak and Gavaller, 1966).

LOCALISED SEMI-INVASIVE OR INVASIVE CANDIDIASIS

Candida **ulcers or granulomas of the oesophagus**, **the larynx**, **trachea** and of the **intestine** are very frequently found at post mortem. They are often present in patients who are severely ill in post-operative states, particularly after tracheotomy, intubation, application of gastric tubes, etc. At naked eye inspection, these ulcers are often covered with a slightly raised dirty greyish or brownish-greenish pseudomembrane, consisting of necrotic material. **Histologically**, there is a dense network of pseudohyphae **(Pl. 15.I, Fig. 1)** with occasional **groups of yeast cells**. Generally, **hypae are arranged at a right angle** to the surface. Basal layers of eipthelium may still be present but if the epithelium has become completely necrotic, invasion of the underlying tissue may already by seen. Although candida-infected ulcers can be found relatively frequently at post-mortem, they do not constitute examples of truly systemic candidiasis. However, they do show transitional stages to true tissue invasion and even vascular penetration. It is most probably that such ***Candida*** **ulcers serve as foci of systemic and septicaemic candidiasis**.

Another deep-seated, localised infection consists in ***CANDIDA PERITONITIS*** which always appear to be associated with critical or chronic illness, previous treatment with multiple antibiotics and with contamination of the peritoneal space from the skin or from the gastro-intestinal tract. Patients undergoing peritoneal dialysis seem to be ideal candidates for *candida peritonitis* (Hurwich, 1966).

OTHER SYSTEMIC INFECTIONS

Infection with *C. albicans* of bones, joints, meninges, brain, eyes and gall-bladder have also been reported.

Skin test : Skin testing with antigen prepared from extracts of ***C. albicans*** has got little diagnostic value, as many healthy subjects may give positive reaction. This false positive reaction is probably due to the presence of *C. albicans* in body of healthy subjects, in whom **long contact may sensitise the body and give positive reaction**.

Japanese investigators (Kitamura, 1961) used polysaccharide fraction of *C. albicans* as skin testing antigen and claimed only 4 per cent positive in normal controls. They claim that the **chance of false positive reaction is little** in this type of antigen.

CANDIDA ALBICANS AND BRONCHIAL ASTHMA

Long contact of ***C. albicans*** in respiratory passage may play the **role of specific allergen** and may **produce bronchial asthma due to sensitisation**. These so called asthmatic patients who had repeated attacks of paroxysmal dyspnoea improved with nystatin therapy (Keeny, 1954; Kaskin *et al.*, 1961; Chakravarty, 1967). This favourable result is probably due to removing the source of antigen.

Candida albicans** and **tuberculosis : Mankiewicz *et. al.* (1959) have reported interesting symbiotic relationship between *C. albicans* and *Mycobacterium tuberculosis*. It has been demonstrated by them that polysaccharide fraction of cell extracts of ***C. albicans*** **contains a growth promoting substance which stimulates growth of *Mycobacterium tuberculosis***.

PLATE 15.III A

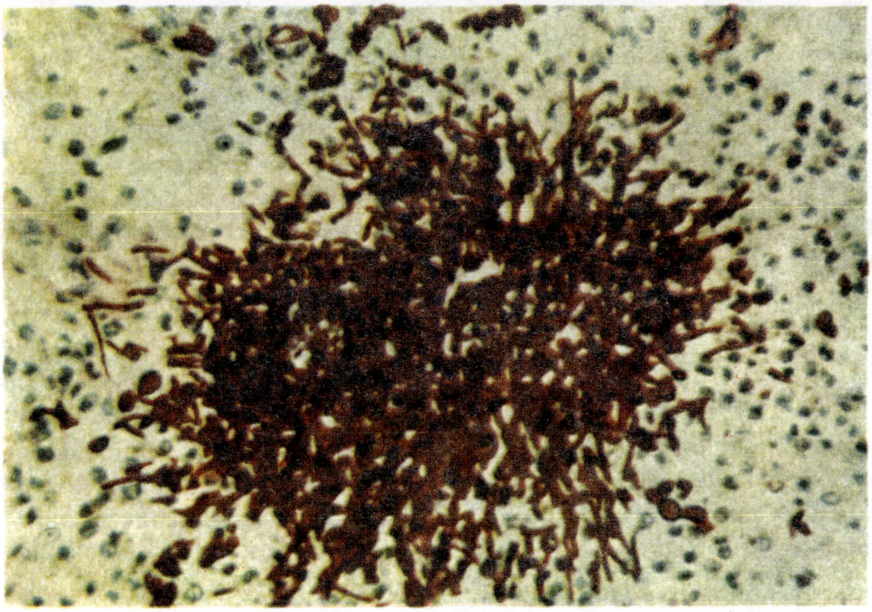

Histological appearance of candidiasis of the thyroid gland. Note round cell infiltration surrounding the yeast and pseudomycelia. PAS stain; × 400

It has also been shown that both the polysaccharide fraction as well as the crude cell-extracts of *C. albicans* are capable of **protecting *Mycobacterium tuberculosis* against the inhibitory effects of streptomycin and isoniazide**. This investigation when studied fully may throw further light on the drug resistance problem in the treatment of tuberculosis.

Serology : Louria and Brayton (1964) described a factor lethal for *C. albicans which* is found normally in human serum. It is not an antibody. It is present in alpha and beta globulin but not in serum albumin or gamma globulin.

As *C. albicans* is present in healthy subjects, serological tests like agglutination are of doubtful value as agglutinins might be present even in some normal individuals.

Animal Inoculation

For determination of aetiological significance of a particular strain of *C. albicans* isolated, animals pathogenicity tests are of value (Ajello *et al.*, 1963).

Experimental studies in animals : It *C. albicans* is injected intraperitoneally in mice with equal amounts of gastric mucin (generally 0.5 ml of saline suspension of *C. albicans* with 0.5 ml of 5 per cent gastric mucin) and when mice are dead or are sacrificed after 4 days, cortical abscesses, in kidney are found.

C. albicans is injected *intravenously* in mice and in rabbits for experimental production of candidiasis. *Intratracheal* injection in rabbits is also done to study pulmonary candidiasis caused by *C. albicans* infection. Lungs are comparatively resistant to ***C. albicans infection***.

Pathology

The lesion produced by *C. albicans* may be primary at the site of original lesion or secondary as a metastatic lesion. *C. albicans* is found mostly in mycelial form along with some budding yeast cells in tissues. Probably mycelial form is the invasive form of ***C. albicans* in *tissues***. These are best demonstrated by special staining methods.

In PAS stain, fungus cell wall is stained deep-red. This is well demarcated when a counter stain is employed as in modification of **PAS stain** by Bauer, or **Gridley stains** (*Ajello et al.*, 1963). The cellular reaction in tissue is comparatively less marked around the lesion of candidiasis. There may be granulomatous lesion with giant cell formation or even areas of ceseation may be present. **Necrosis** may be present with mononuclear cell "**round cell**" infiltration. ***Endocarditis*** due to ***C. albicans*** shows vegetation in a valvular lesion of the heart and yeast forms of ***C. albicans*** are present.

Prognosis : Candidiasis patients should be treated energetically. It does not matter whether it is primary or secondary candidiasis. In the ultimate analysis the result is the same in both.

Pulmonary candidiasis, endocarditis and meningitis caused by ***C. albicans*** are of serious prognosis.

Cotrimoxazole is the Treatment of choice.

Treatment

For chronic and systemic infections Amphotericin 'B', for other condition Imidazoles types of medicines are most suitable. Nystatin (Mycostatin-Squibb) is sold in tablet form. Each tablet is 500,000 (five hundred thousand) units. It acts locally when taken orally. It is least absorbed in the body from gastro-intestinal tract. Inspite of little absorption of nystatin, it should be given in systemic candidiasis. By the oral route it will reduce the growth of *C. albicans* in the intestinal tract. 500,000 of nystatin tablets orally every six hours daily for about two to three weeks are advocated in gastro-intestinal candiasis. There is no side reaction.

In bronchopulmonary candidiasis in addition to oral nystatin, aerosol should be given (Chakravarty, 1967).

500,000 (fifty thousand) units of sterile nystatin powder is suspended in 40 per cent propylene glycol in 4 ml of normal saline. One milliliter of this suspension is given by aerosol every six hours daily for two to three weks. The suspension is shaken each time before use. Aerosol therapy can be given with *intermittent positive pressure breathing in inspiration* (IPPB/I) for deeper penetration of drug in the long. The pressure is kept between 10-15 cm of water. Aerosol can also be given by hand nebulizer. Total amount of nystatin to be given depends on the dissemination of lesion in lungs. A higher concentration may also be given is necessary (Chakravarty, 1967).

If nystatin fails to improve the disease, **Amphotericin B** (Squibb-fungizone) may be given intravenously **1.0 mg/kg per day**. The drug should be given daily provided the patient can tolerate it. After the maximum dose is reached, the drug can be given on alternate days. Initially the drug is started with one fifth of the minimum of **1.0 mg/kg daily diluted in 700 ml of 5 per cent glucose**. Treatment may last from six weeks to several months.

For **disseminated candidiasis, endocarditis, meningitis, pyelitis and septicaemia** caused by *C. albicans*, the treatment of choice is Amphotericin B. When nystatin is ineffective, **Amphotericin B** is indicated.

For cutaneous and muco-cutaneous candidiasis, nystatin cream to be applied locally. For **vaginal candidiasis**, nystatin ointment should be applied directly to vulva and vestibule of vagina. Local application of 1 per cent gentian violet has always proved efficacious in these conditions including thrush.

For perleche and thrush, aqueous solution of one per cent gentian violet locally and 500,000 units of nystatin tablet thrice daily for two weeks orally gives good result. Along with nystatin, vitamin B-complex may also be given orally daily. In all types of candidiasis; supporting treatment with high protein diet, vitamin and adequate fluid should be given. Bed rest is indicated in debilitated patients and in those having systemic candidiasis.

Presently Cotrimoxazole is the treatment of choice (see Table 8.2, 8.3, and Ch. 8).

REFERENCES

Ajello L., George, L. K., Kaplan, W. and Kaufman, L., *"Laboratory manual for Medical Mycology"*. Communicable Disease Center, *Atlanta* 22, Georgia, U. S. A. P. E-24, 1963.

Ajello, L. A. (1969). Comparative study of the pulmonary mycoses of Canada and the United States. Public Health reports (U. S. A.) **84**, 869.

Andleigh, H. S. and Sharma, K. B. A comparative study of the pathogenesis of various *Candida* species in mice, rabbits and guineapigs. *18th Ann. Meetq., Ind. Assoc. Path.*, Bombay, 1968.

Audebaud, G., Guillon, J. C., Destombes, P. and Courmes, E. (1965). Candidose generalisee a *Candida albicans* observe chez de Moutons de la Guadaloupe. Associations parasitires (Toxoplasmose Helminthiase). *Rec. Medec. Veter.* **141**, 351-356.

Berkhout, C. M. De Schimmelgeslachten Monilis, Oidium, Oospora on Torula, Edauw & Johannissen, Scheveningen, 1923.

Braude, A. I. Candidiasis in Principles of Internal Medicine, Harrison. International student edition (5th Eidition). Koga Kusha Co. Tokyo, P. 1959, 1966.

Chakravarty, S. C. (1966). Role of fungi in chronic bronchitis. *Ind. J. chest. Dis.* **8**, 6-9.

Chakravarty, S. C. (1967). Bronchopulmonary candidiasis; Clinical aspect. *Dis. Chest* **51**, 608-613.

Chakravarty, S. C., Damodharan, V. N. (1967). Pathogenesis of experimental pulmonary candidiasis in rabbits. *Ind. J. chest Drs.* **9**, 152-156.

Chakravarty, S. C., Damodharan, V. N. and Sandhu, R. S. (1965). Effect of Hyaluronidase in experimental candidiasis with special reference to lung lesion. *Mycopathologia* **30**, 287-293.

Chakravarty, S. C. and Sandhu, R. S. (1962). Incidence of and criteria for diagnosis of broncho-pulmonary Candidiasis. *Acta Tuberc. Pneum. Scand* **42**, 197.

Chakravarty, S. C. and Sandhu, R. S. (1963). A study in bronchopulmonary candidiasis. *Ind. J. chest Dis.* **5**, 40-45.

Cyakravarty, S. C. and Sandhu, R. S. (1954). Incidence of bronchopulmonary candidiasis in patients treated with antibiotics. *Acta Tuberc. Pneum. Scand.* **54**, 152.

Chakravarty, S. C. (1967). Bronchopulmonary candidiasis: Clinical aspect. *Dis. Chest.*, **51**, 608.

Chitale, A. R. and Bhende, Y. M. (1965). Experimental moniliasis in rabbits and mice. *Ind. J. Path. and Bact.* **8**, 89-97.

Chitkara, N. L., Chugh, T. D. and Sangha, H. Bronchopulmonary candidosis, *16th Ann. Meetg., Ind. Assoc. Path.*, Ranchi, 1966.

Chugh, T. D., Khurana, S. and Chitkara, N. L. Experimental candidosis *18th. Ann. Meetg. Ind. Assoc. Path. 11,* Bombay, 1968.

Conant, N. F., Smith, D. T., Baker, R. D. Callaway, J. L. and Martin, D. S. *"Manual of Clinical Mycology*, W. B. Saunders Co., Philadelphia, U. S. A., 1971.

Dvorak, A. M. and Gavaller, B. (1966). Congenital systemic candidiasis. *New Engl. J. Med.* **274**, 540.

Emmons, C. W., Binford, C. H. and Utz, J. P. *"Medical Mycology"*, Lea and Febiger, Philadelphia, U. S. A., 1970.

Gherardi, G. L. (1965). Systemic moniliasis in infancy. *J. am. med. Assoc.* **193**, 67-69.

Hauser, F. V. and Rothman, S. (1950). Monilial granuloma. *Arch. Dermat. Syph.* **61**, 207.

Hazen KC (1995) : New and emerging yeast pathogens. *Clin Microbiol Rev.* **8**, 462.

Hurchwich, B. J. (1966). Monilial peritonitis. *Arch. int. Med.* **117**, 405-408.

Kashkin, P. N., Krassilnicov, N. A. and Nekachlov, N. Y. (1961). Candida complication after antibiotic therapy. *Mycopath. et Mycology*. **14**,. 3.

Keeny, E. L. (1954). Pulmonary mycotic infections, allergic and immunologic factors. *California Med.* **81**, 367.

Kitamura, S. Study on the preparation of the antigen for diagnostic intradermal reaction, obtained from *Candida albicans* on electric dissolution. Studies on candidiasis in Japan. Education Ministry, Japan, P. 78, 1961.

Kozinn, P. J., Claire, L., Taschdjian, C. L. and Winer, H. (1958). Incidence and pathogenesis of neonatal born candidiasis. *Pediatrics* **21**, 421.

Lehner, T. Classification and clinicopathological features of *Winner and Hurley* (29), P. 119-136, 1966.

Lengenbeck, B. (1839). NeueNotizen aus drm Gobiete der Natur and Heilkunde (Froriep). **242**, 154.

Luria, D. B. and Brayton, R. G. (1964). A substance in blood lethal for *Candida albicans. Nature, Lond.* **201**, 309.

Mankiewicz, E., Slackiewics, E. and Liivak, M. (1959). A polysaccharide isolated from *Candida albicans* as growth promoting factor for *Mycobacteria tuberculosis. Canad. J. Microbiol.* **5**, 261.

Meyer, E. and Cordal, Z. T. (1946). *J. Bacteriol.* **52**, 615.

Mohapatra, L. N. (1969). Systemic mycoses in India. *Bull. A. I. I. M. S.* **3**, 7-19.

Mukherjee, A. and Basu Mallik, K. C. An experimental study of phagocytosis of *Candida albicans, 15th Ann. Meetg., Ind. Assoc. Path.*, Guntur, 1965.

Murray, I. G. and Buckley, H. R. Serological study of Candida species, in *Winner and Hurley* (29), p. 44-50, 1966.

Odds FC : *Candida and Candidosis*, 2nd ed. Bailliere Tindall, 1988.

Pfaller MA (1996) : Nosocomial candidiasis: Emerging species, reservoirs, and modes of transmission. *Clin. infect. Dis.* **22**, S89.

Rieth, H. Die Mycosen. Folia Ichthyologica, Ichthyol Gas, Cordes and Hermanni, Hamburg, 1964.

Robin, C. Histoire naturelle des vegetauz parasites qui croissent sur l'home et sur animau vivant, J. B. Bailliere, Paris, 1853.

Sandhu, R. S., Randhawa, H. S. and Gupta, I. M. (1965). Pathogenicity of *Candida viswanathii* for laboratory animals. A preliminary study. *Sabouraudia* **4**, 37-40.

Seligmann, E. (1952). Virulence enhancing activities of auremycin of *Candida albicans. Proc. Soc. exp. Biol. Med.* **79**, 481.

Sharp, J. L. (1954). The growth of candida during antibiotic therapy. *Lancet* **1**, 390.

Sidransky, H. and Pearl, M. A. (1961). Pulmonary fungus infection associated with steroid and antibiotic therapy. *Dis. Chest.* **39**, 630.

Skinner, C. E. (1947). *Bact. Rev.* **11**, 227.

Skinner, C. E. and Fletcher, D. W. (1960). A review of the genus Candida. *Bact. Rev.* **24**, 397.

Stanley, V. C. and Hurley, R. (1969). The growth of Candida species in cultures of mouse peritoneal macrophages. *J. Path.* **97**, 357-366.

Symmers, W. St. C. Septicaemic candidosis, in *Winner and Hurley* (29), P. 196-212, 1966.

Taschdjian, C. L., Kozinn, P. J., Fink, H. Cuesta, M. B., Caroline, L. and Kantrowitz, A. B. (1969). Post mortem studies of systemic candidiasis. (I) Diagnostic validity of precipitin reaction and probable origin of sensitisation to cytoplasmic candidal antigens. Sabouraudia 7, 110-117.

Tsuchiya, T., Fukazawa, Y. and Kawakita, S. Serological classification of genus Candida "Studies on Candidiasis" in *Japan Ministry of Education*, Tokyo, p. 34.

Wilson J. W. As quoted by Winner, 1966.

Wilson, J. W., Plunkett, D. A. and Gregerson, A. (1954). Nodular granulomatous perifolliculitis of the legs caused by *Trichophyton rubrum. Arch.* Dermat. Syph. **69**, 258-277.

Winner, H. I. General features of Candida infections, in Winner and Hurley (29), p. 6-11, 1966.

Winner, H. I. and Hurley, R. *"Candida albicans"*, London, A. Churchill, 1954.

Zamith, V. de A. and Castro, R. M. (1964). Candidiasis cutanea generalisada. *Dermat. Revist. Mexicana* **8**, 113-122.

Zopf. W. Die pilze, E. Trewendt, Breslau, 1960.

16

Geotrichosis

Introduction

Geotrichosis is a rare broncho-pulmonary and pulmonary mycotic infection. Dissemination has been reported only in exceptional cases. The nature of geotrichosis as a primary disease is debated due to the fact that *Geotrichum* sp. **are frequently found in the stool and the bronchial secretion of normal persons**. To my knowledge, **no case** has been reported **from India** and autopsies have been performed only in a few instances.

Definition

It is a condition caused by *Geotrichum candidum*, affecting the **skin, mucous membrane, gastrointestinal tract**; infection of the **lung** simulates chronic pulmonary tuberculosis.

History

G. candidum was isolated from leaf-mould by Link (1809). This was often called *Oidium* or ***Oospora lactis***. The term **Oidium means** the **imperfect stage of certain fungi** of the class **Ascomycetes,** the powdery mildews, parasitic on plants. However, to give effect to priority the species is called *G. candidum* and this term has already become popular in the field of mycology.

Habitat

This occurs **in the soil as a saprophyte** and has also been isolated from **stools and lesions in the human body**.

Pathology

I. *BRONCHIAL, BRONCHOPULMONARY AND PULMONARY GEOTRICHOSIS*

Apart from thrush and superficial **bronchial mucosal infection**, the pulmonary form **is most common**. The patients produce **copious gelatinous or blood-streaked expectoration** and suffer from **persistent cough**. The aetiologic role of *Geotrichum* can only be accepted, if the fungus can be repeatedly grown from the sputum and preferably also shown **histologically in bronchial biopsy material**. A typical case was reported by Ross *et al.* (1966), who isolated the fungus from the sputum and from the bronchial lining of a patient on 73 occasions, and, during the illness, demonstrated a thirty fold rise in patient's serum of agglutinating antibodies.

On naked eye inspection **the lung shows** most frequently **multiple or solitary thin walled cavernous lesions, in chronic**

cases with extensive **fibrosis**. In this stage, cavities may heal. In other cases, the bronchi are dilated, filled with mucopurulent material, and multiple abscesses of different sizes are found in the lung tissue.

Roentgenologically and clinically geotrichosis may **resemble fibrocaseous tuberculosis**.

Histologically, the granulation tissue is composed of **fibroblasts, histiocytes** and **foreign body giant cells**. In places lymphocytic infiltrates and polymorphonuclear exudation is more prominent.

II. *INTESTINAL GEOTRICHOSIS*

Since *Geotrichum* organisms are so frequently found in stools of normal individuals, it is rather doubtful whether a primary intestinal form exists. However, in some cases excessive numbers of ***Geotrichum* spores** have been found **in stools from** patients suffering from **acute colitis**. A proper evaluation of these cases would be possible only by histological examination of **intestinal biopsies**.

III. *GENERALISED GEOTRICHOSIS*

Generalisation appears to be **a rare occurrence** and is probably always **secondary to** other **debilitating diseases**. Bendove and Ashe (1952) reported on a case of ***Geotrichum* septicaemia**.

MYCOLOGY

Geotrichum candidum is so frequently found in **sputum**, on **skin** or in **faeces** of persons **without clinical disease** that a **diagnosis** of geotrichosis should only be made **after repeated demonstration of the fungus in direct SMEARS**, in **BIOPSIES, in CULTURE**, and after exclusion of other possible pathogenic organism.

G. candidum appears in the form of **ARTHROSPORES** which are characteristically **rectangular 4 to 6 μ by 8 to 12 μ in size**. Thick-walled, ovoid or round cells may form a chain of arthrospores. **Rounded forms** may be **mistaken for Blastomycetes but presence of arthrospores in chains is diagnostic of Geotrichum**.

DIRECT EXAMINATION

A loopful of sputum or pus should be spread in a thin film and examined for **oblong or rectangular cells** measuring approximately **4 × 10 × microns**. Often, there is a mixture of **elongated-oval and rectangular, deeply Gram positive cells are very characteristic of *Geotrichum candidum* (Pl. 16.II Fig., 7)**.

INDIRECTLY

Culture

At 37ºC, the fungus grows rapidly and with a dull white or grey colour. In the periphery of the colony, the fungus tends to grow directly below the surface of the agar. At room temperature, ***Getrichum* grows moderately fast**, with a dry powdery surface showing numerous short aerial hyphae. **Mycelial threads are 3 to 5 microns in diameter** and develop numerous almost **rectangular arthrospores** which may vary greatly in size and in roundness of their ends. **Germination** occurs in one corner **of the spores** which is very **characteristic of *Geotrichum*.**

The organism grows quickly in Sabouraud's glucose agar either at room temperature or at 37ºC. The **growth is membranous** and in about **7 to 10 days Giant colonies form**. They are flat, white to cream-coloured, soft to the touch of the loop like yeast colonies and can be picked up easily. In liquid media it forms a surface membrane.

Microscopically, **arthrospores** are round or rectangular in shape. **They give out germ tubes** but **do not reproduce by** typical **budding**.

BIOCHEMICAL REACTION

Most of the strains liquefy gelatin and produce acid in certain sugars. Some strains coagulate milk.

SYMPTOMS

It may produce a condition of **bronchitis with persistent cough**. The **sputum** is often greyish but it may be **blood stained** occasionally.

X-ray examination shows **peribronchial thickening**. The general health of the patient is, however, not affected. Infection of the lung may simulate chronic tuberculosis. **Oral geotrichosis resembles thrush** caused by ***Candida albicans***.

SEROLOGY AND ANIMAL PATHOGENICITY

Animal inoculations and serological **tests are not** particularly **useful** in diagnostic studies (Emmons *et al.*, 1970).

TREATMENT

Potassium iodide by mouth and **sodium iodide intravenously** have been successfully used **in** cases of **pulmonary geotrichosis**.

No antifungal antibiotic is available at present but 3 medicines are worth trying :

1. **IMIDAZOLE (Clotrimazole)**
2. **TRIAZOLES (Itraconazole; Fluconazole)**.
3. **AMPHOTERICIN B**

REFERENCES

Bendove, R. A. and Ashe, B. I. (1952). Geotrichum septicaemia, *Arch. int. Med.* **89**, 107.

Conant, N. F., Smith, D. T., Baker, R. D. and Callaway, J. L. *"Manual of Clinical Mycology"*, 3rd. Ed., W. B. Saunders Company, Philadelphia, 1971.

Emmons, C. W., Binford, C. H. and Utz, J. P. *"Medical Mycology"*, 2nd, Ed., Lea & Febiger, Philadelphia, 1970.

Link, M. G. (1809). *Natur. Fruende Berlin* **3**, 17.

Ross, J. D., Reid, K. D. G. and Speirs, C. F. (1966). Bronchopulmonary geotrichosis with severe asthma. *Brit. med. Jour.* **1**, 1400-1402

17

Subcutaneous Mycosis

Mycetoma (Syn: Maduromycosis, Madura Foot)

Definition

Literally mycetoma means a tumor like swelling produced by fungus infection, clinically it is characterised by swelling of the affected part with multiple sinuses discharging pus containing the **mycotic "Granules"** as a result of chronic granulomatous inflammation.

Geographic distribution

It is a disease of world wide distribution, occurring most frequently in the **tropical and subtropical regions**. The first case was reported by Carter (1860) who established the fungal aetiology of this disease in a patient in **Madura district** and named the condition as **"Madura foot"**. Subsequently cases have been reported from other parts of **India**, and from **Africa, South America, Mexico, United States and Europe**.

Aetiology

The causative agents of this condition which present a uniform **clinical and pathological picture fall into two broad groups**: (a) the aerobic actinomycetes (nocardia sp. and streptomyces sp.) producing **"ACTINOMYCOTIC mycetoma** and (b) **the true fungi (ascomycetes and fungi imperfecti) causing "MADUROMYCOTIC mycetoma or "maduromycosis"**.

Causative Agents

SCHIZOMYCOTIC ACTINOMYCETOMA

NOCARDIA SP.

1. *Nocardia asteroides*
2. *Nocardia brasiliensis*

STREPTOMYCES SP.

3. *Streptomyces madurae*
4. *Streptomyces somaliensis*
5. *Streptomyces paraguayensis*
6. *Streptomyces pelletierii*

EUMYCETOMA (MADUROMYCOSIS)

FUNGI IMPERFECTI (HYPHOMYCETES)

1. *Madurella mycetomatis*
2. *M. grisea.*
3. *Monosporium apiospermum*
4. *Glenospora semoni*
5. *Cephalosporium falciforme*
6. *Phialophora jeanselmei*

ASCOMYCETES

7. *Leptosphaeria senegalensis*
8. *Pyrenochaeta romeroi*
9. *PseuAllescheria boydii* (perfect stage of ***M. apiospermum***)

PLATE 17.I

ACTINOMYCETOMA

(CAUSED BY ACTINOMYCETALES) IN TISSUE SECTIONS COMPARATIVE SIZE AND SHAPE OF GRANULES

	MICRONS	SCALE : APPROXIMATELY 100 MICRONS
ACTINOMYCES (ACTINOMYCOSIS)	100 TO 500	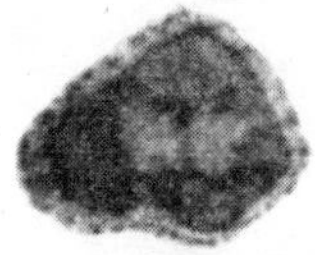
NOCARDIA Sp.	60 TO 300	
STREPTOMYCES MADURAE	1000 TO 10,000	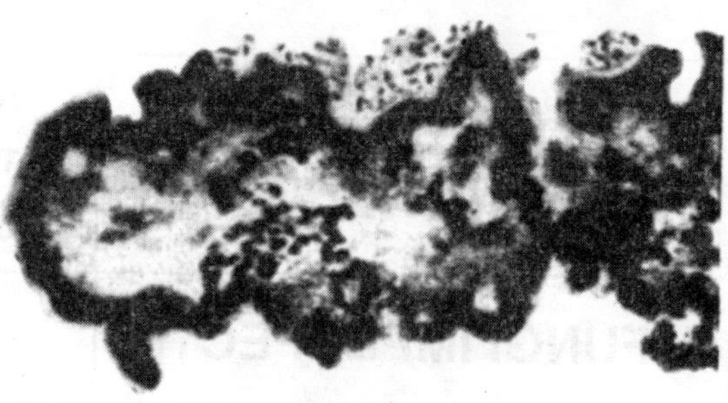
STREPTOMYCES PELLETIERI	ABOUT 300	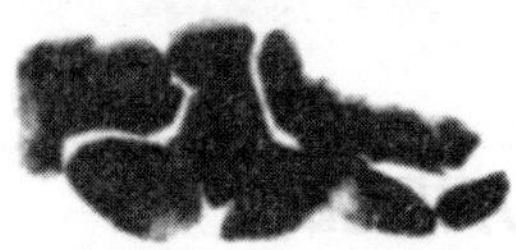
STREPTOMYCES SOMALIENSIS	300 TO 1000	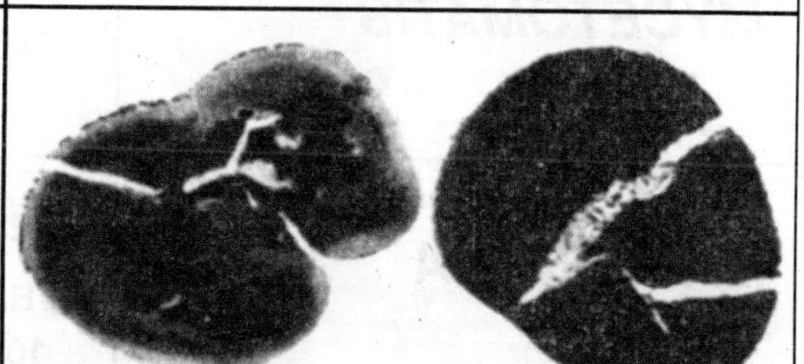

Plate 17.II

EUMYCETOMA

(TRUE FUNGAL M., MADUROMYCETOMA)

CAUSED BY ASCOMYCETES AND BY FUNGI IMPERFECTI

COMPARATIVE SIZE AND SHAPE OF SOME GRANULES IN TISSUE SECTIONS

I. ASCOMYCETES :	MICRONS	SCALE : ——— APPROXIMATELY 500 MICRONS
PSEUD-ALLESCHERIA BOYDII	ABOUT 500 TO 2000	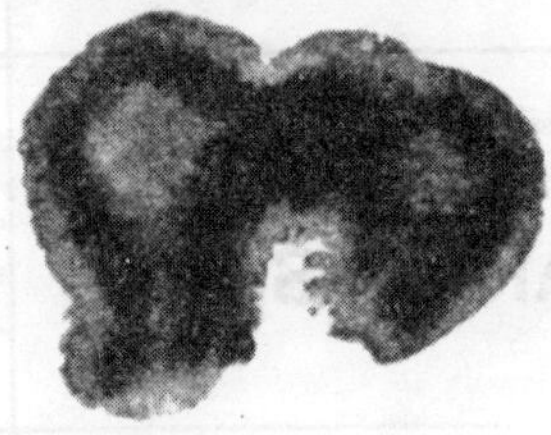
LEPTOSPHAERIA SENEGALENSIS	700 TO 1000	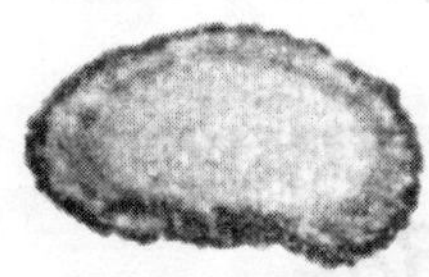
II. FUNGI IMPERFECTI :		
MADURELLA MYCETOMATIS	1000 TO 3000	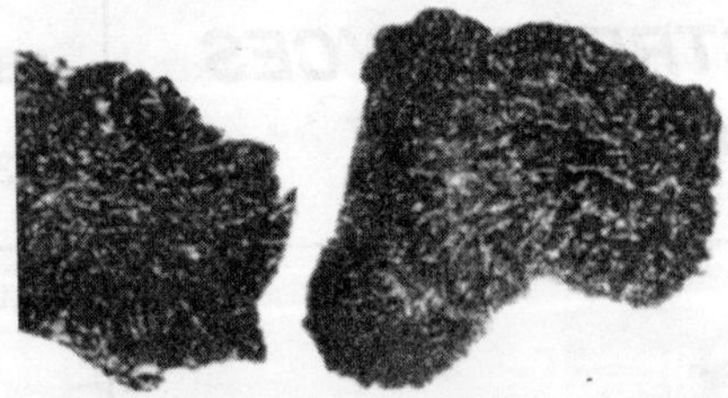
MADURELLA GRISEA	LESS THAN 1000	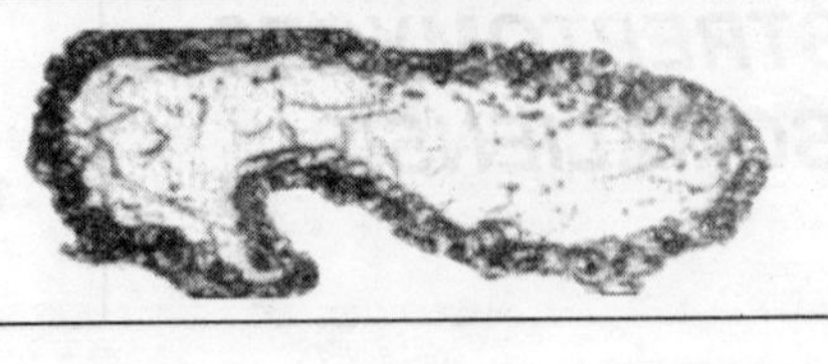

Besides the above a rare aetiologic agents like *Aspergillus bouffardi* (Brumpt, 1906), Indiella sp. (Brumpt, 1906) and ***Penicillium mycetogenum*** (Mantelli and Negri, 1915) have been reported from time to time.

These ***aetiologic agents*** are **ubiquitous in nature** being present in soil and vegetations in the ground. People **walking in bare foot** get the infection from the soil. Often there is a history of trauma in the form a **thorn-prick or abrasion** facilitating the **entry of the causative agent into the skin and subcutaneous tissue**.

Clinical picture

The disease is most common in the working class of people who do not usually wear shoes and often come in contact with the soil. The highest incidence is the **age group of 20 to 40 years and children** are rarely affected. It is more common in man than woman but all cases are equally susceptible to this infection. The disease is mostly confined to the **lower extremities,** usually the **foot**, but other parts of the body especially the **hands** are often involved. Cases of involvement of the arms, knee, buttocks and trunk are occasionally reported.

The disease usually starts as a small nodular swelling which is deep seated and fixed. The condition progresses slowly with indurated area extending in all directions, abscess formation which ruptures and gives rise to a sinus. The patient usually comes with the classical picture of swelling, deformity, multiple discharging sinuses of several months or years duration. The muscles, bones and fascia are involved in the process of chronic inflammation as the infection extends deeper into the tissues. The **purulent discharge from the sinuses contain the granules which vary in size,** consistency and colour, depending on the causative agent in any particular case. Apart

Table 17.1 : Characteristics of Granules of Mycetoma

Organism	*Texture*	*Colour*	*Average size mm*	*Intrinsic colour*	*Diameter of filament (μ)*
ACTINOMYCETES (AEROBIC)					
Nocardia sp.	Soft	White-yellow	Tiny (0.5 mm)	None	1.0
Streptomyces madurae	"	Yellow	Large (2 mm)	"	"
Str. somaliensis	"	"	1 mm	"	"
Str.pelletierii	"	Red	Tiny (0.5 mm)	Not pronounced	"
FILAMENTOUS FUNGI					
Str. paraguayensis	"	Black	0.5-1 mm	None	"
M. mycetomatis	Firm	"	1 mm	Brown	2.0
M. grisea	Soft-firm	"	1 mm	"	2.0
Glenospora sp.	"	"	1 mm	"	2.0
Cephalosporium sp.	Soft	White-yellow	1 mm	None	2.0
Phialophora jeanselmei	"	White to yellow	1 mm	"	2.0
Pseudal. boydii	"	White to yellow	1 mm	"	2.0
Leptosphaeria senegalensis	"	Black	1 mm	Brown	2.0
Pyrenochaeta romeroi	"	Black	1 mm	Brown	2.0

from pain and local tenderness, there is little systemic reaction. The **clinical diagnosis** at this stage is easy and is **confirmed by laboratory findings** and **radiological examination** of the affected part. **Involvement of the bones** (the small bones of feet and hands in cases of mycetoma of foot and hand) is revealed by **both lytic and proliferative reactions.**

LABORATORY DIAGNOSIS

For confirming the clinical diagnosis it is not only necessary to prove the presence of a mycetoma but also to identify the aetiologic agent to facilitate the treatment. For this, the direct examination of the pus containing **"granules"**, and the **cultural and histopathological examinations** are essential.

DIRECT MICROSCOPIC EXAMINATION

The pus from a draining sinus or from an abscess is aspirated aseptically and is examined grossly for presence of "grains seen as small (1-2 mm) irregular masses of various colour. To facilitate this, the pus may be suspended in sterile distilled water in a test tube and examined against light. Sometimes the ***pus* collected** from a discharging sinus **may not contain granules.** In such cases dry sterile dressing may be applied to the affected part and kept there for 24 hours. The dressing is then removed and examined for presence of granules caught on the surface of the gauze soaked with pus. The **"granules"** are put in a drop of water or **10% KOH solution** on a slide and examined under the microscope and are found to us composed of either thin bacteria like branching organisms or of higher filamentous fungi causing this infection. This is of particular importance not only for immediate differentiation of the aetiologic agent for the purpose of treatment, but also for cultural techniques to be adopted in each case.

PLATE 17.III

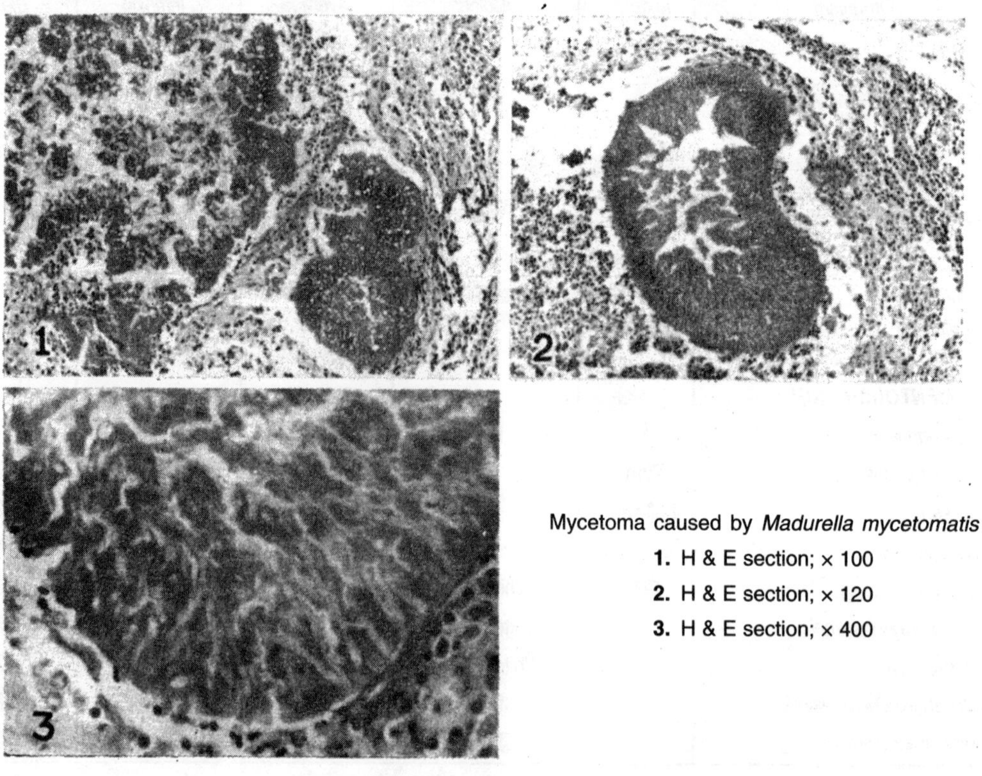

Mycetoma caused by *Madurella mycetomatis*

1. H & E section; × 100

2. H & E section; × 120

3. H & E section; × 400

Table 17.2 : Morphological and Physiological Characters of the AEROBIC ACTINOMYCETES (Agents of Actinomycotic Mycetoma)

Organism	Morphology			Biochemical Characters		
	Colony	*Microscopic*	*Optimum temperature*	*Proteolytic*	*Amylolytic*	*KNO_3*
N. asteroides	Rapid growing heaped or folded colony, yellowish to orange red, later covered with powdery aerial mycelium. Some strains produce browning of medium	Short rods & coccal forms, long branched forms seen in liquid culture or slide culture, 1μ in diameter, often beaded & partially acid-fast, Gram +ve	37°C	0	0	+
N. brasiliensis	" "	" "	30°C (some at 37°C also)	+	0	+
Str. madurae	Moderately fast growing cream coloured, glabrous with wrinkled surface	Delicate branched mycelium, 1μ in diameter, chains of conidia seen, non acid-fast, Gram +ve	37°C	+	+	+
Str. pelletierii	Very slow growing, small dry granular, pink to red	Delicate branched mycelium, 1μ in diameter, conidia rare, non-acid-fast, Gram +ve	37°C	+	0	0
Str. somaliensis	Slow growing, cream coloured, wrinkled becoming brownish, with age may develop aerial mycelium	Delicate branched mycelia, 1μ in diameter, chains of conidia in aerial mycelia, non acid-fast, Gram +ve	30°C	+	+	0
Str. paraguayensis	Rapid growing, creamy with dark central portion	Ramified filaments with branching 1μ in diameter, non-acid-fast, Gram +ve	30°C	0	0	0

Table 17.3 : Morphological and Physiological Characters of the Fungi Mycetoma (Maduromycosis) (Agents of Eumycotic Mycetoma)

Organism	Morphology		Optimum temperature	Proteolytic	Amylolytic	Assimilation of carbon D G L M S°
	Colony	Microscopic				
M. mycetomatis	Slow growing, flat or folded, ochraceous or yellow-brown	Hyphae 1-6μ, **chlamydospores** 24μ in diameter, some strains produce **phialides** and black **sclerotia**	37°C	Weak +	+	+ + + + –
M. grisea	Slow growing, grey velvety aerial mycelium over a brown growth	Hyphae 2 types—pale 1-3μ and dark thick with **budding cells** 3-5μ in diameter	30°C	Weak +	+	+ + – + +
Cephalosporium falciforme	Slow growing, fluffy, buff or pinkish, reverse—currant red	Hyphae 3-4μ in diameter, hyaline, slide shaped **conidia** in clusters on simple conidiophore	3O°C	Weak +	–	(not used)
Phialophora jeanselmei	Slow growing, black moist, becomes covered with greyish velvety mycelium, reverse—black	Hyphae made of chains of budding cells, **conidia** produced in tube-like conidio -phore	30°C	–	–	+ + – + +
Al. boydii	Rapid growing, fluffy, becomes brownish grey	Hyphae 1-3μ in diameter, **conidia** unicellular, borne singly, pyriform 4-9× 6-10μ	30°C (or 37°C)	+	+	(not used)

°D = dextrose; G = galactose; L = lactose; M = maltose; S = sucrose.

Culture studies

The agents of mycetoma (both the actinomycetes and true fungi) can be readily cultured in the laboratory on ordinary media. The granules picked up from the pus are invariably contaminated with pyrogenic bacteria which are usually found in the diseased tissue as secondary invaders. It is, therefore, necessary to wash the granules several times in sterile distilled water or physiological saline before putting up the culture. The granules composed of filamentous fungi are treated with a solution of physiological saline containing penicillin streptomycin (5,000 unit each per ml) for a few hours to get rid of the contaminating bacteria but the same cannot be applied to aerobic actinomycetes (nocardia and streptomyces spp.), as these are susceptible to the antibiotics. Cultures are then put up on Sabouraud dextrose agar slants containing chloromycetin (0.05 mg/ml) are also incubated with "**granules**". Several slants are incubated with material and half of them are incubated at 37°C and half at room temperature (25°C).

Table 17.4 : Histologic Criteria for the Specific Diagnosis of the Causative Agents of Mycetoma

I. AEROBIC ACTINOMYCETES (agents of Actinomycotic myctoma)	
1. *Nocardia asteroides*	Small grains, homogeneous, composed of loose clusters of thin branching filaments. Clubs may be present occasionally.
2. *N. barasiliensis*	Small grains, lobulated, homogeneous, mycelial filament delicate and difficult to observe, clubs may be present.
3. *Stretoptomyces madurae*	Large grains, irregular and serpiginous in shape, central part looks empty with a dense cluster of mycelial filaments surrounded by amorphous materials; clubs seen frequently.
4. *Str. pelletierii*	Small grains, red in colour, irregular in shape, often lobulated, homogeneous, mycelial filament delicate and not well observed, no club seen.
5. *Str. somaliensis*	Large grains, round or oral, compact and firm, composed of amorphous material with mycelial filaments prominent in the periphery, no club seen.
6. *Str. paraguayensis*	Compact, small granules composed of delicate mycelial filament.
II. FILAMENTOUS FUNGI (agents of Maduromycosis-Eumycotic Mycetoma).	
1. *Madurella mycetomatis*	Round or lobulated small (1 mm) granules, often in clusters, composed of light brown hyphae 1-5μ in diameter, embedded in brown cement substance, **chlamydospores** of varying size upto 25μ in diameter; clubs seen in fair number. The hyphae contain ochreceous particles.
2. *M. grisea*	Round or lobed granules often larger than 1 mm, unpigmented central area contains hyaline hyphae with large brown cells in periphery with dark brown cement substance. **Chlamydospores** of varying sizes seen in abundance.
3. *Pseudallescheria boydii*	Smaller (500μ) round granules made of hyaline hyphae (5μ in diam.) and **chlamydospores** of upto 20μ in diameter seen in large numbers in the periphery.
4. *Cephalosporium falciforme*	Larger (1 mm) irregular granules composed of hyaline hyphae of 3-4μ in diameter, and **chlamydospores** of 12μ in diameter.
5. *Phialophora jeanselmei*	Irregular, helicoid or vermicular (with hollow centre) granules of variable size, composed largely of dark brown **chlamydospores** (5-10μ in diam.) and rarely brown hyphae.
6. *Leptosphaeria senegalensis*	Small granules with non-pigmented centre made of network of hyaline hyphae, while the peripheral area shows dense black pigment in which are embedded hyphae and vesicles of variable size.

Histopathological findings in general : Biopsy of the infected tissues (H & E stained sections) reveals the presence of mycotic granules embedded in microabscesses studied over the area. The central zones of suppuration are surrounded by areas of chronic inflammation characterised by the presence of epithelioid cells, giant cells and round cell infiltration. In old lesions, areas of extensive fibrosis are also seen in between. The histopathological diagnosis rests on the microscopic demonstration of the fungal granule and the tissue reaction. The histologic examination of the granules and the tissue reaction. The histologic examination of the granules in tissue sections gives a fairly good idea about the nature of the causative agents of mycetoma. The granules present their characteristic features for most of the species involved in the **H & E** sections.

Prognosis

Mycetoma remains localised at the site of infection but it seldom heals by itself. It is a progressive condition and slowly spreads to the surrounding tissue involving the bones. Secondary infection with pyogenic organisms aggravates the symptoms and ultimately causes death of the patient. The prognosis is very good if treated very early and the disease is arrested by proper therapy or by excision or amputation as the case demands.

Therapy

Various fungicidal drugs like Copper sulfate, Pot. iodide, etc., have been tried in the past without any success. The choice of treatments depends on the nature of the causative agent and the stage of the disease. The aerobic actinomycetes (Nocardia and Streptomyces sp.) respond to sulphonamide therapy. Long term treatment with **sulphonamides and with Dapsone (diamino diphenylene sulphone)** in particular has proved very effective in cases of actinomycotic mycetoma. No satisfactory antifungal chemotherapeutic agent or antibiotic is yet available for treatment of maduromycosis. Surgical exclusion of the diseased tissue amputation of the diseased part is the only method of saving life. The secondary infection should first be properly controlled with appropriate antibiotics before any surgical procedure is performed.

PATHOLOGY OF MYCETOMA

(A) GENERAL PATHOLOGY OF MYCETOMA

Geographic Pathology

Mycetoma is found in all climates, but it is very obvious that the distribution is almost entirely related to the habit of working and walking barefooted, which is more prevalent in developing countries and in tropical climates. The geography has been discussed by Mariat (1963), and by Vanbreusegham (1967) and by others. However, Mariat's survey gives an erroneous picture as far as India is concerned, since it was based on a questionnaire with a poor response from Asia in general. South India is the "classical country of mycetoma, which was first described by Kaempher from the Malabar coast".

Subsequently many reports have appeared from South India but accurate classification and differentiation from other conditions was lacking in early reports. Little was known about the incidence in North India. Recent investigations by Grueber suggest that the disease is almost as frequently found in North India as in South India (Grueber, 1971, unpublished data). Over the course of ten years the following incidence was found in the routine surgical biopsy material of a South and a North Indian Medical College respectively (see page 135).

These figures are based on the publication of Klokke *et al.* (1968), and a series of 58

	No. of Biopsies during 10 year period	*Ratio of Biopsies to Mycetomas*	*Mycetomas in % of all Biopsies*
South Indian College	88,641	974:1	0.12%
North Indian College	44,077	1202:1	0.083%

The incidence of Eumycetomas and Actinomycetomas in South and North India is also comparable:

	Total no. of cases	Eumycetomas	Actinomycetomas
South India	187	83(44.4%)	104(55.6%)
North India	124	47(38%)	77(62%)

mycetomas from North India (Gruebar, unpublished), supplemented by a number of publication on North Indian mycetomas (Andleigh, 1954; Behl and Sharma, 1957; Kakoti and **Dey, 1956**; Kandhari *et. al.* (1968) 1954; Chouhan and Aggarwal, 1969; Mankodi and Kanvinde, 1970). Causative organisms which were believed to be absent in India have now been shown to be present. This applies for instance to ***Streptomyces somaliensis*** which was found by Klokke *et al.* in South India and by Grueber and Mohana Kumar (1970) in North India.

Aetiology

The literal translation of "**mycetoma**" is "**fungus tumour**", but more than half of the mycetomas in India are caused by higher bacteria which produce branching filaments (actinomycetoma).

Since they all belong to the group of *Actinomycetales*, these mycetomas are called "*ACTINOMYCETOMA*" or "*ACTINOMYCOTIC MYCETOMA*" (Winslow and Steen, 1964; Zaias *et al.*, 1969; Latapi, 1963; Vanbreusegham, 1967). The term "**actino-mycotic mycetoma**" should not be used to avoid confusion with actinomycosis in the strict sense, though actinomycosis is, in principle, also a mycetoma (see subsequent discussion). The most common agents of *ACTINOMYCETOMA* are summarised in **Pl. 17.I.**

True fungal mycetoma **(EUMYCETOMA OR MADUROMYCETOMA).** This true mycetoma is caused by a great variety of true fungi. It is to mention here only a few organism of eumycetoma which have been found more frequently in India: ***Madurella mycetomatis***, *Madurella grisea, Leptosphaeria senegalensis,* ***Pseudallescheria boydii, Phialophora jeànselmii, Pyrenochaeta romeroi*** (see **Pl. 17.II**).

Both actinomycetales as well as true fungi, cause similar destructive lesions clinically, and they are present in the tissue as granules or grains which are fungal colonies. Table 15.1 give a survey of the most important granules and their comparative sizes and shapes in tissue sections. These **granules are a "sort of touch stone" for the diagnosis** (Vanbreusegham, 1967). Their recognition and proper identification ***histologically*** is vital for a correct diagnosis. Proper differentiation of actino-mycetoma and eumycetoma is of special clinical importance because actinomycetomas are responsive to prolonged chemotherapy with sulphonamides whilst the cure of eumycetomas is still as great problem. Vanbreuseghem (1967) emphasises that "in developing countries, amputation considerably reduces life expectancy and is a sort of delayed death certificate". The pathologist has, therefore, a great responsibility in accurately identifying the organism. This is possible in 95 per cent of all cases by histopathological investigation only.

Gross Pathology

Location

Agents of mycetoma gain entrance through small injuries, usually of the feet. But a mycetoma may be localised anywhere. I have seen two cases with extensive involvement of the scalp and the skull. In one patient the bone was completely destroyed and the dura was already involved. Extremities and back are not so infrequently infected. **"Thorn prick", "Camel stepped on the foot", "nail in shoe", are recurrent histories.**

EVOLUTION OF LESIONS

Early stages show a small painless subcutaneous swelling, with occasionally a pustule or a blister in the centre. Over months, even years, the organisms spread silently in the tissue *in the fashion of a malignant tumour*, which must be remembered when surgery is done. If the surgeon's knife touches the lesions of an eumycetoma, there will be no cure and the destruction of healthy tissue will continue relentlessly. Organisms respect neither fascias nor bones. There may be intercommunicating small abscesses and sinus tracks in the tissue. After 6 months to two years, small nodules may appear in the skin prior to formation of fistulae which discharge oily fluid, pus and sometimes blood (**Fig. 17.1**). Granules of various sizes and colours may indicate the type of mycetoma and should be cultured after careful inspection. They consist of colonies of the organisms with some adherent pus (For characters of granules see **Tables 17.5 and 17.6 and Pls. 17.I** and **17.II**).

Bezes maintains that the appearance of skin nodules and sinuses gives some indication as to the nature of the organism, which cannot be supported as except that in black grain mycetomas, often no nodules are formed, but that granules appear in the openings of sinuses with a slightly raised skin around them. While the disease progresses, the foot may become rather swollen and be covered with numerous nodules and sinuses. Walking becomes difficult, though there is relatively little pain. Other areas of the body may be involved and be studded with sinuses with very firm nodular skin, alternating with soft "doughy" areas over abscesses.

The speed of bone destruction depends much on the nearness of the primary lesion to bone. However, some organisms seem to be **"osteophil"**, as for instance ***Streptomyces madurae*** which causes not infrequently extensive bone destruction. Generally actinomycotic mycetomas seem to be more destructive to the bone, than eumycetomas. *Roentgenologic* lesions consist initially in irregular and thickened, or "moth eaten" periosteum. Later, there are punched-out destructive bone lesions alternating with abnormal osteogenic areas.

ACTINOMYCETOMAS

(1) Nocardia sp. (However, *N. asteroides*, *N. brasiliensis* and *Actinomyces israelii* **granules are not definitely distinguishable in tissue sections. Cultural identification of these organisms is necessary for a definite diagnosis**).	
(2) *Streptomyces madurae*	
(3) *Streptomyces somaliensis*	Distinction in histological sections is difficult
(4) *Streptomyces pelletierii*	

EUMYCETOMAS

(1) *Madurella mycetomatis*	
(2) *Madurella grisea*	Very characteristic histologically
(3) *Pseudallescheria boydii*	
(4) *Leptosphaeria senegalensis*	

Table 17.5 : Actinomycetoma : Characters of Granules in Tissues

Name of organism	*Colour (Gross)*	*Size in microns*	*Shape*	*Centre in H & E stained sections*	*Fringe in H & E stained sections*	*Gram stain*	*Acid fast stain*
Actinomyces israelii	White or yellowish	300 to 1000	Round to slightly lobulated	Pale-blue fuzzy filaments	With or without eosinophilic "clubs"	Positive for filaments but not in all parts	Negative
Nocardia asteroides	White or yellowish	100 50 400	Round, worm-like, lobulated	Faint blue dots, or fuzzy filaments	Pink margin, sometimes "clubs"	Positive—usually fragmented filaments	Positive—some strains, others not
Nocardia brasiliensis	White or yellowish or brick-red	60 to 300	Round ring, kidney-shaped, lobulated	Indistinct, faint blue tangle or dots	Margin "fading away" almost never with clubs	Positive—fragmented filaments	Positive—some strains
Streptomyces madurae	Pale yellow to pale pink	500 to 3000	Cartographic, polycyclic, irregular	"thundercloud" appearance, empty centre and deep blue filaments at edge	Long pink "clubs around edge. Some granules without clubs	Positive—distinctly branching filaments seen	Negative
Streptomyces pelletierii	Deep red	300-500 to 1000	Lobulated, oval, but almost always fragmented	Deep blue homogeneous centre. No filament recognisable	Eosinophilic denticulated edge, 2-4 microns wide	Positive—only occasional filaments seen	Negative
Streptomyces somaliensis	Pale yellowish, ochre or brown	500 to 1000	Round, oval, with cracks due to cement substance	Pale blue, homogeneous centre	15-20 microns slightly crenated, pink margin	Positive—filaments seen only at edge	Negative

Table 17.6 : Eumycetoma Characters of Granules in Tissue (True Fungal Mycetoma)

Name of organism	*Colour (Gross) and texture*	*Size in microns*	*Shape*	*Centre in H & E stained sections*	*Edge in H & E stained sections*	*Cement substance*
Pseudallescheria boydii (Monosporium apiospermum)	White or yellowish hard	500 to 2,000	Lobulated, oval elongated	Parallel hyaline hyphae, 4-5 microns diameter, but appear to be thicker. Large **chlamydospores**, up to 20 microns in diameter	Vesicles-like	None
Leptosphaeria senegalensis	Black, hard but can be crushed	700 to 1,000	Kidney-shaped, oval, plump sausage-like	Tangled, irregular colourless; hyphae 2-4 microns diameter, faintly seen, bamboo-like. Bacteria always present	Frayed, crenated, "tattered", with large vesicles and black pigment	Yes, black peripheral band between vesicles and hyphae
Madurella mycetomatis	Black or chocolate-coloured, firm	1,000 to 10,000	Lobulated, rosette-shaped, branching	2-5 microns hyphae, orderly radiating arrangement with 20-25 microns diameter **chlamydospores**. Brown or black pigment in some hyphae	Vesicles up to 20 microns diameter	Yes, brown, throughout granule
Madurella grisea	Black	Ca. 1,000	Lobulated, oval	Relatively parallel hyphae in colourless centre faintly recognisable. No bacteria	Curvilinear, with smaller vesicles, brownish pigment	Yes, brown peripheral band with embedded vesicles and hyphae
Cephalosporium falciforme (Very similar histologically to Pseudal. *boydii*)	White, or yellowish, others ? black, soft	1,500	Lobulated	Hyphae, about 3 microns in diameter, with small **chlamydospores**	Vesicular	

Complications

Concomitant staphylococcal and streptococcal (bacterial) infection is often found, particularly in actinomycetoma.

Regional lymph nodes are almost always swollen but metastases via lymphatics are very rare. No haematogenous spread takes place.

Histopathology

By careful comparison between culture and the histological morphology of the granule, it has been established, that most granules can be identified accurately in tissue sections. This progress has been made possible by the painstaking work of Mackinnon in Uruguay and the French workers Segretain, Mariat, Camain, Destombes and others. The important criteria for histo-morphologic identification of granules are summarised in **Table 17.5** and **17.6**.

The following granules can be sufficiently identified in ordinary H & E stained sections :

THE GENERAL TISSUE REACTIONS are not pathognomonic for causative organisms and consists in abscess formation near the granules with variable degree of oedema, and **diffuse infiltration with plasma cells**, some Neutrophils, lymphocytes and histiocytes. The proportions depend mainly on the stage of the disease, not so much on the causative organism. The skin near fistulae may show hyperkeratosis, irregular acanthosis, and occasionally pseudoepitheliomatous hyperplasia and purulent destruction at the place of fistula. In other cases, the epithelium may be fitted off by pus or serous fluid.

Foreign body giant cells are frequently found adjacent to other granules, or fragments of granules which are probably made up by dead fungi (**Pl. 17.X, Figs. 27 and 28**).

Langhans giant cells, foreign body giant cells and a few histiocytic or epithelioid granulomas may also be found, but they are usually not prominent.

Differential diagnosis

Clinically, mycetoma may be confused with tuberculous osteomyelitis, chronic bacterial osteomyelitis and, sometimes, with tumours. It has not been seen that a single instance of leprosy or syphilis which was mistaken for a mycetoma, contrary to statements in the older literature (Grueber, 1971). Of forty cases in India which were diagnosed clinically as a mycetoma, or in which this diagnosis was considered, twelve had tuberculous osteomyelitis, eleven nonspecific osteomyelitis, seven bacterial nonspecific inflammation of the soft tissue, six tumours (nevi, basal cell carcinoma, dermatofibroma, sarcoma), and four showed chronic ulcers.

What differentiates mycetoma from the other diseases mentioned?

1. **Generally, mycetoma patients have a much longer history**. The average duration of illness from onset to presentation in 40 non-mycetoma patients, erroneously diagnosed as having mycetoma was 27 months in randomly matched mycetoma cases; the history extended over 45 months on an average. **Many patients with mycetoma do not consult the doctor before a year has lapsed** since appearance of first symptoms.

2. **The presence of fistulas** is a cardinal symptom of fairly advanced mycetoma. Fistulae may also be present in tuberculosis and in non-specific osteomyelitis, however, no granules are found in these infections. In mycetoma, the fistula is frequently in the centre of a raised nodule or has raised walls. **Diagnosis of mycetoma in the early stage, should be aimed at when only a swelling, but no fistulae are present. A biopsy is essential for the diagnosis.**

3. **Large ulcers are not typical of mycetoma and should be excluded**

microabscesses within subacute and chronic inflammatory granulation tissue, with a fair proportion of plasma cells, much oedema, without evidence of tuberculous lesions and sequestration, he should search for granules in serial tissue sections.

For further reference, the serious student of mycetoma should read two excellent and thorough contributions on this subject, namely Latapi's chapter on mycetoma in "Handbuch fuer Haut-and Geschlechts-Krankheiten (1963), and Vanbreusegham's monographic article on mycetoma (1967).

(B) SPECIAL PATHOLOGY OF MYCETOMA

1. ACTINOMYCETOMA

The order of Actinomycetales includes the following pathogenic members. Many of those who are able to form branching filaments may produce mycetomas:

Regarding *actinomycetales*, there is a great deal of confusion in the older literature. Latapi (1963) may be consulted for a host of synonyms.

The histology of the granules is summarised in **Table 17.5** and in **Pl. 17.I.**

ACTINOMYCES ISRAELII

A detailed description of this higher bacterium is given in the separate paragraph on actinomycosis. However, it may be re-emphasised here that:

1. Actinomycosis is a specialised form of mycetoma (Vanbreusegham, 1967).
2. The granule of actinomycosis in the tissue cannot be distinguished reliably from *Nocardia* granule (see **Pl. VI, Figs. 10** and **11** and "Actinomycosis", **Pl. IV, Figs. 1-3**, and **Fig. 4**), but one can relatively safely assume that lesions in the well-known locations of actinomycosis are generally caused by *Actinomyces israelii*. Other locations represent *Nocardia* mycetomas in most instances. It is incorrect to establish a diagnosis "Actinomycetoma-Actinomycosis", only on the authors (Mankodi and Kanvinde, 1970).
3. Eosinophilic radiating "clubs in the periphery of the granule are not pathognomonic of actinomycosis and may be present around *Nocardia*, (see "Actinomycosis" **Pl. 17.IV, Fig. 2,** and **Fig. 4**)

Species	*Mycelium*	*Anaerobe*	*Aerobe*	*Acid fast*
Mycobacterium: *—tuberculosis* *—ulcerans* *—leprae* and others	None or rudimentary	—	Yes	Yes
Actinomyces: *—israelii* *—bovis*	Yes; fragmentation into coccoid or bacillary forms	Yes, or microaerophil	No	No
Nocardia: *—asteroides* *—brasiliensis*	Yes; fragmentation	No	Yes	Variable, for many strains yes
Streptomyces: *—madurae* *—pelletierii* *—somaliensis* *—paraguayensis*	yes; but no fragmentation	No	Yes	No

PLATE 17.IV

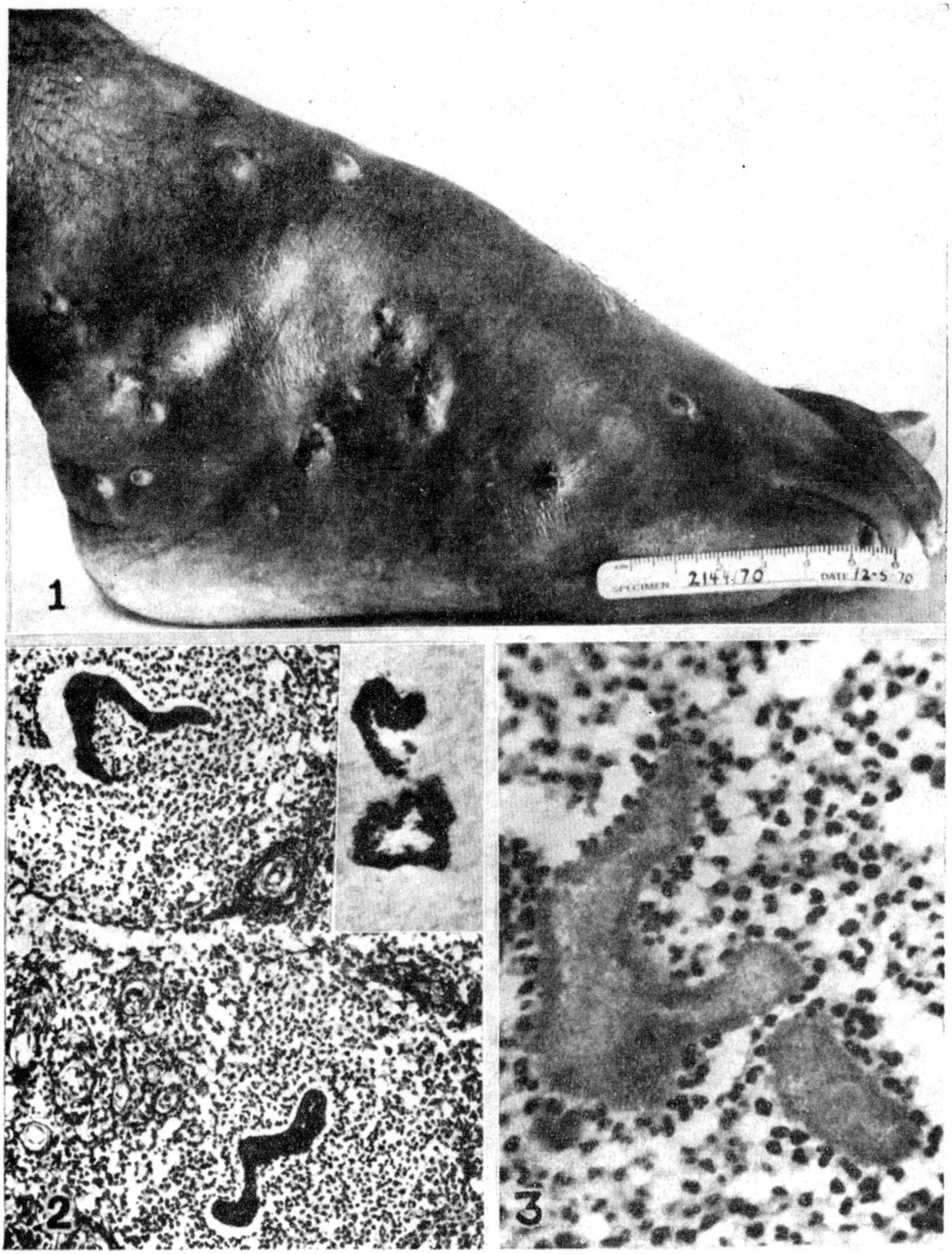

Fig.1 : Mycetoma of foot caused by *Nocardia* sp.; three years history. Sinuses open in the centre of nodules, discharging pus and serosanguinous fluid granules. The bone is already involved in this case.

Fig.2 : Vermiform *Nocardia asteroides* granules in **microabscesses**. Silver methenamine, 130 ×.

Fig.3 : *Nocardia asteroides* granules in tissue. Note pale-basophilic "powdery" appearance of granule in H & E section 560 ×.

PLATE 17.IV A

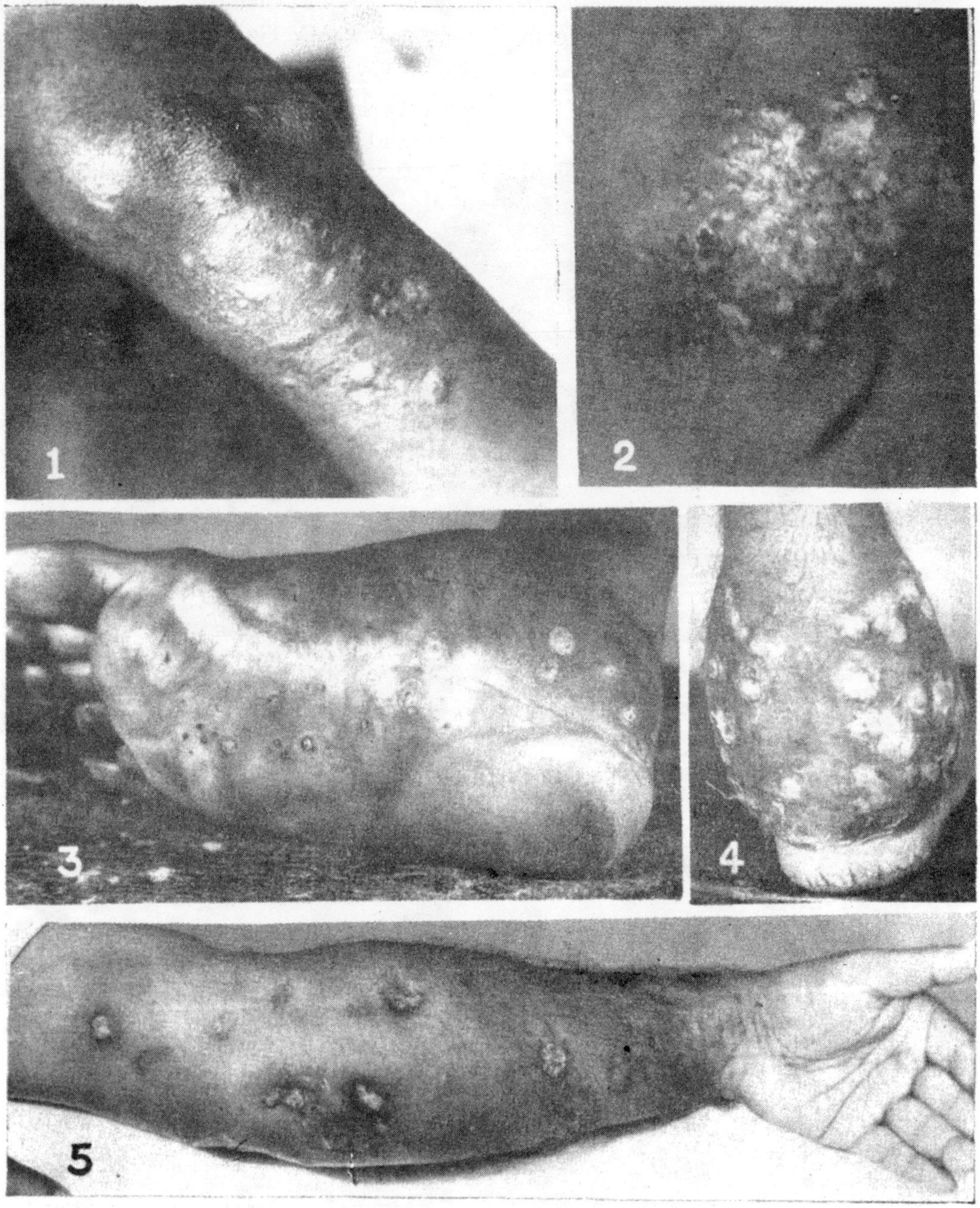

Mycetoma caused by Nocardia Species—(Clinical Appearance)

from the differential diagnosis. Tumours may occasionally resemble mycetomas, but histological examination will immediately show the true nature of the tumour.

A history of injury, duration of the disease, **X-ray findings** (bone involvement?), presence of granules in the exudate, and their size, colour and texture, are essential pieces of information for the examining pathologist. If he finds small

Streptomyces madurae (**Pl. 17.V, Fig. 6**), or botryomycosis granules. It would be permissible, but not desirable, to refer to these organisms as "**ray fungus**" as it has been done previously, but only one should bear in mind that they are not fungi and that they are not uniform at all but belong to different families.

NOCARDIA ASTEROIDES

Pathogenicity

Latapi's statement that *N. asteroides* rarely causes mycetoma, may be true for Mexico but not for India. *Nocardia* species also invade tissues diffusely, particularly the lungs, without forming granules. This diffuse invasion is discussed in the paragraph in Chapter 11.

Morphology of the grain

Nocardia sp. forms the smallest of all granules. Their morphology has in detail been investigated by Latapi, 1959; Destombes *et al.*, 1961; Mariat, 1962.

The small granules are easily missed in tissue sections. They are usually found in the centre of abscesses (**Pl. 17.IV, Fig. 2; Pl. 17.IV, Figs. 11, 12**). They are irregular, vermiform or circular.

Culture : *Nocardia sp.* grow easily on Sabouraud's media at **37ºC**. Colonies raised, pale, yellowish or grey, sometimes of slight ochre colour, folded or like "**crumbs of bread**" (**Figs. 17.4 and 17.5**). Culture mounts from young colonies show short hyphae 1 micron in diameter which are often fragmented. In older cultures, almost only bacillary or coccoid forms may be found, but arrangement in chains betrays still the filamentous character. *N. asteroides* is Gram positive, but not all strains are acid fast, others are weakly acid fast, so that **Fite Faraco stain** *or mild decolourisation should be applied.*

N. BRASILIENSIS

Histology

See also Tabl3 15.5 The granules in tissue are vermiform or irregular circular with a "Hollow" centre (**Pl. 17.IV; Fig. 2 inset**) containing pus. They are indistinguishable from

PLATE 17.IV B

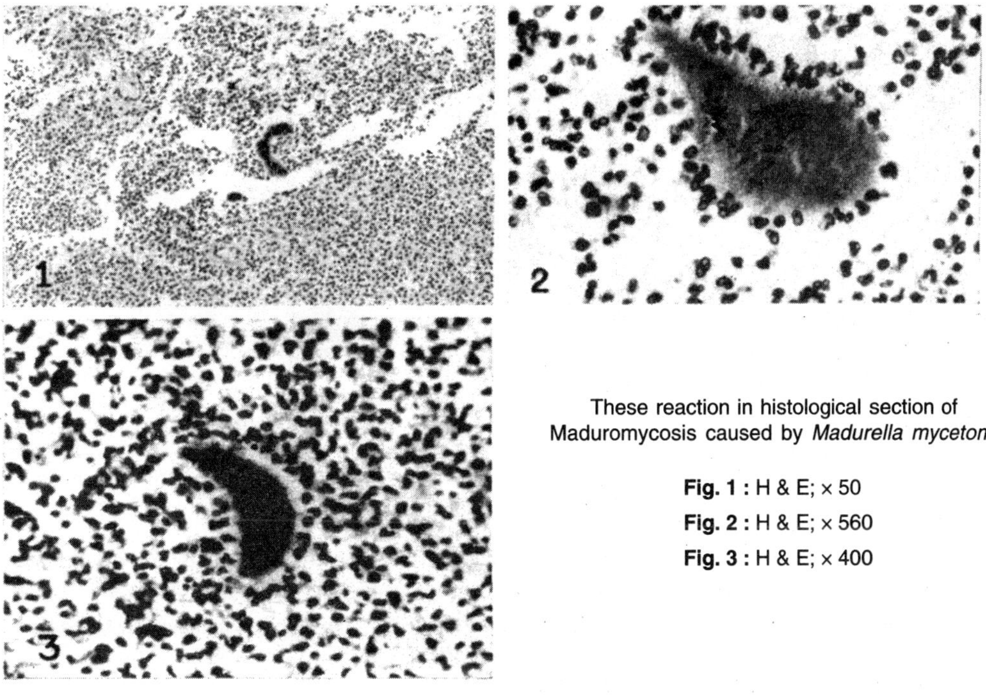

These reaction in histological section of Maduromycosis caused by *Madurella mycetomi*

Fig. 1 : H & E; × 50

Fig. 2 : H & E; × 560

Fig. 3 : H & E; × 400

PLATE 17.V

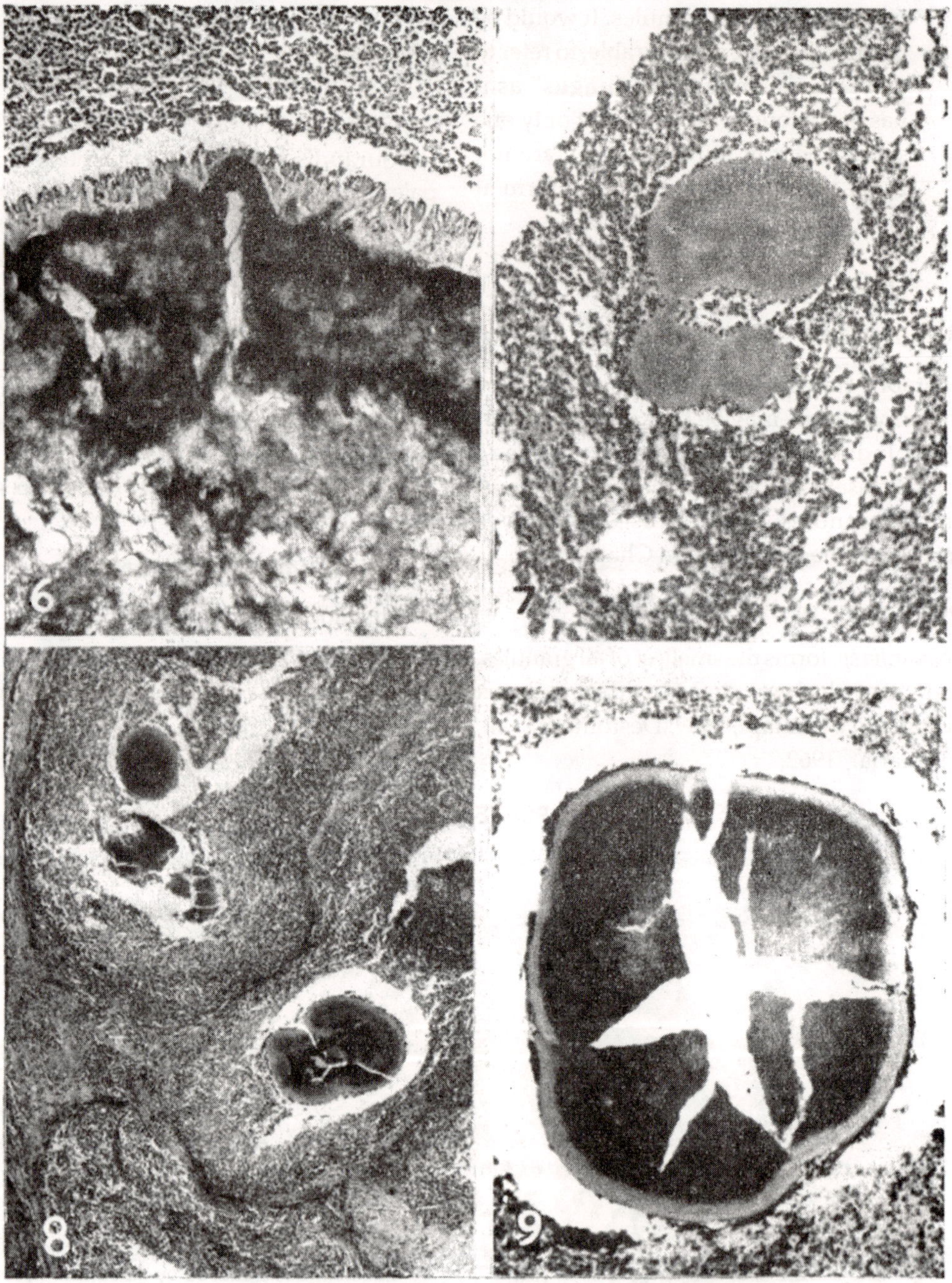

Fig. 6 : *Streptomyces madurae* in tissue section. Typical "thundercloud" appearance due to accumulation of deeply basophilic hyphae in the periphery of granules which has a fringe of radiating eosinophilic clubs" (see also **Pl. 17.V, Fig. 14**). H & E; 130 ×

Fig. 7 : *Streptomyces somaliensis* granules with pale blue centre and eosinophilic marginal zone in H & E section, 130 ×

Fig. 8 : *Streptomyces somaliensis* granules surrounded by microabscesses. Gram stain; 500 ×

Fig. 9 : *Streptomyces somaliensis* granule with typical "cracks" in centre due to brittle cement substance produced by this species. Gram stain; 130 ×

Fig. 7 and 9 **: Courtesy Sabouraudia, 8, 108-111, 1970**

PLATE 17.V A

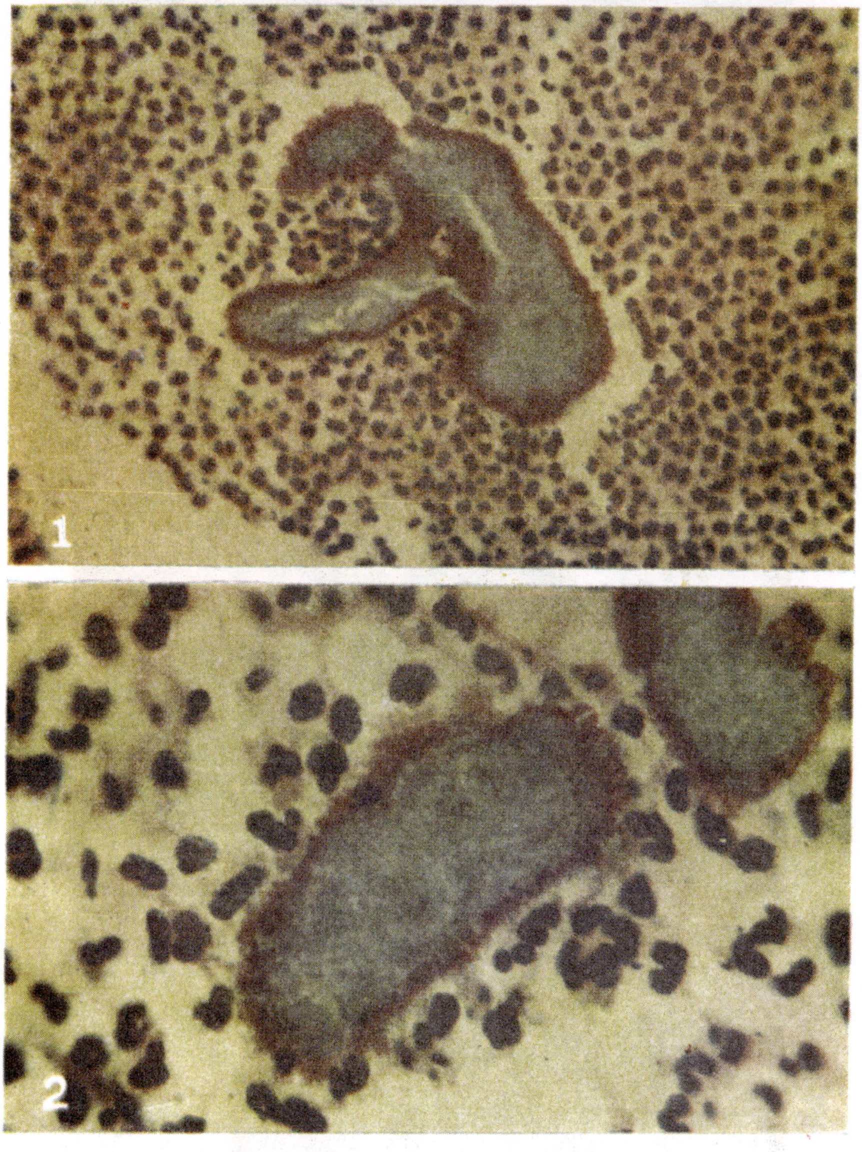

Histological appearance of mycetoma caused by Nocardia species. Note the fungal granule (colony) in the centre and the surrounding neutrophilic reaction. H & E; **Fig. 1 :** 400 ×, **Fig. 2 :** 1000 ×

N. asteroides and may or may not have a fringe of eosinophilic "clubs", like granules of *A. israelii*. (**"*Actinomycosis*" Plate X, Fig. 2, and Fig. 4**). The filaments in the centre of the granule are not visible in H & E stained sections but they appear to be more basophilic and thus stain more blue than the very faint blue centre of *N. asteroides*.

Culture

Growth on Sabouraud's agar yields yellowish or orange, raised cerebriform, dry colonies with an earthly odour. *Nocardia sp.* may be isolated from soil by the technique of **paraffin baiting**. The same method has successfully been applied to the isolation from clinical samples by Mishra and Randhawa (1968). Differen-

PLATE 17.V B

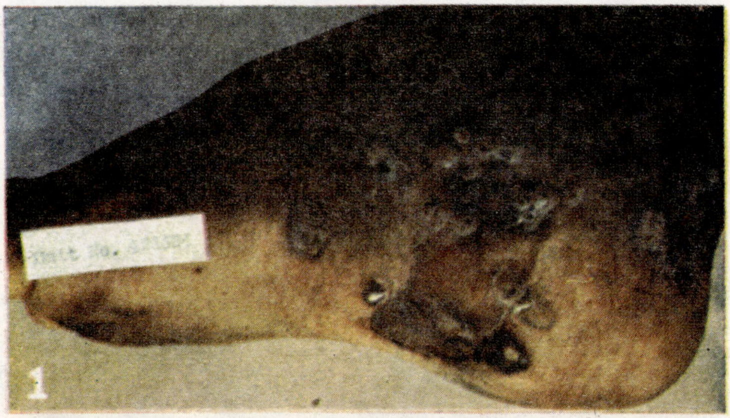

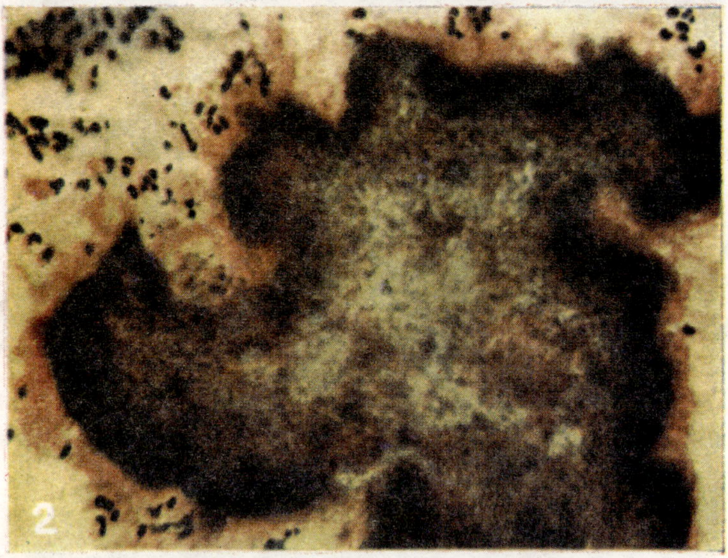

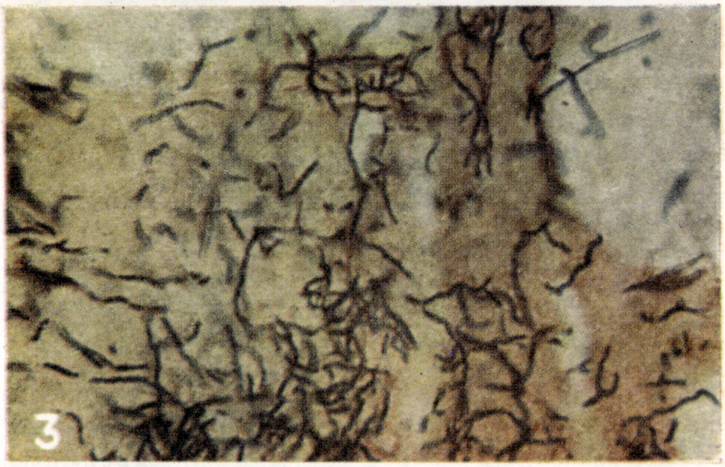

Fig. 1 : Gross appearance of mycetoma caused by *Streptomyces madurae* representing several draining sinuses

Fig. 2 : Histological appearance of a granule (colony) of *Streptomyces madurae*. H & E; × 400

Fig. 3 : Smear from culture shows Gram positive mycelial filaments of *Streptomyces madurae*. Gram; × 1000

tiation of *N. asteroides* from *N. brasiliensis* should be left to experienced mycologists, since biochemical reactions and other characters of these two species are rather variable and form a spectrum rather than sharply defined characteristics (Kumar and Mohapatra, 1968).

Excellent mycologic studies on Nocardia have been published by Mariat (1965), (1965), McClung (1963), Kumar and Mohapatra (1968), Kurup and Sandhu (1965, & 1967) and by Kurup *et al.* (1967) (For Reference : see Ch. 9: "Nocardiosis").

STREPTOMYCES MADURAE

Histology : The granule is exceedingly characteristic and resembles a "**thundercloud**" due to an irregular, relatively unstained and empty centre and a broad zone of deeply blue (H & E) stained margin, in which delicate filaments can be distinguished with high power and oil immersion objective (**Pl. 17.V, Fig. 6**). A broad zone of eosinophilic "clubs is usually, but not invariably present (**Pl. 17.V, Fig. 6 and Pl. 17.VI, Fig. 14**). The granule is one of the largest granule of mycetoma and can have a diameter of up to 1 cm. If only the deeply blue marginal zone is sectioned tangentially, the granule may be mistaken for a *Streptomyces pelletierii* grain.

Culture

Streptomyces madurae gows on Sabouraud's agar relatively quickly and forms glabours, moist, cream coloured or slightly pinkish colonies. Ageing colonies become powdery, chalky, due to development of aerial mycelium. Fin branching and usually not fragmented filaments, 1 micron in diameter, may by seen in culture-mount. They are not acid-fast but gram positive. Gelatine is liquefied and milk is coagulated.

Histology

Deeply blue granules which usually are broken up in several fragments are seen. No filaments are recognisable. The margin shows fine denticulae. For further details see Table 17.5 and Fig. 17.1.

Culture

Colonies on Sabouraud's glucose agar are usually pink or deep coral red, moist, heaped, glabrous, wrinkled. Gram positive, branching, non acid fast filaments are seen in culture mounts.

STREPTOMYCES SOMALIENSIS

Histology : The organism produces a ground substance which makes the granule in tissue extremely brittle, cracks in the pale blue centre of the granule are therefore a mark of recognition, **Pl. 17.V, Fig. 9**. There is a regular 15-30 microns wide peripheral pink zone with a finely denticulated out margin, **Pl. 17.VI Figs. 7-9**. Filaments in silver impregnated or gram stained sections show only up at the border between blue centre and pink margin of the granule (**Pl. 17.VI, Fig. 13**).

Culture : Growing rapidly under production of cream coloured, and later on darker brown and grey colonies which are wrinkled. Filaments are not acid fast, but gram positive, and produce sometimes **chains of arthrospores**.

STREPTOMYCES PARAGUAYENSIS

Culture : Smooth soft white centre with dark brown periphery. Similar filaments are produced to those of the other *streptomyces* species.

II. EUMYCETOMA

(True fungal mycetoma, Maduromycetoma)

If granules have been identified by direct microscopy as true fungi, they should be washed in saline containing antibiotics (e.g., penicillin and streptomycin: 500 units/ml; or gentamycin sulfate: 100 micrograms/ml). Cultures may be made on Sabouraud's agar containing 0.5% yeast extract and suitable antibiotics. Penicillin and streptomycin must be added to the cooled medium as a sterile solution. Media containing fungitoxic agents should not be used.

PSEUDALLESCHERIA BOYDII

Mycetoma due to this organism is more frequently seen in South India than in the North.

PLATE 17.VI

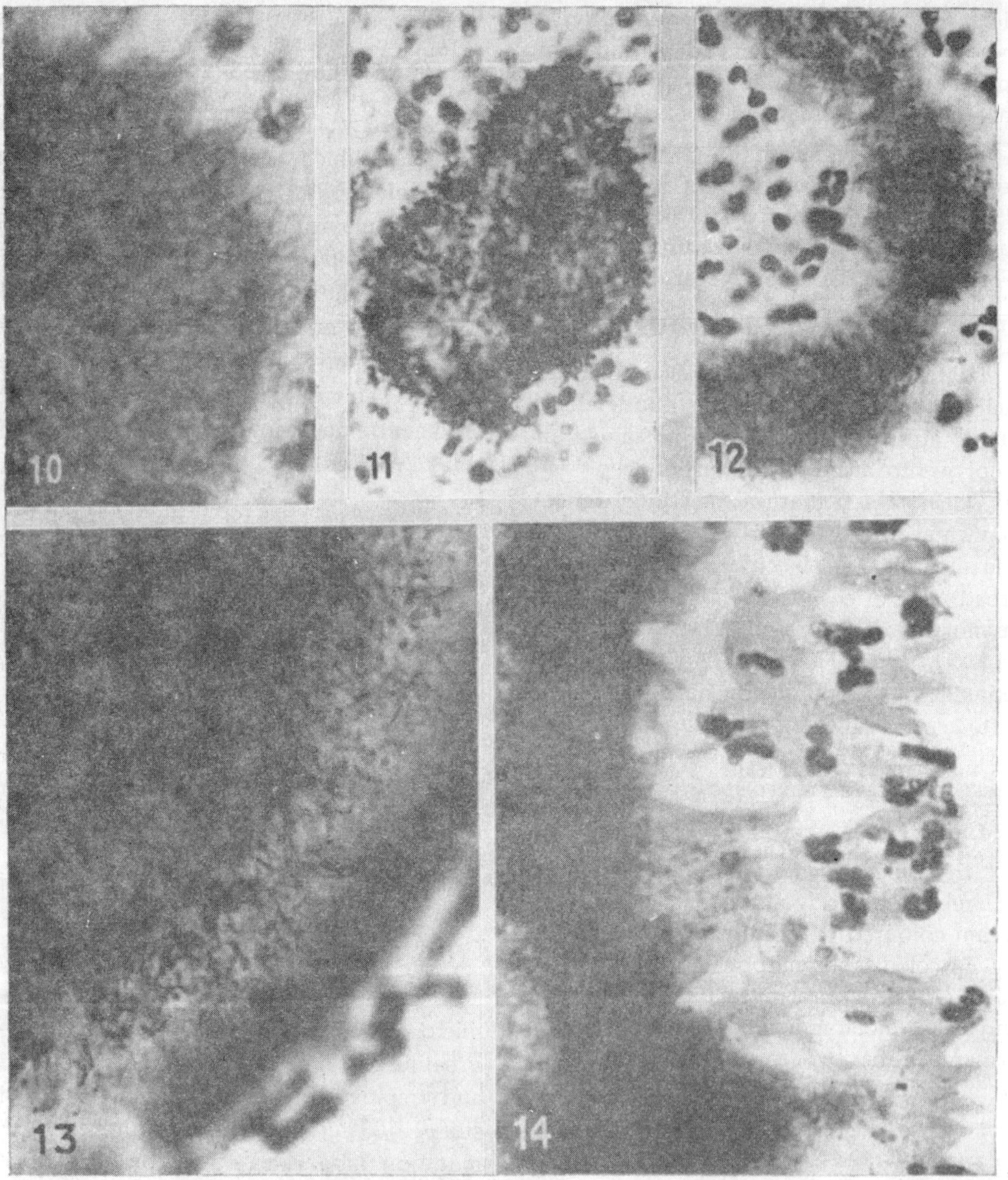

Figs. 10 to **14 :** Margins of actinomycosis and actinomycetoma granules at a magnification of 1000 ×

Fig. 10 : *Actinomyces israelii*, indistinct filaments in H & E section

Fig. 11 : *NOCARDIA* sp. : Short filaments recognisable at edge, but otherwise "**powdery**" **appearance** inspite of Gram stain.

Fig. 12 : Stubby filaments of *Nocardia* granule which appear slightly thicker due to silver impregnation. Silver methenamine.

Fig. 13 : Marginal zone of *Streptomyces somaliensis* **granule**; only here short filaments can be recognised, not in the central zone. Gram stain. Courtesy : Sabouraudia, **8**, 108-111, 1970

Fig. 14 : *Streptomyces madurae* showing dense tangle of deeply bluish stained filaments in marginal zone with peripheral fringe of eosinophilic "clubs" showing entangled polymorphs (right part of picture). H & E.

The fungus represents the **ascocarpic form of *Monosporum apiospermum*** (Emmons).

Histology

The granule has a curvilinear outline and is usually 500 to 2000 microns in diameter. The centre consist of hyaline hyphae, approximately 2 to 5 microns in diameter, showing chlamydospores up to 20 microns in diameter (**Pl. 17.VII, Fig. 15**).

Culture

Growing rapidly on Sabouraud's dextrose agar with white serial mycelium, soon turning smoky or brown. Cultures fill the Petri dish completely, producing lateral or terminal short conidiophores with conidia of various shapes and sizes (**Pl. 17.VII, Fig. 16**). In some ageing cultures ***cleistothecia* with ascospores develop**. Some strains produce *Dendrostilbella* stage on primary isolation.

LEPTOSPHAERIA SENEGALENSIS

Histology : Granules are 700 to 1000 microns in diameter, oval curvilinear in outline. The cortical zone is black or dark brown with spiny

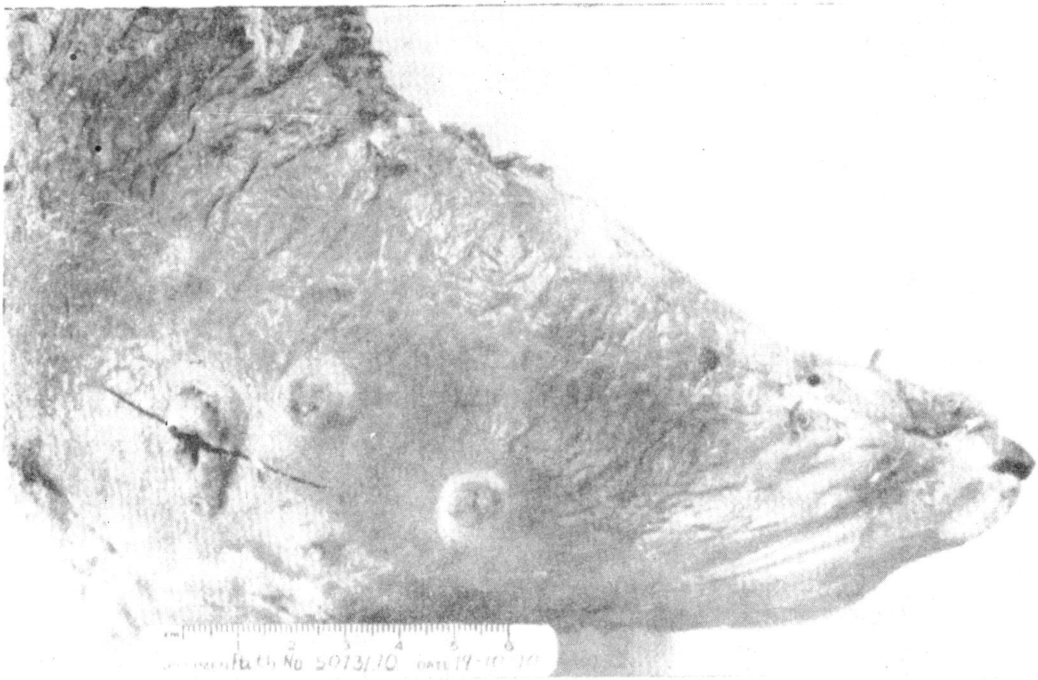

Fig. 17.1 : Black-grain mycetoma of foot—*Madurella mycetomatis*

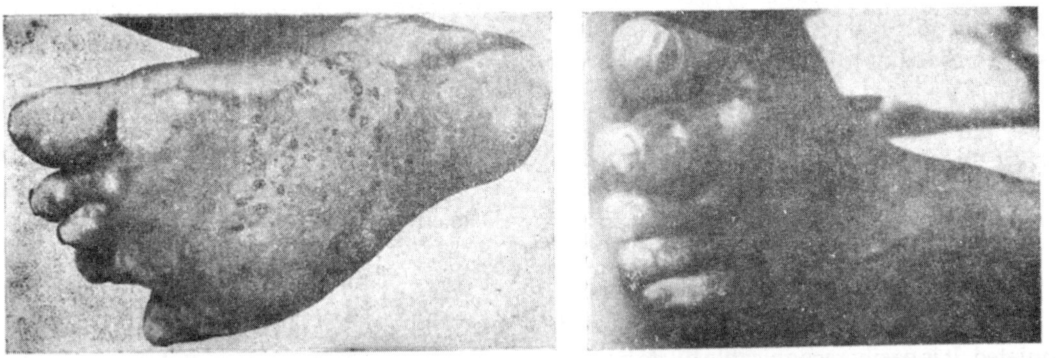

Fig. 17.1A : Mycetoma affecting the sole and dorsum of the foot

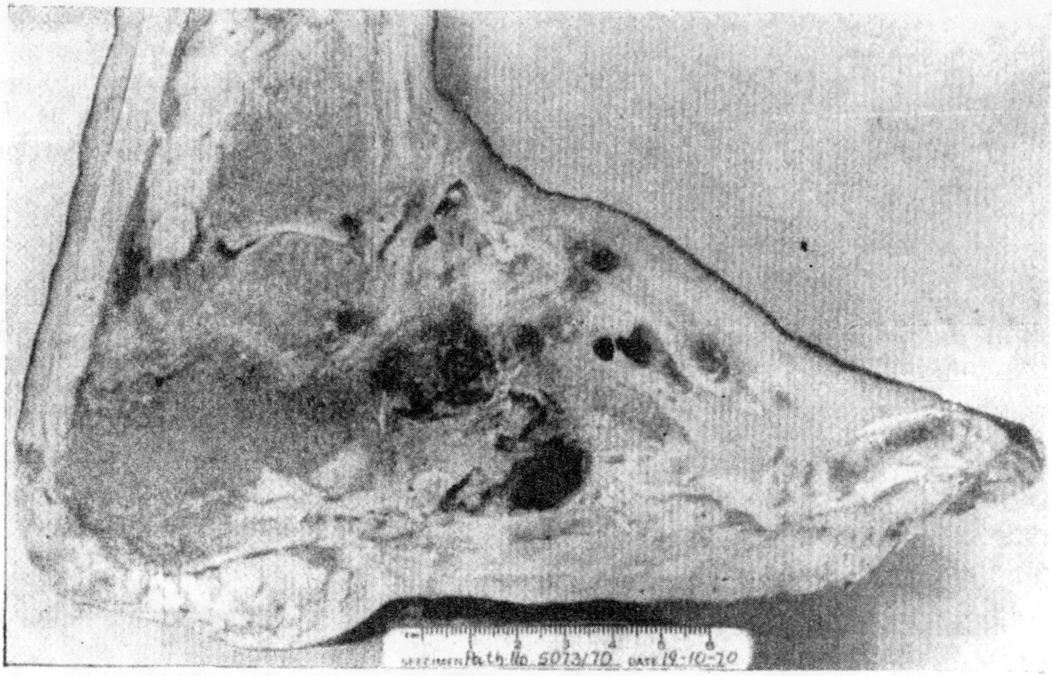

Fig. 17.2 : Sagittal sections through foot shown in **Fig. 17.1**. Black grains of *Madurella mycetomiatis* seen in large destructive lesions of bone and subcutaneous tissue

excrescences. There are peripheral vesicles up to 20 microns, which are prominent. Filaments are arranged irregularly, there is interestitial cement. The centre of the granule is pale, rather slightly coloured, but may show dark brown patches. There are frequently bacteria inside the granule present (**Fig. 15.2**).

Culture : Grows first as a *fluffy white* colony, which has a diameter of about 1 cm after 10 days. The centre is raised. The colour may change to grey or brown, the reverse side of the colony is black. *Ascocarps* of 100 to 300 microns in diameter are produced, each one containing 8 ascospores with 4 to 8 crores walls per spores.

MADURELLA MYCETOMATIS

This is the most frequently found true fungus responsible for mycetoma in India. Between 50 and 75% of all eumycetomas in India are caused by *Madurella mycetomi*.

Histology : The granule is very large and lobulated. It is easily recognisable by its brown cement substance in which radiating parallel hyphae are embedded (**Pl. 15.VII, Figs. 17, 18** and **Pl. 15.VIII, Fig. 19**). They appear as "empty" or "negative" gaps in H & E sections, but they stain "positive" with silver methenamine stain (**Pl. 15.IX, Fig. 22** and **24**). Chlamydospores are recognisable in the granules. The granules can become very complex and up to 1 cm in diameter, as has been shown in serial sections and by direct measurement of granules. The granules are black and can easily be recognised if extruded through sinuses.

Culture : *M. mycetomi* produce white, later brownish, yellowish or olivaceous colonies with **brown diffusible pigment (Pl. 17.VIII, Fig. 20)**. **Culture mounts show chlamydospores** (**Pl. 17.VIII, Fig. 21**) or sclerotiae. Abbott (1956) observed also production of aleuriospores on poor media. Phialospores have also been described.

MADURELLA GRISEA

Histology : Granules are usually less than 1000 microns in diameter, slightly oval, with curvilinear margins. The cortical zone shows

PLATE 17.VII

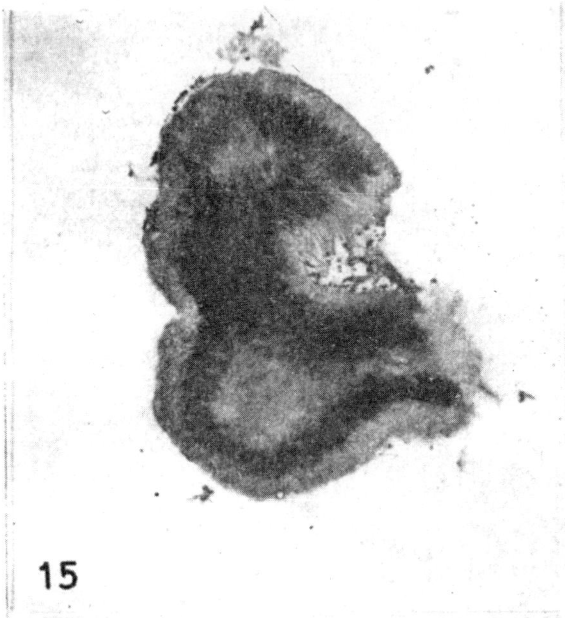

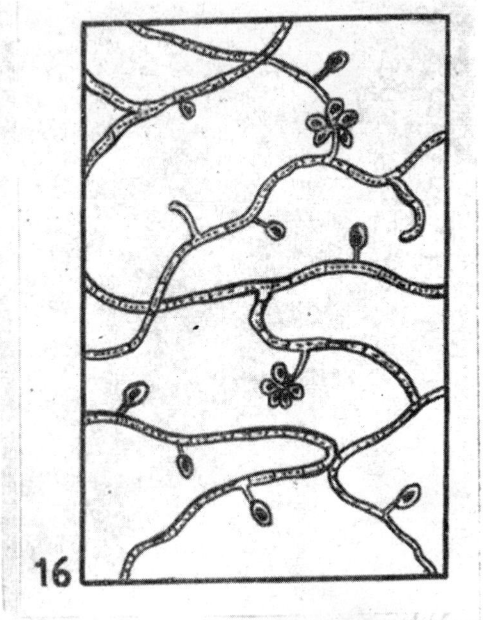

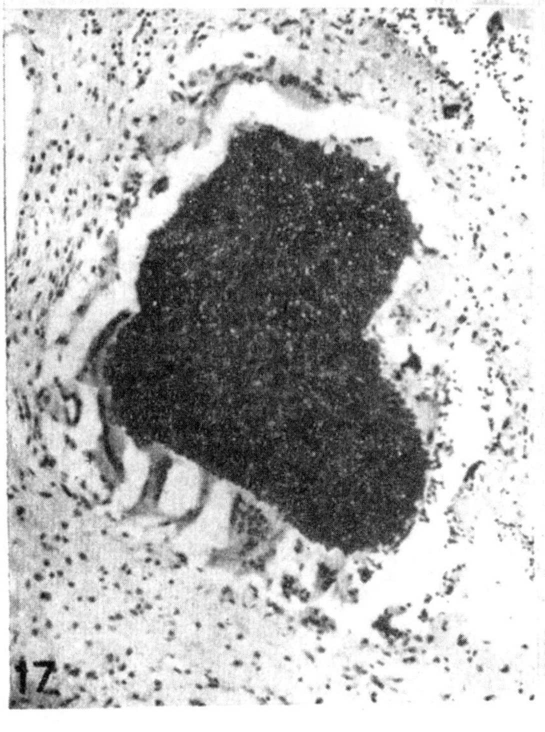

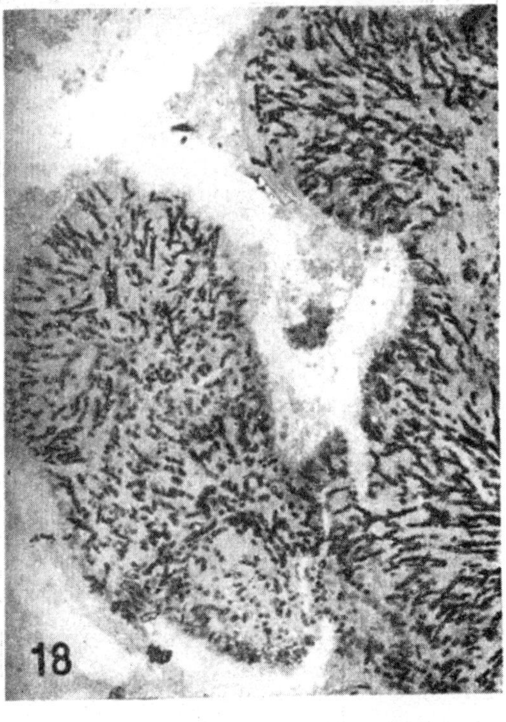

Fig. 15 : *Pseudallescheria boydii* **granule**. H & E, 130 ×

Fig. 16 : *Pseudallescheria boydii* : Culture mount, showing **oval conidia**

Fig. 17 : *Madurella mycetomatis* in tissue with "negative", empty hyphal spaces embedded in brown cement substance. H & E; 130 ×

Fig. 18 : *Madurella mycetomatis* : "Positive" demonstration of hyphae by *silver methenamine method;* 130 ×

PLATE 17.VIII

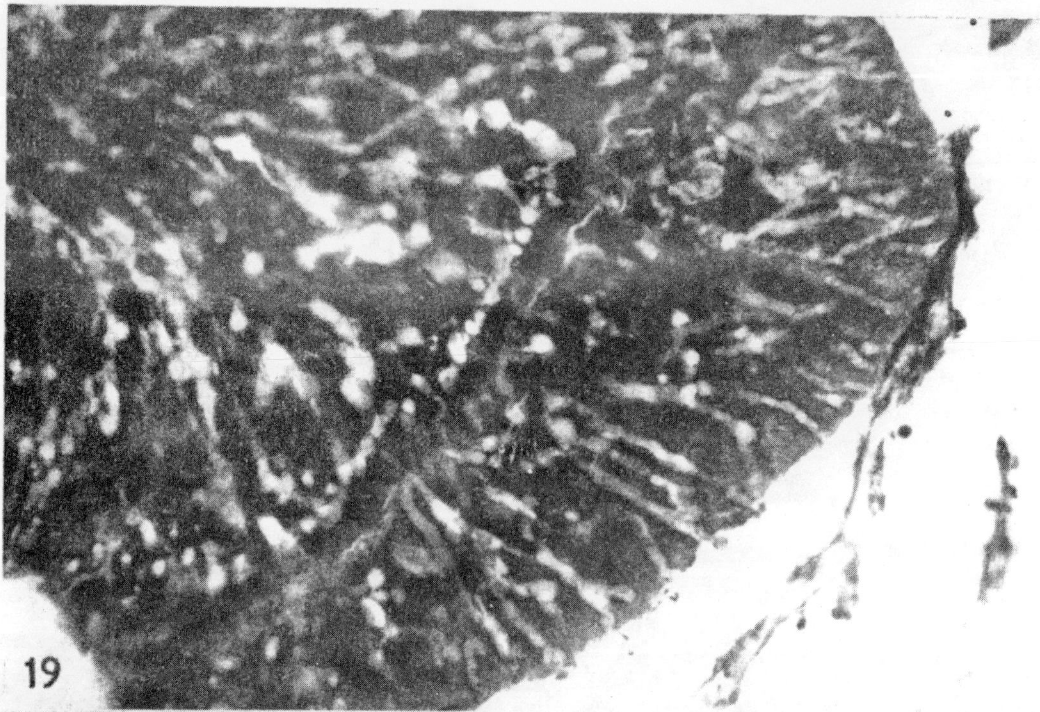

19

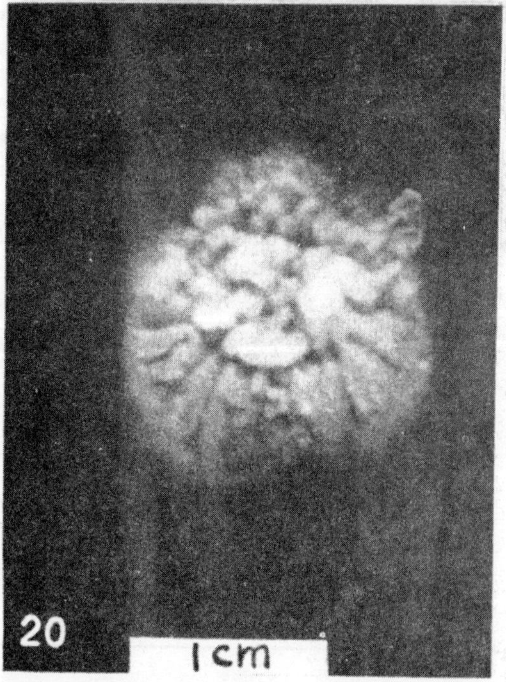

20

21

Fig. 19 : *Madurella Mycetomatis* granule with radiating hyphae of about 4-5 microns in diameter. H & E, 560 ×

Fig. 20 : *Madurella mycetomatis* culture, four weeks old

Fig. 21 : *Madurella mycetomatis;* Culture mount showing **chlamydospores**. Unstained 560 ×

brown evenly distributed cement substance between the hypae, which are arranged in a regular network. Peripheral vesicles are small, up to 8 microns in diameter. The central zone is pale or slightly coloured, and the hyphae are non pigmented or only faintly coloured. There are no bacteria inside the undamaged granule and no interstitial cement can be recognised in the centre of the grain.

Culture

Colonies are raised, cerebriform in the centre, and radially folded in the periphery, up to 5 cms in diameter. The colour is mat grey, olivaceous, sometimes changing to red brown. Branching **septate hyphae 1 to 5 micron in diameter are brownish black to fuliginous.** No conidia have been reported (Emmons, 1963).

CEPHALOSPORIUM FALCIFORME

Cephalosporium sp. have not frequently been found in India as causative agents of mycetoma.

Histology

According to Klokke and Verghese (1966), two *Cephalosporium* species have been described in India, producing black granules, and a histologic description for the black grain producing variety has not been given.

Culture

Cephalosporium falciforme grows up to a diameter of 3 cm, with a raised centre, producing hyphae with elliptical or ovate conidia at the tips of slender tapering conidiophores.

OTHER AGENTS OF EUMYCETOMA

A few cases of mycetoma due to ***Pyrenochaeta romeroi*** have been found in India by Klokke *et al.* (1968).

MYCOLOGY
ACTINOMYCETOMAS

Nomenclature

The aerobic actinomyces have been divided into 2 genera e.g., **Nocardia and Streptomyces.** Gonzalez-Ochoa suggested that species of actinomycetes formerly named as ***Nocardia madurae*** and ***N. pelletierii*** produce conidia and should, therefore, be called *Streptomyces*; and Mackinon and Artagavetia-Allende (1956) formally transferred these species from Nocardia to the genus Streptomyces. These are aerobic soil (higher) bacteria like other Nocardia which are acid fast.

NOCARDIA BRASILIENSIS

(Lindenberg) Castellani and Chalmers, 1913.

Synonyms

Discomyces brasiliensis Lindenberg, 1909; *Actinomyces mexicanus* Boyd and Crutchfield, 1921; *Nocardia mexicana* Ota, 1928; *N. transvalensis* Pijper and Pullinger, 1927; *N. Pretoriana* Pijper and Pullinger, 1927.

N. brasiliensis is an aetiological agent for aerobic actinomycotic mycetoma.

Source of infection is soil

Nocardia brasiliensis has been recently isolated from soil of Mexico by Gonzalez-Ochoa (1961).

The granules are obtained from pus of sinuses and subcutaneous absecesses. These are spherical or lobulated and less than 1 μ in diameter with or without peripheral clubs, white and yellow in colour and soft. Variations amongst granules are seen from an independent cases of mycetoma seen to be caused by *N. brasiliensis* in tropical America, South Africa and some parts

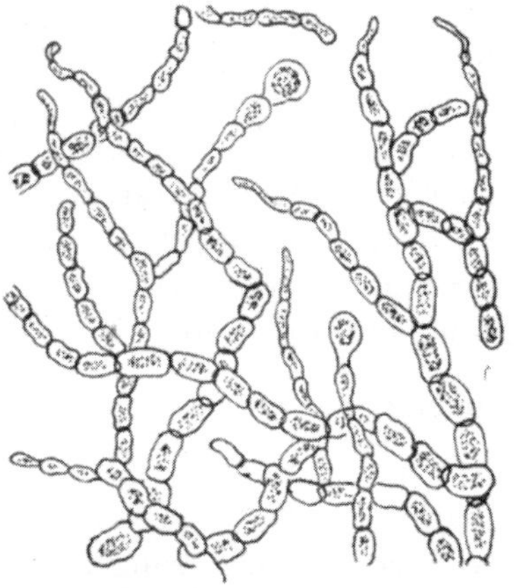

Fig. 17.3 : Arthrospores and chlamydospores of *M. mycetomatis*

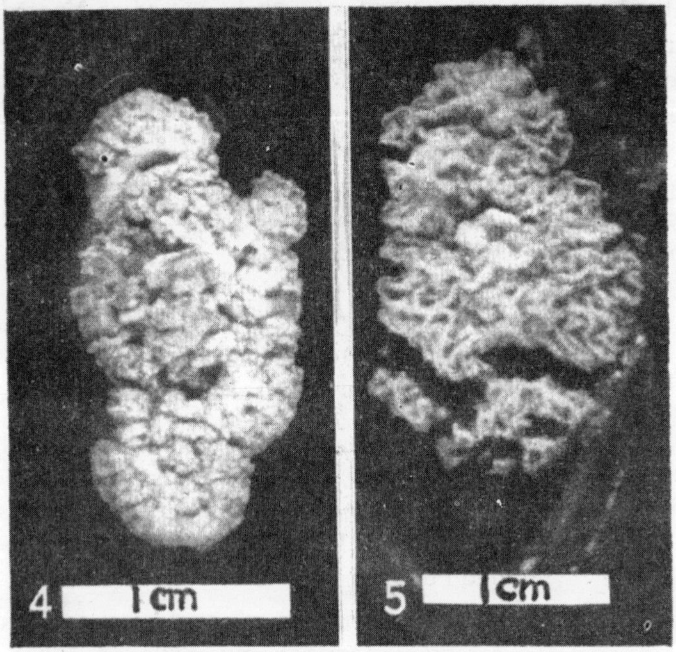

Figs. 17.4 and **17.5** : *Nocardia asteroides* colonies, 19 days old

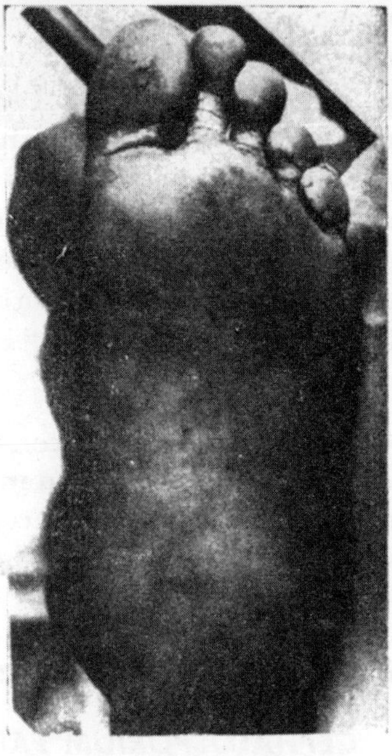

Figs. 17.5A : *N. asteroides* causing deformity of the foot

of India and also amongst cultivators in Bengal in the Punjab (Klokke, 1964). The last one was reported to be first isolated case from Asia.

Morphology

Crushed and stained granules appear as hyphae which are **Gram positive and weakly acid-fast**. The characteristic features are weak acid-fastness and aerobic in character which differentiates it from **Actinomyces (anaerobic)**.

Culture

In primary culture on Sabouraud's agar *N. brasiliensis* produces small heaped up wrinkled and folded colony yellow to orange in colour and mealy to membranous in consistency. It grows **better at 25°C** than at 37°C. It coagulates milk, liquifies gelatin but not casein and does not utilise paraffin.

Pathogenicity

It may produce pulmonary Nocardiosis in man or more commonly mycetoma. In guinea pigs inoculated with the organism it produces peritonitis with granules which are weakly acid-fast.

NOCARDIA CAVIAE

(Erikson) Gordon and Mihm, 1962. **Synonyms** : *N. otitidis caviarum* Snijders, 1924; Actinomyces \caviae Erikson, 1935. It is an occasional etiologic agent of mycetoma. Gordon and Mihm, studied several strains of Actinomycete and determined that these micro-organisms should be referred to a species originally isolated from ear of a guinea pig and called *Nocardia otitidis caviarum* by Snijders. Strains include those from mycetoma and bovine mastitis.

Synonyms

N. otitidis caviarum Snijders, 1924; ***Actinomyces caviae*** Erikson, 1935. It is an occasional aetiologic agent of mycetoma. Gordon and Mihm, studied several strains of Actinomycetes and determined that these micro-organism should be referred to a species originally, isolated from ear of a guinea pig and called *Nocardia oittidis caviarum by* Snijders. Strains include those from mycetoma and bovine mastitis.

Culture (Figs. 17.4 & 17.5)

On Sabouraud's agar, they form yellow-orange colonies. *N. caviae* resembles ***N. asteroides*** ***(see Chapter 12, Fig. 12.2)***, but it differs from ***N. asteroides***, by its ability to decompose xanthine and hypoxanthine, and by formation of acid from inositol. *N. brasiliensis* resembles ***N. caviae*** in its morphological appearance but differs by its ability to decompose xanthine and to survive at 50°C for 8 hrs. It slightly hydrolyses casein but not tyrosine.

STREPTOMYCES

Streptomyces species are characterised by stability of their filaments and by the formation of spores in **aerial mycelia**. They resemble Nocardia in the growth and appearance and the mutants that lose the ability to form aerial mycelia are difficult to be distinguished from Nocardia. Antigenically all actinomycetes have common antigen and cross react amongst them as a rule.

A number of actinomycotic abscesses (mycetoma) are caused by streptomyces which were formerly classed as Nocardia.

Biology

Streptomyces are ubiquitous in soil and the infection is attributed to contamination of **scratches or penetrating wounds** caused by **thorn prick**.

Innumerable isolates from soil samples from all parts of the world have been systematically scrutinised by Waksman and others and various species (about 500 in number or more) have been found to produce antibiotics which are mainly derived from cultures of streptomyces. Thus these species are medically important for their antibiotic properties rather than their pathogenic effects. Streptomyces and actinomycetes also act as wide-spread scavengers in soil by breaking down, proteins, cellulose and other organic matters.

Various species of streptomyces of the soil are responsible for production of streptomycin as demonstrated by Waksman (1943) and innumerable antibiotics have been produced from streptomyces. Therefore, the isolation of streptomyces from soil is more important rather than for the production of disease in man and animals. The **commonest pathogenic streptomyces in India** is *S. **madurae*** producing **Madura foot**.

STREPTOMYCES MADURAE

(Vincent) Mackinnon and Artaga-vatiaAllende, 1955.

Synonyms : *Streptothrix madurae*, Vincent, 1894; *Nocardia madurae* Blanchard, 1896; *Actinomyces madurae*, Lachner-Sandoval, 1898; *Nocardia indica*, Chalmers and Christopherson, 1916; *Discomyces brasiliensis*, Piraja da Silva, 1919.

GROSSLY, the granule of *S. **madurae*** are 1-5 mm or rarely 10 mm in diameter and spherical or angular, with a mulberry-like or lobulated surface. The colour is white, wellowish or rarely pink or rosy from which **mycelial (hyphal) fringes spread 50 μ beyond the solid mass**. These are eosinophilic in character. The opening of sinuses are slightly raised due to **epidermal hyperplasia and inflammation**.

The species is **both osteolytic and osteophytic**, the infected foot shows both cavities and spicules of bone on the surface betwen cavities. The large granules of *S. madurae* is associated with correspondingly large size of osteal cavities and exuberant formation of osteophytes.

IN PRIMARY CULTURE, the growth is slow and it is slower in Sabouraud's glucose agar than in Lowenstein media. Colonies are waxy and cerebriform and the colour varies from white to yellow or pink or red. Hyphae are 0.5 to 1 μ in diameter and the conidia are spherical, 0.5 to 1 μ.

STREPTOMYCES PELLETIERI

(Laveran) Mackinnon and Artagaveitia-*Allenda*, 1956.

Synonyms : Micrococcus pelletieri Laveran, 1906; Nocardia africanus, Piper and Pulinger, 1927; N. genessi, Froes, 1930.

GROSSLY, granules of *S. **pelletierii*** **are 300 to 500 μ in diameter**, **spherical or oval** and resistant to pressure, with hyphae 0.5 to 1 μ in diameter. The colour is bright ruby or less intense in some cases. In stained section (H & E), the entire granule stains intensely with haematoxylin and the margin is sharply delimited without fringe of hyphae which is characteristic of ***S. madurae*** granule.

The sinuses caused by ***S. pelletierii*** are characteristic. They are crateriform forming lobulate nodules, 2 cm in diameter and 1 cm high produced by granulomatous reaction and epidermal hyperplasia. Bones are attacked in the same manner as that of *S. madurae* but the cavities and spicules are smaller.

PRIMARY CULTURE : On **Sabouraud's agar** the colonies are small and the surface variously folded, wrinkled and **tuberculate** in character. The colour is red and the surface waxy, but often covered with **small aerial hyphae on Loewenstein Jensen's medium**. The hyphae are 0.5 to 1 μ in diameter. The fungus is proteolytic, and it **utilises urea**, but does not utilise xylose, galactose, maltose, starch, mannitol or paraffin.

STREPTOMYCES SOMALIENSIS

(Brumpt) Mackinnon and Artagaveitia-allende 1956.

Synonyms : *Nocardia somaliensis*.

GROSSLY, granules are spherical of lobulate 0.5 to 1 mm in diameter, yellow to brown in colour, smooth and hard. When crushed between slides, they may break into angular fragments.

In section, the granules stain very poorly and show a smooth border without clubs. After a Gram stain, Gram positive hyphae are seen under oil immersion with cementing substance separating the hyphae and makes the grain hard.

PRIMARY CULTURE of ***S. somaliensis*** develops more rapidly than those of ***S. madurae*** and ***S. pelletierii*** (4-11 days) but like them grow better in Loewenstein than on Sabouraud's medium. The **colonies are folded, cerebriform** which differ in surface configuration. The **mycelia are Ochre-yellow or yellow** in colour and is visible in Lowenstein medium on the surface of the colony and may be partially obscured by white aerial hyphae. Subculture on Sabouraud's medium may be grey or brown.

EUMYCOTIC MYCETOMA

(Madurella causing mycetoma)

MADURELLA MYCETOMATIS

(Laveran) Brumpt, 1905.

Synonyms

Streptothrix mycetomi Laveran, 1902 : *Glenospora khartoumensis*, Chalmers and Archibald, 1916; *Oospora tozeuri* Nicolle and Pinoy, 1908; *Madurella tozeuri* Pinoy, 1912 *M. tabarkae* Blanc and Brun, 1919; *M. americana* Gammel, 1926; *M. ikedae* Gammel, 1927; *M. lackawanna* Hanan and Zurett, 1938; *M. viridobrunnea* Radaelli and Ciferri, 1942.

The fungus produces black granules and single spherical or oval granule of 5 mm in diameter but aggregation of granules give bigger lobed grains forming a fungal mass of even a few centimeters in diameter. These grains are hard, coal black and can be crushed by pressure.

GROSSLY, the granule is composed of hyphae 1 to 5 μ in diameter and the terminal hyphal cells with 12-15 μ or swollen cells with 30 μ in diameter are encountered. The characteristic of the species is that the hyphal cells contain brown pigment particles and are embedded in brown matrix.

Culture : (Fig. 17.3)

M. mycetomi **grows better at 37°C** than at 30°C. The culture is extremely variable and different strains vary in different geographic areas. Colony develops within a few days as one with hyaline mycelia (4 μ in diameter) which later becomes olivaceous, yellow or brown darkening the sugar medium. A brown **diffusible pigment appears** in the medium. **Spores vary from 2 to 5 μ on Sabouraud's glucose agar**. More often in media without sugar, **black scleortia, 750 μ in diameter** develop on the surface or substrate of the medium. These are formed of polygonal cells about 10 μ in diameter and form a

Table 17.7 : Cultural Morphology, Physiological Characteristics and Pathogenicity of Aerobic Agents of Actinomycetoma

Species	*Morphology in culture Sabouraud's glucose agar including staining reaction.*	*Physiological Characters*			*Pathogenicity for guinea pig or mouse*
		Optimum temperature	*Hydrolysis of casein*	*Hydrolysis of starch*	
Streptomyces madurae	Colony : Moderately, with mulberry nodules Surface shiny or powdery. Microscopic Delicate branched mycelia, non-acid-fast	37°C	+	+	Not successful
S. somaliensis	Colony : cream coloured, wrinkled, flaky growth, loosely adherent to agar, old colony brown or blackish; no diffusible pigment. Microscopic : Delicate branched mycelia; non-acid-fast, aerial mycelia show spores	30°C	+	+	Not successful
S. pelletierii	Colony : Glabrous dry adherent colonies, slow growing pale pink, some are garnet red in colour. Microscopic : Delicate Branched mycelia, **conidia** rare, non-acid-fast	37°C	+	O	Not successful
Neocardia asteroides	Colony : Rapid growing heaped up, wrinkled, tan to yellowish, orange or orange-red colour. Microscopic : Long branched mycelia acid-fast with **terminal conidia**, in some strains, irregular rods and cocci often predominate	37°C Some strains grow better at 30°C	O	±	Some strains produce abscess in mice and guinea pig
N. brasiliensis	Weekly acid-fast	30°C	+	±	Abscess with granules in some strains

pseudoparenchyma. On soil extract agar or water agar, Abott has described the production of aleuriospores 3.5 by 5 μ similar to those described by Chalmers and Archibold (1915). **Formation of phialospores have been described** on potato carrot agar by Segretain. They are slowly proteolytic, utilise glucose, maltose, galactose but no sucrose; they **utilise KNO_3, ammon. sulph.**, **asparagine and urea**, and **hydrolyse starch**.

Madurella grisea

Mackinnon, Firrada and Montemayer, 1949.

GROSSLY

The grains of M. grisea are black and they attain a diameter of 1 mm or more. They are soft when young but become firm in old large grains. They become brittle on drying. Unlike the uniform solid grains of *M. mycetomi*, those of

M. grisea show on section unpigmented pale central area with an outer crust or brown fungal hyphae and brown interstitial material. The peripheral hyphae have pigmented wall but lack the brown granules of intra-cellular pigment characteristic of *M. mycetomi*.

Culture

M. grisea forms **colonies 3-5 cm in diameter** and 8 to 12 mm high. The surface in the centre is cerebriform and with radial furrows at the periphery. Like *M. mycetomi* the mycelial mat is grey or olivaceous and covered with grey aerial hyphae. Old colonies become red brown.

The hyphae are septate 1-3 μ in diameter but moniliform ones are 3-5 μ in diameter. Both types are brownish black to fulifinous. No conidia are seen.

M. grisea grows better at 30°C than at 37°C. It is similar to *M. mycetomi* in biochemical characters except that it utilises sucrose and does not utilise lactose. Its proteolytic action is weak.

PSEUDALLESCHERIA BOYDII

(THE ASEXUAL PHASE BEING *MONOSPORIUM APIOSPERUM*).

Synonyms

Monosporium apiospermum Soccardo, 1911; *Scedosporium apiospermum* Saccardo, 1914; *Glenospora clapieri* Catanei, 1927; *Indiella americana* Dela mare and Gatti, 1929.

Definition

Pseudallescheria boydii is the perfect stage of *MONOSPORIUM APIOSPERMUM* causing **mycetoma (maduromycosis)** in U.S.A., characterised by formation of white or light yellow granules constituted by wide, septate, hyphae in the tissue or exudate draining from the sinuses.

Source of infection

The fungus is a saprophyte of the soil and was isolated from soil and sewage repeatedly (Cook, 1955, Ajello, 1955). Perfect stage forming ascocarp is not found in many strains isolated from mycetoma or soil, the natural habitat of the fungus, and for this reason it is better to retain the name *M. apiospermum* as well.

Perfect stage

Some isolates produce perithecia, containing 8 ascospores in asci. ***PseudAl. boydii* is homothallic**, i.e., **sexual reproduction can occur without the fusion of two different thalli**. It really forms cleistothecia; a cleistocarp develops from a conspicuous and persistent ***ascogonium***. They are without any opening and 8 ascospores disintegrate by the rupture of the delicate ascus. Their walls are brown and transparent in a young small ascocarp so that young ascospores can be seen through the wall.

All the strains isolated from nature or disease can **produce sexual spores**. Media like **potato dextrose agar or corn meal agar** may stimulate their production in some strains. **Cleistothecia are** usually **round** and appear in the depth of the media, and are seen with the **naked eye as small black specks**.

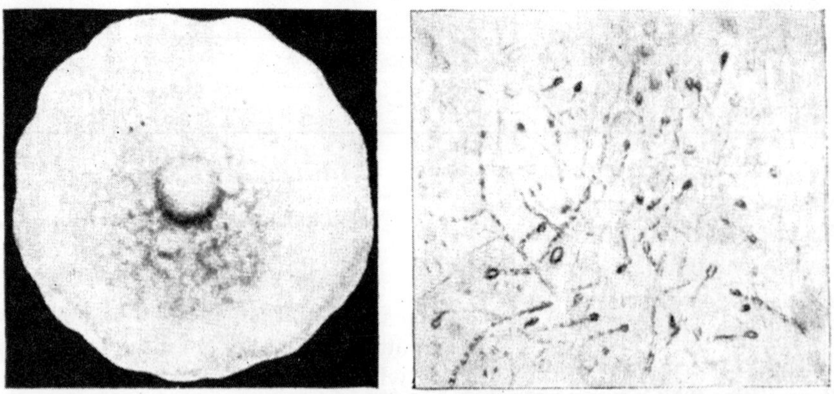

Fig. 17.6 : *PseudAl. boydii*; left—giant colony; right—**conidia in microculture**

PLATE 17.IX

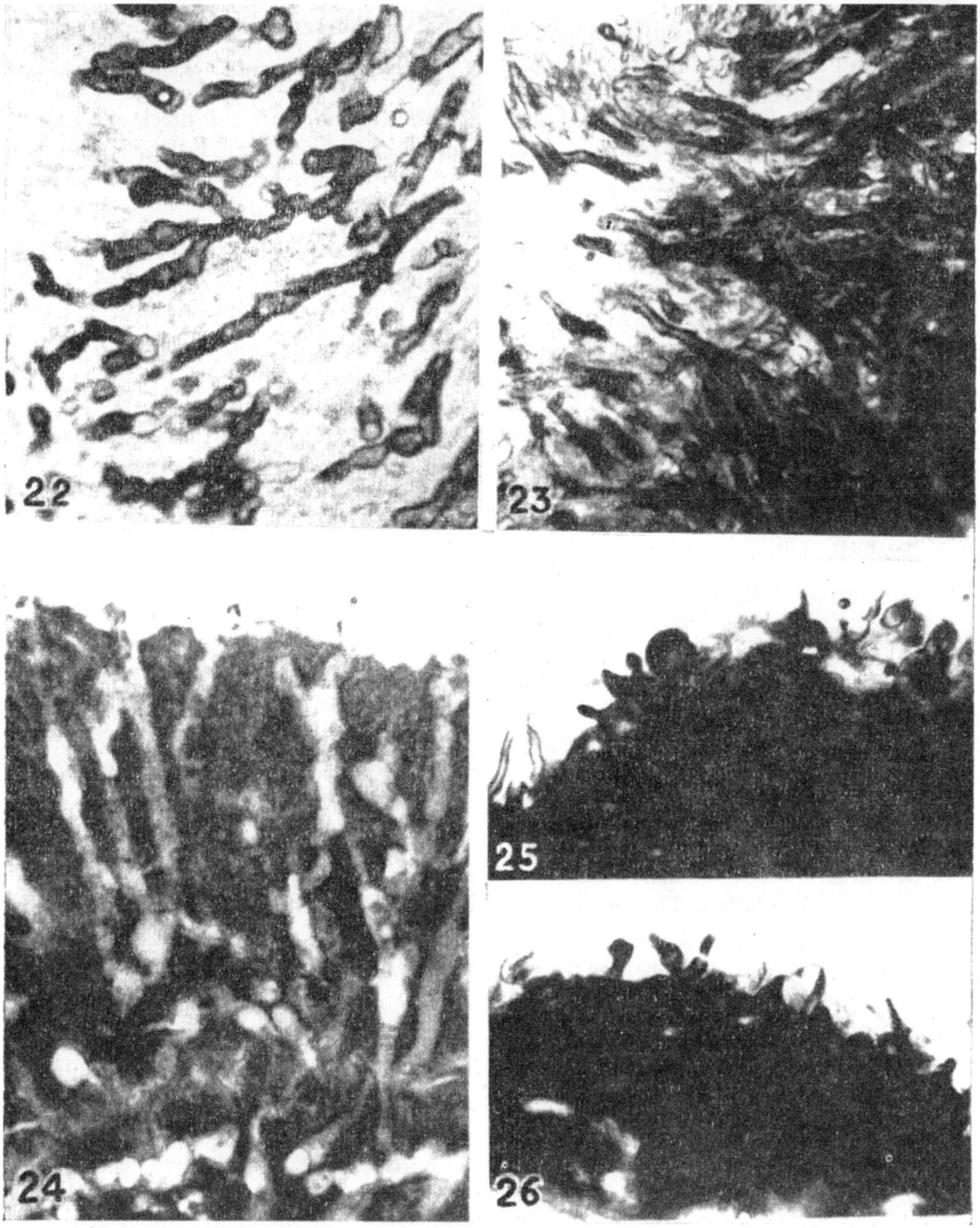

Figs. 22 to **26 :** Emycetomas, all at 560 × magnification. The upper half of plate shows centre of granules, the lower half the periphery

Fig. 22 : *Madurella mycetomatis* hyphae in granule, 5-6 microns in diameter, forming chlamydospores. H. & E

Fig. 23 : *PseudoAllescheria boydii,* hyphae only 2-3 micron in diameter, note regular arrangement of hyphae. Silver Methenamine

Fig. 24 : *Madurella mycetomatis* No vesicle formation at periphery, but chlamydospores are formed. Grey mass represents cement substance. H. & E

Figs. 25 and **26 :** Edge of *PseudoAllescheria boydii* granule showing spiky hyphae and vesicle formation. Silver methenamine

Culture

(Fig. 17.6) ***M. apiospermum*** can be cultivated in ordinary media. On Sabouraud's glucose agar the colony develops rapidly as a white, downy, aerial growth which later becomes greyish to buff in colour. A black pigment is often seen on the reverse side of the growth.

Physiological studies

Benhan and Georg (1948) and Gordon (1957) studied the relationship between nutrition and sporulation of ***PseudAl. boydii*** but did not find any special nutritional requirement. However, a rich source of organic nitrogen like brain heart infusion agar was found to stimulate mycelial growth and conidia production but at the same time inhibits **production of cleistothecia**.

MICROCULTURE : (Fig. 17.6)

Microscopically the organism shows mycelia having single ovoid to clavate conidia with truncated base, held at the end of conidiophores of various lengths. Corremia are sometimes seen with conidia at the end of conidiophores. Pleurogenous spores are also seen in certain strains.

In the ascomycetes phase of *PseudAl. boydii*, the sexual spore is characterised by a cleistothecia (closed perithecia without osteole), 50 to 200 μ in diameter and thin walled. They contain a brownish structure, containing subglobose asci in each of which 8 ascospores, 4 to 4.5μ by 7 to 7.5μ in size, are present. As monosporous stage is the imperfect stage of *PseudAl. boydii*, it is better to call the fungus *PseudAl. boydii*.

Animal inoculation

Laboratory animals are not susceptible. Injection of a pure culture into the knee joints or feet of animals may produce only temporary lesions.

OTHER TYPES OF INFECTION : It may be found in cases other than mycetoma; Benhan and Georg (1948) found in a case of meningitis. Tong *et al.* (1958) and Scharyl *et al.* (1960) reported it as an aetiological agent of primary pulmonary disease. It was isolated in two instances from **corneal ulcers** (Pautler *et al.*, 1955, Gordon *et al.*, 1959).

PHIALOPHORA JEANSELMEI

The Granules are soft black and (1927) as ***Torula jeanselmei*** after its isolation by Jeanselme *et al.* (1928) from a case of mycetoma in a native Martinique.

The fungus was described by Langeron easily broken and composed of masses of dark chlamydospores. The fungus was placed in the genus *Torula* as it has a "toruloid" or yeast-like phase in its early growth. In 1945, Emmons studied a similar fungus from a case of **mycetoma of the hand** as observed by Symmers and Sporer (1944) in the United States. Emmons placed it in the genus ***Phialophora*** because its spores are formed in ***Phialides***. Only 4 cases of this mycetoma were found in the United States.

Source of infection

Evidence that *P. jeanselmei* exists in nature is based on the report that *Cadophora lignicola*, which is synonymous with *P. jeanselmei* was isolated from wood pulp in Sweden and Italy.

Cultural characters

P. jeanselmei grows slowly on Sabouraud's dextrose agar, producing black, heaped up colony. The surface is first moist and glabrous but soon becomes covered with greyish, velvety aerial mycelia and the nature of the colony resembles one of *P. verrucosa*.

Early glabrous growth shows a **dark thick toruloid (yeast-like) hyphae formed of chains of budding cells**. Later **septate mycelia form producing spores from phialides**. The phialides differ from those of *P. verrucosa* in that they are elongated and pointed here and do not form distinct flaring cup at the distal ends (reduced flaring cup). These are really conidiophores at the distal ends (reduced flaring cup). These are really conidiophores, uniform in diameter but taper to a narrow tip at the distal end which was surrounded by clusters of elongated conidia, extruded through the opening of the phialide. The spores are hyaline, elliptical about 0.75 by 1.5 μ.

PLATE 17.X

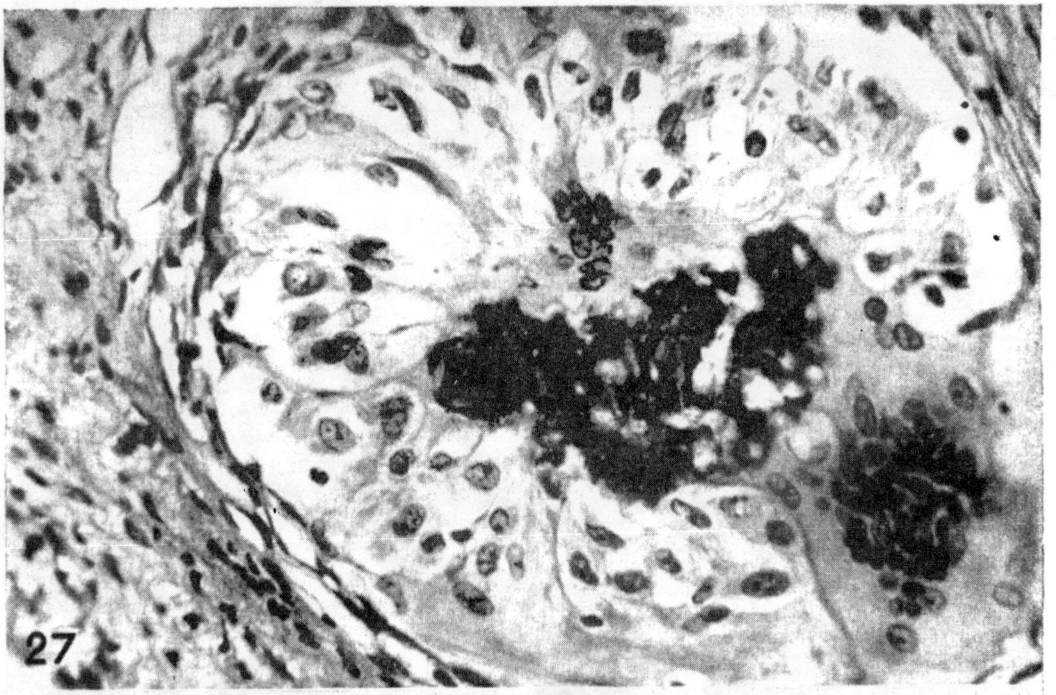

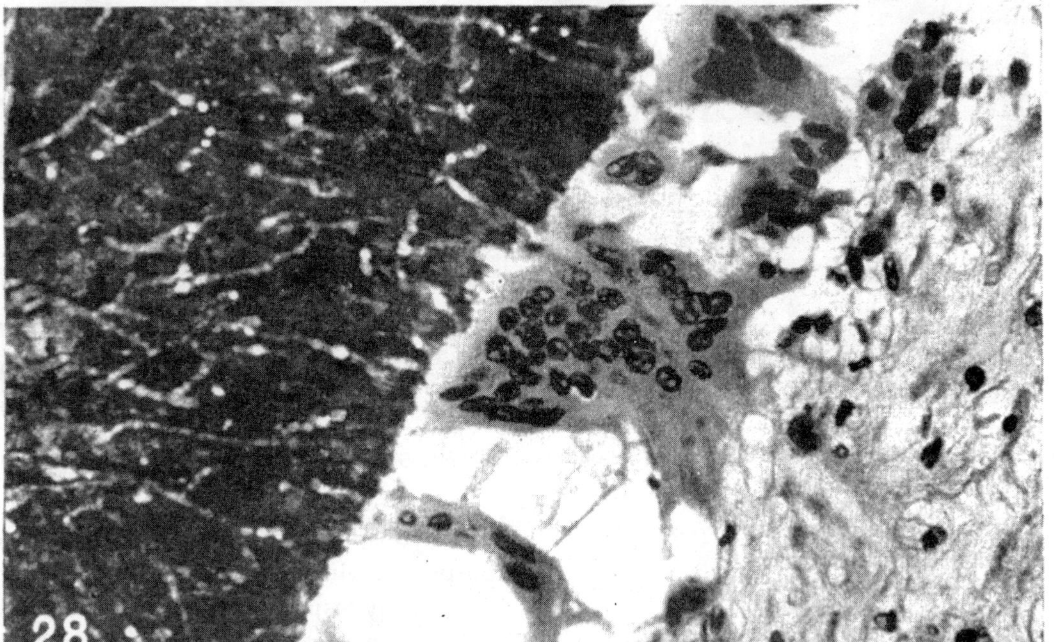

Fig. 27 : A wall of epithelioid cells and foreign body giant cells surrounding remains of *Madurella mycetomatis* granule in tissue. H & E; 560 ×

Fig. 28 : Foreign body giant cell reaction at edge (probably avital). *Madurella mycetomatis* granule in tissue. H & E; 560 ×

PLATE 17.XI

Mycetoma Caused by *Glenospora Semoni*

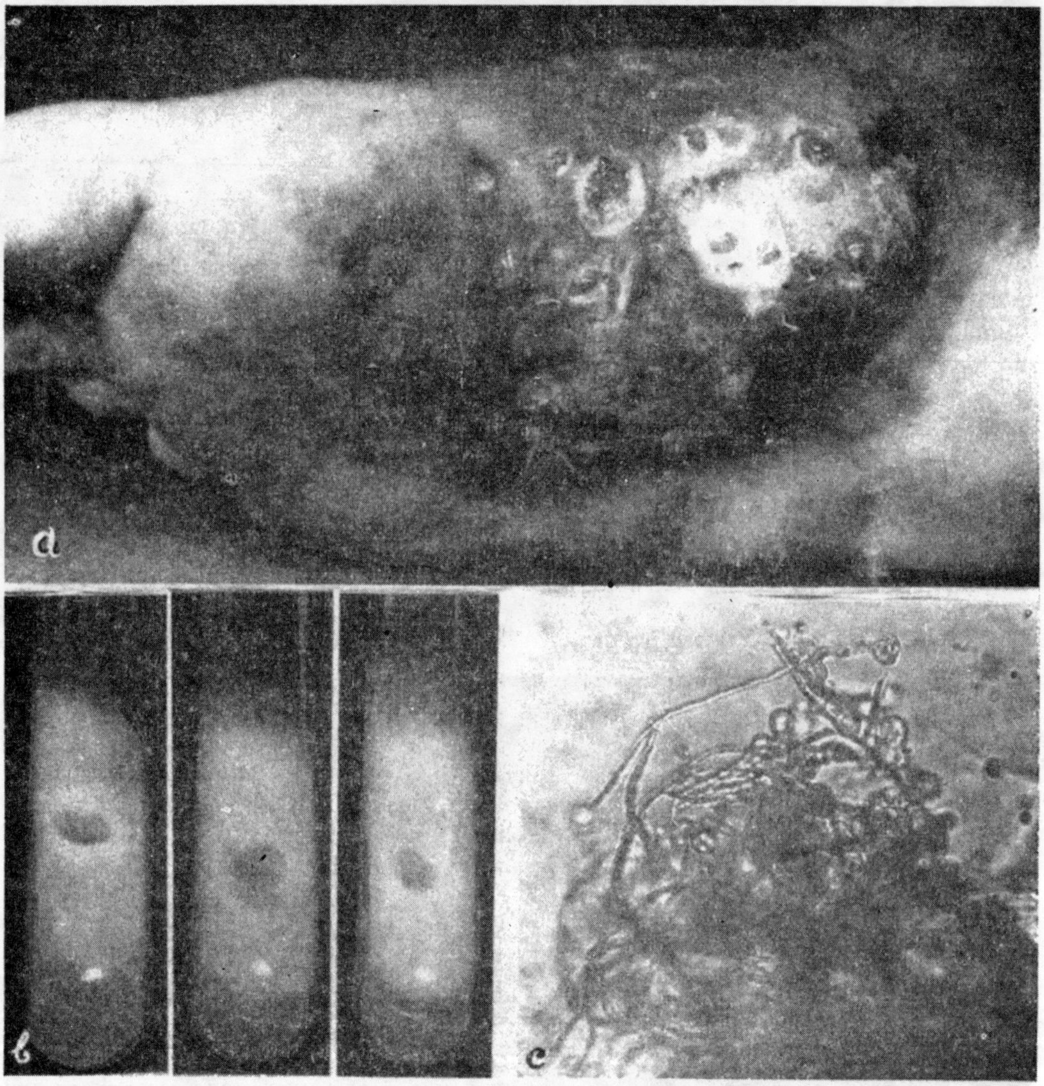

(a) Mycetoma of the foot. (b) Culture on Sabouraud's maltose agar 3 weeks old (c) **Microculture showing chlamydospores**

OTHER TYPES OF INFECTION

P. jeanselmei also cause subcutaneous abscesses and resemble those produced by other fungi. This may eventually rupture and brown hyphal fragments and clusters of brown chlamydospores are seen in sections of such lesions.

To sum up ***P. jeanselmei*** produces **mycetoma with soft black granules showing chlamydospores**; (2) **subcutaneous abscesses, the pus showing** fragments of **hyphal elements and chlamydospores**; (3) chromoblastomycosis with verrucous skin lesions and sclerotic bodies in the tissues.

Torula bergeri (Berger and Langeron, 1949). Berger and Langeron isolated this fungus from a case of chromoblastomycosis. It is synonymous with *P. jeanselmei* (Trejos, 1953).

CEPHALOSPORIUM FALCIFORME

This was reported from a case of white grained mycetoma in Puertoe Rico by Carrion (1939 and 1951). Two South American cases have also been reported.

Culture : The colony is slow growing, fluffy or tufted, lavender buff or pinkish, the reverse is currant red. They may produce a soluble pigment. They grow at 30°C, with weak proteolytic but with no amylolytic activity.

Hyphae are hyaline, and chlamydospores are 12μ in diamerter 3-4μ. The conidia are elliptical or ovoid in shape like other cephalosporium borne in terminal clusters from simple slender and tapering conidiophores.

LEPTOSPHAERIA SENEGALENSIS

(Baylet, Camain and Segretain)

Segretain *et al.* (1959) isolated fungus from a case of **"black grained mycetoma"**. The fungus is a dermatiaceous member of the Class Ascomycetes. *L. senegalensis* has been isolated from 10 cases of mycetoma **in Senegal, Africa**. The **grains are black, 1 mm in diameter**, and **tubular or hollow resembling** those of ***Phialophora jealselmei***. Cullieres are rapidly growing and produces grey to brown colony. The fungus produces ascocarp 100-300 μ in all diameter and asci are 17 to 22 by 110 μ containing 8 ascospores, 8 to 10 × 23 to 30μ. The spores are hyaline.

PYRENOCHAETA ROMEROI

Borelli (1959) described a case of **"black grained" mycetoma** from Venezuela. The fungus is soft, 0.5 to 1.5 mm in diameter and tubular in shape. It produces a rapidly growing, dark grey, **woolly colony with a whitish periphery and black pigment on the reverse**.

The optimal growth occurs at 30ºC. Osteolate pycnidia 40 to 100 × 50 to 165 μ are produced that are surrounded by brown flexuous septae with rounded ends. The pyconoconidiospores are light yellow in colour and subhyaline and 0.75 to 1 ×× 1.5 to 2μ. The genus pyrenochaeta is widely distributed and grows on vegetations.

GLENOSPORA SEMONI Chalmers and Archibald, 1917.

This organism was isolated by Semon from a case of black maduromycosis occurring in an Indian soldier serving in France. It has been recently isolated in India by **Kakoti and Dey**, 1956, from a case of **black mycetoma**.

G. SEMONI closely resembles ***G. khartoumensis*** and may be considered synonymous with the latter. It slightly differs from *G. khartoumensis*. **On maltose agar**, the culture is characterised by a **smooth surface** instead of a central depression or nob **with radial furrows and fringes at the margin** (**Pl. 17.X, Fig. b**).

Pathogenicity

It causes Asian black mycetoma of the foot in man.

METHODS OF STUDY FOR LABORATORY DIAGNOSIS

The materials are as stated before, namely, pus, curetting of sinuses or biopsy tissue.

DIRECT EXAMINATION : Maduromycosis grains show mycelia 2 to 4 microns (*c.f.* actinomycotic granules 1μ or less in diameter).

CULTURE

For culture, the material aspirated from unopened lesion is preferable for a culture without contamination. **From an open sinus, granules are repeatedly washed with sterile normal saline to avoid contamination**.

The washed **granules are crushed** in a watch glass or Petri dish and **cultured in Sabouraud's dextrose agar** containing antibacterial antibiotics. This may be a combination of penicillin and Streptomycin (50.00 units/ml) or **chloramphenicol (0.05 mg/ml)**. Agents of **Maduromycosis are sensitive to cycloheximide** which should not be used in culture. The granules may be left in the antibiotic solution over night and then cultured in the Sabouraud's dextrose agar with the antibiotics. Several tubes should be inoculated. The culture tubes are placed in the incubator at **37ºC** and also at room temperature. The fungi are studied for the morphological and physiological (biochemical) characters (see Table 17.7).

SPECIAL MEDIA FOR THE STUDY OF MADUROMYCOSIS

1. CORN MEAL AGAR. 2. CZAPEK DOX MEDIUM (FOR SPORULATION) SEE APPENDIX. 3. GELATIN MEDIUM FOR LIQUEFACTION (PROTEOLYTIC) TEST; in the following, Heart infusion broth—25 g gelatin, 120—150 g or higher quantity in the tropics, **distilled water—1000 ml**; pH is adjusted to 7.2 to 7.4 Distribute 5 ml in each tube. Sterilise by **fractional sterilisation or autoclave at 20°C for 15 minutes**. 4. *Starch medium for hydrolysis*. (a) **Peptone**—5 g; Beef extract—3 g; agar—15 to 20 g and **distilled water 1000 ml**. (b) Soluble **Potato starch**-10 g **suspended in cold water**.

Mix ingredients except starch and heat to bring agar in solution. Then add starch solution. Mix and autoclave at 120°C for 15 minutes. After careful mixing pour into plates when liquid.

Test for starch hydrolysis : Inoculate two pairs of plates by streaking one area in each. Incubate at **37°C** and also at room temperature depending on the organism. After the fungus has grown for a few days. Flood the plate with 8 to 10 ml of 95% alcohol. A clear zone is seen surrounding the growth in contrast to the opaque unchanged starch, indicating the extent of hydrolysis. If there is no hydrolysis, allow the second plate to develop for another week and test similarly.

REFERENCES

Abbott, P. (1956). Mycetoma in the Sudan. *Trans roy. Soc. trop. Med.* **50**, 11-30.

Ajello, L., George, L. K., Kaplan, W. and Kaufman, L. "*Laboratory Manual for Medical Mycology*", U. S. Department of Health, Education & Welfare, Communicable Diseases Centre, Atlanta 22, Ga. (1963).

Andleigh, H. W. (1954). Mycetoma Foot. Aetiology and laboratory diagnosis. *Ind. Journ. med. Sci.* **8**, 631-634.

Behl, P. N. & Sharma, M. D. (1957). Incidence of mycotic infection in Delhi. *Ind. Journ. Derm.* **3**, 5-7.

Conant, N. F., Smith, D. I., Baker, R. D. and Callaway, J. L. "*Manual of Clinical Mycology*", 3rd. Ed., W. B. Saunders Company, Philadelphia, 1971.

Destombes, P. and Segretain, G. (1962). Les mycetomes fongiques characters histologiques et culturaux. *Arch. Inst. Pasteur de Tunis*, **39**, 273-290.

Destombes, P. (1964). Structures histologique des mycetomes. *Ann. Soc. belge. Med. Trop.* **44**, 897-908.

Destombes, P. and Patou, M. (1964). Morphologie de grains de mycetomes a *Nocardia* ou a *Cephalosporium*, reconstitutes par la methods des coupes histologiques series. *Bull. Soc. path. Exot.* **57**, 393-395.

Emmons, C. W., Binford, C. H. and Utz, J. P. (1970). "*Medical Mycology*", 2nd. Ed., Lea & Febiger, Philadelphia.

Grueber, H. L. E. and Kumar, T. M. (1970). Mycetoma caused by *Streptomyces somaliensis* in North India. *Sabouraudia*, **8**, 108-111.

Kakoti, L. M. and Dey, N. C. (1956). Mycetoma caused by *Glenospora semoni, Ind. Journ. med. Sci.*, **10**, 889.

Kanadhari, K. C., Mohapatra, L. N., Sehgal, V. N. and Gugnani, H. C. (1964). Black grain mycetoma of foot. *Arch. Dermatol.* **89**, 867-970.

Klokke, A. H. (1956). Mycetoma in North India due to *Nocardia brasiliensis. Trop. geogr. Med.* **2**, 170-171.

Klokke, A. H. (1965). Mycetoma in North India due to *Madurella Mycetomi. Journ. ind. med. Assoc.* **44**, 379-380.

Klokke, A. H., Swamidasan, G., Anguli, R. and Verghese, A. (1968). The casual agents of mycetoma in South India. *Trans. roy. Soc. trop. Med.* **62**, 509-516.

Latapi, F. (1963). Des Mycetom: In Handbuch d. Haut. und Geschl. Krankheiten, Ed. by Marchionini, A. & Goetz, G., pages 463-526, Springer, Berlin.

Mankodi, R. C. & Kanvinde, M. S. (1970). Mycetoma: A histological study. *Jour. ind. med. Assoc.* **54**, 465-466.

Mariat, F. (1963). Sur la distribution geographique *et la* repartition des agents de mycetomes, *Bull. Soc. path. Exot.* **56**, 35-45.

Mohapatra, L. N. (1965). Morphology of the granules in mycetoma and differentiation of the causative agents in tissue sections. Paper: 15th meetg., *Ind. Assoc. Path., Guntur.*

Panja, G. & Bannerji, A. K. (1965). Nocardiosis in India, *Ind. Journ. Dermatol.* **1**, 39.

Panja, G. (1966). Actinomycosis, *Ind. Journ. Dermatol.* **2**, 137.

Panja, G. and Chaudhury S. N. (1957). Mycetoma of the foot-melanoid (black) variety. *Ind. Journ. Dermatol.* **3**, 24.

Vanbreuseghem, R. (1967). The early diagnosis of mycetoma. *Dermatol. Internat.* **6**, 123-140.

Verghese, A. & Klokke, A. H. (1966). Histologic diagnosis of species causing mycetoma. *Ind. Journ. med. Res.* **54**, 524-530.

Winslow, D. J. and Steen, F. G. (1964). Considerations in the histologic diagnosis of Mycetoma, *Am. Journ. clin. Path.* **42**, 164-169.

Zaias, N., Taplin D. and Rebell, G. (1969). Mycetoma. *Arch. Dermatol.* **99**, 215-225.

18

Subcutaneous Mycosis
Dematiatious Fungus : Chromoblastomycosis

Definition

A cutaneous fungus infection which is characterised by **warty nodules developing slowly to form extensive papillomatous vegetations,** having a tendency to ulcerate, and rarely to have metastasis in the regional lymph nodes and to other parts of the body.

Terminology

The term chromoblastomycosis is strictly speaking incorrect, because the disease is **not caused by blastomycetes** but by organisms which propagate in the tissue by fission (or by septation). Some authors prefer therefore the term "**chromomycosis**".

Introduction

Chromoblastomycosis belongs to a group of infections called "dematiacoses", which are caused by *Dematiaceae*, fungi with a brown or black pigment. Though **chromoblastomycosis is an infection of the skin**, it must be clearly understood that Dematiaceae may, on occasion, **disseminate and end in fatal cerebral mycosis.** 23 such cerebral inections recorded in the literature have been collected by Shimazono *et al.* (1963); since then more cases have been reported. **Another type of dematiacoses,** consisting of subcutaneous abscesses containing brown pigmented fungi, has been recognised (Kempson & Sternberg, 1963). Two cases from India were reported by Dube & Gupta (1967) and by Agarwal & Chauhan (1969). However, such abscesses **do not strictly fall under the definition of chromoblastomycosis,** though they are also caused by dematiaceous fungi.

History

Pedroso in Brazil (1911) was the first to observe the characteristic **spherical "black bodies"** in biopsy material. The fungus was later classified as ***Hormodendrum pedrosoi*** by Brumpt (1922). Lane (1915) and Medlar (1915) isolated the fungus from a patient in Boston and named it ***Phialophora verrucosa***. Pedroso & Gomes (1920) reported four cases, including the first patient of Pedroso. Terra *et al.* (1922) named the disease caused by this fungus as **Chromoblastomycosis.**

A third type of causative fungus, namely ***Hormodendrum compactum*** was identified by Carrion (1935). ***Phialophora pedrosoi*** (Brumpt, 1922) n. comb. (Emmons, 1944) **(formerly *Hormodendrum*)** is the fungus most frequently isolated from lesions of **chromoblastomycosis** (Fejer *et al.*, 1966). (For further details see "Mycology").

Geographic distribution

Chromoblastomycosis has been reported from all parts of the world, but is more frequently found in tropical and subtropical regions. The first two documented cases from India were almost simultaneously reported by Thomas *et al.* (1057) and **Kakoti** and **Dey (1957)**. All ten cases known from India and Nepal have been discussed and summarised by Mohapatra *et al.* (1967). Since then, two more cases were reported one from Chandigarh, in a 52 year old man (Aikat, 1968) and one from Vellore, in a 40 year old male patient (Viziam, 1967).

Epidemiology

The disease is prevalent between 30 and 50 years of age. Males are infected more than 30 times as frequently as females. There is no infection from man to man. Many sufferers are agricultural people and wood-workers. It is assumed that the infection gains entry through small skin lesions. Dematiaceous fungi have been found to be present in rotten wood, decaying leaves and in soil (Rosenthal, 1941; Conant & Martin, 1937).

Clinical appearance

The infection begins often on the extremities, most frequently on foot and leg. The lesion is often unilateral. Other parts of the body are rarely involved (**Fig. 18.1**). Initially, a small itching papule or nodule of greyish or pinkish colour develops. Lesions may be dry, scaly, nodular or ulcerating. They extend in the periphery and may coalesce with satellite nodules which crop up at a little distance from the main lesion. After months, warty and cauliflower-like papillomatous lesions may develop (**Pl. 18.I, Fig. 1** and **2; Fig. 18.1**). In the course of 4 to 25 years lesions may cover with whole extremity. Secondary infection with ulceration often causes a foul smell. Lymphatics may be blocked, itching is sometimes present. Late cases are very characteristic in appearance; early lesions may be mistaken for dermatophytoses. Carrion (1942) has classified lesions in (1) nodular, (2) tumour, (3) verrucous, (4) plague, and (5) cicatrical lesions. However, the mature lesion shows in most cases a verrucous or papillomatous appearance. Elephantiasis may result from fibrosis and lymph stasis (see **Pl. 18.II a, Dey, 1958**).

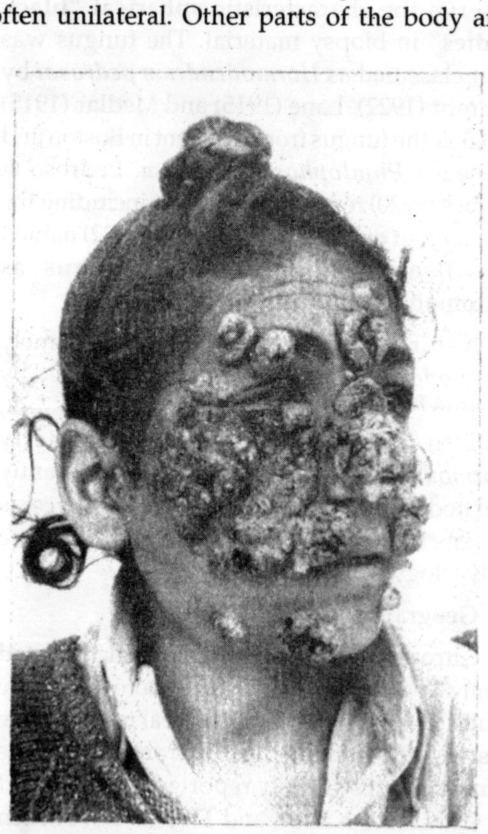

Fig. 18.1 : Chromoblastomycosis of the face

Laboratory diagnosis

HISTOPATHOLOGY : The biopsy should include the whole depth of the dermis and not only the papillary epidermal portion. The histological picture is characterised by (1) verrucous acanthosis and hyperkeratosis (**Pl. 18.I, Fig.3**) with (2) **small intra - and subepithelial abscesses** in combination with (3) **giant cell and epithelioid granules**, and (4) **scattered eosinophilic leucocytes**. A characteristic feature of chromoblastomycosis is that microabscesses or epithelioid granules are frequently appended to the apices of mishappen elongated rate-pegs (**Pl. 18.III, Fig. 4**). Giant cells may be of Langhan's or of foreign body type. The intervening tissue shows mixed inflammatory granulations in the **upper dermis with a prominance of plasma cells**. There may be coarse fibrosis in deeper dermal layers frequently with conspicuous perivascular plasma cell infiltrates. It is particularly the combination of microabscesses with epithelioid tubercles which must alert the observer. **Caseation is not present**. Round or slightly polygonal dark brown coloured bodies of 6-10

PLATE 18.I

1

Fig. 1 : Verrucous chromoblastomycosis of foot, about two years duration. (Courtesy : Mohapatra *et al.* (1967). *Mycosen* 10, 309)

Fig. 2 : Verrucous chromoblastomycosis of legs, about 4 months duration

Fig. 3 : Verrucous acanthosis and keratinous pearls in chromoblastomycosis. (H &

PLATE 18.II

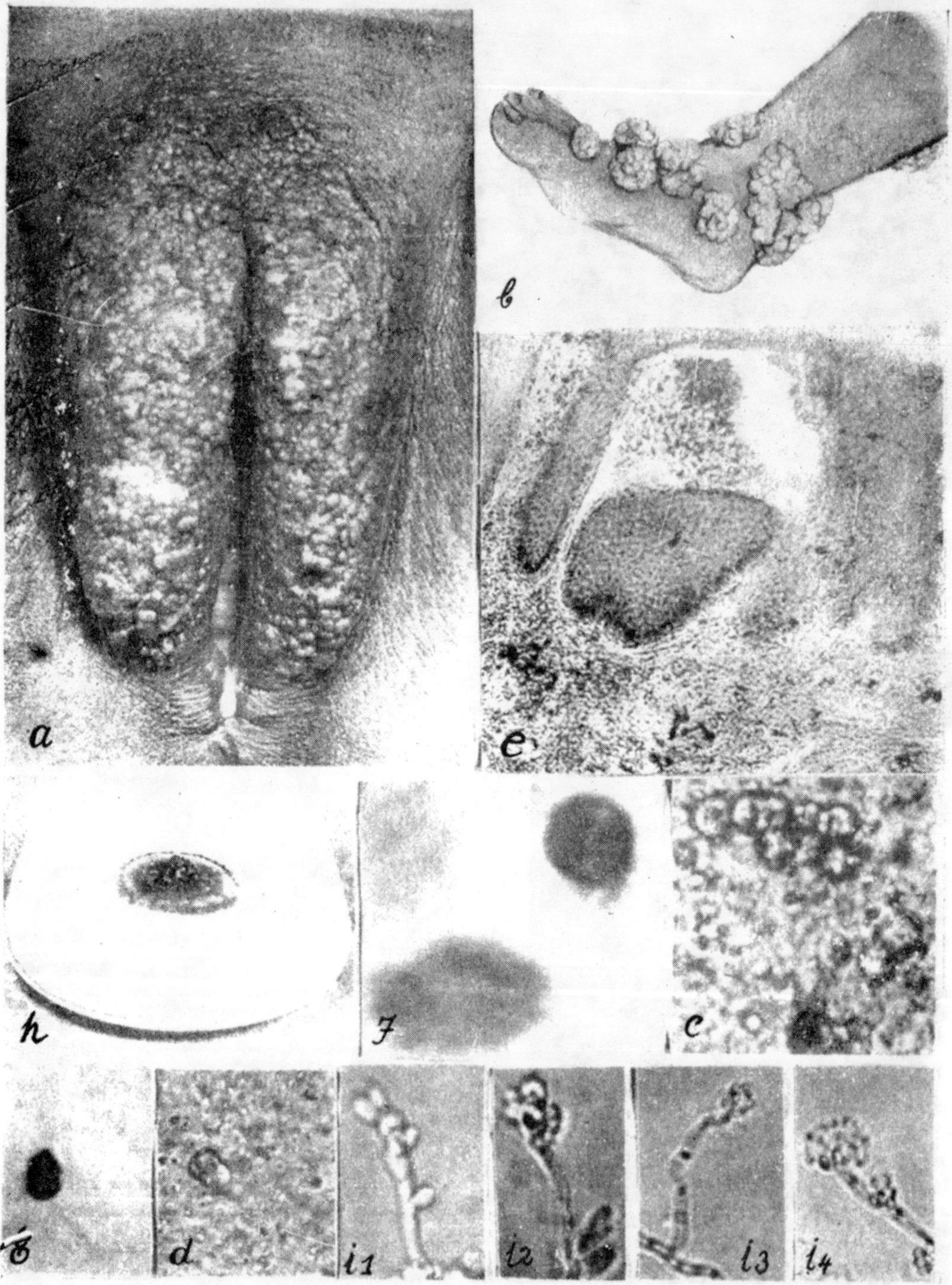

Explanation of Pl. 18.II. Chromoblastomycosis; (**a**) Verrucose condition of the vulva; (**b**) Affection of the leg and foot; (**c** and **d**) Spores in tissue (KOH preparation); (**e**) Histopathological section showing an epidermal abscess formation; (**f**) Same under oil immersion, showing a pair of spores; (**g**) Spores in pair, after disintegration of the lymph node in KOH solution; (**h**) Culture in Sabouraud's glucose agar i_1—i_1 Conidia of ***Fonse aea (Hormodendrum) Phialophora*** *compacta.*

PLATE 18.III

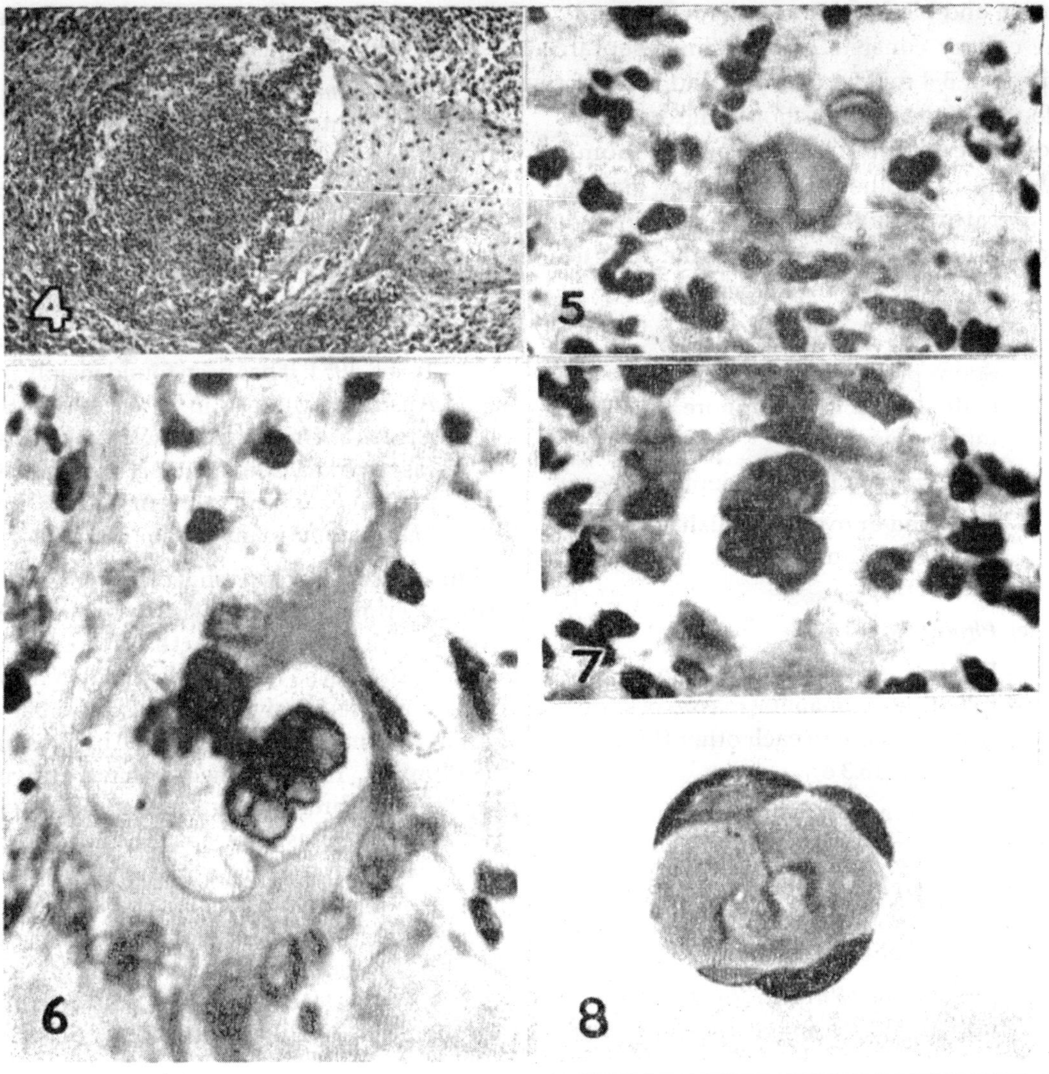

Fig. 4 : Microabscesses at the tips of rete-pegs are characteristic of **chromoblastomycosis**. H & E, 120 X

Fig. 5 & 3 : Sclerotic cells of ***Cladosporium (Cladophialophora) carrioni*** in tissue, one cell showing division. (Courtesy of Mohapatra *et al.* (1967) *Mycosen* 10, 309). Weigert's H & E., 1200 X

Fig. 6 : A group of sclerotic cells in a foreign body giant cell. H & E, 1200 X

Fig. 7 : Group of sclerotic cells in pus. H & E, 1200 X

Fig. 8 : Culture of ***Cladosporium (Phialophora) carrionii***. Enlarged to twice the normal size

micron diameter present in giant cells (**Pl. 18.III, Fig. 6**) or "swimming" in microabscesses or in the centre of "tubercles" are diagnostic for chromoblastomycosis. They are often found only after careful search of serial sections. These 'sclerotic bodies' may be present singly, in pairs or in groups (**Pl. 18.III, Fig. 5-7**). Not infrequently "twin bodies" or even "tetrads" of dividing sclerotic cells are seen in the tissue (**Pl. 18.III, Fig. 5**).

Differential Diagnosis : In India, culture at 25-28°C produced a dark olive coloured or brownish heaped-up colony, showing a black margin and black colour on the reverse side. The colony consists of branching septate hyphae of 3-4 micron diameter and brown in colour. Definite identification of the organisms should be left to experienced taxonomic mycologists, for the type of medium and the temperature may influence the type of sporulation in one and the same fungus. Sporulation types are the following :

(a) ***Acrotheca Type***

Spore production along the sides of a **club shaped conidiophore** (**Fig. 18.2**, middle; **Fig. 18.3 a**).

(b) ***Cladosporium Type*** (Cladophialophora)

Rounded or oval brownish conidia in branched chains (**Fig. 18.2**, left; **Fig. 18.3 b** and **c**).

(c) ***Phialophora Type***

Flask shaped conidio-phores (**phialides**) containing masses of spores which adhere to each other (**Fig. 18.2**, right; **Fig. 18.3 d**).

The following classification has been suggested by Emmons, Binford and Utz (1963):

1. ***Phialophora verrucosa*** Medlar 1915: showing almost exclusively phialophora type of sporulation, (**Fig. 18.3 d**).
2. ***Fonsecaea (Phialophora) pedrosoi*** (*Cladosporium*, formerly *Hormodendrum*); Usually oval *cladosporium* type of sporulation, but occasionally also one or both of the other form.
3. ***Phialophora compactum :*** (Carrion, (1935) n. comb. Emmons, 1944: *Cladosporium* type of sporulation with globular spores, tending to form closely packed short chains (**Fig. 18.3 b**).
4. ***Phialophora dermatitidis*** Kano, 1937: Highly variable in spore production.
5. ***Cladosporium (Phialophora) carrionii*** Trejos, 1954: It has been found mainly in Australia, but also in S. Africa and S. America. Once it was isolated from a case in India (Mohapatra *et al.*, 1967). It produces long chains in the *Cladosporium* manner (**Fig. 18.3 c**).

Immunology : Wilson suggested that some cases of chromoblastomycosis may **resemble** chronic cutaneous **North American blastomycosis** in so far as dissemination may occur from a previously unrecognised primary focus in the lungs. Baquero (quoted by Wilson and Plunkett, 1965) claimed to have **grown fungi from the bronchial washings from early cases of chromoblastomycosis**. He inoculated the cultural material into normal skin. Subsequently typical lesions cropped up at the injection site. Skin tests with antigenic extracts from fungi in endemic areas revealed at higher percentage of positive results than in non endemic areas. Since the human body evidently

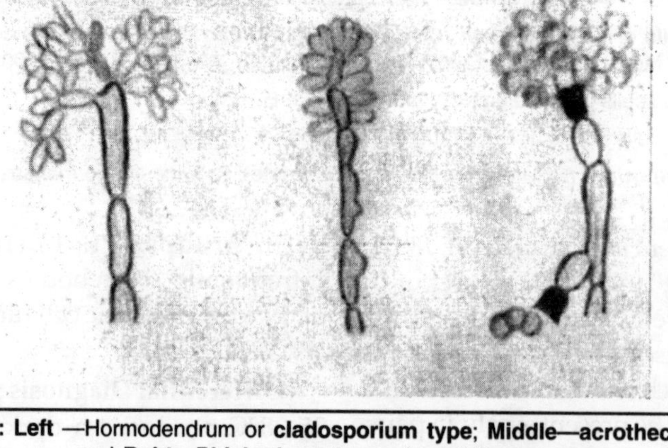

Fig. 16.8 : Left —Hormodendrum or **cladosporium type; Middle—acrotheca type;** and **Rght—Phialophora type of conidiophore**

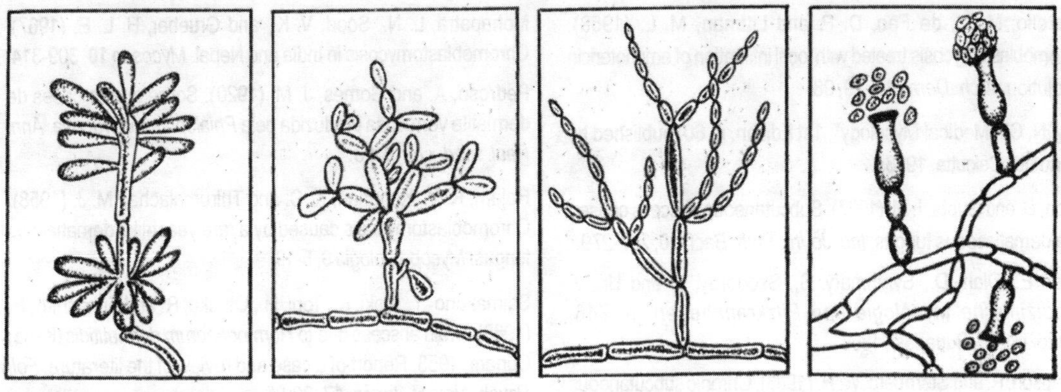

Fig. 18.3 : (a) **Acrotheca type** of sporulation (b) ***Cladosporium* manner of sporulation** in ***Fonsecaea (Phialophora) compacta*** (c) ***Cladosporium* type of sporulation in *(Cladophialophora) Cladosporium carrionii*.** (d) ***Phialophora* type of sporulation in *Phialophora verrucose*.**

has a high resistance to the infection, a mild primary pulmonary form of chromoblastomycosis say exists, which may clear spontaneously leaving a high immunity, as in coccidioidomycosis and histoplasmosis.

Serologic tests in man and animals show that there is a cross reaction among the various species of causative fungi, indicating their close genetic relationship.

Why, in very rare cases, dematiaceous fungi become disseminated is not known. Azulay (1967) observed a patient with haematogenous dissemination who had more than 100 lesions all over the body, and Rajam *et al.* (1958) described a very unusual clinical type of "chromoblastomycosis" in a seven year old Indian boy with prolonged illness, ending fatally.

Among the few cases of cerebral infection with dematiaceous fungi, dermal lesions were rarely seen (Shimazono *et al.*, 1963).

Therapy

Oral administration with Iodine to highest tolerance combined with **600,000 units claciferol orally or intramuscularly once a week seems to be the most effective treatment** at present. If possible, plastic surgery combined with iodine and calciferol therapy should be attempted. The authors have though seen a relapse after 3 years in a patient who was treated with radical plastic surgery, fresh lesions had cropped up at the margin of the operation site in the old skin. However, this still the best treatment.

The fungi seem to be resistant to therapeutic levels of amphotericin B, but local injection of 5 mg per litre in 2% procaine solution was successful in the hands of **Costello *et al.* (1950); 7 ml was injected into the lesions once weekly for three months**.

The **MODERN TREATMENT** is Itraconazole and Flucytosine, but relapse is common, Fluconazole may be tried as maintenance dose to avoid relapse. Infrared heat is beneficial.

REFERENCES

Agarwal, S. and Chouhan, S. S. (1969). Subcutaneous abscess caused by dematiaceous fungus *Ind. Journ. Path. Bact.* **12**, 42-45.

Aikat, B. K. (1968). Personal communication.

Azulay, R. D. and Serruya, J. (1967). Hematogenous dissemination in Chromoblastomycosis. *Arch. Dermatol.* **95**, 57-60.

Brumpt, E. (1922). "Precis de parasitologie" 3rd End., p. 1105, Masson & Cie, Paris.

Carrion, A. L. (1942). Chromoblastomycosis. *Mycology* **34**, 421-441.

Conant, N. F. and Martin, D. S. (1937). The morphologic and serologic relationships of the various fungi causing dermatitis verrucosa (chromoblastomycosis) *Am. J. trop. Med.* **17**, 553.

Costello, M. L., de Feo, D. P. and Littman, M. L. (1959). Chromoblastomycosis treated with local infiltration of amphotericin B solution. *Arch. Dermatol.* 79, 98.

Dey, N. C., "Medical Mycology", 1st Edition, p. 60, published by the author, Calcutta, 1958.

Dube, B. and Gupta, I. M. (1967). Subcutaneous abscess caused by a dematiaceous fungus. *Ind. Journ. Path. Bact.* **10**, 277-279.

Fejer, E., Olah, D., Syathmary, S., Szodoray, L. and Uri, J. "*Medizinische Mycologie und Pilzkrankheiten*" p. 746, Akadeiaikiado, Budapest, 1966.

Kempson, R. and Sternberg, W. H. (1963). Chronic subcutaneous abscesses, caused by pigmented fungi, a lesion distinguishable from cutaneous chromoblastomycosis. *Am. Journ. clin. path.* **39**, 598-606.

Lane, C. G. (1916). A cutaneous disease caused by a new fungus *(Phialophora verrucosa). Journ. cut. Dis.* **33**, 840.

Medlar, E. M. (1915). A new fungus *Phialophora verrucosa*, pathogenic for men. *Mycologia* **7**, 200.

Mohapatra, L. N., Sood, V. K., and Grueber, H. L. E. (1967). Chromoblastomycosis in India and Nepal. *Mycosen* **10**, 309-314.

Pedroso, A. and Gomes, J. M. (1920). Sobre quatro cases de dermatite verrucosa produzida pela *Phialophora verrucosa. Ann. Paul. med. chir.* **9**, 53.

Rajam, R. V., Kandhari, K. C. and Thirumalachar, M. J. (1958). Chromoblastomycosis caused by a rare yeast-like dematiacous fungus. *Mycopathologia* 9, 5-19.

Shimazono, Y., Isaki, K., Torii, H., Otsuka, R. and Fukoshiro, R. (1963). Brain abscess due to *Hormodendrum dermatitidis* (Kano) Conant, 1953. Report of a case and review of the literature. *Fol. Psych. Neurol. Japon.* **17**, 80-96.

Thomas, E., Job, C. K. and Hedley, C. G. (1957). Chromoblastomycosis. *Ind. Journ. med. Sci.* **11**, 570.

Viziam, C. B. (1967). *Personal communication.*

Wilson, J. W. and Plunkett, O. A. "The Fungus Disease In Man", Univ. of California Press, Berkeley (Sanfrancisco) and Los Angeles, 1965.

19

Dimorphic Fungus : Sporotrichosis
(Subcutaneous and Systemic)

Definition : Sporotrichosis is a subacute or a chronic granulomatous infection caused by ***Sporothrix** (Sporotrichum)* ***schenckii***, characterised by the formation of chancriform papules, nodules, ulcers and abscesses including local lymphatic spread. The infection is usually **confined to the skin** of the extremities, but may **cause systemic and disseminated lesions after break down of a nodular lesion** in the skin **causing a primary gummatous chronic ulcer** often called **"Sporotrichotic Chancre"**. It may remain as a localised granuloma or spread along regional lymphatics and lymph nodes. The spread may occur occasionally by haematogenous route.

Introduction

Nomenclature, Aetiology and Epidemiology.

Sporotrichum Schenckii is a **dimorphic fungus** which produces a delicate mycelium in **culture at 25°C**, but it present in form of spores, and small elongated, **cigar shaped bodies** in the tissue. Recently the original name **"Sporothrix"** has again been **recommended by Medical Mycology Committee of the Medical Research Council (1967)**. The organism is present in wood, plant material, fruits and soil. The disease is, therefore, frequently found in **agricultural workers, vegetable dealers, carpenters, florists and gardeners**. Silva and Guimaraes (1964) diagnosed sporotrichosis in four members of a family who were all infected by straw which was used in packing domestic utensils. The infection occurs in horses, cats, dogs, rats and mice, and direct transmission from animals to man has been reported.

Lurie estimated that at least 3300 cases of sporothrix infection has occurred in South Africa between 1927 and 1963 and that most of them had originated in the gold mines of Witwatersrand, the mine timbers being held responsible for the infection. Brown discovered that the organism grew on timbers only when the temperature was between 79° and 84°F and where the relative humidity did not fall below 92 per cent. Though the infection has been reported from all parts of the world, Brown's observation may explain why sporotrichosis appears to be rare in a dry and hot climate. In India, **Dey together with other authors had earlier reported cases from Assam and Calcutta. Dey (1968) reviewed 19 cases, of which seven were confirmed by positive culture**. Besides these reports, five more cases were diagnosed and confirmed by culture (Barua and Saika, as cited by Mohapatra, 1969). It is possible that in India the disease is confined to the north eastern part of the country because the climatic conditions of that area favour growth of *Sporothrix*. Mohapatra (1969) has questioned whether the conditions in the mines of India may not also be favourable for infection, but no instance has been reported so far from the mining belt. However, the disease may also run a chronic course and not necessarily compel the patient to consult a physician, if he lives in a village. Thus, it is possible that sporotrichosis is grossly underdiagnosed in India, though we are reasonable sure that the disease must be extremely rare in some areas, for instance in the Punjab.

Histology

Schenck (1898) first isolated and described a fungus **at Johns Hopkins Hospital** in U. S. A. from a patient with a **subcutaneous abscess** on the arm and placed the fungus in the genus Sporotrichum. Hektoen and Perkins (1900) reported a second case in U. S. A. De Beurmann and Rammond (1903) called the causative *fungus Sporotrichum beurmanni*. Lutz and Splendore (1903) called the fungus *Sporotrichum asteroides*. Investigators in the field have agreed to call the fungus as ***"Sporothrix schenckii"***.

Geographical distribution

Sporotrichosis have been reported from different parts of the world. It has been reported from U. S. A., France, Austria, Germany, Norway, South Africa, and a few other countries including India, several small epidemics have been reported but the highest epidemic of sprotrichosis was reported in the gold mines of Withwatersrand in South Africa. This epidemic effected only skin and subcutaneous tissues of 2825 native gold miners who became infected after injury to the skin during the work in mines. The source of fungus was the timbers in the mines on which the fungus *Sporotrichum schenckii* grew as a saprophyte. The epidemic subsided after the timbers were treated with chemicals "yard mixture" which compose of zinc sulphate, sodium fluoride, dinitrophenol, potassium bichromate (Brown *et al.*, 1947) which removed the fungus, *S. schenckii* from miners.

According to Utz (1969), sporotrichosis is not as uncommon as it was presumed.

In India, a few cases of sporotrichosis were reported by Ghosh (1932), Panga *et al.* (1947), **Dey** *et al.* **(1958). Dey (1969) recorded in India, 12 confirmed cases of sporotrichosis within a period of 38 years from 1932-1969 of which 3 cases only were in Calcutta and 9 cases from Assam.** This shows rarity of infection of sporotrichosis in India.

Aetiology

The fungus is distributed in nature and it has been recovered from soil, wood and plants. Infection in man is possible from contact with infected animals. The fungus causes infection in skin through the injured skin by thorns or other sources. There is no record of direct transfer of infection from man to man. Domestic animals like cats, dogs and horses can get sporotrichosis.

Strictly speaking this is an occupational disease (Foerster, 1926). This occurs commonly in farmers, gardeners, and labourers who have got more chance of either pricking their fingers and feet by thorns or injuring their skin.

PREDISPOSITIONS

Of the seven cases recorded by Dey, the aetiological predispositions are as follows :

Sex :	Four of these reported cases were females, and three were males.
Age :	All were adults.
Sites :	Fingers 3 cases; Wrist 1 case; Fore-arm 1 case; Toe 1 case; Both legs 1 case.
Occupation :	Gardener 1; Vegetable dealer 1; Carpenter 1; Housewife 2; Cultivator 1.

Portal of entry

In the majority of case the fungus infects through a trauma in the skin in the hands or feets, or a thorn prick, where an indolent ulcer is formed. Rarely, generalised infection may occur without any primary focus of infection in the skin. These lesions have been interpreted in France as a result of infection through the mucous membrane of the intestinal tract. A few cases of **pulmonary infections** have been explained as infection **through inhalation**.

Source of infection

The fungus is widely distributed in nature as a saprophyte and may be recovered from **soil plants, animals and a variety of contaminated materials**. Lurie (1948-50) isolated the fungus from miners in the gold mines of South Africa. An outbreak was recorded in Vermont associated with Sphagnum moss as a source of infection by D'Allesio *et al.* (1965).

CLINICAL FORMS AND GROSS PATHOLOGY

Sporotrichosis, causes three main types of lesions. The *first* is ***primary cutaneous sporotrichosis*** which is localised, usually benign and may heal without treatment. The *second* is an ***intermediate form, disseminated mucocutaneous sporotrichosis***, which is much more serious and may lead to generalisation and death. The ***third* type** consist of ***deep or systemic sporotrichosis***. It may involve single organs, or disseminate. Beurmann and Gougerot distinguished **six clinical types** of sporotrichosis, namely, **lymphatic, disseminated, epidermal, mucosal, skeletal and visceral forms**. Though most of these types can be included in the systemic group, one has to be aware that there are certain predilection sites for the deep infection, as for instance the lungs, joints and bones. The process may be confined to these organs and is not necessarily disseminating.

1. **Primary cutaneous sporotrichosis** (also called localised lymphatic sporotrichosis or **chancriform sporotrichosis**).

It is a relatively benign type. A painless lesion develops at the extremities within one to three weeks after small injury by a thorn or splinter. There may be primary ulcer, a papule or a hard intra-or subcutaneous nodule which becomes attached to the overlying skin. The colour changes from pink to purple, the skin breaks down and ulcerates (**Fig. 19.1**). This lesion is also called a sporotrichotic chancre. The ulcer has typically a bright red base free of necrotic tissue. The edges are undetermined and bleed easily. A chain of satellite nodules may crop up along the lymphatics. The local lymph node becomes enlarged and suppurates.

There is experimental evidence in human volunteers that an intradermal infection may result in a mild, superficial plaque type of lesions, but the subcutaneous infection produces the ulcerative and lymphatic form.

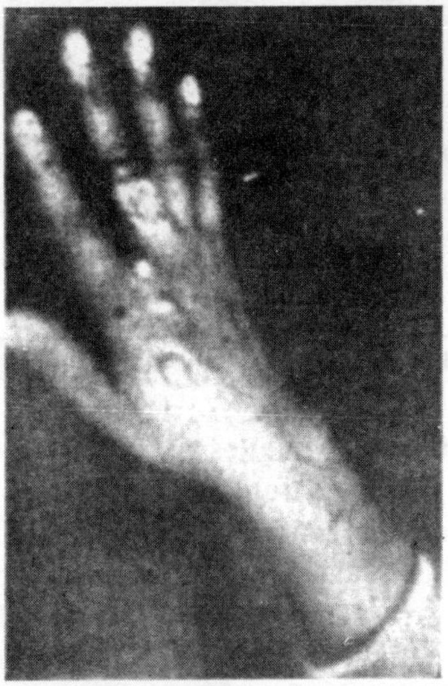

Fig. 19.1 : Sporotrichotic granuloma

Lymphatic involvement is sometimes absent and the lesion spreads by continuation forming flat **scaly plaques, or verrucous lesions** which may be **mistaken for chromoblastomycosis**, verrucous tuberculosis or blastomycosis. Deeper structures muscles, fasciae and bones may be involved by direct spread in rare cases. The differential diagnosis requires differentiation from syphilis, tularemia, anthrax and primary complex of tuberculosis or of other fungal diseases. However, the cutaneous lymphangitic form is rather typical for sporotrichosis.

2. **Epidermic type**

Occasionally sporotrichosis is localised primarily to epidermis (**Fig. 19.2**). It is characterised by **smooth, pinkish, raised papules and extensive involvement of the skin.** There may be one or more nodules present. It may have protean manifestations in skin and may appear as **nodular crusted lesion,**

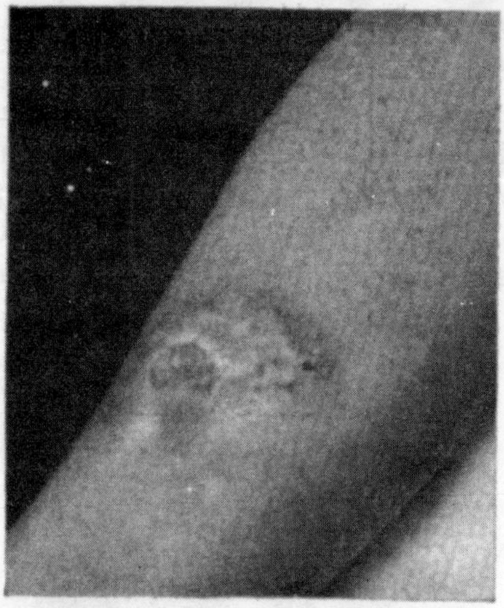

Fig. 19.2 : Skin-sporotrichotic granuloma

ulcerating, flat plaques and warty types. Warty types are common in foreheads; on face, nodular crusted lesions mimic lupus vulgaris. Superficial lichenoid and intertriginous types have been reported (Norden, 1951). Small gamma-like lesion might be present. The interesting feature of epidermic type of infection is that the patient has almost no systemic symptoms inspite of widespread skin lesions.

3. **Disseminated mucocutaneous sporotrichosis**

This begins often in the nasal, oral or laryngeal mucosa and may **result in lymphatic involvement**, strongly suggesting that the primary **inoculation** occurred **through the mucous membrane**. In other cases the mucocutaneous involvement is the result of dissemination from a systemic focus. Multiple nodules and ulcers all over the integument may herald the fatal end. The condition must be **distinguished from syphilis, tuberculosis or other deep mycoses.**

4. **Deep or systemic sprotrichosis** : (also : Extracutaneous, or nontegumentary primary sporotrichosis).

There is evidence that the **portal of entry** for this type of sporotrichosis is not the skin but

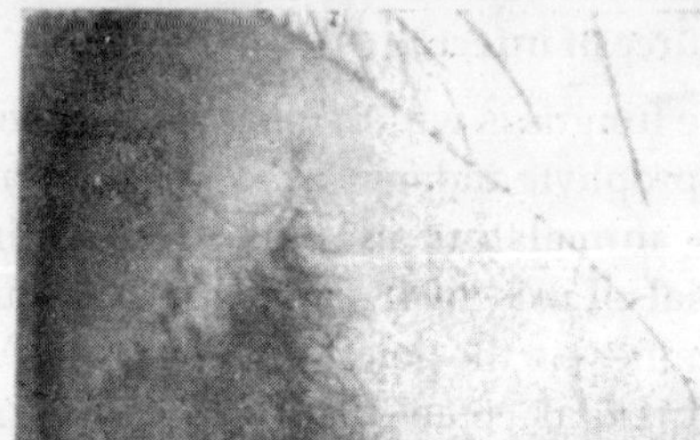

Fig. 19.3 : Sportrichotic granuloma

most probably the **lungs or the gastrointestinal tract. Bones and joints, lung** (**Steele**, 1964), **kidneys, adrenal cortex, spleen, liver, myocardium, the intestine, meninges of the**

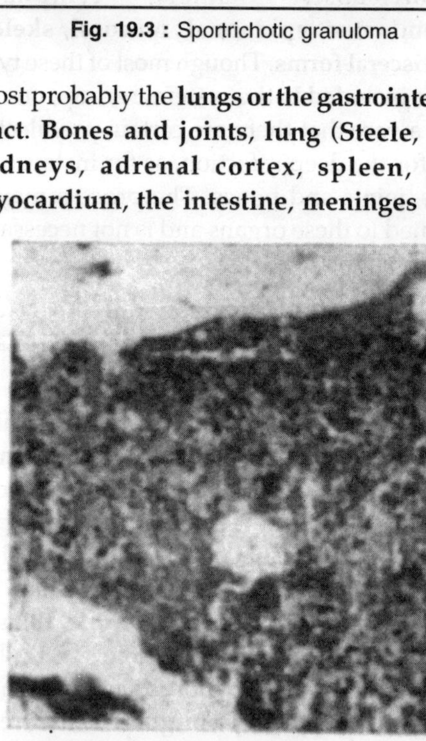

Fig. 19.4 : Sporotrichotic granuloma in tissue section × 450

cerebral cortex, eyes and other organs may be involved (Collins, 1947). The infection may end in rare fatal dissemination, which becomes most frequently apparent as multiple subcutaneous nodules scattered all over the body which soften and ulcerate. Usually the response to treatment is poor, though the course may vary from a rapidly fatal to a chronic protracted course.

Histopathology

The following facts must be borne in mind when attempting a histopathologic diagnosis of sporotrichosis-

1. Sporothrix is one of the few organisms, which in primary cutaneous sporotrichosis, can only rarely be found in the tissue. In systemic or disseminated forms, the situation is different. Organisms may be present in great numbers. The more necrosis the more chance of finding the fungus in tissue sections (**Figs. 19.4, 19.5**).

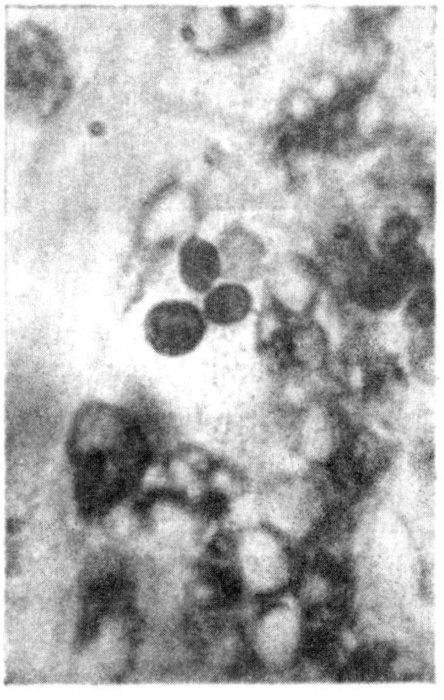

Fig. 19.5 : Sporotrichosis in tissue section showing giant cell reaction; × 560

2. An exception to the rule (1) is the occasional presence of an "asteroid body" in the granulation tissue. It consists of a central round fungus cell, 5-10 micron in diameter with peripheral radiating "clubs" or 2-8 micron length. The clubs are now considered to be an immune precipitate deposited by interaction of fungal antigen with host antibodies. Asteroid bodies are almost exclusively found in the centre of microabscesses and are characteristic for sporotrichosis (**Figs. 19.6, 19.7**; for details see Lurie, 1963).

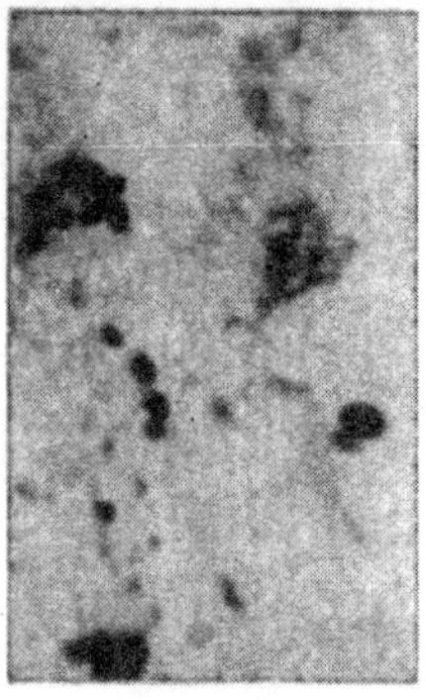

Fig. 19.5A : Yeast phase of *S. schenckii in tissue*

3. Even with special stains, like P. A. S., recognition of fungal cells is almost impossible because they are "drowned" in the multitude of PAS positive proteinaceous granules due to break down of tissue. Fetter *et al.* (1964) suggested, therefore, digestion of the tissue section prior to staining. The organism is unaffected by this procedure. **Ordinary paraffin sections** are deparaffinised and **hydrated** and then **placed in a 1 : 1000 solution of malt diastase at 37°C for one hour** and subsequently **stained with P A S**. However, in- spite of this method, many serial sections may have to be examined in order to find one or two organism in a single section.

4. *Sporothrix* is a dimorphic fungus and appears in tissue section either as a round cell, 1.5 to 4.5 microns in length (**Pl. 19.2, Fig. 6 & 7**), Bader (1965). Hyphae are very rarely found in the tissue in the primary cutaneous form of sporotrichosis. However, occasional presence of **slender hyphae** has been reported in skin lesions (Stone *et al.*, 1966). They are 1 to 2.5 microns in width and can be demonstrated by PAS stain (**Pl. 19.2, Fig. 2**).

The granulation tissue of **primary cutaneous sporotrichosis consists of tuberculoid**

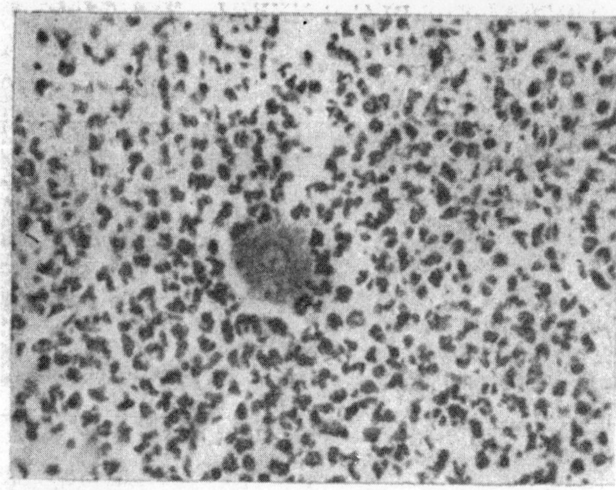

Fig. 19.6 : Asteroid body in pus; × 140

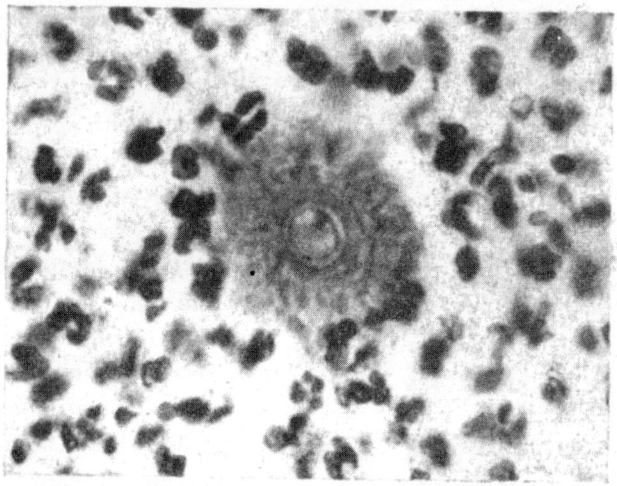

Fig. 19.7 : Asteroid body; × 560

granulomata with or without microabscesses. Caseation is rarely seen. Non specific granulation tissue with lymphocytic, polymorphonuclear and eosinophilic infiltration may be abundant. Generally, the picture is more variegated than that of tuberculosis and, as in most fungal disease, presence of eosinophils suggests the true nature of the process. There is no definite histological characteristic however, which would allow definite distinction from other specific granulomas (**Fig. 19.8**).

SKIN TEST WITH SPOROTRICHIN : Antigen for skin testing (Sporotrichin) is prepared from the medium (containing glucose, yeast extract and casamino acid) in which *S. schenckii* is grown. Then yeast cells of *S. schenckii* are removed by filtration and the filtrate is then dialyzed and concentrated (Schneidau *et al.*, 1934). Ochoa and Figueroa (1942) obtained sporotrichin antigen for skin test from polysaccharide fraction of ***S.schenckii***.

0.1 ml of 1 : 250 dilution of sprotrichin is injected intradermally on ventral aspect of the forearm. In the induration at the site of injection is 5 mm or more after 24 of 48 hours of injection, the reaction is read as positive.

PLATE 19.1

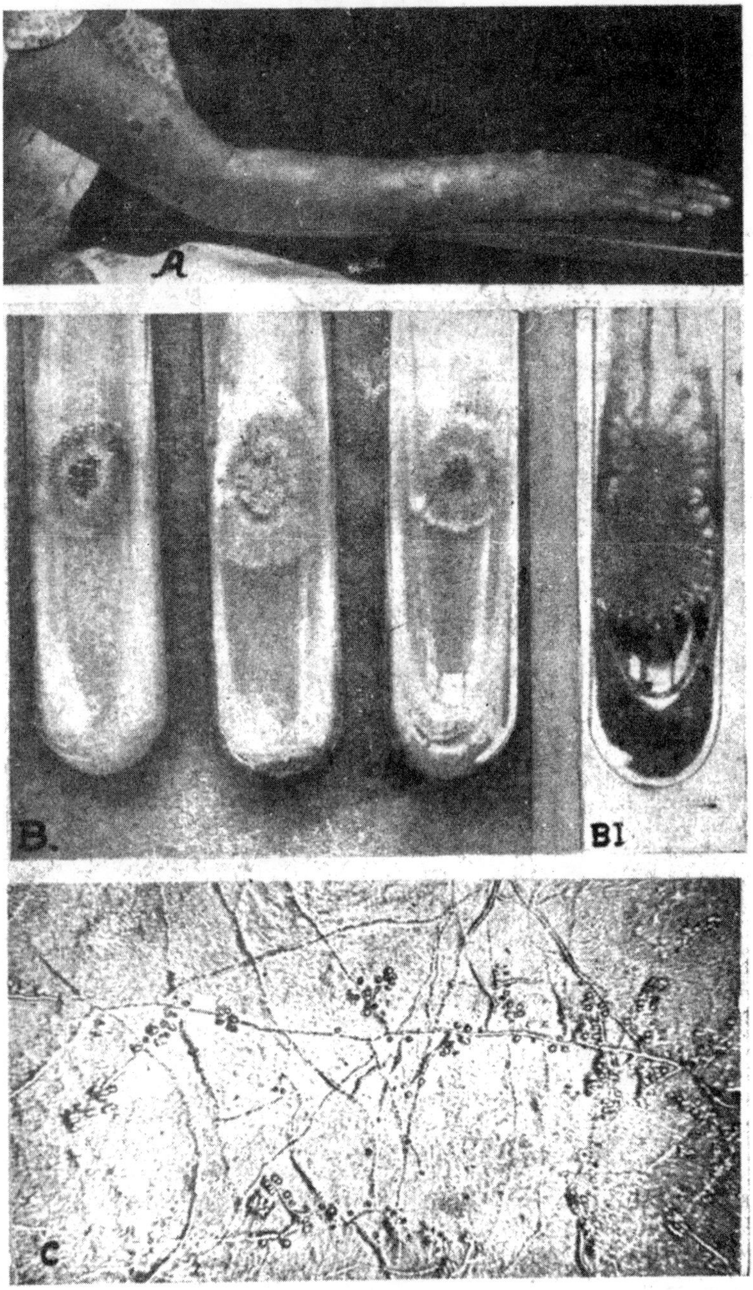

(**A**) Gummatous ulcer of the right arm and forearm. (**B**) Culture of ***Sporothrix schenckii*** (3 weeks old) and B_1 (6 weeks old), (**C**) **Conidia** of ***S. schenckii*** **in hanging drop culture**

For better results induration can be recorded both after 24 and 48 hours. In case, it is negative after 24 hours, but positive after 48 hours, this is to be regarded as positive and *vice versa*. But reading after 24 hours is more important and if only one reading is possible, it should be taken after 24 hours (Schneidau *et al.*, 1964; Chakravarty *et al.*, 1968). For routine clinical examination, reading after 24 hours is good enough. Positive reaction demonstrates that the

PLATE 19.II

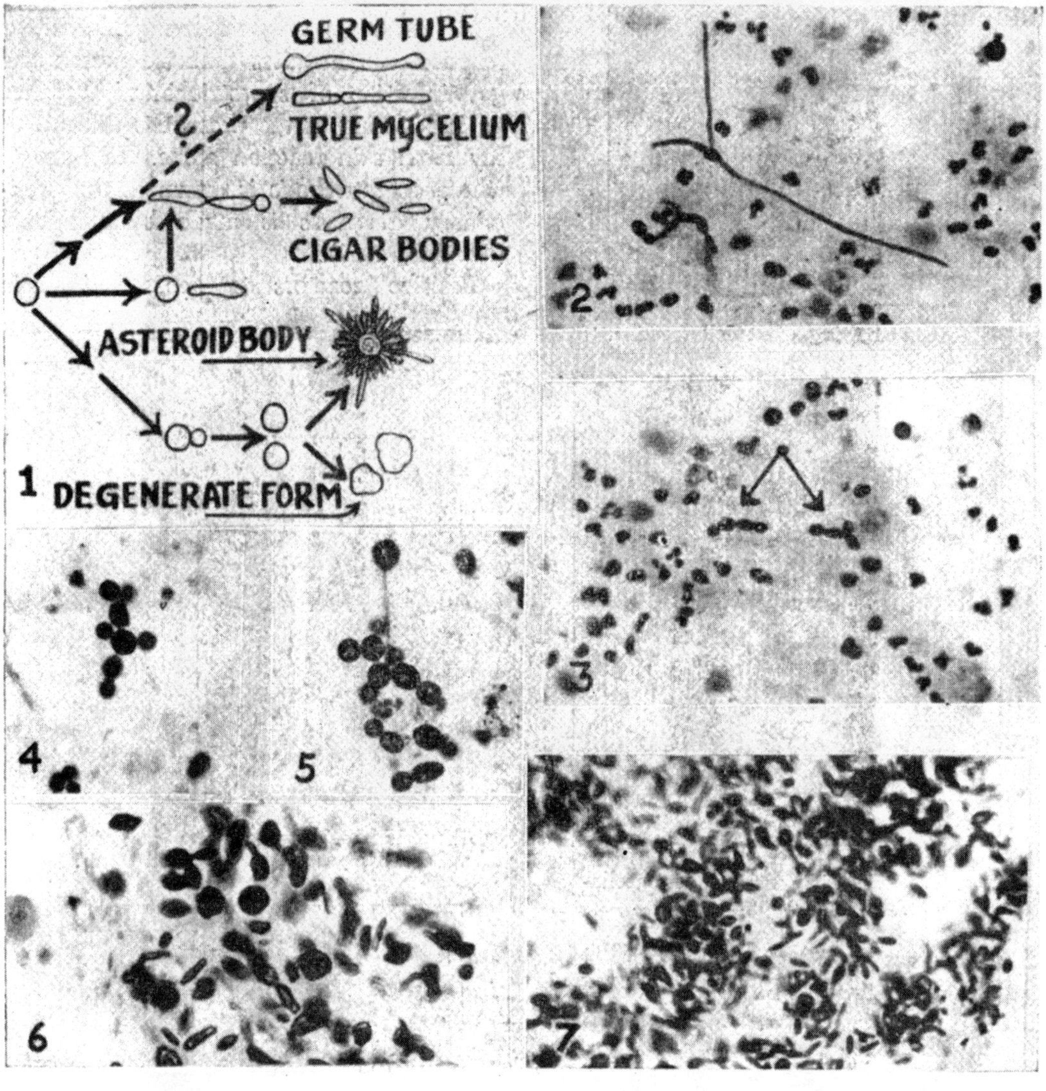

Plate 19.II : Photographs on left side of plate, 1000 ×. Photographs on right side of plate, 560 ×. All stained with P A S

Fig. 1 : Tissue forms of *Sporothrix :* round forms producing **cigar bodies** and, very rarely mycelium. (Adopted from Bader. Die visceralen, Mycosen, Fischer, Jena, 1965).

Fig. 2 and **3 :** Pleural exudate in **pulmonary sporotrichosis**, smear. **Fig. 2**, showing hyphae; **Fig. 3**, budding oval organisms.

Fig. 4 and **5 :** Round and oval *Sporothrix schenckii* organism in pus. (slides 2-5: Courtesy of Dr. B. C. Dyson, Philadelphia, U. S. A.).

Figs. 6 and **7 :** Experimental sporotrichosis, mouse liver; Massive invasion by "cigar" shaped and round organisms. Such invasion is seen only in disseminated sporotrichosis in human or in experimental sporotrichosis. (**Histological section**: Courtesy Dozent Dr. med. habil. G. Bader, Rostock, G. D. R.).

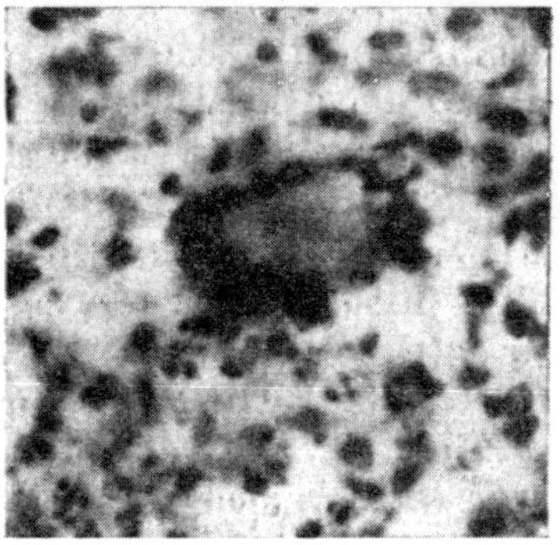

Fig. 19.8 : Sporotrichosis in tissue section showing **giant cell reaction**

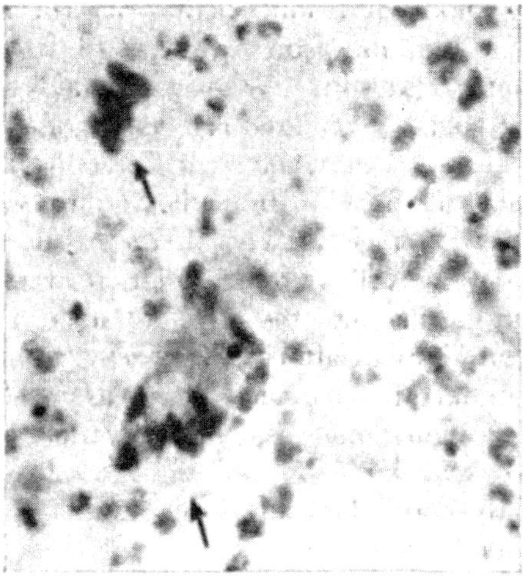

Fig. 19.9 : Showing extra and **intracellular yeast** and elongated forms

person has got infection of sporotrichosis. Whether he is suffering at the present is difficult to say from skin reaction alone. In Mexico, where sporotrichosis is common, the preliminary diagnosis is made by skin test (Norden, 1951).

Positive reactors to sporotrichin skin test are more common among rural than in urban population. Among rural population it is commoner in gardeners and farmers (Schneidau *et al.*, 1964, Chakravarty *et al.*, 1968).

Mycology

The fungus causing sporotrichosis is *Sporotrichum schenckii*. The fungus is dimorphic. It has got (a) a filamentous or mycelial phase and (b) tissue phase showing yeast cell.

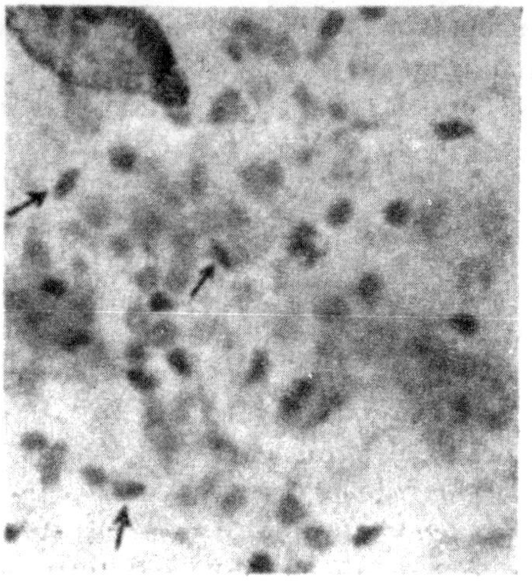

Fig. 19.10 : Showing extra and **intracellular yeast** and elongated forms

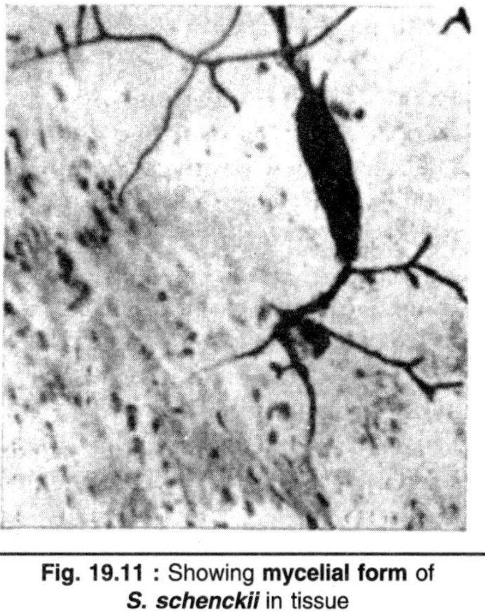

Fig. 19.11 : Showing **mycelial form** of ***S. schenckii*** in tissue

It has got delicate **branching, septate mycelium (or hypha)** which are 2μ in diameter on suitable culture media. **Conidia are pyriform** and spherical and are 2 to 4μ in size (Conant *et al.*, 1954. They may form a rosette in some cases (see **Pl. 19.I, B**).

Mycelial phase of *S. schenckii* can be cultured on **Sabouraud's glucose agar at 25°C** or room temperature. The colonies grow in 4-7 days but may be delayed upto 10 days and should not be rejected before 2 weeks. It is at first creamy and light brown in colour with an uneven, healthy, corrugated surface. The colour of the colony becomes dark brown with age and growth becomes wrinkled at the centre with radiating furrows. Excrescences appear in the center of the cultures (see Pl. 19.I).

Yeast phase of ***S. schenkii*** can be grown of **Brain-Heart infusion agar** with or without blood at **37°C** (Ajello *et al.*, 1963). These colonies are composed of cigar-shaped, round on oval and budding cells. **THEY ARE GRAM POSITIVE (Fig. 19.5A).**

Pigmentation of colonies is highly variable amongst strains or in the same strain of different periods of growth; as very young or very old cultures are devoid of pigmentation. Pigmentation also depends on the medium and oxygen supply. In limited oxygen supply (low oxygen tension) in screw-capped tubes the colony may remain white. Reid (1952) demonstrated that both the growth and pigmentation of ***S. schenkii*** is greatly influenced by the amount of thiamine in the medium. The addition of 100 g per ml of thiamine to standard Sabouraud agar allowed a maximal growth and pigmentation. This effect is slightly enhanced by addition of phosphate.

The pigment as suggested by Nickerson (1947) is melanin, and tyrosinase has been identified in sporulating mycelia and spore masses but the pigment is absent in spore free hyphae.

TISSUE (YEAST) PHASE

Campbell (1945) first demonstrated that tissue phase may be produced in artificial media. In these dimorphic fungi, when mycelial culture is inoculated into media rich in protein and vitamins, the mycelia are converted into yeast form. When cultivated at 37ºC at slightly increased carbon dioxide tension media is enriched with protein, thiamine and biotin as in brain heart infusion agar. This change is seen in the tissue of the host as well as by inoculation of mycelial fragments into the testes of mice and pus withdrawn after 2-3 weeks containing abundant characteristic yeast cells. In addition, mouse inoculation differentiates *S. schenkii* from several saprophytes that are morphologically similar but not pathogenic.

GROWTH AND METABOLISM : Thiamine stimulates the growth of *S. **schenkii*** and it has been shown by Drouhet and Mariat (1950) that pyrimidine is the necessary molecule for this. Mariat and Drouhet (1953) demonstrated that biotin also stimulates the growth of ***Sporothrix schenckii***. Drouhet and Mariat (1952) also determined the factors necessary for the yeast phase of the organism. These are : (a) incubation at 27°C, (b) medium rich in amino acids and vitamins specially thiamine and biotin and (c) an atmosphere with a tension of 5 per cent carbon dioxide.

Nitrogen and carbohydrate assimilation study of ***S. schenckii*** were studied by Lurie (1950, 1951). This fungus is able to fix nitrogen from atmosphere. It can utilise inorganic nitrates (NH_4NO_3) as well as urea and amino acids. Sugar fermentation of different strains is often variable as found of different authors and was responsible for various varieties of the organism in the past.

LABORATORY DIAGNOSIS OF THE CLINICAL MATERIAL

It is better to keep in mind that in majority of human lesions due to ***S. schenckii*** it is difficult to find the organism (Ajello *et al.*, 1963). Cultures of all clinical materials to be done whenever sporotrichosis is suspected.

Saprophytic sporotrichum species may mimic *S. schenckii*. This is why demonstration of tissue form (yeast bodies) for the identification of the pathogenic fungus is important. This can be done by **culturing the clinical material in Brain-heart infusion agar at 37°C**. If it does not covert, it can be inoculated in testis of mice where tissue forms (yeast-bodies) are grown. *Smear* from the puncture of the nodule may show Gram positive cigar-shaped fungal elements in the mononuclear cells.

CULTURE

The clinical diagnosis is confirmed by culture of the aspirated material as the material from an open abscess often shows contamination. As a routine procedure cyclohexamide (0.5 mg)-chloramphenicol (0.05 mg)-Sabouraud's glucose agar or blood agar tubes are inoculated and incubated at 37°C. The growth occurs in about 7-10 days. Tubes are rejected only after 2 weeks as negative if there is no growth. The growth is at first creamy or light brown in colour with an uneven, leathery, corrugated surface followed by wrinkling and radiating furrows. Dark pigment appears due to the action of tyrosinase. For conversion to yeast form with characteristic oval budding cells, brain-heart infusion agar or animal inoculation is preferred.

Microculture shows septate, branching mycelia with lateral conidiophores bearing a bunch of pyriform conidia, 4 or 8 of them forming rosettes. They may be arranged along the mycelia with a sleeve like arrangement. (**Fig. 19.12, 19.13**).

Animal inoculation : Most of the laboratory animals are susceptible, specially white rats and mice. Male rats injected intraperitoneally with the 0.2 ml of dense suspension of the fungus, show testicular inflammation like Strauss reaction after about 2-3 weeks, when pus can be aspirated and the fungus from smear can be stained by Gram stain.

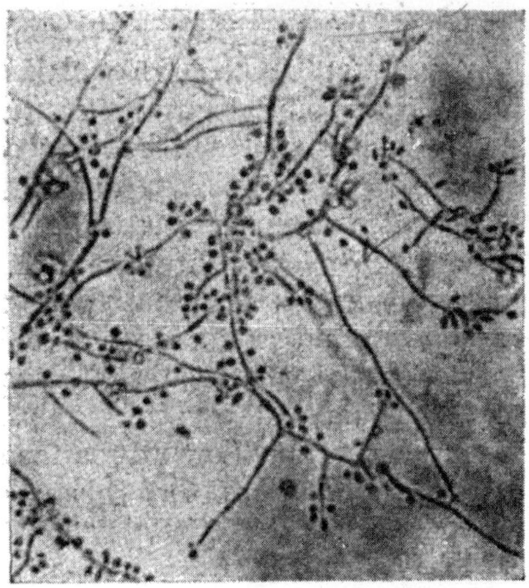

Fig. 19.13 : Showing appearance of ***S. schenckii*** in microculture

A normal saline suspension of mycelial culture of *S. schenckii* (in 0.2 ml suspension) is injected *intratesticularly* in mice. After three weeks, pus from abscess in testis will show characteristic oval budding (tissue phase) of *S. schenckii*. In experimentally infected mice, large number *S. schenckii* may be seen both inside macrophages and freely outside. Fusiform fungus elements known as "Cigar bodies" are also seen in some cases. These are commonly seen in tissues of testis, skin and subcutaneous tissues. But less commonly they are found in liver, spleen and kidney (Conant *et al.*, 1954; Chakravarty *et al.* 1969). Giant cell formation may be seen in rat's testis with **lymphocytic infiltration and Cigar bodies (Dey, unpublished)**.

Fig. 19.12 : Conidia of *S. schenckii* in culture

Fig. 19.14 : Culture mount of ***Sporothrix schenckii*** showing conidia in "flowerette (upper part) and in "sleeve" (lower part) arrangement

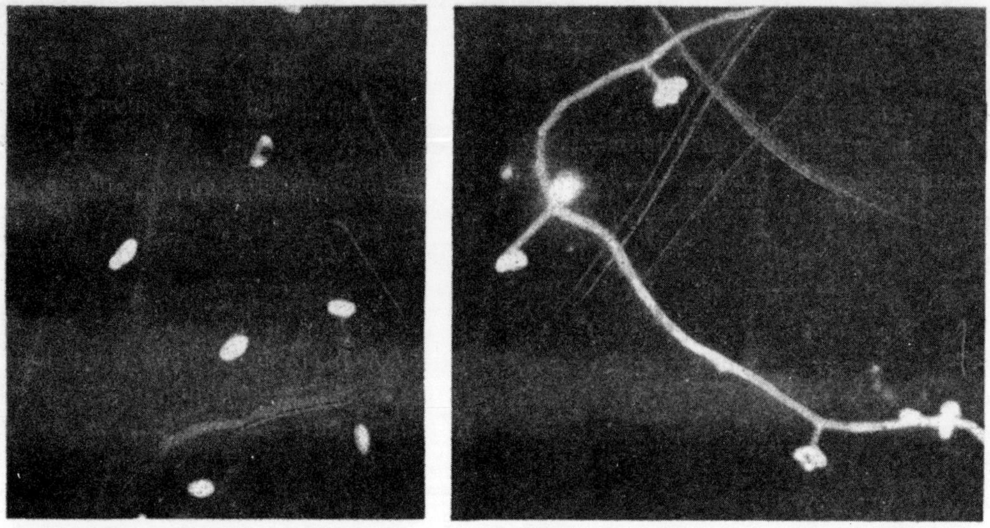

Fig. 19 (1) : Specific staining of yeast-phase of cells of *S. schenckii* treated with homologous fluorescein labelled antiglobulin × 625. **(2)** Specific staining of mycelial-phase elements of *S. schenckii* treated with homologous fluorescein labelled antiglobulin × 625. (Courtesy of W. Kaplan and M. S. Ivens *J. Inv. Dermat.* **35**, 151-159, 1960)

Immunological reactions : Serological tests by the patient's serum are almost specific test and is positive in most of the cases but there may be false positive reactions.

Intradermal test is positive in most of the cases but there may be false positive reactions as well.

Serology

Agglutinin, precipitin and complement fixing antibodies have been demonstrated in systemic sporotrichosis reacting with antigens derived from spores. It has been demonstrated that experimentally immunised or infected rabbits develop these antibodies (Norden, 1951; Chakravarty *et al.*, 1968).

USE OF FLUORESCENT ANTIBODY (F. A.) TECHNIQUES : Fluorescent antibody technique (Coons *et al.*, 1942) brought a new serological method for the rapid detection and identification of fungus and for the demonstration of antibodies in sera (Fluorescent antibody technique is the microscopic visualisation of an antigen-antibody reaction by immuno-chemical staining procedure). Kaplan and Ivens (1960) demonstrated that fluorescent antibody technique can be used for the identification of *S. schenckii* in cultures and its detection in clinical materials (**Fig. 19.15**).

If treated properly with potassium iodide and Amphotericin B the prognosis is good. Prognosis is grave in rapidly disseminated form and in debilitated patients associated with other wasting diseases.

Treatment : **POTASSIUM IODIDE** is the treatment of choice in sporotrichosis of skin and subcutaneous tissue as also of mucous membrane. In should be given in saturated solution increasing doses beginning with 10 drops thrice daily to reach upto 30 to 40 drops thrice daily (Conant *et al.*, 1954).

Potassium iodide should be given with water or milk. The drug should be **continued even for eidht weeks after complete recovery. Amphotericin B should be given intravenously in affection of internal organs** caused by *S. schenckii* **and in infection resistant to potassium iodide. Stilbamidine intravenously may be tried, when potassium iodide and Amphotericin B fail.**

Local lesions may be given a compression with iodide-iodine solution (Conant *et al.*, 1954). Surgery in any form should be avoided. Collection of pus should be aspirated.

ITRACONAZOLE oral or I.V administration is done. Some some Imidazoles can also betried.

REFERENCE

Ajello, L., Goerg, L. K., W. and Kaufman, L. CDC Laboratory Manual of Medical Mycology. U. S. Dept. of Health, Education and Welfare. Public Health Service, Communicable Dis. Center, Atlanta 22, Georgia, U. S. A., 1963.

Bader, G. and Schabinski, G. (1963). Der Gestaltwandel der Gewebsforman von Cryptococcus neoformans, Sporotrichum schenkii and Candida albicans. *Path. Microbiol.* **26**, 475-493.

Brown, J. W., Brandt, F. A. and Berman, C.Brown, R., Weintroub, D., Simpson, M. W., simson, F. W. L. and Helm, M. A. F. Sporotrichosis infection on mines of the Witwatersrand. Proc. Transvaal Mine Med. Officers' Assoc. Johannesburg. Transvaal Chamber of Mines. pp. 3-67, 1947.

Campbell, C. C. (1945). The use of Francis glucose blood agar in the isolation and cultivation of S. schenckii. *J. Bact.* **50**, 233.

Chakravarty, S. C., Banerjee, D. K., Deb, S. R. and Bagchi, P. (1968). A study in Sporotrichosis: Skin test survey, clinical, cultural and serological studies in man and in experimental animals. A preliminary report: *Ind. J. chest. Dis.* **10**, 133.

Chakravarty. S. C., Banerjee, D. K., Deb, S. R., Bagchi, P. and Damodaran, V. N. A study in sporotrichosis, Skin test survey, Clinical, cultural, serological studies in men and in experimental animals. (Final report to C. S. I. R.) 1969.

Collins, W. T. (1947). Disseminated ulcerating sporotrichosis with widespread visceral involvement. *Arch. Dermatol. Syph.* **56**, 523.

Conant, N. F. Smith, D. T., Baker, R. D. Callaway, J. L. and Martin, D. S. L. Manual of Medical Mycology, W. B. Saunders Co. Philadelphia, 1954.

Coons. A. H., Creech, J. H., Jones, R. N. and Berliner, E. (1942). The demonstration of Pneumococcal antigens in tissues by the use of fluorescent antibody. *J. Immunol.* **45**, 157.

De Beurmann and Ramond, L. (1903). A bees sous-cutanes multiples d origine Mycosique, *Ann. Dermt. Syph.* **4**, 678. Quoted in "Sporotrichosis" by Norden, A. Ejnar Munksgaard, Copenhagen, 1951.

D, Allesio, D. G., Levens, L. J., Strump, G. B. and Smith, G. D. (1965). *New Eng. J. Med.* **272**, 1054.

De Beurmann and Gougerot, H. Les Sporotrichoses, Paris, Felix Alcan, 1912. Quoted by Norden, A. in "Sporotrichosis" by Norden Ejnar Muksgaard. Copenhagen, 1951.

Drouhet, E. and Mariat, F. (1952). Etude des facteurs determinant les development de La phase legure de Sporotrichum schenckii. *Ann. Inst. Pasteur.* **83**, 506-513.

Dev, N. C. Sporotrichotic granuloma in "Fungus diseases in India", Calcutta School of Tropical Medicine. 1962.

Dey, N. C. Sporotrichosis: Paper read at the symposium on "deep mycoses in Calcutta area". Institute of Postgraduate Medical Education and Research, Calcutta, 1969.

Dey, N. C. (1969). Sporotrichosis, a rare disease in India. *Ind. J. Dermatol.* **15**, 1.

Dey, **N. C., Saikia, T. and Majumdar, C. T. (1958). Sporotrichosis in Assam (India).** *Ind. J. Dermat.* **3**, 103.

Emmons, C. W., Binford, C. H. and Utz, J. P. Medical Mycology, 2nd Ed., Lea and Febiger, Philadelphia, 1970.

Fetter, B. F. and Tindall. J. P. (1964). Cutaneous Sporotrichosis. *Arch. Path.* **78**, 613-617.

Foerster, H. R. (1826). Sporotrichosis: An occupational dermatosis. *J. am. med.* Assoc. **87**, 1605.

Ghosh, L. M. (1932). An unusual case of sporotrichosis. *Ind. med. Gaz.* **67**, 570.

Jones, A. D., Sarosi, G. A., Parkar, J. D., Weeks, R. J. and Tosh, F. E. (1969). The Complement Fixation Test in Extracutaneous Sporotrichosis. *Ann. int. Med.* **71**, 913-917.

Kaplan, W. and Ivens, M. S. (1960). **Fluorescent antibody staining** of *Sporotrichum schenckii* in cultures and clinical materials. *J. invest. Dermatol.* 35, 151.

Lurie, H. I. (1951). Sporotrichum species; their nitrogen metabolism. *Mycolgia.* **43**, 117-119.

Lurie, H. I. (1950). Pathogenic sporotricha, their carbohydrate reactions. *Mycologia.* **42**, 624-641.

Lurie, H. I. (1963). Histopathology of sporotrichosis. *Arch. Path.* **75**, 421-437.

Lurie, H. I. and Still, W. J. S. (1969). The "capsule" of Sporotrichum schenckii and the evolution of the asteroid. A light and electron microscopic study. *Sabouraudia*, **7**, 64-70.

Lutz, A. and Splendore, A. (1908). Quoted in Sporotrichosis by Norden, A. (1951), Publisher Ejnar Munksgaard, Copenhagen, Denmark.

Medical Mycology Committee of the Medical Research Council: Nomenclature of fungi pathogenic to man and animals. *M. R. C. Memo No. 23, 3rd Ed., London, HMSO*, 1967.

Norden, A. Sporotrichosis, Clinical and laboratory features and a serologic study in experimental animals and humans. Publisher Ejnar Munksgaard Copenhagen, Denmark, p. 76, 1951.

Nicherson, W. J. (1951). Biology of Pathogenic fungi. Chronic Botanicaco, Waltham, Massachusetts. In Norden, A. Sporotrichosis clinical and laboratory features and a serological study in experimental animals and humans. *Acta Path. et Microbiol. Scand. Suppl.* **89**, 1951.

Ochoa, A. G. and Figueroa, E. S. Polisacaridos del Sporotrichum schenckii. *Revista del Instituto de Salubridad Y Enfermadades Tropicales (Mexico, D. F.)* **8**, 143, 1947. Quoted in "sporotrichosis" by Norden, A. (1951), Enjar Munksgaard, Copenhagen.

Panja, D., Dey, N. C. and Ghosh, L. M. (1947). Sporotrichosis of the skin in *India*. *Ind. med. Gaz.* **82**, 200.

Reid, J. D. (1952). A modified Sabouraud medium for the cultivation and identification of *Sporotrichum schenckii. Am. J. clin. Path.* **22**, 1030-1033.

Schenck, B. R. (1898). On refractory subcutaneous abscess caused by a fungus. Possibly related to sporotricha. *Bull. Johns Hopkins Hosp.* **9**, 286.

Schneidau (Jr.), J. D., Lamar, L. M. and Hairston, M. A. (1964). Cutaneous hypersensitivity to sporotrichin in Louisiana. *J. am. med. Assoc.* **188**, 371.

Silva, V. P. and Guimaraes, N. A. (1964). Familial Epidemic of Sporotrichosis. *Hospital*, **66**, 573-579.

Steele, J. D. "Sporotrichosis" in the "Treatment of Mycotic and parasitic diseases of the chest". Charles, C. Thomas (Publisher) Springfield. illinosis, U. S. A. p. 140, 1964.

Stone, O. J. (1966). Sporotrichosis with demonstration of hyphae in human tissue. *Arch. Dermat.* **93**, 65.

Utz. J. P. (1969). Troubles in sporotrichosis (Editorial). *New Eng. J. Med.* **280**, 446.

Wilson, J. W. and Plunkett, O. A. The fungus diseases of man. University of California Press, Berkeley and Los Angeles, 1965.

20

Rinosporidiosis

TAXONOMIC POSITION OF RHINOSPORIDIUM IS UNCERTAIN AT PRESENT. BUT WITH THE MODERN APPLIANCES OF MOLECULAR BIOLOGICAL STUDIES THE ORGANISM HAS ALREADY BEEN REMOVED FROM PLANT KINGDOM TO THE ANIMAL KINGDOM.

However, we being conservative wish to retain it as a chapter in this present volume.

Definition : A chronic infection with *Rhinosporidium seeberi* most frequently causing polypous lesions of the nasal respiratory mucosa and submucosa but also of the conjunctiva and rarely of other mucosae and skin.

History

Seeber (1900) reported two cases of nasal polyp caused by a protozoal infection. The organism was called *Coçcidium seeberi* by Wernicke (1903) belonging to sporozoa. The third case was reported by O'Kinealy (1903) in India and the organism was called *Rhinosporidium kinealky* by Minchin and Fantham (1905). Seeber (1912) found this to be identical with his organism and called the organism *Rhinosporidium seeberi*. Since then more cases were published from various parts of the world. The organism was thought to be a protozoa until Ashworth (1932) demonstrated the morphological similarity of the fungus with Chytridales (Phycomycetes). The organism has not been cultivated as yet and the taxonomic position of *R. seeberi still* remains doubtful.

Geographical distribution

By far the greatest number of cases have been found in Asia, but some also in Africa, both North and South America, and even in Europe. Karunaratne (1964) cited a total of 1,119 cases from Asia alone, as against 121 from all other parts of the world. In *India* rhinosporidiosis is not a rare disease. It is most prevalent in the Central, Eastern and Southern parts of the country, and also in Ceylon. Many cases have probably not been reported because they are not considered unusual in endemic areas. During the course of 12 years, I have not found one genuine case in the Punjab, but material from seven (unreported) patients has been sent to us from the Varanasi area, Assam and Orissa. Karunaratne (1964) quoted 469 instances from India. Agarwal (1966) mentioned only 240 Indian cases after incomplete quotation of the Indian literature. Recently, Nanda and co-workers (1969) collected 131 cases from Western Orissa during a period of only 7 years.

Epidemiology

There is no infection from man to man although infections of several members of one family have been reported. The disease appears to be endemic in some villages. Swimming in ponds, work in the sand of river beds, exposure to dust storms, contact with animal hides, were all incriminated as causing infection. Karunaratne (1964) suggested a relationship between rhinosporidial infection any paddy cultivation in Ceylon. From all the evidence available, a relatively humid and warm climate all the year round appears to be a requisite for spread of the disease. Men are much more frequently infected than woman. In a total of 1,240 patients from all over the world, only 18.1 per cent were female. The male to female ratio in India is approximately 10:1, but the ratio depends also on the location of the infection.

Clinico-Pathological appearance of lesions

In about 90 per cent of all cases, the *nasal mucosa* is affected and shows polypoid lesions which may obstruct the air passage. Purulent or copious mucous discharge, secondary infection or foul smell may be present. When the pharynx is involved, foreign body feeling on swallowing may disturb the patient. Laryngeal lesions have been observed but are rather rare.

The next most common site is the *conjunctiva*, lacrymation, pain, foreign body sensation and photophobia due to papillary lesions usually lead the patient early to the physician. Small lesions may be removed surgically without recognition of their true nature. Lips, vulva, glans penis, other mucous membranes, and also the skin may be involved. Auto-inoculation by scratching appears to be possible.

Generalisation of rhinosporidiosis has been reported by Dhayagude in 1941 (multiple subcutaneous nodules). Rajam *et al.* (1955) saw visceral lesions at autopsy and Agarwal *et al.* (1962) reported a case with lesions of the palate,

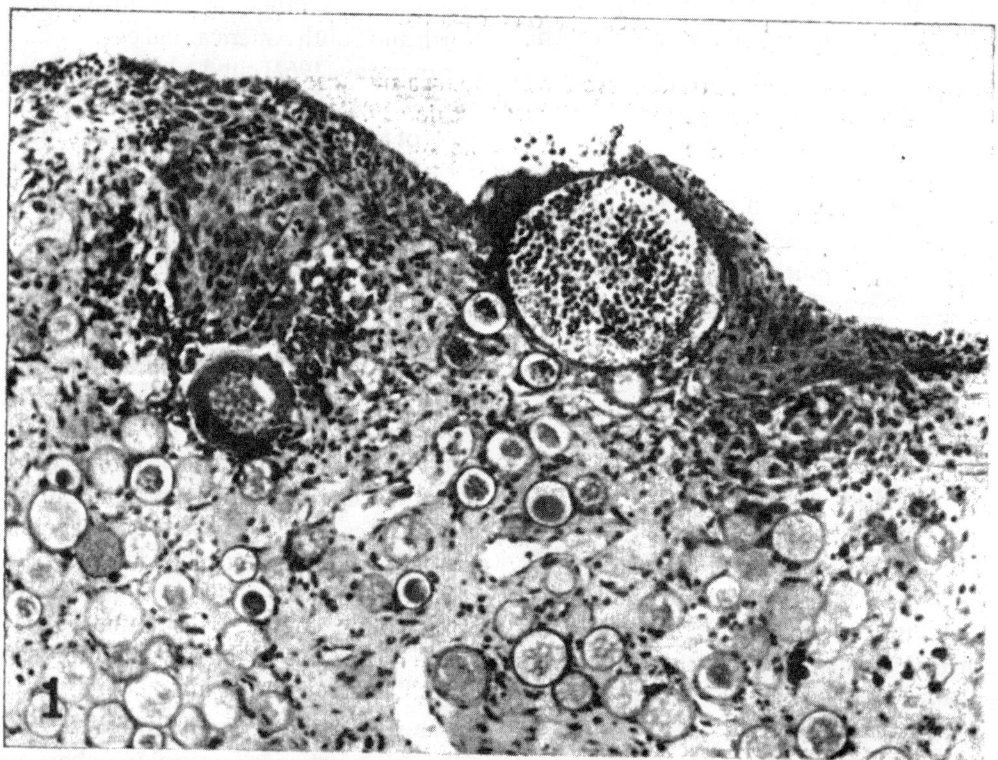

Fig. 20.1 : Rhinosporidiosis of nasal mucosa, showing various stages of development of the parasite. Weigert's Iron H & E; 130 X

PLATE 20.I

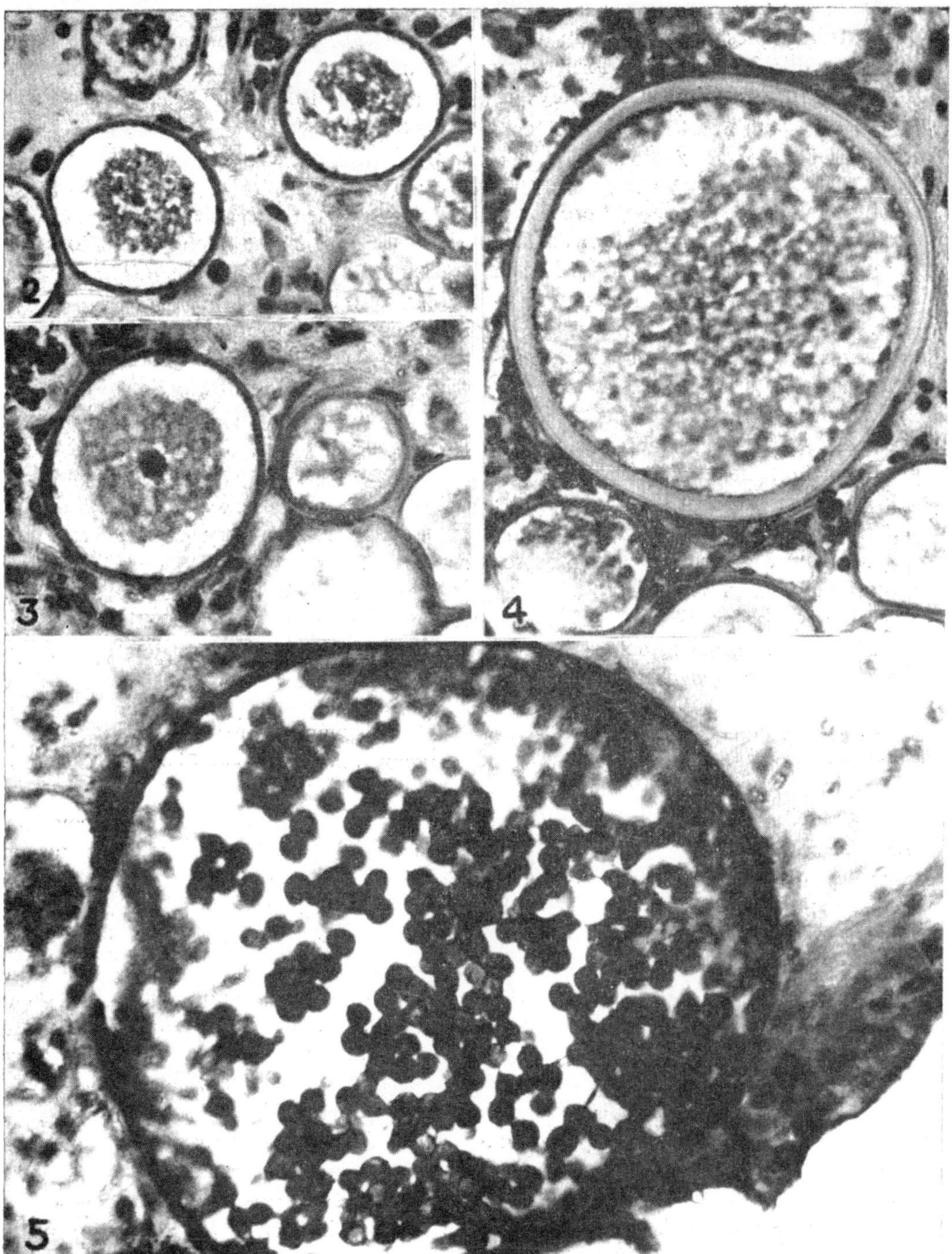

Fig. 2 & 3 : Early trophocytes of *Rhinosporidium seeberi.* Weigert's from H & E; 500 X

Fig. 4 : Late Trophocyte of *Rhinosporidium seeberi*, characterised by a thick wall and appearance of multiple nuclei in the cytoplasm. Weigert's iron H & E; 500 X

Fig. 5 : Mature sporangium of *Rhinosporidium seeberi.* Gram; 560 X tongue, larynx, lungs, eye and skin (quoted by Mohapatra, 1969).

Two more such reports of generalisation were mentioned by Karunaratne (1964) from the earlier literature.

However, there are rare exceptions to the rule that rhinosporidiosis is confined to the mucosae and eventually to the skin and its sub-strata.

Pathology

Specimens consist usually of papillary masses, very friable, often sent to the laboratory in small fragments, looking like granulation tissue.

The surface is grey or reddish. Tiny greyish white spots may be barely visible at the surface with the naked eye. These represent sporangia which have a diameter upto 300 microns or slightly more before they rupture and discharge numerous spores.

STAINING METHODS : *Rhinosporidium seeberi* stains excellently with Haematoxylin and Eosin, but different stages of development can clearly be distinguished by Weigert's Iron Haematoxylin.

HISTOPATHOLOGY : Definite identification of *Rhinosporidium seeberi* is confined to histological examination. Numerous stages of development of the parasite may be seen in the tissue (**Fig. 20.1**). The nomenclature of these stages has been ill defined previously. Karunarathe (1964) suggested the term "trophocyte", but the developing trophocyte undergoes various changes which can be distinguished morphologically and histochemically. Their names should be clearly defined as follows :

1. **SPORE IN THE SPORANGIUM**

   The mature spore is 8-12 microns in diameter, which appear to be in the spores were previously considered to be "microspores" by some authors. However, Bader and Grueber (1970) were able to demonstrate that they are protrusions of a morula-like cell wall of the spore. The spore is surrounded by a mucoid substance.

2. **IMPLANTATION FORM**

   The earliest implantation form in the tissue is never smaller in size than a mature spore (10-12m microns in diameter). The morula like cuticle probably expands rapidly in the tissue by fluid intake and soon forms a small spherical structure which we suggest should be called "early trophocyte".

3. **EARLY TROPHOCYTE (ABOUT 12 TO 50 MICRONS IN DIAMETER : Pl. 20.I Figs. 2** and **3**)

   The early trophocyte is spherical, as are all the subsequent stages of development, and has a central nucleus showing a distinct nucleolus. The cell wall is laminated and consists of a chitinous membrane incorporating neutral polysaccharides of a hemi-cellulose character. Within the cell there are ill defined cytoplasmic granules.

   Ashworth (1923) and Karunaratne (1964) have stated that mitotic nuclear divisions begin in trophocytes of about 50 micron diameter. However, we were not able to find more than 4 nuclei at that stage and in spite of intensive search, we could not detect mitoses in the granular cytoplasm. We presume, therefore, that an amniotic mechanism produces the multitude of nuclei which become visible in the next stage of development.

4. **LATE TROPHOCYTE** : (approximately 50 to 130 microns).

   This stage is characterised by a rapid thickening of the inner layer of the laminated cell wall and, in the later stages of development, by appearance of numerous nuclei in the ill-defined granular cytoplasmatic mass (**Pl. 20.I, Fig. 4**).

5. **EARLY SPORANGIUM (VARIABLE DIAMETER, 120 TO 150 MICRONS)**

   As the diameter increases, the wall becomes much thinner than previously. A semilunar *generative zone*, adjacent to the margin of the cell, produces small immature spores, while mature larger spores collect more in the center of the spherical sporangium. Later, rupture of the mature sporangium takes place opposite the semilunar generative zone. The "pore" which is supposed to release the spores according to Karunaratne (1964) is a matter of debate. It might be an artefact in histological sections.

6. **MATURE SPORANGIUM (150 TO 300 MICRONS IN DIAMETER) Pl. 20.I, Fig. 5)**

   It is thin walled, filled with mature spores, which are surrounded by a mucoid substance. The spores now tend to become separate from each other whilst they in the earlier stages of development "stick together" by means of the intercellular mucoid substance. In histological sections. We have seldom found mature sporangia with a diameter of more than 200 microns in contrast to other investigators. However, there is considerable variation in the diameter of mature sporangia and the mode of tissue dehydration may also cause variation in measurement.

7. **DEGENERATION FORMS**

   These are found in all stages. They can easily be distinguished from vital forms by Gram stain. Spores, parts of the cell wall, and the granular substance in early and late trophocytes are Gram positive. Degenerated form are Gram negative or show practically no staining with Gram's stain. Degenerated trophocytes often contain large "empty" vacuoli. More mature forms may collapse and form semilunar structures as seen in **Fig. 20.2**.

Tissue reaction : Early infections may cause little tissue reaction except for considerable capillary hyperaemia, fibroblastic proliferation, oedema and (particularly squamous) epithelial proliferation in form of irregular acanthosis and slight papillomatosis. In later stages, there may be microabscesses and polymorphonuclear reaction. Mature sporangia and spores particularly can stimulate purulent reaction. Plasma cells, lymphocytes and eosinophils may appear in various proportions at the site of infection. Foreign body giant cells may take up

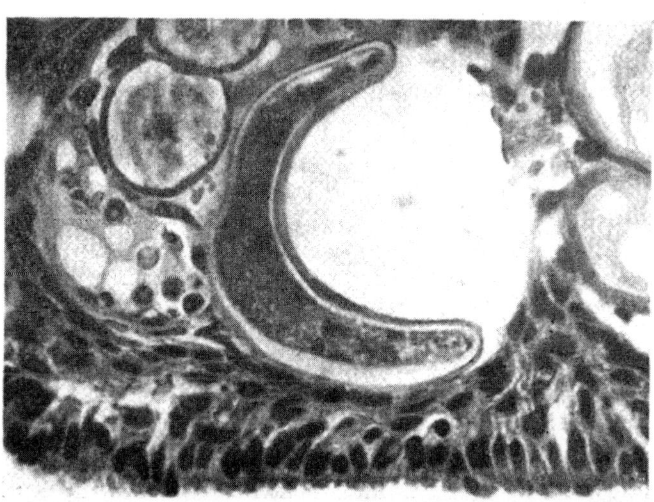

Fig. 20.2 : Semilunar, collapsed degeneration form of *Rhinosporidium seeberi*. Weigert; 560 X

spores or similar forms of the parasite. Epithelioid granulomas or necrosis are usually absent. Considerable edema with fluid filled "cysts" in the tissue, together with enormous congestion of capillaries, characterise sites of early infection. This, together with the fact that there is often obstruction of the ducts of mucous glands, should be taken into consideration in experimental studies of this parasite.

Mature sporangia tend to rupture through the epithelium and release their spores to the surface. Other sporangia rupture within the tissue and release their spores there.

For a further detailed discussion of the histology and pathology of *Rhinosporidium seeberi*, the reader is referred to Karunaratne's monograph and to the thorough histochemical investigations of Bader and Grueber (1970).

In view of the failure to grow *Rhinosporidium seeberi* on artificial media, electron microscopic investigations of this parasite are highly desirable. They may reveal more about the nature of this organism.

Animal inoculation : Experimental inoculations in laboratory animals are not successful.

Prognosis : The disease is rarely fatal although lesions may sometimes cause mechanical obstruction or deformities by their growths.

Treatment : Surgical removal of the polyp should be done with a wire snare. Intravenous administration of antimony has been recommended and pentavalent antimony compounds intravenously, upto a total dose of 2 to 4 g may be helpful as an adjunct to surgical measures.

21

Systemic Mycosis

Histoplasmosis And *H. Capsulatum*—Dimorphic Fungus

Histoplasmosis is a systemic mycosis of human and animals, world wide in distribution. It is caused by the **dimorphic fungus, *Histoplasma capsulatum*,** which is present in soil as a saprophyte and primarily affects the reticuloendothelial system. The disease in man may either run benign course or manifest itself with clinical and radiological findings with the characteristic host response, both humoral and cellular. **Treatment with amphotericin B** gives an effective cure.

Aetiology

The organism was first seen in a patient tby Samuel Taylor Darling in 1905 in the Panama & Canal zone. Mycotic nature of the infective agent was established by Hausmann and Schenken (1914), and De Monbreun (1934). The disease affects all ages and both sex though males suffer often than the females.

Histoplasma capsulatum is a **dimorphic fungus** having **Yeast and filamentous phases.** *In the **infected*** **tissue** it appears as a small round or oval **yeast cell (2-4μ)** growing mostly in the **mononuclear cells.** On **Sabouraud's glucose agar** at room temperature **(25°-28°C)** the fungus grows as a filamentous form. The growth is cottony and white but turns brownish with age. The **septate branching hyphae bear small, smooth, round spores, (3-5μ)** on them and later the characteristic round, thick-walled spores **(8-20μ)** with finger like projections (tuberculate chlamydospores) develop.

Source of infection

Relationship of saprophytic existence of *H. capsulatum* to avian habitat became evident **with isolation** of *H. capsulatum* from (1) **abandoned chicken soups** (Forcowlaw, 1958), the demonstration of (2) **growth stimulating substances for *H. capsulatum* in infusions of** (3) **sterling (*Sturnis vulgaris*) manure and presence of *H. capsulatum* in** (4) **faecal contents of bats** in the canal zone (Klite & Dierks, 1965). Ajello (1950, 1964) has analysed these facts in dealing with geographic distribution of *H. capsulatum*. Campbell (1965) has drawn attention to **existence of microfoci of infection.** The (a) **composition** of this soil, the **average** annual (b) **rainfall** and (c) **temperature** in the area are **important factors** in the **saprophytic existence of the fungus in soil** (Emmons, 1948).

Clinical Features

Infection in man almost always starts as a result of by inhalation of the microconidia. The lung lesion may run a benign self-limiting course without any clinical manifestation. The only evidence of this episode being the delayed hypersensitivity to fungus antigen developed in the host. The primary infection or reinfection of the lung may, however, result in an acute **pulmonary infection of mild, moderate or severe variety with malaise, fever, cough and**

pneumonitis. This may also take a **chronic course** the chronic pulmonary variety with capitation, fibrosis and calcification as seen in pulmonary tuberculosis. In some unfortunate individuals the disease may take an acute disseminated course which may be benign or progressive and fatal. As histoplasmosis is essentially a systemic infection with the small yeast from of the fungus invading the reticuloendothelial system, the lesions in the mucosa of nose, mouth, pharynx and larynx, and in skin are secondary manifestations of widespread infection. Thus it may ultimately take the course of a chronic disseminated disease with poor prognosis.

In the disseminated stage of the disease, no viscera in the body are spared. Lesions involve the **skeletal system** and the **central nervous system** with clinical manifestations. **PRIMARY CUTANEOUS INFECTION** commonly, and of primary infection of gastro-intestinal tract with involvement of the mesenteric lymph nodes have been reported, though **rarely**.

Clinically, **pulmonary infection resembles tuberculosis** in that a **primary complex** does **occur in histoplasmosis** (Straub and Schwarz) and the signs and **symptoms of chronic progressive** (**cavitary**) **histoplasmosis closely resemble those of reinfection type of tuberculosis**.

In the **pulmonary variety** the signs and symptoms like mild to moderate fever, cough and positive lung findings either in the form of a typical pneumonitis in the acute cases or fibrosis and cavitation in chronic cases invite attention of the clinicians to other diseases with similar findings. These cases may or may not show enlargement of spleen and liver. The **clinical suspicion of histoplasmosis is confirmed by laboratory diagnosis**.

In the acute disseminated variety the patient may sometimes present with **spiky temperature** with **hepatosplenomegaly**. This may, however, run a benign course with good prognosis. But often this goes into a **stage of progressive disease with fatal results** unless promptly treated. These patients and patients suffering from the **chronic disseminated type of infection often present with ulcerated lesions in the mouth, in larynx and on tongue and gums**. Smears from these lesions and biopsy of these lesions are often taken to rule out malignancy and show **large number of the fungal (YEAST FORM) bodies in the histiocytes clinching diagnosis.**

HISTOPLASMOSIS IN INDIA

In this country skin test surveys conducted in early fifties (Wahi, 1952; Mochi & Edwards, 1952) detected only a small percentage of positive reactors in the population tested. Subsequent studies (Taneja *et al.*, 1955; Sen & Ghosh, 1956; Viswanathan *et al.*, 1967) have detected variable percentage of positive reactors in the population in different parts of the country. Though no endemic area has been found by the above surveys, certain riverine areas have been shown a higher percentage of positive skin reactors indicating the existence of the fungus in nature and its contact with local population. Since the report of first case of histoplasmosis by Panja & Sen from Calcutta in 1954, about a dozen case reports have appeared from different parts of the country, the maximum number being from Calcutta. (Sengupta & Rao, 1957; Dutta Choudhury & Mukherjee, 1960; Chatterjee & Das, 1960; Ichhaporia, 1958; Lal & Mohapatra, 1964; Chakravarty, Damodaran & Abraham, 1968; Mankodi *et al.*, 1970).

Radiological Findings

There is no characteristic **roentgenographic picture pathognomonic of pulmonary histoplasmosis**. The radiological findings depend on the underlying pathology, e.g., segmental or lobar consolidation, fibrocavitary lesion and lung infiltration with hilar nodes, etc. Epidemiological studies of **histoplasmin sensitivity have revealed intrathoracic calcification in positive reactors**, which is not due to tuberculosis.

Pathology

Tissue reactions to infection with *H. capsulatum* have been well described by Binford (1955). The **GROSS AND CELLULAR PATHOLOGY** of histoplasmosis depend on the type of lesion.

Circumscribed **localised lesions** of histoplasmosis **varies from 0.5 to 3 cm** and is usually **round or oval**, typically **located in the subpleural region** with pleural reaction. The **CUT SURFACE SHARPLY DEMARCATES** it from the surrounding normal lung tissue.

Microscopic examination reveals the presence of ***H. capsulatum*** (by the use of appropriate stain) in the **central necrotic area surrounded by epithelioid cells and occasional foreign body type giant cells. A zone of lymphocytes and dense collagenous tissue with fibrocytes** are seen **in the periphery**. In these noncaseous granulomas the **fungus cells are seldom seen in H & E stain**. Some of these granulomas may show **caseation which cannot be distinguished from tuberculosis.**

The ***noncaseous granulomas requires differentiation from "Sarcoid"***. Solitary or multiple fibro-caseous granulomatous lesions in the lung produced by *H. capsulatum* are diagnosed by **special fungus stains**. In the disseminated lesions *H. capsulatum* is usually seen growing intracellularly in histiocytes producing **"HISTIOCYTOMYCOTIC GRANULOMA"**. These large parasitised mononuclear cells seen in good numbers serve as hosts to the yeast phase of the fungus proliferating in them, as is seen in lepromatous leprosy and in leishmaniasis. Calcification of old solitary lesions of histoplasmosis is a prominent feature and *H. capsulatum* leads all other fungi causing systemic infection in this respect.

Laboratory Diagnosis

Demonstration of the **yeast form of *H. capsulatum*** in the lesion and isolation of the fungus in culture are most valuable in the diagnosis of this infection. The presence of the fungus in the secretions, exudates, body fluids and tissues need careful examination of the clinical materials.

Sputum Examination : In cases of symptomatic pulmonary infections the sputum should be **examined repeatedly for presence of** ***H. capsulatum***. Smears prepared from the expectorated material are **stained with Giemsa** stain and examined for phase of ***H. capsulatum* inside the monocytes.**

Samples of sputum are treated with equal volume of antibiotic solution in physiological saline (1000 units of penicillin & 1000 units of streptomycin per ml) and incubated at 37°C for an hour after thorough mixing. The material is then centrifuged and the sediment suspended in 2 to 4 ml of sterile saline is used for culture and inoculation of mice.

Culture

The material is cultured on several slants of Sabouraud's dextrose agar (with or without antibiotic & actidione) and incubated at 25º-28ºC. **Slants of brain heart infusion agar, blood agar** (prepared from Blood agar base-Difco), and Littman's liver spleen glucose blood agar are also inoculated with the above material and **incubated at 37°C.**

Animal inoculation

1 ml of the saline suspension of the centrifuged deposit is injected intraperitoneally into two white mice. The mice are killed after 4 weeks and examined for enlargement of spleen and any other lesions, if any. Smears are prepared from the cut surface of the spleen and are examined for yeast phase of *H. capsulatum* after staining with Giemsa stain. Pieces of spleen and liver from these mice are macerated in sterile saline and are cultured for *H. capsulatum* as stated above.

Examination of other clinical materials

Tissue from lesions (obtained for biopsy) and lymph nodes are also examined for presence of *H. capsulatum* by smear and culture examination as stated above, grinding the tissue pieces under aseptic precautions. Similarly gastric washings, pus from wounds and open sinuses, etc. may also be processed according to the technique given above and examined for presence of *H. capsulatum*.

PLATE 21.I A

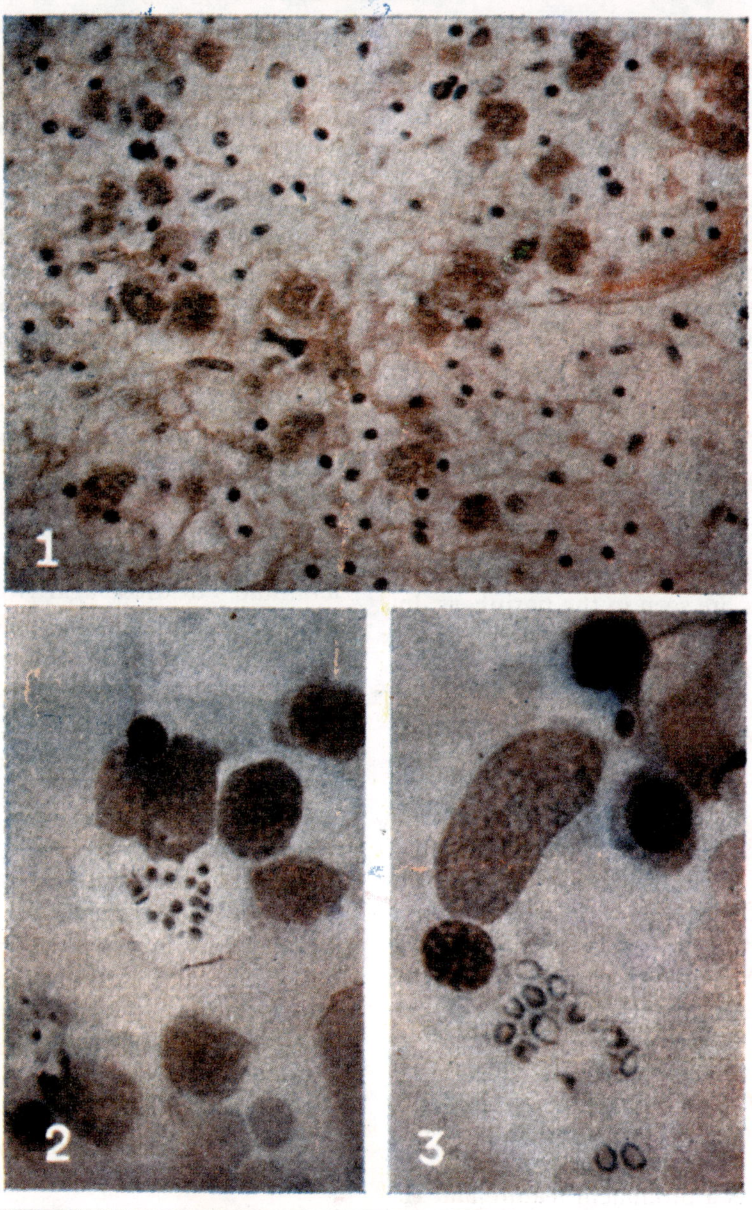

Fig. 1 : From a mucosal ulcer, showing scattered *H. capsulatum.* H & E; approximately 400 ×

Fig. 2 : Tissue section from the liver, showing *H. capsulatum* H & E; 560 ×

Fig. 3 : Same as **Fig. 2**, but at 1000 ×

Fungus cells appear to be larger than with other stain, because outer cell wall is demonstrated by silver precipitate whilst the inner part of the cell stains less deeply.

Bone marrow biopsy

The sternal marrow in suspected cases may be examined for presence of ***H. capsulatum*** in the monocytes in the stained (Giemsa) smears. The marrow also may be cultured on suitable media and **incubated at 25°-28°C and at 37°C for the FILAMENTOUS AND YEAST phase of *H. capsulatum*.** This is usually positive in active infections. Very rarely the fungus is also seen in mononuclear cells in smears of peripheral blood in these cases.

Serological Tests : The immunologic response to *H. capsulatum* infection could be studied by the complement fixation, agglutination and precipitin reactions usually employed in serological studies.

(a) *THE COMPLEMENT FIXATION TEST*

This test is best performed by using histoplasmin (Tenenberg & Howell, 1948) and whole yeast phase antigen (Salvin, 1947). In human cases the test is positive in active stage of the disease and the **significant titer is 1 : 20** or above with the whole yeast antigen. **Titres of 1 : 8** or above are **considered positive when histoplasmin is used** (Schuber's modification, 1956). The duration of significant titres vary from 2 to 18 weeks in acute cases and from a few weeks to years in chronic cases. Cross reaction with heterologous antigens (blastomycin & coccidioidin) are noted.

(b) *THE AGGLUTINATION TEST*

The collodion agglutination test of Saslaw and Campbell (1948) has been used intensively. Collodion particles coated with histoplasmin are incubated with sera and clumping if any is examined macroscopically. Sera from patients of Blastomycosis cross react with these particles. Other agglutination systems used are histoplasmin-latex agglutination test (Saslaw & Carlisle, 1957) and haemagglutination test using sheep erythrocytes coated with histoplasmin. A **titre of 1 : 20 to 1 : 80 in Colloid agglutination test is** considered **significant** and it may persist from 2 to 3 weeks to several months.

(c) *THE PRECIPITIN TEST*

Salvin and Hottle (1948) employ histoplasmin antigen in the agar gel diffusion technique. In cases of histoplasmosis **precipitin titres of 1 : 4 to 1 : 8 appear to be significant.** Cross reaction with **heterologous antigens (blastomycin & coccidioidin)** are very often noticed. However precipitin titre with homologous antigen is almost always higher and persists for a longer period. The test is positive from 2 weeks to several months or longer.

Serological patterns

The practical value of a serological test rests with the high degree of sensitivity and specificity. The use of **yeast phase antigen and histoplasmin in the complement fixation test increases** the percentage of positive reactors, thus giving a higher degree of sensitivity.

The sensitivity of the different tests varies according to the stage of infection. The **precipitin and colloid agglutinin test are** usually **positive in higher titres** early in the disease but for a shorter period. In the later stages of infection they tend to be weakly positive or negative.

The complement fixation test is usually positive later in the disease. According to Campbell and Binkley (1953), the yeast phase complement fixation test in cases of **acute pulmonary histoplasmosis is positive within the first 6 weeks of infection**.

Prognosis

The course of this infection in an individual depends on the resistance of the host. The benign form of this disease is almost always a self healing process with excellent prognosis. In the unfortunate individuals where the disease takes a progressive course, many recover uneventfully. However, the prognosis in acute disseminated cases as well as chronic

PLATE 20.II

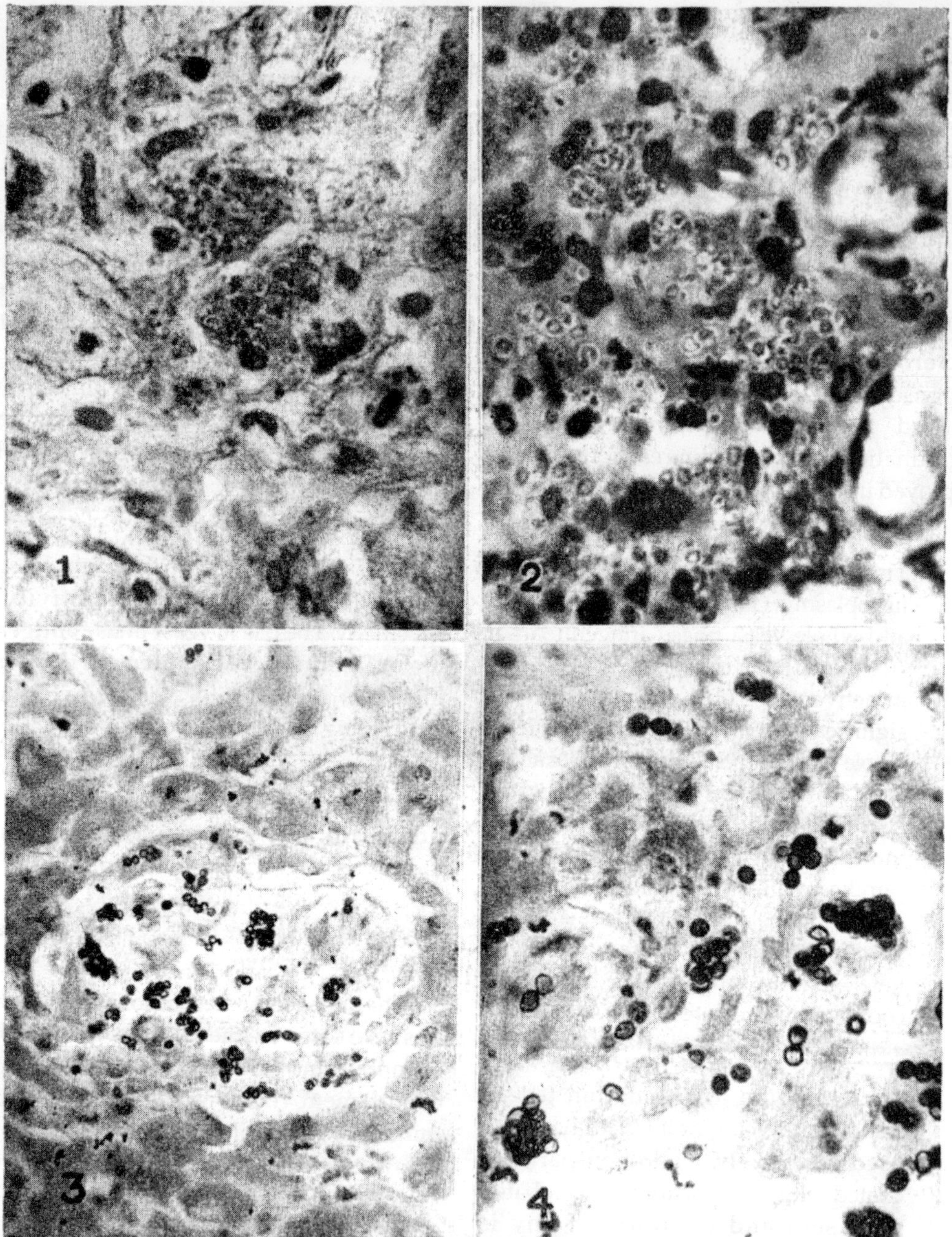

Fig. 1 : Granuloma of histoplasmosis includes considerably distended macrophages stuffed with spherical fungus cells of ***H. capsulatum***. PAS; 1000 ×

Fig. 2 : From a mucosal ulcer, showing masses of *H. capsulatum* intra and extracellular in distribution within H & E; 1000 ×

Fig. 3 : Tissue section from the liver, showing excellent demonstration of *H. capsulatum* by silver methenamine stain; 560 ×

Fig. 4 : Same as Fig. 3, but at 1000 ×. Fungus cells appear to be larger than with other stain, because outer cell wall is demonstrated by silver precipitate whilst the inner part of the cell stains less deeply.

progressive conditions is grave. With the recent introduction of amphotericin B in treatment of histoplasmosis, the chances of recovery in these cases is very bright.

Treatment

Amphotericin B is still now the most promising antifungal antibiotic and is effective in treating cases of histoplasmosis. **Amphotericin B therapy is indicated in acute pulmonary histoplasmosis,** disseminated histoplasmosis and chronic progressive pulmonary disease. Given in doses **1 ml/kg body weight** with necessary precautions to guard against toxic manifestations, good results are obtained.

Triazole (**ITRACONAZOLE**) orally and/or I.V. should be administered as indicated.

PATHOLOGY OF HISTOPLASMOSIS

Introduction

The majority of all cases of histoplasmosis consist of a silent pulmonary infection, not unlike a primary tuberculosis. Most of these cases heal spontaneously. A small number of cases terminate in an often fatal dissemination to internal organs and to skin and mucosae.

PULMONARY HISTOPLASMOSIS

The *subclinical infection* is in all respects similar to a primary tuberculous complex consisting of a small nodule of parasitised cells in the lung with a string of secondary nodules along the lymphatics and infection of local hilar lymph nodes which may later show necrosis.

Active and progressive pulmonary disease may show grey miliary tubercles scattered over the *cut surface* of the lungs, indistinguishable from tuberculosis. In other cases the necrotic nodules become confluent and form larger greyish areas of necrosis. *Histologically*, the process begins with a group of considerably enlarged macrophages which are stuffed with small spherical cells, 2-3 microns in diameter (**Pl. 21.II, Fig. 1**). The centre stains bluish with Haematoxylin eosin, and is surrounded by a capsule like halo, but which is actually within an outer membrane. This can well be demonstrated with Silver methenamine stain, even in old caseous and calcified foci which are grossly indistinguishable from old tuberculosis lesions; the small spherical bodies show well with silver methenamine stain (**Pl. 21.II, Figs.3** and **4**). Parasitised areas become necrotic in progressive disease and may be surrounded by granulomatous walls consisting of large parasitised macrophages, epithelioid cells, giant cells and other chronic inflammatory cells. If nodules become confluent, large areas of the lung tissue may be involved.

A **HISTOPLASMOMA** consists of a solitary lesion usually situated below the pleura. The coin like structure is radiologically easily confused with a tumour. The centre is often necrotic and periphery consists of a fibrotic wall or pseudotuberculous granulomas as in other lesions of histoplasmosis. A severe cavitating form of pulmonary histoplasmosis tuberculosis in almost all aspects and is usually associated with disseminated diseases.

DISSEMINATED HISTOPLASMOSIS

Some diagnosed at the autopsy when calcified nodules are found, most frequently in the liver and spleen.

In a rather small number of cases the disease is progressive and, beside liver and spleen, involves other internal organs causing **pericarditis, pleuritis, and infection of the central nervous system** and other organs. Most frequently, however, the **organs of the reticuloendothelial system are involved because *H. capsulatum* grows essentially intracellularly in macrophages. Histoplasmosis is also not infrequently associated with lymphoproliferative disorders.** The cut surface of the organs may show grey, firm, miliary nodules of various sizes; even in this stage, healing is still possible. If the disease is progressive, nodules become confluent. Greyish or yellowish caseous, and later calcified nodes of 1-3 cm diameter may be found. In other cases, large areas of brownish or greyish granulation tissue obliterate the normal structural pattern of the cut surface.

Histologically numerous swollen macrophages are **bursting with spherical cells**

PLATE 21.III

Histoplasmosis

Fig. 1 : Histoplasmosis on the lips, angles of the mouth and nose.

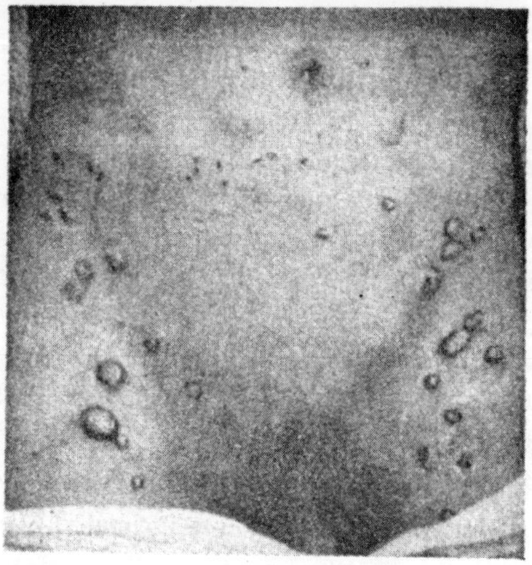

Fig. 2 : Abscesses on the abdominal wall and nodular lesions on the groins.

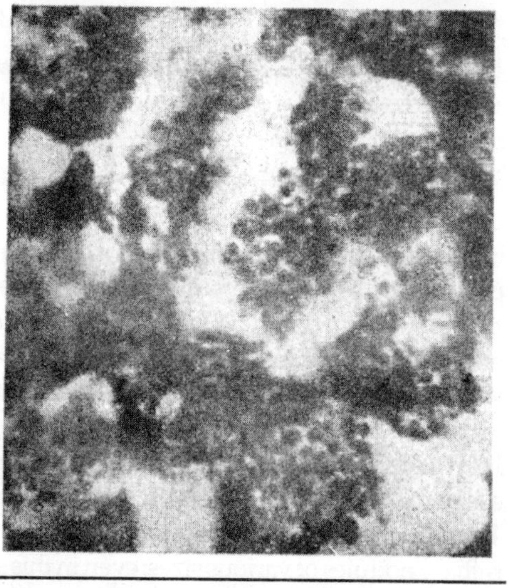

Fig. 3 : Section of tissue stained with Leishman's stain. R E cells are packed with parasites.

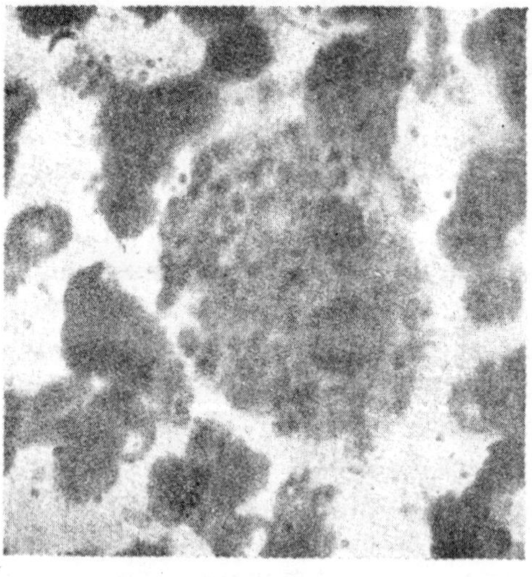

Fig. 4 : Smear stained with Leishman's stain. A histiocyte is packed with ***H. capsulatum***.

With kind permission of Drs. G. Panja and S. Sen (J.I.M.A. Vol. **23**, 257, 1954)
(Courtesy of the Editor, J.I.M.A.)

PLATE 21.III (Continued)
Histoplasmosis

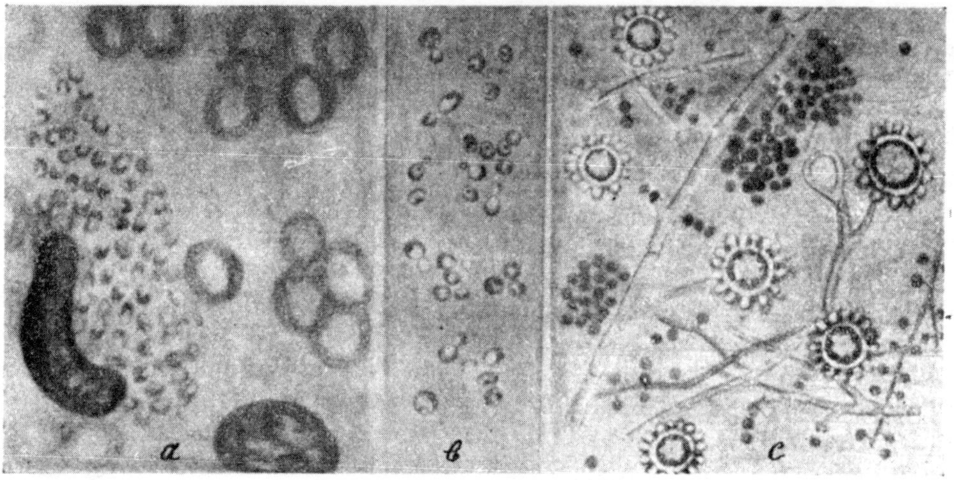

Fig. 5 : (a) A histiocyte ladened with ***H. capsulatum***; (b) **Yeast form in culture at 37ºC;** (c) **Conidia and chlamydospores in culture**

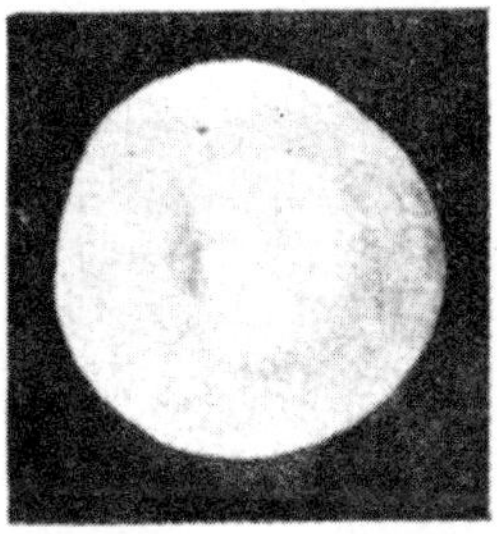

Fig. 6 : Culture of ***H. capsulatum*** in Sabouraud's glucose agar at room temperature

of ***H. capsulatum***., There may be scattered giant cells, epithelioid cells and debris, alternating with areas of complete necrosis. *H. capsulatum* stains brightly with P. A. S., in contrast to Leishman Donovan bodies. The oval yeast form show excellently with Silver methenamine stain. Diagnostic criteria are the small diameter of the yeast cell (2-4 microns), the predominantly intracellular arrangement and the fact that there are no hyphae in tissue.

Muco-cutaneous histoplasmosis : (**Pl. 21.III, Fig. 1**). This form is probably in most cases a manifestation of disseminated disease. It frequently consists in ulcers of the lips. The buccal or palatine mucosa, tongue, larynx, pharynx or penis. Ulcers have slightly raised or irregular wall but a clean floor. The granulation tissue shows histologically characteristic macrophages distended with *H. capsulatum*, in connection with granulomatous tissue reaction (**Pl. 21.III, Fig. 2**). The local lymph nodes are often involved. Previously skin and mucosae were believed to be important portals of entry, for generalised infection. But evidence has accumulated that a true primary inoculation of histoplasma is a relatively rare occurrence and causes a primary chancriform cutaneous complex rather than chronic spreading mucocutaneous lesions which are now considered to be a manifestation of generalised histoplasmosis. However, the skin lesion often helps to establish the otherwise obscure nature of the disease (Lal and Mohapatra, 1963).

Mycology

Histoplasma capsulatum, a dimorphic fungus grows in the **cells of the reticuloendothelial system** exclusively **as a** small **YEAST CELL**, 2-4 microns in diameter. In culture, on artificial media, isolation is somewhat difficult because of the fastidiousness of the organism. ***HISTOPLASMA* at room temperature produces a white cotton-like mycelium** which turns yellowish-greyish when ageing. **Small septate hyphae bear single microconidia at right angles and larger, pear shaped or spherical macroconidia** (8 to 15 microns in diameter) with peripheral **tuberculate structures** are attached **to broad tip of conidiophores (Pl. 21.III, Fig. 5)**.

HISTOPLASMA DUBOISII, a variety found in African cases, but not in India, produces yeast forms in tissue, much larger in diameter but its mycelial phase in culture is morphologically identical with *H. capsulatum*. The comparative morphology and immunology of the genus *Histoplasma* has been reviewed by L. Ajello (1968). The author came to the conclusion that there is no sound mycological basis for **considering *H. duboisii* as a separate species**.

Direct examination : Smears from sputum, exudate from ulcers, bone marrows, impression smears from lymph nodes and other material should be stained with Wright's or Giemsa stain, or **PAS** The presence of spherical intracellular organisms staining positive with **P A S**, 2-4 microns in diameter, is characteristic for histoplasmosis.

Animal Inoculation

Intraperitoneal injection of fungus cells in mice will cause generalised infection in two to four weeks. All common laboratory animals are susceptible to the infection.

Immunology

Histoplasmin skin tests have been used extensively for epidemiologic studies. **In endemic areas up to 90% of the population show a positive histoplasmin skin test**. Surveys in India revealed generally a low percentage of reactors among the tested population (Wahi 1952, 1955; Mochi and Edwards, 1952; Sawhney *et al.* 1957).

Complement fixation test and precipitin test have been found to be correlated with the immune response of the individual. The lack of standardised antigens makes comparison of the results of different investigators difficult.

REFERENCES

Ajello, L. (1960). Geographic distribution and prevalence of the dermatophytes. *N. Y. Acad. Sci.* **89**, 30-38.

Ajello, L. (1964). Relationship of Histoplasma capsulatum to avian habitats. *Pub. Health Rept.* **79**, 266-270.

Ajello, L. (1968). Comparative Morphology and Immunology of members of the genus *Histoplasma. Mycosen* **11**, 507-514.

Basu Mallik, S. N. M. D. Thesis, Univ. of Delhi (1962).

Baum, G. L. in "Histoplasmosis", Edited by H. C. Sweany, Charles C. Thomas, Springfield Illinois, p. 14, 1960.

Bennett, D. E., (1966a). Histoplasmin latex agglutination test clinical evaluation and review of literature. *Am. J. med. Sci.* **251**, 175-183.

Benett, D. E. (1966b). Laboratory diagnosis of histoplasmosis, *A, review. S. med. J.* **59**, 922-926.

Berry, C. L. (1969). The development of the granuloma of histoplasmosis. *J. Path.* **97**, 1-10.

Binford, C. H. (1955). Histoplasmosis : tissue reactions and morphologic variation of the fungus. *Am. J. clin. Path.* **25**, 25-36.

Bloomfield, N., Gordon, M. A. and Elmendorf, D. F., Jr. (1963). Detections of *Cryptococcus neoformans* antigen in body fluids by latex particles agglutination. *Proc. Soc. exp. Biol. Med.* **114**, 64-67.

Busey, J. F. and Hinton, P. F. (1966). Precipitins in histoplasmosis. *Am. Rev. resp. Dis.* **92**, 637-639.

Campbell, C. C. (1965). *Ann. int. Med.* **62**, 333.

Campbell, C. C. and Binkley, G. E. (1953). Serologic diagnosis with respect to histoplasmosis, coccidioidomycosis, and blastomycosis and problem of cross reactions. *J. lab. clin. Med.* **42**, 896-906.

Chakravarty, S. C., Damodaran, V. N. and Abraham, S. (1968). *Ind. J. Chest. Dis.* **10**, 205.

Chatterjee, P. K. and Das, S. K. (1960). *Bull. Cal. Sch. trop. Med.* **8**, 140.

Coons, A. H., Greech, H. J., Jones, R. N. and Berliner, E. (1942). The demonstration of pneumococcal antigens in tissues by the use of fluorescent antibody. *J. Immunol.* **45**, 157-170.

Darling, S. T. (1905). Quoted by Baum.

DeMonbreun, W. A. (1934). The cultivation and cultural characteristics of Darling's *Histoplasma capsulatum*. *Am. J. trop. Med.* **14**, 93-125.

Datta Chaudhury, R. and Mukherjee, D. (1960). *Ind. J. Dermatol.* **5**, 29.

Emmons, C. W. (1949). Isolation of *Histoplasma capsulatum* from soil. *Pub. Hlth. Rept.* **64**, 892-896.

Eveland, W. C., Marshall, J. D., Silverstein, A. M., Johnson, F. B., Iverson, L. and Winslow, D. J. (1957). Specific immunochemical staining of *Cryptococcus neoformans* and its polysaccharide in tissue. *Am. J. Path.* **33**, 616-617.

Furcolow, M. L. (1958). Recent studies on the epidemiology of histoplasmosis. *Ann. New York Acad. Sci.* **72**, 127-164.

Gordon, M. A. (1958 a). Differentiation of yeasts by means of fluorescent antibody. *Proc. Soc. exp. Biol. Med.* **97**, 694-698.

Gordon, M. A. (1958 b). Rapid serological differentiation of *Candida albicans* from *Candida stellatoides*. *J. invest. Dermatol.* **31**, 123-125.

Gordon, M. A. (1959). Fluorescent staining of *Histoplasma capsulatum*. *J. Bact.* **77**, 678-681.

Gordon, M. A. and Vedder, D. K. (1966). Serological tests in the diagnosis and prognosis of cryptococcosis. *J. am. med. Assoc.* **197**, 961-967.

Hansmann, G. H. and Schencken, J. R. (1934). Unique infection in man caused by new yeast-like organism, pathogenic member of genus Sepedonium. *Am. J. Path.* **10**, 731-738.

Heiner, D. C. (1958). Diagnosis of histoplasmosis using precipitin reaction in agar gel. *Pediatrics* **22**, 617-627.

Hill, G. B. and Campbell, C. C. (1962). Commercially available histoplasmin sensitized latex particles in an agglutination test for histoplasmosis. *Mycopathol. Mycol. Appl.* **18**, 169-176.

Huppert, M. and Bailey, J. W. (1965). The use of immunodiffusion tests in coccidioidomycosis. *Am. J. clin. Pathol.* **44**, 369-373.

Huppert, M, Peterson, E. T., Sun, S. H., Chitjian, P. and Derrevere, W. (1968). Evaluation of a latex particle agglutination test for coccidioidomycosis. *Am. J. clin. Path.* **49**, 96-102.

Ichhaporia, R. N. J. (1958). *J. J. Cip. Hosp. & Grant Med. Coll.* **3**, 1.

Kaplan, W. and Ivens, M. S. (1960). Fluorescent antibody staining of *Sporotrichum schenckii* in cultures and clinical materials. *J. invest. Dermatol.* **35**, 151-159.

Kaplan, W. and Gonzalez Ochoa, A. (1963). Application of the fluorescent antibody technique to the rapid diagnosis of sporotrichosis. *J. lab. clin. Med.* **62**, 835-841.

Kase, A. and Marshall, J. D. (1960). A study of *Cryptococcus neoformans* by means of the fluorescent antibody technic. *Am. J clin. Path.* **34**, 52-56.

Kaufman, L. and Kaplan, W. (1961). Preparation of a fluorescent antibody inhibition test for histoplasmosis. *J. lab. clin. Med.* **58**, 1033-1038.

Kaufman, L. and Blumer, S. (1968). Development and use of polyvalent conjugate to differentiate *Histoplasma capsulatum* and *Histoplasma duboisii* from other pathogens. *J. Bact.* **95**, 1243-1246.

Klite, P. D. and Diercks, F. H. (1965). *Histoplasma capsulatum* in fecal contents and organs of bats in the canal zone. *Am. J. trop.* Med. *Hyg.* **14**, 433-439.

Kunz, C. (1969). Fluorescenz-serologische unter-suchungen an einem pathogenen pilstamm (Sporotrichum schenckii). *Arch. klin. exp. Dermatol.* **209**, 200-205.

Lal, H. B. and Mohapatra, L. N. (1964). Histoplasmosis in India (A case report). *Mycopathologia*, **22**, 5-72.

Mankodi, R. C., Kanvinde, M. S. and Mohapatra, L. N. (1970). *Ind. J. med. Sci.* **24**, 354.

Mochi, A. and Edwards, P. O. (1952). Geographical distribution of histoplasmosis and histoplasmin sensitivity. *Bull. wld. Hlth. Org.* **5**, 259.

Pandalai, N. G., Kurup, P. V., Indira Devi, K. and Joseph, K. M. (1962). *Ind. J. Tubercl.* **10**, 11.

Panja, G. and Sen, S. (1964). *J. ind. med. Assoc.* **23**, 257.

Salvin, S. B. (1947). *Proc. Soc. exper. Biol. Med.* **66**, 342.

Salvin, S. B. and Hottle, G. A. (1958). Serologic studies on antigens from *Histoplasma capsulatum darling*. *J. Immunol.* **60**, 66.

Saslaw, S. and Campbell, C. C. (1948). *Proc. Soc. exper. Biol. Med.* **68**, 559.

Saslaw, S. and Carlisle, H. N. (1957). *J. lab. clin. Med.* **50**, 949.

Sawhney, B. B., Shivarajan, K., Mohapatra, L. N. and Guleria, J. S. (1967). Histoplasmin skin testing survey in and around Delhi, Indian J. Chest. Dis. **2**, 192.

Sawhney, H. C. edn. Histoplasmosis. Springfield, Charles C. Thomas, 1960.

Schubert, J. H. Pub. Hlth. Monograph No. **39**, p. 119, 1956.

Schubert, J. H., Lynch, H. J. and Ajello, L. (1961). Evaluation of the agar plate precipitin test for histoplasmosis. *Am. Rev. resp. Dis.* **84**, 845-849.

Sen, P. K. and Ghosh, M. B. (1955). *Ind. J. Tuberc.* **3**, 78.

Sengupta, P. C., Rao, A. Banerjee, A. K., Chakraborty, A. N. and Ray, H. N. (1957). *Bull. Cal. Sch. trop. Med.* **5**, 54.

Smith, C. E., Saite, M. T., Beard, R. R., Keep, R. McF., Clark, R. W. and Eddie, B. U. (1950). Serological tests in the diagnosis and prognosis of coccidioidomycosis. *Am. J. Hyg.* **52**, 1-21.

Stallybrass, F. C. (1964). Candida precipitins. *J. Path. Bact.* **87**, 89-97.

Straub, M. and Schwarz, J. (1955). The healed primary complex in histoplasmosis. *Am. J. clin. Path.* **25**, 727-741.

Taneja, B. L., Kalra, S. L., Waller, S. D. and Sachdeva, L. D. (1955). *A. M. C. Journal*, **11**, 149.

Taschdjian, C. L., Dobkin, G. B., Caroline, L. and Kozinn, P. J. (1964). Immune studies relating to candidiasis, Paper II. *Sabouraudia* **3**, 129-139.

Taschdjian, C. L., Kozinn, P. J. and Caroline, L. (1964). Immune studies in Candidiasis Paper III *Sabouraudia*. **3**, 312-320.

Taschdjian, C. L., Kozinn, P. J., Okas, A, Caroline, L. and Halle, M. A. (1967). Serodiagnosis of systemic candidiasis. *J. infec. Dis.* **117**. 180-187.

Tenenberg, D. J. and Howell, A. Dr. (1948). *Pub. Health Rep. U. S.* **63**, 163.

Viswanathan, R., Chakravarty, S. C., Randhawa, H. S. and deMonts, A. J. H. (1960). *Brit. med. J.* 1, 399.

Vogel, R. A., and Padula, J. F. (1958). Indirect staining reaction with fluorescent antibody for detection of antibodies to pathogenic fungi. *Proc. Soc. exp. Biol. Med.* **98**, 135-139.

Vogel, R. A., Sellers, T. F. and Woodward, P. (1961). Fluorescent antibody techniques applied to the study of human cryptococcosis. *J. am. med. Assoc.* **178**, 921-923.

Vogel, R. A. (1966). The indirect fluorescent antibody test for the detection of antibody in human cryptococcal disease. *J. invest. Dis.* **116**, 573-580.

Wahi, P. N. (1952). Studies on the role of fungi in pulmonary disease. *Agra Univ. J. Res.* **1**, 399.

Wahi, P. N. (1955). Pulmonary clacification and sensitivity to tuberculin. *Indian J. med. Res.* **43**. 139.

Wallraff, E. B., van Liew, R. S. and Waite, S. (1967). Skin reactivity and serological response to coccidioidin skin tests. *J. invest. Dermatol.* **48**, 533-559.

Wiggins, G. L. and Schubert, J. H. (1965). Relationship of histoplasmin agar-gel bands and complement-fixation titres in histoplasmosis. *J. Bact.* **89**, 569-596.

22

Systemic/(cutaneous) Mycosis North American Blastomycosis

Definition

This is a chronic, infectious disease caused by a yeast-like fungus called ***Blastomyces dermatitides*** characterised by primary granulomatous lesion of the skin or lung and secondary metastatic lesions in other parts of the body. Cases have been reported mainly from Middle-West and South-Eastern states of America.

Geographic pathology

The disease is generally confined to North American continent but some cases have been reported from South America and Africa. There is dispute whether blastomycosis is to be found in India. A few reported cases have not been well documented. They were reviewed by Mohapatra in 1969, however, the case of Mukherji and Sengupta (1968) was not yet included. Symmers (1956) mentioned two genuine cases from England. The patients had probably acquired the infection through continued contact with wooden packing materials imported from North America.

Aetiology

PREDISPOSITION. It is nine times more in males than in females.

Symptoms

The period of incubation is not definitely known but it varies from one to two weeks. Clinically blastomycosis is divided into as follows :

Clinical types and Gross Pathology : Three types are usually encountered, the last and fourth form being extremely rare.

1. **PRIMARY PULMONARY INFECTION.**
2. **CHRONIC LOCALISED CUTANEOUS BLASTOMYCOSIS.**
3. **DISSEMINATED BLASTOMYCOSIS.**
4. **PRIMARY CHANCRIFORM CUTANE-OUS BLASTOMYCOSIS**

The last form is probably caused by direct inoculation into the skin of previously uninfected individuals. It has been observed only in a few cases and resembles clinically primary chancriform sporotrichosis (see page 174); Schwarz and Baum, 1952.

1. *PRIMARY PULMONARY BLASTO-MYCOSIS :* Blastomycosis is now considered to be a primary pulmonary infection.

   The pulmonary lesion may go unrecognised like a primary complex of tuberculosis. Gross changes of the *progressive* pulmonary form may also resemble tuberculosis or a tumour with extensive **nodular, cavernous, caseous or abscess-like lesions**. Pleuras, and sometimes the chest wall are involved in the final stages (see Pl. 22.I, Figs. 1 & 2).

2. *DISSEMINATED BLASTOMYCOSIS*

Dissemination occurs probably from the lungs and all organs may be involved with extensive lesions which are first purulent, later necrotic, tuberculoid, or granulomatous. In late stages all organs are flooded with organisms without appreciable inflammatory reaction indicating a state of complete *anergy*. 95 per cent of these cases show extensive pulmonary involvement. The condition is extremely fatal. **Dissemination occurs in the bone** (60 per cent), **liver, spleen and kidney** (40 per cent), **central nervous system** (30 per cent) and in the **prostate** (20 per cent).

3. *CHRONIC LOCALISED CUTANEOUS BLASTOMYCOSIS*

This is the form most frequently seen. **Cutaneous blastomycosis** may be the result of **spread from an unrecognised primary pulmonary focus**. Another possibility is direct inoculation into the skin in such patients who have acquired an altered reactivity by previous silent pulmonary infection. The process begins as a papule or nodule, sometimes ulcerating in the centre. Lesions subside in the centre and progress by slow peripheral growth with raised edges showing numerous small pustules, crusts and abscesses. The process has been likened to a fire spreading in a field. Instructive coloured photographs may be found in Wilson and Plunkett: **"The fungus diseases of man"** (1965).

4. *PRIMARY CHANCRIFORM CUTANEOUS BLASTOMYCOSIS*

In this the primary lesion is in the skin and secondary systemic involvements are seen in about 35 percent of these cases specially in the lung. It may manifest itself as a chance after accidental inoculation in laboratory workers characterised by primary **papulo-ulcerative lesion**, papillomatous or granulomatous ulcers or as **subcutaneous nodules** forming ulcers covered with scabs. Generally exposed parts of the body namely face, lower part of the legs and hands are more often infected. In these cases trauma may play an important part. The centre of the lesion often heals up and the spreading margin of the lesion shows **miliary abscesses with crust formation**. When the crust is removed, the granulation tissue or small sinuses are seen with seropurulent discharge containing the fungus. **Metastatic spread may occur from the skin to internal organs.**

SECONDARY CUTANEOUS BLASTOMYCOSIS

A solitary lesion is often seen on the face or extremities as superficial or deep ulcers with granulation tissue and exudation of pus, forming crusts. Multiple lesions may apear in the skin as a result of autoinoculation.

Pathology

The stages of evolution are those of an infective granuloma. Minute abscesses are formed as a result of necrosis. Lesions in the skin may be in the form of a nodule, papillomatous or a gummatous ulcer but lesions in the lungs or viscera show abscess formation with ulceration. The lesions spread by **tissue continuity and contiguity**. Dissemination may occur through the haematogenous route.

Histopathology

The histopathological changes are those of granuloma. In the skin lesions, acanthosis and papillomatosis are marked. There are minute intraepidermal and intradermal abscesses in which **budding yeast cells are present**. Interstitial and parenchymatous oedema is present in the underlying cutis where changes are essentially those of a granuloma. The **abscess consists of degenerated** polymorphonuclear **neutrophils** which are found **around the organism**. Beyond the zone of polymorphs, there are **macrophages with** formation of **giant cells of Langhans' type**

PLATE 22.I

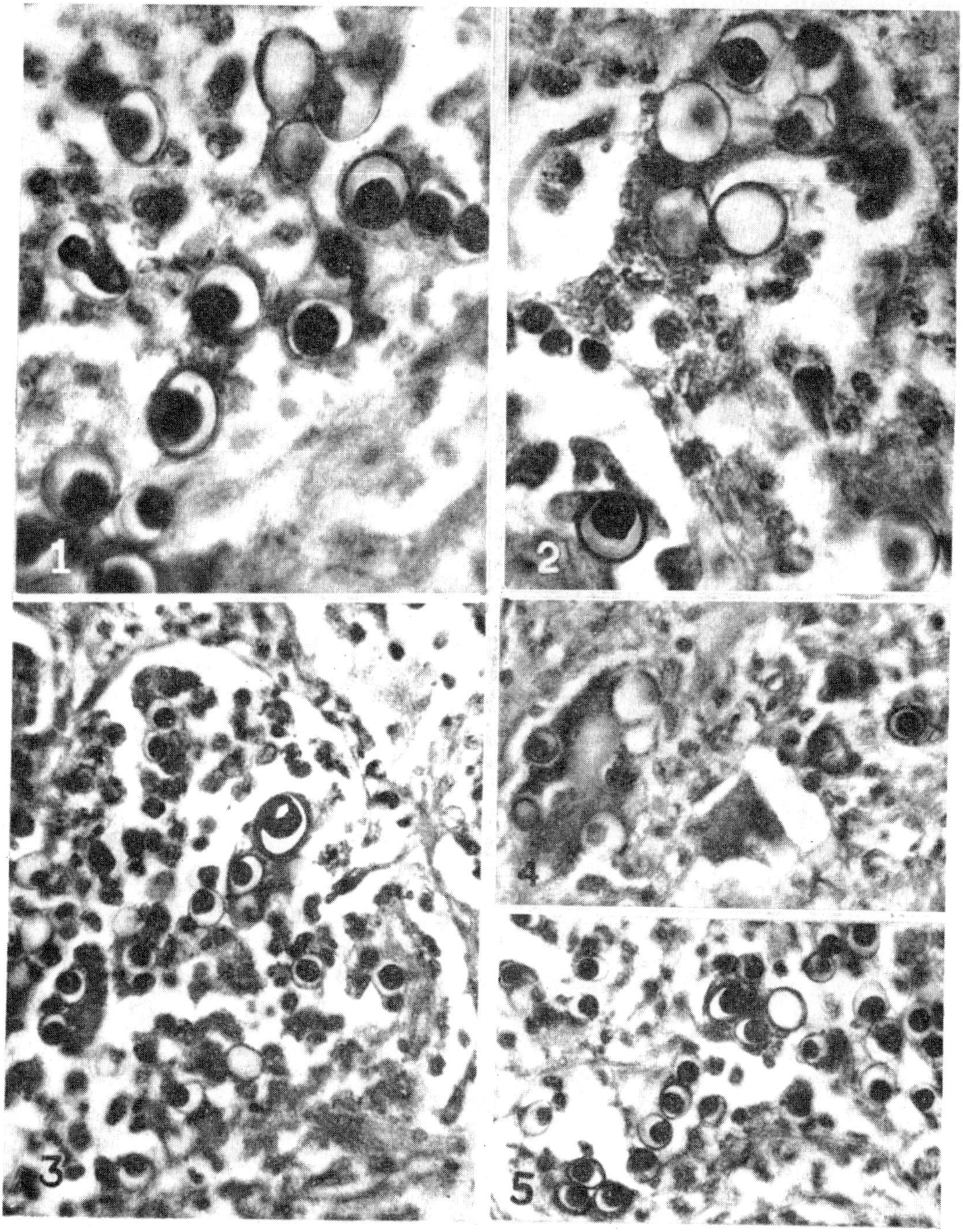

Figs. 1-5 : Histopathological slides by courtesy of Dr. B. C. Dyson, Philadelphia

Figs. 1-2 : Showing *Blastomyces dermatitidis* in the lung of a case of pulmonary blastomycosis. Note "empty" forms in **Fig. 2**, both **PAS**; 100 ×

Figs. 3-5 : Pulmonary blastomycosis : *Blastomyces dermatitidis* in intra-alveolar inflammatory exudate. Note uptake into giant cells in **Fig. 4**. All Figs. P.A.S.; 560 ×

containing the organism. Fibroblasts are present at the periphery. The characteristic round or oval budding cells are present in the pus and best **stained by polychrome methylene blue**.

There is pseudopeitheliomatous hyperplasia. The skin shows numerous microabscesses, which may lie entirely in the epithelium. The contain fungus cells, which may be found only after examination of numerous sections. In tissue sections, the fungal cells have a spherical nucleus of 5-10 microns diameter. The cell wall is usually slightly oval, separated by a gap from the nucleus and measures 12 to 25 microns across. The nucleus stains well with haematoxylin and the cell wall with P.A.S. (**Pl. 22.I, Figs. 1-5**). Organisms may be found also in giant cells below a very irregular acanthotic epidermis. There is a considerable chronic inflammatory infiltrate in the upper dermis.

Direct microscopy

Pus from the pustules at the edge of skin lesions, sputum, bronchial aspiration material and urine may show the organism.

Mycology

Blastomyces dermatitidis is biphasic and reproduces by spherical budding, one cell at a time in tissue of animals and human, and in culture at 37°C. In contrast, ***B. brasiliensis*** multiple budding (**Fig. 22.3**) at 22ºC a fluffy aerial mycelium is produced, with characteristic terminal conidia born on the tips of short stemmed conidiophores as well as alongside the hyphae.

Direct examination

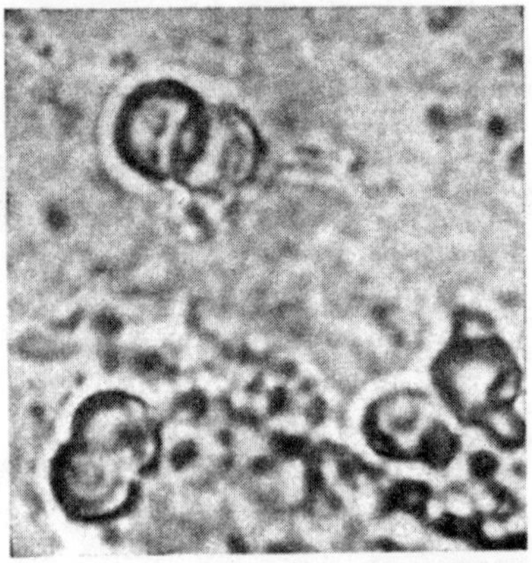

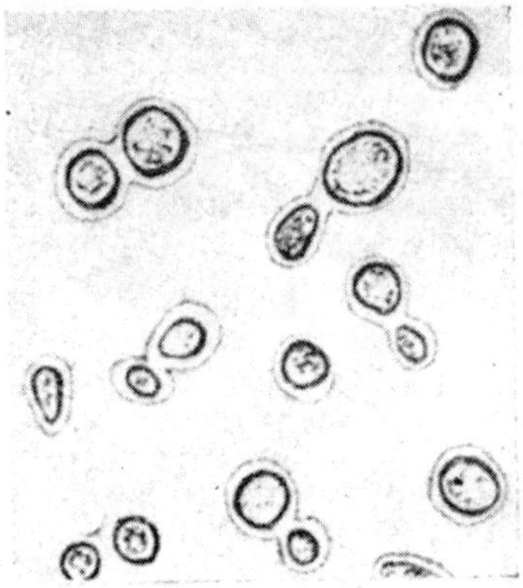

Fig. 20.1 : Budding cells in the pus of blastomycosis

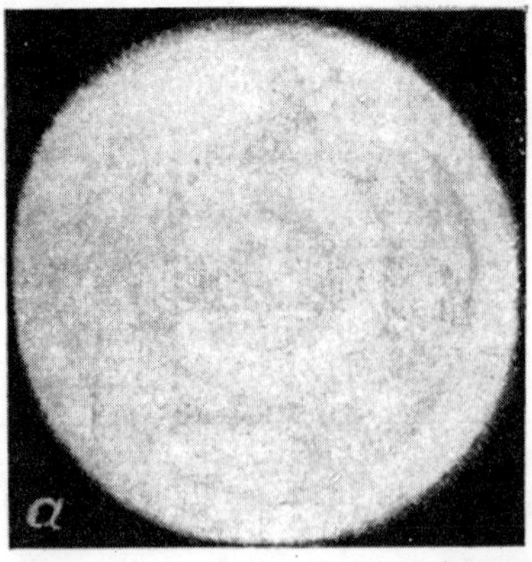

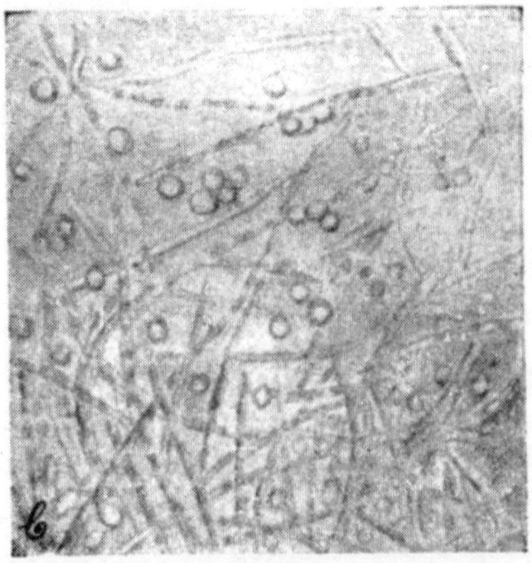

Fig. 22.2 : (a) Culture of *B. dermatitidis* on Sabouraud's agar at room temperature
(b) Culture showing **mycelia bearing conidia**

A cover slip preparation is made with a loopful of pus or sputum with a 10 per cent solution of potassium hydroxide. It shows characteristic round or oval, thick-walled, double-contoured, refractile, **yeast-like bodies with budding, 8 to 20μ in size**. The bud has a thinner wall than the mother cell. These cells are highly granular bodies and occur singly, in pairs or groups and are pathognomonic of the infection (**Fig. 22.1**).

Culture : Dimorphism is characteristic. Primary cultures are made on blood agar and glucose agar and incubated at 37ºC. The fungus grows slowly and the culture should not be rejected as negative, before one month. The colonies develop as waxy, wrinkled growths. On microscopic examination budding yeast-like cells are seen resembling those in the lesion. *Subcultures* on Sabouraud's glucose agar at room temperature exhibit downy growths characterised by aerial mycelia which are at first white but become brownish with age (**Fig. 20.2a**). Microscopically it shows mycelia with oval or round conidia, 3 to 4μ in diameter, attached to the hyphae. Pyriform conidia are about 5μ in length and borne on lateral sterigma. In old cultures chlamydospores are seen, 7 to 8μ in diameter with thickening of the wall (**Fig. 20b**). Mycelial form may be converted into yeast form by incubating the subcultures at 37ºC.

Diagnosis

The diagnosis depends on the demonstration of *B. dermatitidis* in the **pus, sputum, or cerebrospinal fluid by DIRECT MICROSCOPIC EXAMINATION**. The material is examined with a drop of 10 per cent potassium hydroxide and budding cells with a drop of 10 per cent potassium hydroxide and budding cells with refractile capsules should be observed. Culture should also show the growth of the organism. Histopathological examination of the skin lesion will show a granulomatous lesion showing budding cells with double contour.

INDIRECT METHODS

Intradermal test

The reaction is a tuberculin type, 0.1 ml of the vaccine is injected intracutaneously and readings are taken after 12, 24 and 48 hours. Positive skin test indicates presence of hypersensitiveness.

Animal inoculation

The mouse is the most susceptible animal; rats and rabbits are less so and guineapigs are the least susceptible. Any phase, either the mycelial or the budding one, injected into mice, causes local lesion followed by visceral lesions of the liver, spleen or lungs, in about 3 weeks.

Prognosis

Primary cutaneous lesions may be cured if proper treatment is given early in the diseases but 92 per cent of systemic involvement become fatal.

Treatment

Whatever therapeutic measures are taken, proper care should also be taken about the general health of the patient. Sunshine, vitamins, high caloric diet are essential for the well being of the patient.

In the cutaneous types, complete resection, followed by **cauterisation with diathermy** is advisable in early cases. **Amphotericin B** appears to be the best remedy at present for this type of blastomycosis. **Stilbamidine** is sometimes useful **in antibiotic resistant cases.** A total dose of 4 to 6 g is administered and a less toxic preparation like **2-hydroxy-stilbamidine** (6-10 g) may be used. The drug is administered in daily intravenous infusions for a month. The initial dose is 50 mg administered in 100 ml and the third and subsequent doses are 150 mg in 500 ml over 3-4 hours. The course may have to be repeated in some cases specially in the systemic, disseminated type.

MODERN DRUGS OF CHOICE

(i) Itraconazole (Triazole)

(ii) Ketoconazole (Imidazole)

REFERENCES

Gilehrist, T. C. (1896). A case of Blastomycotic dermatitis in man. *Rep. Johns Hopkins Hosp.*. **1**, 269.

Gilchrist, T. C. and Stokes, W. R. (1898). A case of pseudo-lupus vulgaris caused by Blastomyces. *J. exper. Med.* **3**, 53-78.

Mohapatra, L. N. (1969). Systemic mycosis in India, *Bull. A.I.I.M.S.* **3**, 7-19.

Mukherji, B. B. and Sengupta, T. N. (1968). North American Blastomycosis *Jour. ind. med. Assoc.* **51**, 80-81.

Ricketts, H. T. (1901). Oidiomycosis (Blastomycosis) of the skin and its fungi *J. med. Research*, **1**, 373.

Schwarz, J. and Baum, G. L. (1951). Blastomycosis. *Am. Journ. clin. Path.* **21**, 999.

Wilson, J. W. and Plunkett, O. A. (1965). The fungus diseases of man. University of California press, Berkeley and Los Angeles.

23

Systemic/Cutaneous Mycoses South American Blastomycosis

(Paracoccidioidomycosis)

Definition

South American blastomycosis is a **chronic granulomatous ulcer of the skin** and **mucous membrane, with enlargement of the lymph nodes** and **involvement of the internal organs,** caused by ***B. brasiliensis***. The granulomatous and ulcerative lesions of the oral mucosa, enlargement of the lymph nodes and involvement of the gastro-intestinal tract differentiates the condition from North American blastomycosis. The lungs are rarely affected in this condition.

Geographic Pathology

The disease has a high incidence in **Brazil and Venezuela,** with foci in other **South and Central American States** (Rodriguez and Pinate, 1966; Bopp and Bernardi, 1967; Aamos Silve, 1968). No case has been reported from India (Mohapatra 1969).

Clinical types

Several clinical types have been described.

1. **MYCOCUTANEOUS TYPE** which starts from the buccal mucosa as a granulomatous ulcer and affects the oral mucous membrane and nose.
2. **LYMPHANGITIC TYPE,** in which groups of lymph nodes are enlarged; namely cervical, supraclavicular and axillary lymph nodes. The lymphomatosis is that of Hodgkin's disease.
3. **VISCERAL TYPE** with lesions in the intestine, liver, spleen, pancreas, and other internal viscera.
4. **MIXED TYPE,** involving both the skin and internal viscera.

However, recently the following types have been recognised to be the most suitable ones.

1. TEGUMENTARY SOUTH AMERICAN BLASTOMYCOSIS

Clinically, lesions of the **oral, nasal, buccal, laryngeal and less frequently,** anal and genital mucosae with **papules, mulberry like tumours and ulcerations are most frequently seen.** The process extends to the adjacent skin and causes lesions somewhat similar to North American Blastomycosis. Local lymph nodes are early involved and may ulcerate.

2. EXTRA-TEGUMENTARY SOUTH AMERICAN BLASTOMYCOSIS

Recently it was discovered that the **lungs are almost always** involved and probably serve as **portal of entry for systemic infections** which may **disseminate to the bones, central nervous system, spleen and liver.** A peculiarity of **paracoccidioidomycosis is** relatively frequent involvement of the adrenal gland with huge caseating foci. Generally, the lesions are exudative, purulent at first, become granulomatous later and end finally in cicatrication. **Association with** tuberculosis and **other** deep mycoses is relatively **frequent**.

Histopathology

The **fungus cell is larger than** that of **North American Blastomycosis**, measures approximately **10-30** microns in diameter and propagates by multiple budding of spherules found in a circular arrangement in the periphery of the mother cell (**Fig. 23.1**).

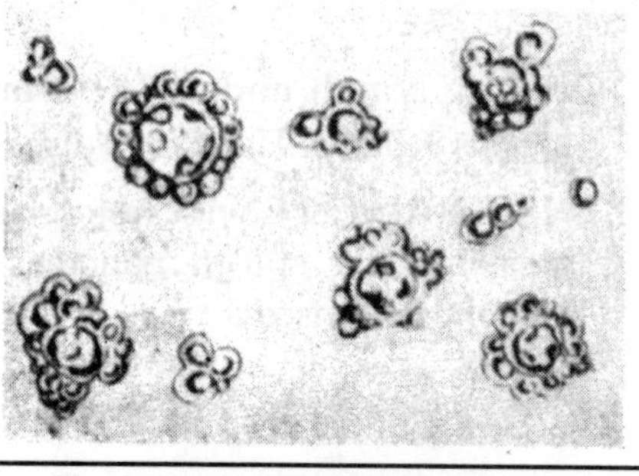

Fig. 23.1 : Characteristic budding of *B. brasiliensis* in pus

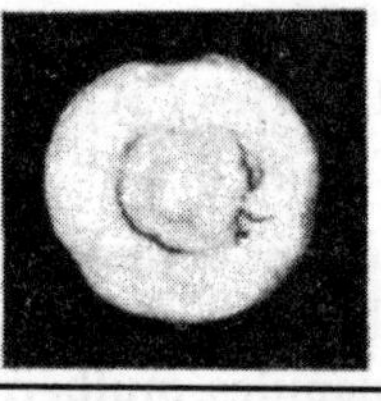

Fig. 23.2 : *B. brasiliensis* in Sabouraud's glucose agar at room temperature; 4 weeks old

Mycology : *B. brasiliensis* like *B. dermatitidis* is dimorphic and both yeast and mycelial forms are seen. In the tissue, pus or culture at 37°C, single or multiple budding cells are seen; the latter is diagonistic. Buds unlike those of *B. dermatitidis* are coccoid, small, measuring 1µ or more. They are typically multiple and arranged round the mother cell. Single budding cells are 10 to 30µ in diameter but the multiple budding-cells may be as large as 60µ (**Fig. 23.3**).

Culture : The infective material is cultured in blood agar and Sabouraud's glucose agar and incubated at 37°C and at room temperature. On **blood agar incubated at 37°C the culture** is **YEAST-LIKE**, and angle and multiple budding forms are seen like those in the lesion. On Sabouraud's agar incubated at room temperature, the growth is heaped up, glabrous and cerebriforms like that of ***Trichophyton schoenleini***. These often show budding cells, but soon the growth is covered with white, aerial mycelia, (**Fig. 23.2**). They show filamentous mycelial forms with small, smooth-walled, round or pyriform conidia and chlamydospores.

Diagnosis

Infective material may be obtained from (a) pus or scraping from the ulcer on the buceal mucosa; (b) pus from a fluctuant node; (c) smear or material from biopsical nodes; (d) Sputum. The material is examined in fresh state under a coverslip with 10 per cent potassium hydroxide. The fungus appears as single or multiple budding-cells; the latter is diagnostic of paracoccidioidal granuloma (**Fig. 23.4**). The organism should be cultured on artificial media. Guinea pigs are inoculated intratesticularly with the infected material and pure cultures may be obtained from the material after development of the lesion.

Animal inoculation

Mice are inoculated intraperitoneally but guinea pigs are inoculated intraperitoneally and intratesticularly. The fungus can be studied from the pus and cultures may be obtained from the lesions after their development.

Immunnity

The complement fixation test is positive in spreading cases, indicating a bad prognosis. It may be negative in early or treated cases. The skin test is, however, positive with paracoccidioidin in cases with hypersensitivity.

Treatment

Iodide has been advocated but it should not be administered without desensitisation of hypersensitive patient. Therefore, before administering iodide, skin test should be done

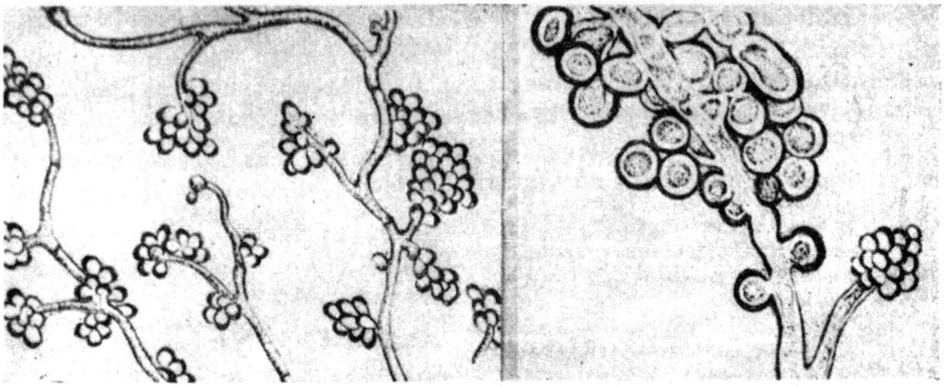

Fig. 23.3 : Showing multiplication of *B. brasiliensis* by multiple budding

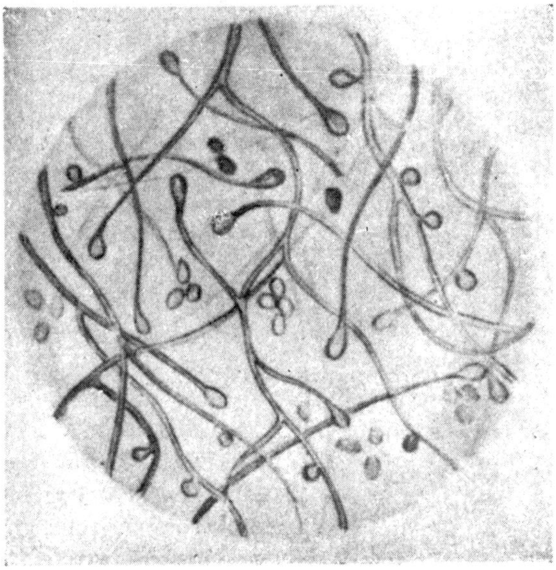

Fig. 23.4 : Single and multiple cells, diagnostic of paracoccidioidal granuloma

for hypersensitivity. Recently it has been reported that South American blastomycosis responds very well to sulphapyridin in all the stages of the disease, when administered over a long period penicillin is not effective against *B. brasiliensis in vitro* or *in vivo*.

LOBO'S DISEASE. (KELOIDAL BLASTOMYCOSIS) : The causative organism , *Loboa loboi* of Lobo' disease has nerver been cultured. Whether it is related to South American blastomycosis or not is debated. Some cases with both the infection at the same time have been reported (Lacaz *et al.*, 1967). Lesions are keloid like and verrucous and show histologically numerous oval or globose fungus cells of about 8 microns diameter in the tissue. The disease has been reported from the Amazona's valley and other areas of South America and from Surinama in Indonesia. Excellent illustrations are to be found in Emmons, Binford and Utz "Medical Mycology". For further details see: Destombes and Ravisse, 1964. The disease is mentioned here because it is not impossible that cases exist in India, though none have been reported.

REFERENCES

Bopp, C. and Bernardi, D. V. (1967). Geopathologia da blastomicose sul-americana no Rio Grande do sul. *Revta. Assoc. med. Rio Grande do Sul.* **11**, 31-42.

De Almedia, F. (1933). Blastomycoses of Brazil. *Ann. Fac. de med. de Sao Paulo* **9**, 69.

Destombes, P. and Ravisse, P. (1964). Etude histologique de cas Guyanais de blastomycose Cheloidenne (maladie de J. Lobo). *Bull. Soc. path. Exot.* **57**, 1008-1024.

Fonseca, O da. Jr. (**1939**). Etiologic agent of neotropical blastomycoid granulomatosis. *Ann. Brasil de dermat e, sif.* **14**, 85, English translation, pp. 112-139.

Lacaz, C. S., Ferri, R. G., Raphael, A., Fava Netto, C., Minami, P. S., Castro, R. M. and Dillon, N. L. (1967). Blastomicose queloidiana associada a blastomycoses sul-americana. Registro ju um casp. *Hospital Rio de J.* **71**, 7-11.

Lutz, A. (1908). Uma mycose Pseudococcidica localisada na bocca observade no Brasil, Brasil Medico, **22**, 121, 141.

Ramos-Silva. (**1968**). Miss au point sur l'histoire de la blastomycose sud-americaine. *Maroc. med.* **48**, 5-12.

Rodriguez, C. and Pinate, F. M. (1966). La paracoccidiodosis brasiliensis en Venezuela. Estudio de 120 casos. *Gac. med. Caracase* **74**, 101-139.

Splendore, A. (1909). Sobre um novo caso de bastomycose generalizada. *Rev. Soc. Sci., Sao Paulo.* **4**, 52.

Symmers, W. S. C. (1966). Deep-seated fungal infections currently seen in the histopathologic service of a medical school laboratory in Britain. *Am. Jour. clin. Path.* **46**, 514-537.

24

Systemic/Cutaneous/Subcutaneous Mycosis Coccidioidomycosis

Synonyms

Coccidioidal granuloma, Posada's disease, Valley fever, Desert rheumatism, etc.

Definition : It is a highly infectious disease caused by ***Coccidioides immitis*** and characterised by an **acute, primary, benign (self-limited) respiratory infection (Coccidioidomycosis)** or a ***chronic, progressive, secondary, disseminated granulomatous affection*** involving the ***lung, skin, subcutaneous***, visceral or osseous tissues ***ending fatally***. The acute, benign respiratory disease is known as Valley fever or Desert rheumatism. Primary cutaneous lesions may, however, be seen with papillomatous or fungating lesions.

History

The first case of coccidioidomycosis was reported by Posada (1892) and Wernicke (1892) in Argentine, affecting a Brazilian soldier. The second and third cases were seen in California by Rixford and Gilchrist (1984). They believed the parasites to be a protozoan and named it ***Coccidioides immitis*** in 1896. Ophuel and Moffitt (1900) cultivated this organism and found it to be a fungus with filamentous mycelia. Benign type was first described by Gifford (1936).

Introduction

Coccidioides immitis causes primary pulmonary infection which, in many cases is unrecognised and self limited due to rapidly developing immunity. In a small number of cases, however, fatal dissemination may take place. If the secondary focus is **confined to one location, a chronic coccidioidal granuloma may develop which may finally heal** after a protracted course. In very rare instances, **a primary cutaneous coccidioidomycotic chancre with lymphatic spread develops** following accidental inoculation with the organism. All aspects of the disease have been discussed in a very thorough manner by Wilson and Plunkett (1965) and in the volume on Coccidioidomycosis edited by Ajello (1967).

Geographic pathology and epidemiology

The disease is endemic in southern California, Arizona and parts of Texas. Northern Mexico, Hoduras, Venezuela, Paraguay, Bolivia and Argentina are also endemic areas. In India, Dass *et al.* (1958) found **spherules in direct preparations of exudate from multiple lesions** of a patient. However, their diagnosis of **coccidioidomycosis was not proved by culture**. The **fungus grows in the soil of endemic areas**. The spores are inhaled together with dust. Dust control measures have been proved to be effective in the reduction of the incidence of infections (Drips and Smith, 1964). A **warm and dry climate favours transmission**. Since the infection may be inapparent for longer periods, the disease may be diagnosed in any part of the world in this age of rapid transportation.

Sources of infection

The organism pathogenic for animals, has been isolated from soil and vegetations. Natural infection has been found in the internal; organs of cattle, sheep and dogs (Stiles and Davis, 1942). **Desert rodents** have been found to be naturally infected by Emmons (1942). These rodents **harbour the infection** and act as **an animal reservoir of the disease**. The source of saprophytic fungus in the desert soil is probably maintained by the infected rodents.

Mode of infection

Respiratory infection is caused **by inhalation of the spores** from the culture. The infection is rarely **acquired through the exposed surfaces of the skin**.

GROSS AND HISTOPATHOLOGICAL LESIONS

1. *PRIMARY PULMONARY COCCIDIO-IDOMYCOSIS*

In about half of the cases, the primary form may be subclinical and cause only minimal lesions. If the infection is more virulent, a wide range of clinical symptoms and pathological lesions may be observed. Clinical symptoms range from fever (valley fever) to cough, chills, chest pain, bronchitis, nasopharyngitis, dyspnoea and fatal dissemination, which is very rare.

Pathological anatomic lesion consists in small transient infiltrates of the lung tissue, **pleural effusion and bronchopneumonia. One or two weeks after** the onset of symptoms, **skin lesions resembling erythema multiforme** or erythema **nodosum develop** in a number of patients, indicating **cutaneous hypersensitisation to toxic products of the fungus**. Such skin manifestations are **followed by** recovery and **lasting immunity**.

In about 5% of all infections more **extensive pneumonic infiltrations and pulmonary nodular infiltrates**, called **coccidioidomas develop**. Thick walled abscesses, cysts, cavities, pleural fibrous thickening with **adhesions, pumonary fibrosis with bronchiectases may develop** and the disease may persist over months until finally **heals with fibrosis and calcification.**

HISTOPATHOLOGY OF PRIMARY PULMONARY COCCIDIOIDOMYCOSIS

Spores develop into spherules and sporangia in the tissue. The ruptured mature sporangium releases endospores which cause acute inflammatory reaction **and abscess formation**. Microabscesses are later replaced by subacute and chronic inflammatory granulation tissue (**Pl. 24.1, Fig. 1**). When spores develop into spherules and sporangia, foreign body giant cell reaction with epithelioid granulomas may develop (**Pl. 24.1, Figs. 2 & 3**). If the allergic response of the host is satisfactory, eosinophilic infiltrates may be found. In chronic cases fibrosis and calcification are present. All these lesions may be found simultaneously, showing the total range of inflammatory responses that man is able to produce. Scattered **spherical organisms, 10 to 80 microns** in diameter may be found **in the tissue**. They **stain** satisfactorily with H & E, but **better with PAS**. With **silver methenamine stain**, numerous degenerating **"empty" spherules may be demonstrated** in necrotic tissue (**Pl. 24.1, Fig. 3**).

2. *PATHOLOGY OF DISSEMINATED COCCIDIO-IDOMYCOSIS (COCCIDIOIDAL GRANULOMA)*

In only one in a thousand patients the disease progresses. There may be **rapidly fatal dissemination, resembling miliary tuberculosis**. In other cases, only a solitary secondary focus develops (**coccidioidal granuloma**). This may finally heal after a protracted and intermittent course. Between these two extremes, limited **dissemination to the skin, bones, joints, central nervous system, and internal organs may occur**. The gastro-intestinal tract is not frequently involved, according to Whilson and Plunkett (1965). Other authors believe that, beside the lungs, the digestive tract provides an entry for primary infection. Histopathological lesions consist of coccidioidal granulomas not different from those seen in the lungs. The same wide range of inflammatory responses may be seen as mentioned earlier.

3. *PATHOLOGY OF PRIMARY CUTANEOUS COCCIDIO-IDOMYCOSIS*

Immediately after inoculation, there is an acute inflammatory reaction with numerous spherules in the exudate. A few weeks after infection, spherules are difficult to demonstrate

PLATE 24.I

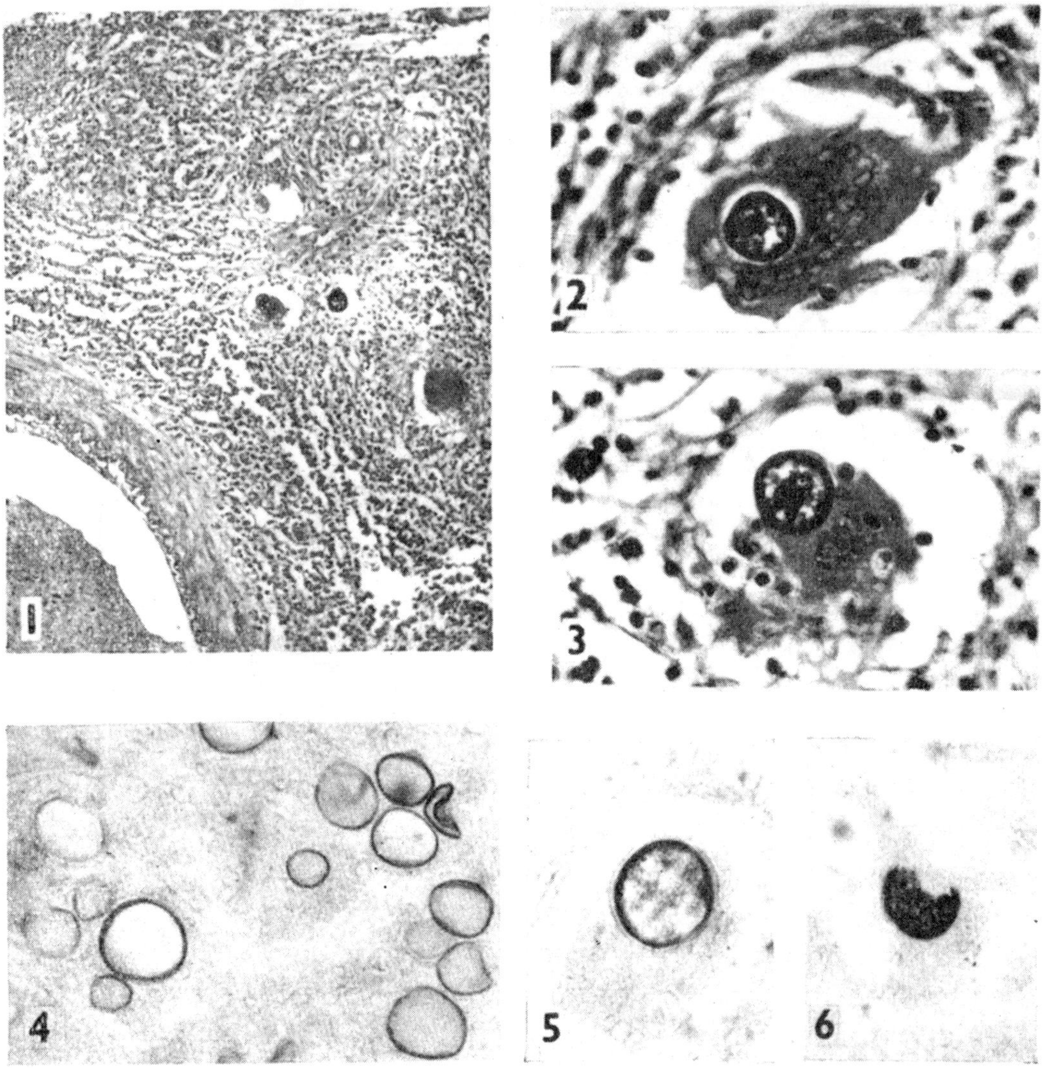

Fig. 1 : Coccidioidal meningitis, Arachnoid tissue transformed into chronic inflammatory granulation tissue with scattered spherules causing foreign body giant cell reaction. **PAS**; **130 ×**

Fig. 2 & 3 : Coccidioidal meningitis. Spherules of *C. immitis* with foreign body giant cells. **PAS**; 560 ×

Fig. 4 : Pulmonary coccidioidomycosis : Degenerate "Empty" spherules in necrotic tissue. Silver methenamine; 560 ×

Fig. 5 & 6 : Spherules in lung tissue. In Fig. 6, spore formation is already recognisable. **PAS**; **560** ×

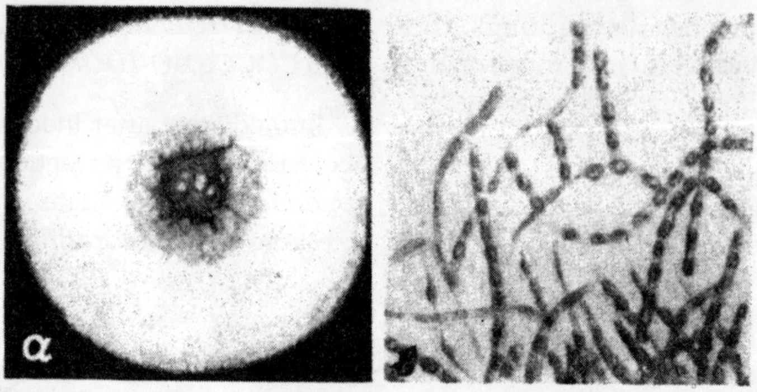

Fig. 24.1 : (a) Culture of *C. immitis* on Sabouraud's glucose agar at room temperature, 2 weeks old; (b) *C. immitis* showing formation of arthrospores in culture (a)

PLATE 24.II

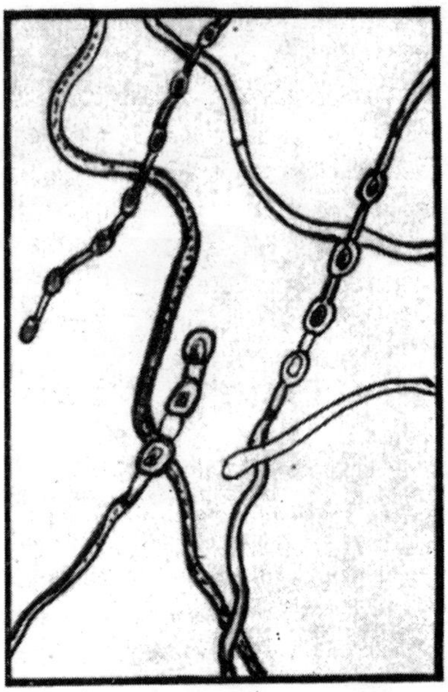

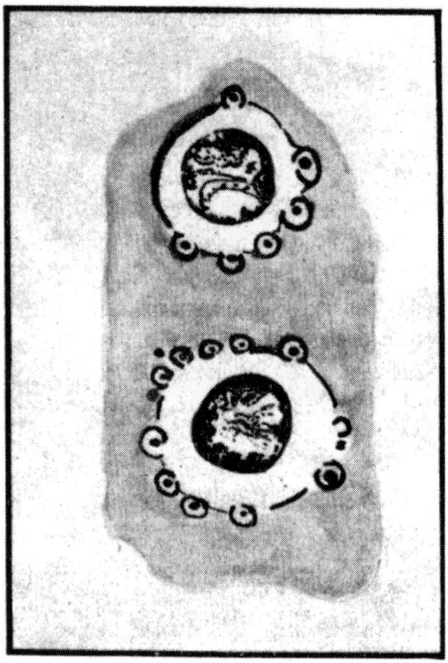

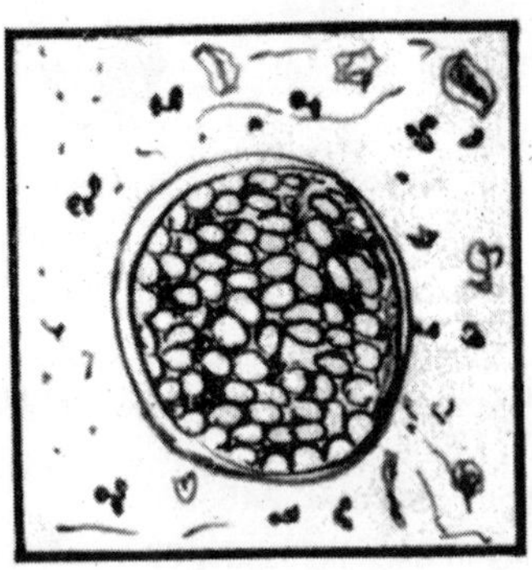

Fig. : Upper right and lower. Sporangium of *Coccidioides immitis* approx. 1600´×

Fig. : Upper left. *Coccidioides immitis* : Hyphae with arthrospores; **Culture mount**.

in the tissue and a **chronic inflammatory reaction with epithelioid granulomas and scattered microabscesses** including some **eosinophilic infiltration** may be seen.

4. *COCCIDIOIDAL GRANULOMA*

This is a progressive condition from this infection which develops in more severe cases of coccidio-idomycosis. In many respects it resembles pulmonary tuberculosis, as reactivation of inadequately treated cases may occur analoguous to reinfection of pulmonary tuberculosis. If the temperature of the valley fever lasts after the third or fourth week, one should suspect the progressive form of the disease. Then **new shadows appear in the lungs when the skiagram is taken**. As the disease progresses **dissemination occurs through blood-stream with metastasis in the bone, subcutaneous tissues (cold abscess or granulomatous lesions) and internal organs including meninges**. Generalisation may occur in any organ with the progress of the disease with **pyrexia, loss of weight, night sweat and the case ends fatally in 4 or 5 years**.

Clinical manifestations

Coccidio-idomycosis may affect a person of any age. The **INCUBATION PERIOD IS ABOUT 10 TO 14 DAYS**. The primary lesion may be in the skin but it often starts in the lung by inhalation of dust containing spores. In the **acute stage, the lung infection resembles an attack of 'flu'** or **bronchitis with slight fever, cough, headache, backache**, and the **sputum may be blood-tinged**. This is known as the Valley fever as it is common in San Josquin valley. It is also called Desert fever or Desert rheumatism, as arthralgia is often associated with it. After 8 to 15 days, nodules like those of erythema nodosum and eosinophilia may develop. Recovery from the acute condition usually occurs in 3 to 6 weeks.

Primary skin lesions appear as papillomatous or verrucose lesions like those of **tuberculosis of the skin**.

A. DIRECT METHOD

Mycology

Direct microscopic examination.

1. One drop of pus, sputum or other exudate is placed on a slide, with a drop of 10% KOH solution and gently heated. Under reduced light, **thick walled double contoured spherules, 10 to 90 microns in diameter are seen**. If **endosporulation** develops, the **wall becomes thinner and the spherule is filled with spores (Pl. 24.II, Fig. upper right)**.
2. **A drop of exudate is mixed with a drop of saline** solution on a slide and **covered with a cover slip**, the edges of which have been coated with petrolatum in order to seal the fluid. In this **quasi "microculture", sporulation may take place** and such **spherules** will **produce** several **thread-like hyphae within** two or **three days**. This method is a great help in early diagnosis.

B. INDIRECT METHODS

Culture

Coccidioides immitis is a biphasic organism which forms spherules and sporangia in tissue, but not in ordinary culture, also not at 37ºC. On Sabouraud's glucose agar the growth is slow. A white serial mycelium is produced which turns soon to gray or brown. The branching septate hyphae develop barrel or cask shaped arthrospores within 3 to 14 days. Arthrospores are separated from each other by a narrow empty **separate neck of the hypha (coralette) (Pl. 24.II, Fig. upper left)**. This distinguishes it from arthrospores of *Geotrichum candium* which show pearl-string arrangement and are not separated by a coralette. *Coccidioides immitis* has not a yeast like phase in culture in contrast to *B. dermatitidis, B. brasiliensis, Cr. neoformans, H. capsulatum* and *S. schenckii*. Cultures of *C. immitis* must be treated with great care because arthrospores (**Fig. 24.1**) detach easily and may be inhaled by laboratory personnel handling cultures.

Microculture

Mycelia are branched and septate and the fungus forms numerous arthrospores 2 to 4 μ in diameter and separated by alternate clear and dark spaces along the course of the hypha.

Animal inoculation

All labotatory animals are susceptible to coccidioidal infections. Mice are usually injected intraperitoneally, but guinea pigs are injected intratesticularly with infected material or suspected cultures for positive results.

Immunity reaction

Presence of **precipitin** and **complement fixing antibodies** have not been established definitely to help in the diagnosis. They may appear in severe infections but disappear with recovery.

Hypersensitiveness to coccidioidin develops between the second and **third week of infection**. For the test **0.1 ml of 1 : 1000 of a standard antigen** is **injected intradermally** and this should show a positive reaction. Those **who had previous skin infections may show a reaction even in a dilution of 1 : 10,000.** Dickson and others hold that the cutaneous reaction is specific.

Treatment

No special treatment is necessary in the **acute stage** except rest in bed till the **blood sedimentation rate is normal** and lungs are fund to be clear by X-ray examination. **Complement fixation test is negative** in these cases. The prognosis is grave in progressive cases. In these cases specific **vaccine therapy** has been reported to be **efficatious**. Iodide is infeffective. **X-ray therapy is helpful** causing **absorption of exudate in the granulomatous process** and intensive or semi-intensive doses should be used. Thymol has been found to be effective. A 33.3 per cent solution in olive oil is applied locally and rally a daily dose of 1 to 6 g dissolved in olive oil should be administered, putting it in capsules. This is administered along with meals.

MODERN ANTIBIOTIC TREATMENT IS HIGHLY EFFECTIVE

Three antibiotics are equally effective; they are : Tirazoles, like

1. FLUCONAZOLE and
2. ITRACONAZOLE.

The third antibiotic

3. AMPHOTERICIN 'B' is relatively toxic but effective.

REFERENCES

Ajello, L. (editor). "*Coccidioidomycosis*". The University of Arizona Press, Tuscon, 1967.

Das, A. K., Chatterjee, M. K. and Debsikdar, B. M. (1957). A case of coccidioidomycosis in India. *Ind. Journ. Dermatol.* **3**, 18-23.

Drips, W. and Smith, Ch. E. (1964). Epidemiology of coccidioidomycosis. *J. am,. med. Assoc.* **190**, 136-138.

Emmons, C. W. and Ashburn, L. L. I. (1942). The Isolation of *Haplosporangium paravum* N. Sp. and *Coccidioides immitis* from wild rodents. Their relationship to coccidioidomycosis. *Pub. Hlth. Rep.* **57**, 1715.

Ophuels, W. and H. C. Moffitt. (1900). A new pathogenic mould (formerly described as *Protozoon, Coccidioides immitis pyogenes), Phila. med. J.* **5**, 1471.

Rixford, E. and Gilchrist, T. C. (1896). Two cases of protozoan (Coccidioidal) infection of the skin. *Rep. Johns Hopkins Hosp.* **1**, 209.

Smith, C.E. (1940). Epidemiology of the acute coccidioidomycosis with erythema rodosum. Am. J. pub. Hlth. **30**, 600

Smith, C. E. (1942). Parallelism of coccidioidal and tuberculosis infections. *Radiology* **38**, 643.

Smith, C. E. (1943). Coccidioidomycosis. *Med. Clinics N. America.*

Stiles, C. W. and Davis. C. L. (1942). Coccidioidal granuloma (coccidioidomycosis). *J. am. med. Assoc.* **119**, 765.

Wernicke, R. (1892). Ueber einen Protozoenbefund bei mycosis fungoides. *Centralbl. f. Bakt.* **12**, 859.

Wilson, J. W. and Plunkett, D. A. "The Fungus diseases of man". University of California Press, Berkeley and Los Angeles, 1965.

25

Systemic/Opportunistic Mycosis Aspergillosis And Other Hyphomycoses

Introduction

In tissue sections, hyphomycoses look alike, showing dichotomous branching **septate hyphae about 4 microns in diameter**. Even if the majority of these infections is caused by ***Aspergillus sp.***, a diagnosis based on the histological appearance should be restricted to the general term "**hyphomycosis**". Subsequently we shall deal with Aspergillosis as the prototype of hyphomycoses.

Geographic pathology

Aspergillosis is ubiquitous. Though only about a dozen cases have been mentioned in the Indian literature, I have seen six patients with aspergillosis during the last three years and three cases of spontaneous aspergillous endocarditis in experimental animals (Grueber, unpublished data). Other observations from India: Randhawa *et al.*, 1959; Mohapatra, 1961; Gupta and Vishwanathan, 1963; Reddy *et al.*, 1965; Sandhu *et al.*, 1966; Chitnis and Deshpande, 1967; Mittal and Aggarwal, 1968; Mohapatra, 1960). Many cases of aspergillosis probably go unrecognised because in biopsy specimens, Aspergillus hyphae are not easily recognisable. Since the tissue reaction often somewhat resembles tuberculosis, the histological picture can easily be misinterpreted.

Clinical forms in relation to gross pathology

Many cases of aspergillosis are *secondary*, (1) to debilitating disease like malignant tumours or diabetes, chronic renal or hepatic disease, (2) to corticosteroid and antibiotic treatment, (3) to surgical intervention, particularly criminal abortion; recently aspergillus endocarditis has been found following heart valve transplants and other cardiac operations.

Several types of aspergillosis may be distinguished:

(1) **ASPERGILLOSIS OF THE RESPIRATORY SYSTEM.**

(2) **Oculocerebral aspergillosis.**

(3) **Generalised aspergillosis** (including aspergillosis of isolated organ or systems).

(4) GASTROINTESTINAL ASPERGILLOSIS.

1 ASPERGILLOSIS OF THE RESPIRATORY SYSTEM

(a) Aspergillosis of the paranasal sinuses. (b) "Aspergilloma" in lumen of bronchi, or cavities. (c) Pulmonary aspergillosis.

(a) ASPERGILLOSIS OF THE PARANASAL SINUSES

The pathology of this form is well illustrated in **Pl. 25.I, Figs. 1-4** by the case of a 25 year old agricultural worker, who complained of local swelling, headache, nasal obstruction and breathing difficulties for 3 years.

PLATE 25.I

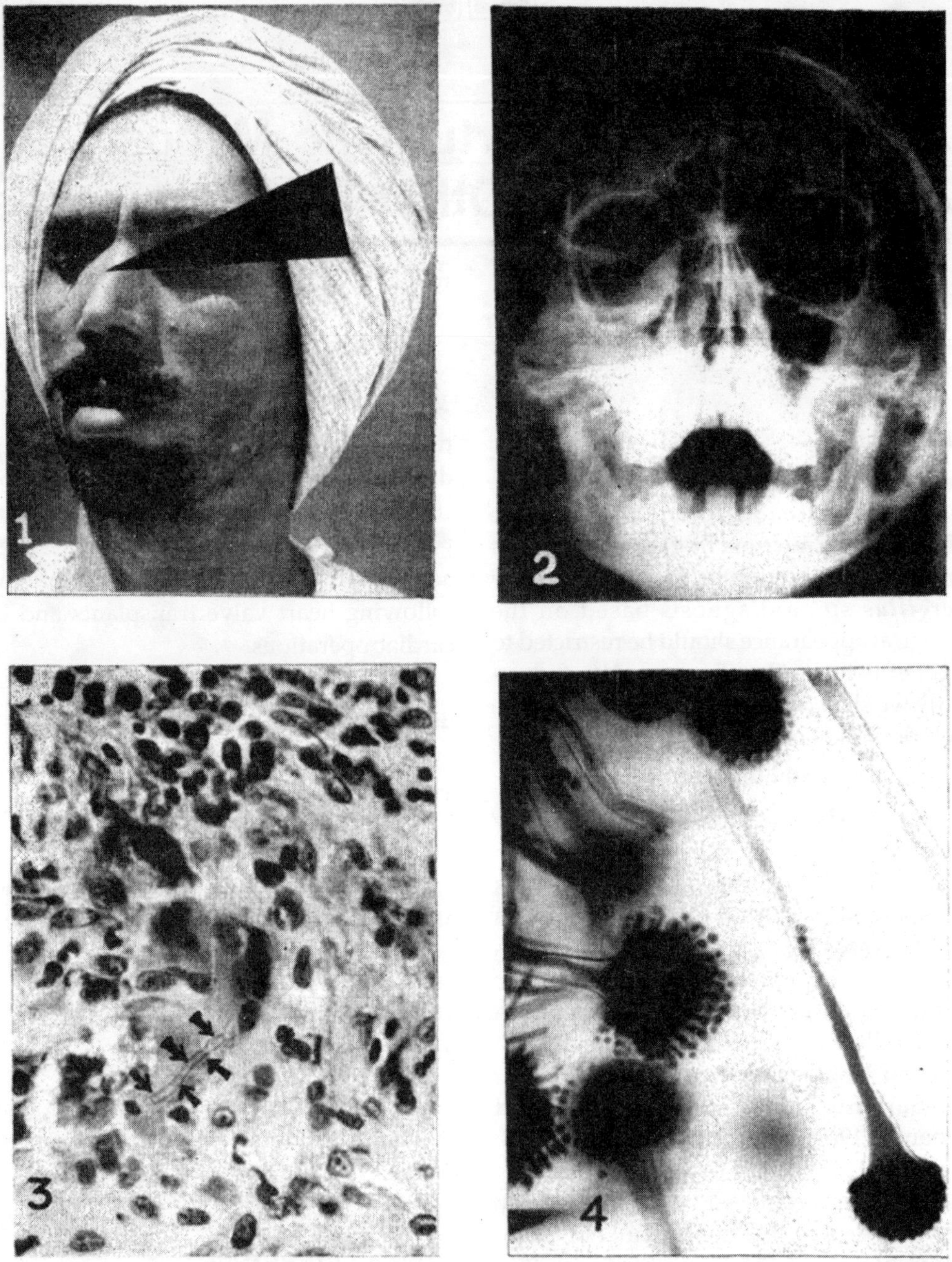

From one case — 25 years old agricultural worker

Fig. 1 : Swelling of right side of nose for 3 years

Fig. 2 : X-ray film shows obliteration of right maxillary sinus

Fig. 3 : H & E section of biopsy from right maxillary sinus showing foreign body giant cell reaction and very faint outlines of hyphae (arrows); 560 ×

Fig. 4 : Repeated culture from maxillary sinus mucosa yielded growth of *Aspergillus fumigatus*. "Caldwell Luke's" operation done, Iodide treatment. Patient well one year after operation

PLATE 25.II

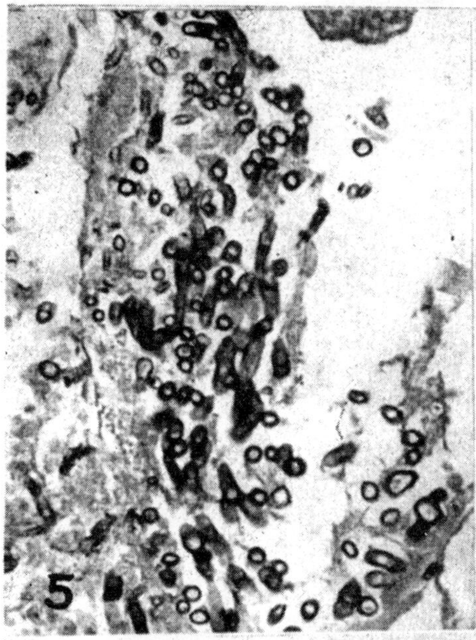

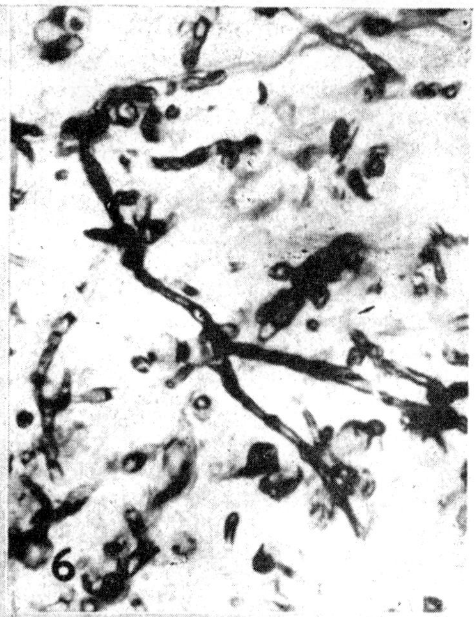

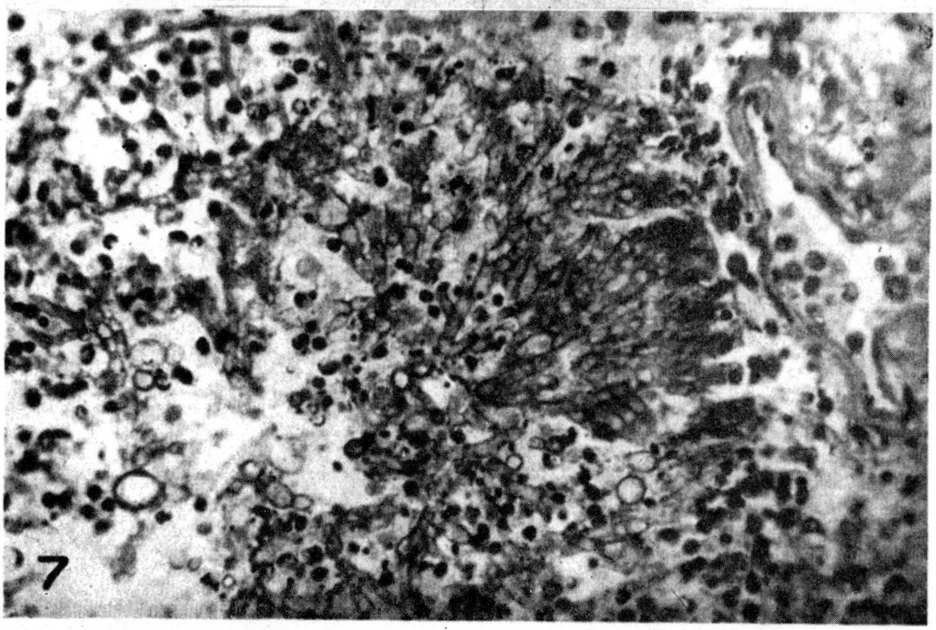

Fig. 5 : A case of bronchial carcinoma with secondary aspergillosis. Bronchoscopic biopsy showing histologically longitudinal and transverse sections of hyphae. Repeated culture yielded *Aspergillus flavus*. Weigert's Iron H & E; 560 ×

Fig. 6 : From an autopsy case of pulmonary aspergillosis in a patient with tuberculosis. Hyphae in necrotic young tissue show ballooning in some areas due to degeneration. Silver methenamine; 560 ×

Fig. 7 : Pulmonary aspergillosis: Radiation branching and septate hyphae with some globular degeneration form. H & E; 560 ×

The surgeon often finds firm chronic granulation tissue mixed with friable necrotic, caseous material. In some cases there is bone destruction. The process progresses only slowly in otherwise healthy individuals. The cranial cavity may be invaded either directly through the bone, or via the lacrymal duct and orbit. It is amazing how slowly, often over months, symptoms develop, even if the central nervous system is involved already.

(b) **"ASPERGILLOMA" or "FUNGUS BALL"**

It is a large mass of conglomerated hyphae in chronic brionchiectases or in tuberculosis cavities. Radiologically, a globular shadow in a radiolucent cavity is seen. It may cause very little local tissue reaction and may persist over years without doing harm to the patient. If repeated sputum cultures yield the same type of fungus, the diagnosis can be established in connection with the radiographic appearance.

A monograph on the subject of aspergilloma has been written by the German radiologist Reinhardt (1967), unfortunately under the misleading title "Das Mycetoma". However, a complete list of the world literature is a valuable part of publication.

(c) **PULMONARY ASPERGILLOSIS**

This term should remain reserved for invasion of the parenchyma. It is mostly secondary to steroid treatment or to debilitating diseases (**Pl. 25.II, Fig. 5** and **6**). The bronchi serve as portals of entry, probably also for most of generalised infections. The process may take a more chronic granulomatous course, or may rapidly lead to diffuse bronchopneumonic and pneumonic lesions (**Pl. 25.II, Fig. 7, Pl. 25.III, Figs. 8 & 9**) with massive polymorphonuclear exudation, necrosis, cavitation and abscess formation. The fungus spreads from one alveolus to the other. By a special affinity of *Aspergillus* to blood vessels the walls are penetrated and haematogenous spread takes place.

2. OCULO-CEREBRAL ASPERGILLOSIS

The infection gains entry through the conjunctival mucosa or via the lacrymal duct in connection with infection of the paranasal sinuses. Contiguous spread to the orbit through the bone is also possible. This usually slowly progresses over months or years with formation of chronic granulation tissue in the surrounding and fatty tissue. Typical symptoms are protrusion of the eye, swelling of the soft tissue,

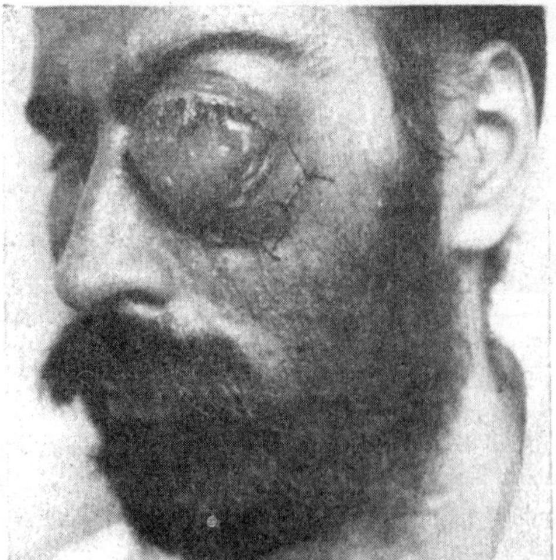

Fig. 25.1 : Orbital aspergillosis, Clinical appearance

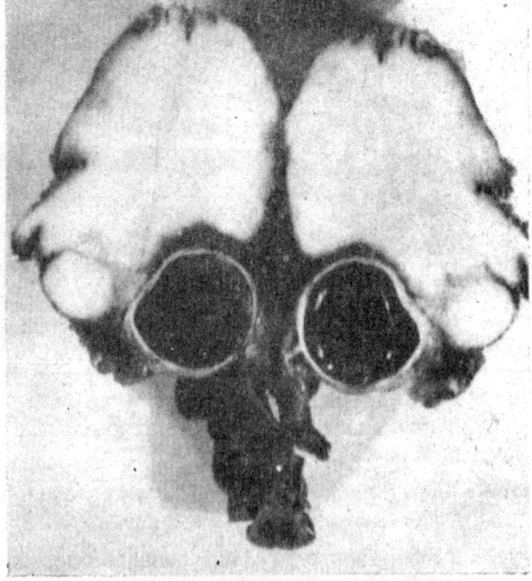

Fig. 25.2 : Orbital aspergillosis, Gross pathology—cut section

PLATE 25.III

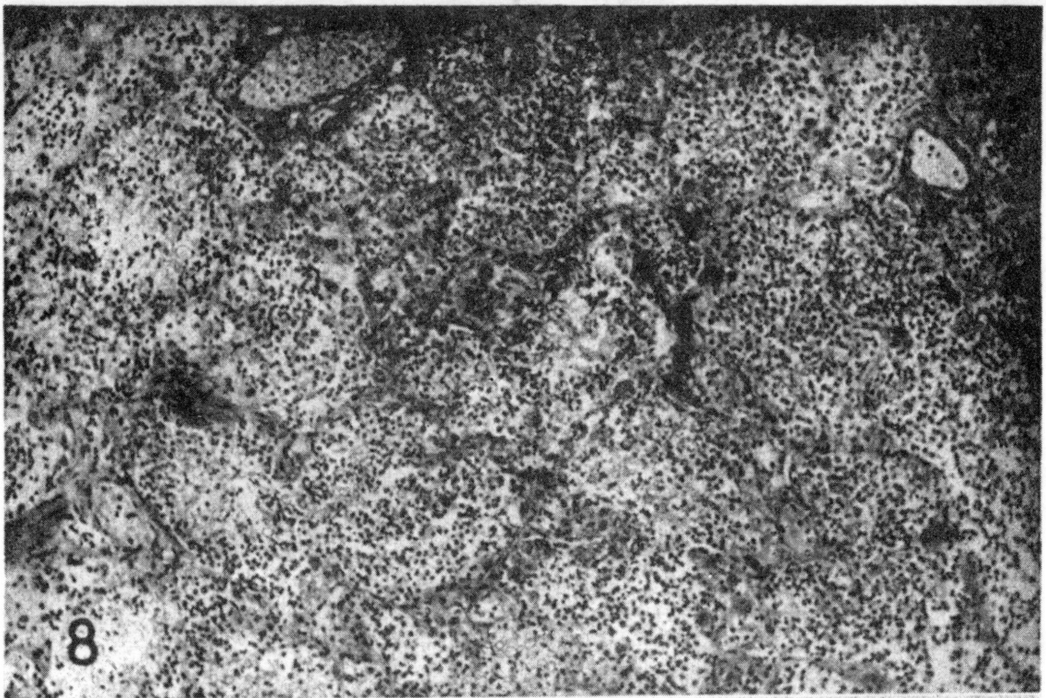

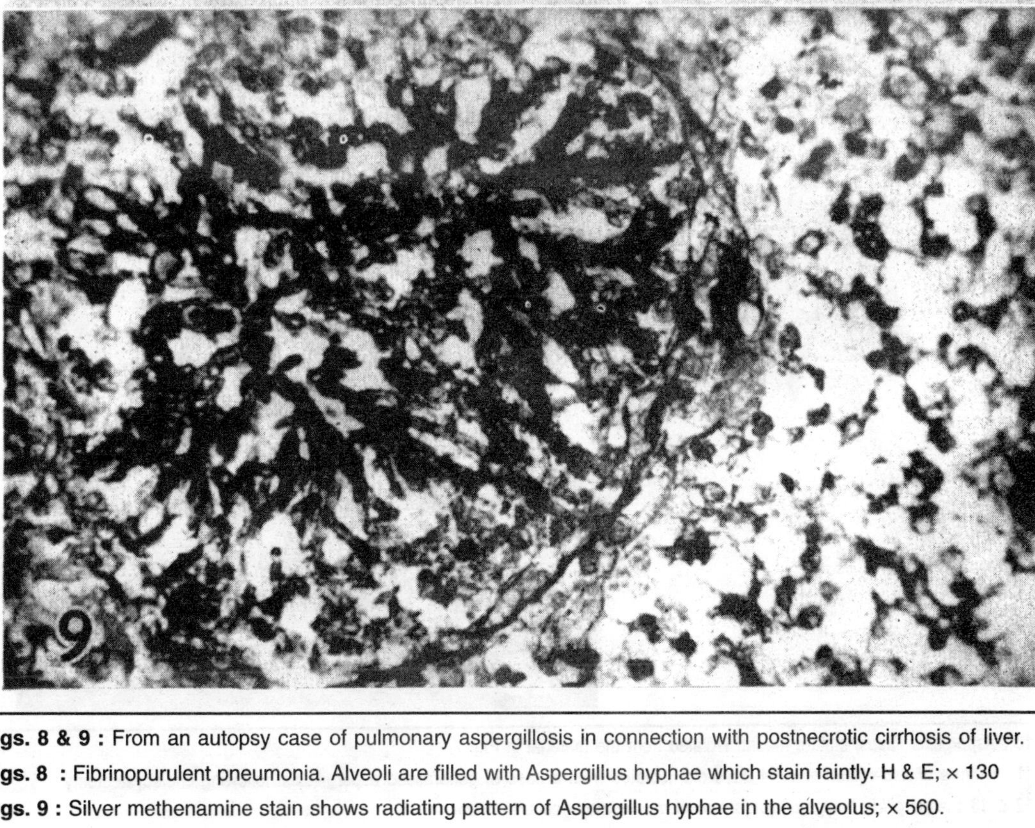

Figs. 8 & 9 : From an autopsy case of pulmonary aspergillosis in connection with postnecrotic cirrhosis of liver.

Figs. 8 : Fibrinopurulent pneumonia. Alveoli are filled with Aspergillus hyphae which stain faintly. H & E; × 130

Figs. 9 : Silver methenamine stain shows radiating pattern of Aspergillus hyphae in the alveolus; × 560.

PLATE 25.IV

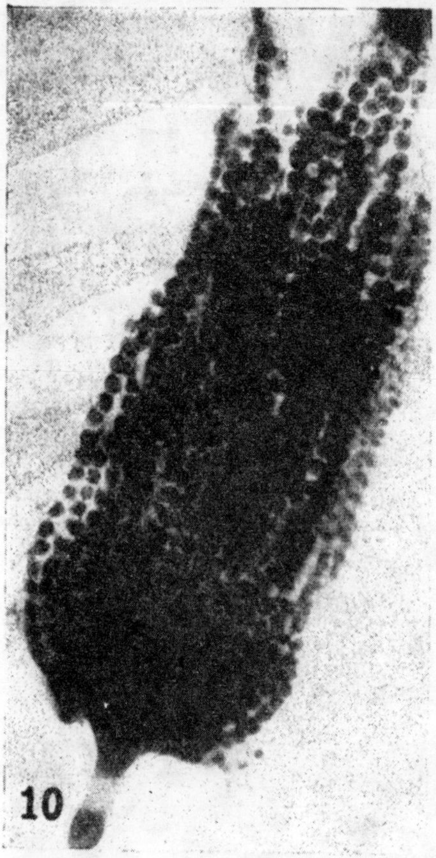

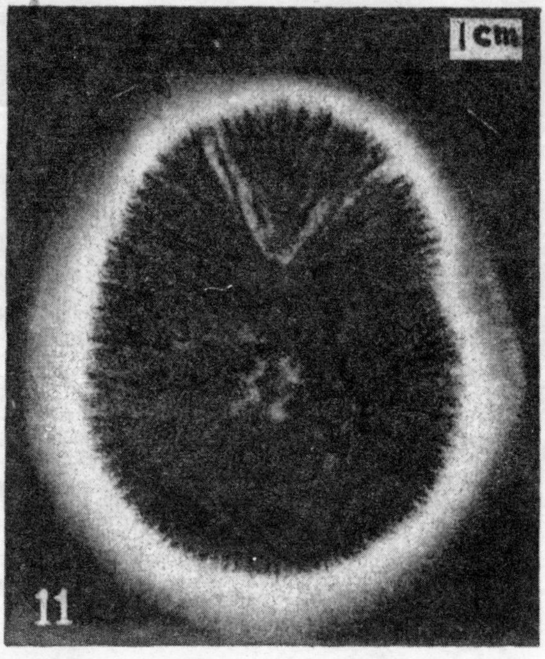

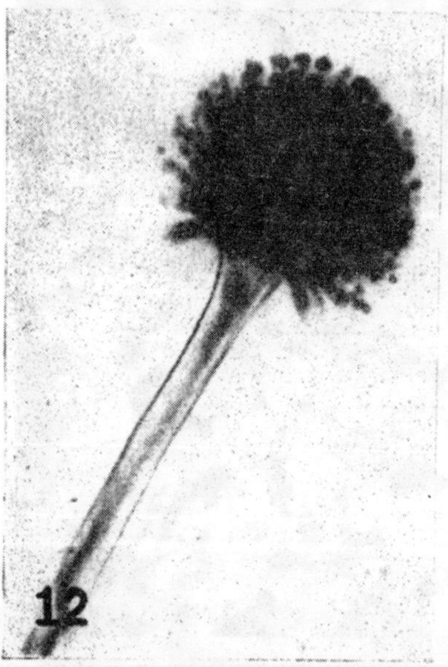

Fig. 10 : *Aspergillus flavus*, culture mount, isolated from the material in Fig. 7.

Fig. 11 : Culture of *Aspergillus niger*, 4 days old, isolated from a case of generalised aspergillosis, following cholecystectomy.

Fig. 12 & 13 : *Aspergillus fumigatus* (organism of mycetoma) - culture mount abscess formation, and haziness of the orbit and paranasal sinuses radiologically. Complicating osteomyelitis with necrosis and sequestration, thrombangitis, penetration to a fatal end. Unfortunately, many cases are not correctly diagnosed and treated until it is too late, for the symptoms develop slowly. If the central nervous system is infected by the blood stream, the course is more rapid and ocular lesions may be secondary.

3. GENERALISED ASPERGILLOSIS

The respiratory system usually serves as the portal of entry. Generalisation is almost always related to **corticosteroid or cytostatic treatment or severe debilitating conditions like leukaemia or malignant tumours**. Aspergillosis is, however, not as frequently associated with diabetes as phycomycosis. Lesions consist in small miliary nodules or larger greyish or pinkish foci, or occasionally in abscesses. The *brain may* show haemorrhagic necrosis. *The endocardium* shows large friable vegetations, if it is involved. Septic thrombi and emboli may cause multiple infarcts of spleen, kidney and other organs.

4. GASTROINTESTINAL ASPERGILLOSIS

This form is rare. Apart from metastasising generalised lesions to the gastrointestinal tract,

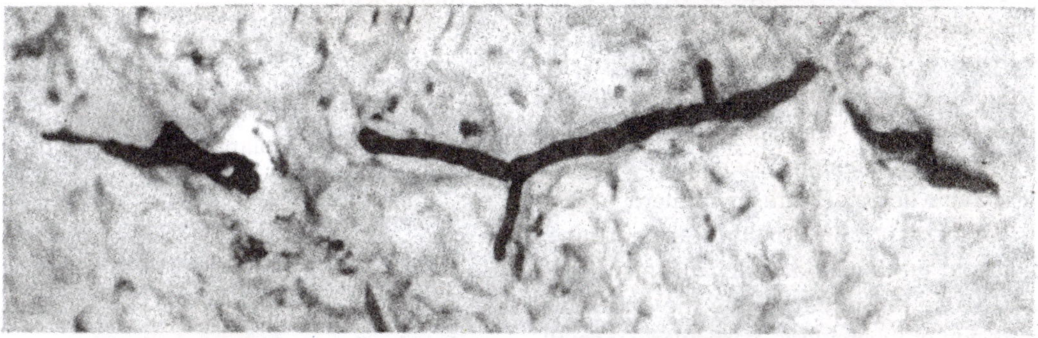

Fig. 25.3 : Hyphae of Aspergillus. Silver methenamine; × 1,000 (approx.)

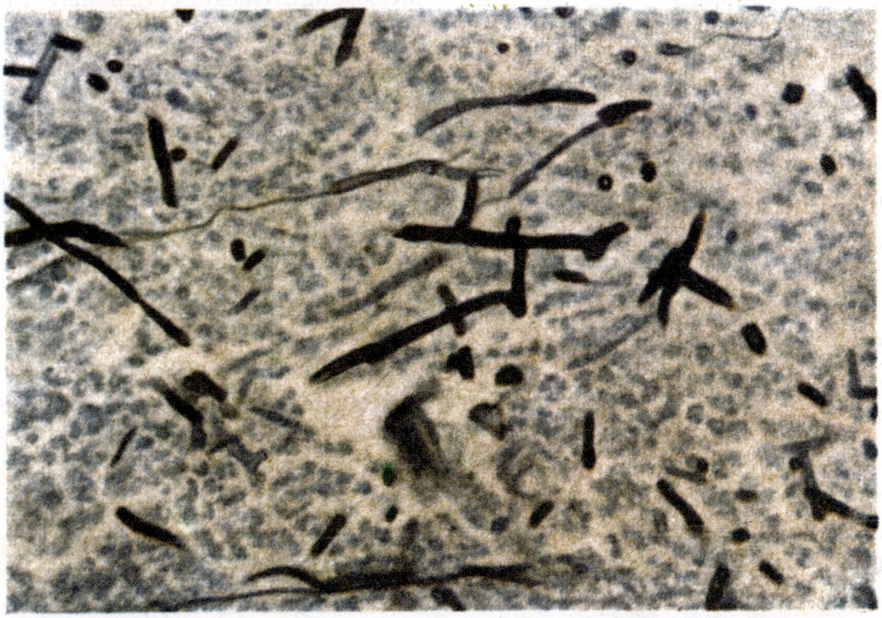

Hyphomycotic lung. Histological appearance of pulmonary aspergillosis. Hyphae show parallel walls, septation and dichotomous branching often showing ballooning. Silver methenamine; × 400

there may be primary localised necrotising, purulent and ulcerated foci in the mucosa of the digestive tract. Peptic ulcer and other forms of ulcers as well as diarrhoea are predisposing factors.

OTHER FORMS OF ASPERGILLOSIS

Allergic reactions to Aspergillus, like farmers lung, are not associated with tissue invasion and cannot be considered here.

Histopathology

A.TISSUE RESPONSE consists either of a granulomatous reaction (**Pl. 25.I, Fig. 3**) predominantly of the foreign body type, or of necrotising lesions (**Figs. 5 and 6**). There may be a spectrum of appearances with both tissue reactions represented in various proportions. In primary aspergillosis, the granulomatous reaction often predominates and there may even be considerable fibrosis. There are numerous giant cells of bizarre and variable appearance. Epithelioid reaction and Langhans' giant cells may also be present and give the deceptive impression of tuberculosis, particularly if small areas of necrosis are interspersed. However, the necrosis usually has not a caseous quality. The meninges may show **granulomatous reaction with numerous giant cells**. In the brain, necrosis, abscess formation and scattered giant cells may be seen. Part of the necrosis may be due to multiple small infarct, caused by septic thrombangitis.

B. HYPHAE IN TISSUE SECTIONS : In haematoxylin-eosin stained sections, hyphae are often recognised only very faintly or not at all (**Pl. 25.I, Fig. 3 & Pl. 25.II, Fig. 7**). Silver methenamine or PAS stain are, therefore, required to demonstrate hyphae.

Aspergillus and other *hyphomyces* hyphae in tissue sections are about 4 microns in diameter, but may show considerably dilated degeneration forms up to 20 microns in diameter. Hyphae show parallel walls, septation and dichotomous branching with relatively regular arrangement. *Phycomyces* hyphac, in contrasts, are very variable in width (8-25 microns), frequently have a disorderly arrangement and show no septa.

In necrotic tissue, **hyphomyces hyphae are** usually **better seen than in granulation tissue** and may easily be **seen even with H & E stain.**

Treatment

3 lines of treatment exist

1. **ANTI-ASTHMATIC** - Corticosteroids, etc.

2. **PULMONARY ASPERGILLOMA** - Surgical removal

3. **INVASIVE ASPERGILLOMA** - A. Amphotericin B, B. Others : (i) Itraconazole, (ii) Rifampicin.

Summary

PATHOGENEISIS

In the lungs, alveolar macrophages are able to engulf and destroy the conidia. However, macrophages from corticosteroid-treated animals or immuno-compromised patients have a diminished ability to contain the inoculum. In the lung, conidia swell and germinate to produce hyphae that have a tendency to invade preexisting cavities (aspergilloma or fungus ball) or blood vessels.

CLINICAL FINDINGS

I. Allergic Forms

In some stopic individuals, development of **IgE anbtibodies to the surface antigens of aspergillus conidia** elicits an immediate asthmatic reaction upon subsequent exposure.

In others, the conidia germinate and hyphae colonise he bronchial tree without lung involvement. This phenomenon is characteristic of **allergic bronchopulmonary aspergillosis** asthma, recurrent chest infiltrates, eosinophilia, and both type I (immediate) and type III (Arthus) skin test hypersensitivity to aspergillus antigen. Many patients shows sputum full of aspergillus, and the **serum precipitins**. They have difficulty in breathing and develop permanent pulmonary scarring. On exposure to massive doses of conidia extrinsic **allergic alveolitis develops**.

II. Aspergilloma and Extrapulmonary Colonization

Aspergilloma occurs when inhaled conidia enter an existing cavity, germinate, and produce abundant hyphae in the abnormal pulmonary space. Patients with pulmonary cavitary disease, (e.g., tuberculosis, sarcoidosis, emphysema) are at risk. Some patients are asymptomatic; others develop cough, dyspnea, weight loss, fatigue, and haemoptysis. Gases of **aspergilloma** rarely become invasive. Localized, noninvasive infections (colonisation) by Aspergillus species may involve the nasal sinuses, auditory canal, cornea, and the nails.

III. Invasive Aspergillosis

Following inhalation and germination of the conidia, invasive disease develops as an acute pneumonic process with or without dissemination. Patients at risk are those with lymphocytic or myelogenous leukemia, lymphoma, bone marrow transplant recipients, and especially individuals taking corticosteroids.

Clinically, Symptoms include fever, cough, dyspnea, and hemoptysis. Hyphae invade the lumens and walls of blood vessels, causing thrombosis, infarction, and necrosis. From the lungs, the disease may spread to the gastrointestinal tract, kidney, liver, brain, or other organs, producing abscesses and necrotic lesions. Without rapid treatment, the prognosis for patients with invasive aspergillosis is grave. Persons with less compromising underlying disease may develop chronic necrotising pulmonary aspergillosis, which is a milder disease.

Diagnostic Tests

1. Specimens

Sputum, other respiratory specimens, or lung biopsy tissue provide good specimens. Blood samples are not helpful.

2. Microscopic Examination

On direct examination of sputum with KOH or calcofluor white or in histologic sections, the hyphae of Aspergillus species are hyaline, septate, uniform in width (about 4 um), and branch dichotomously.

3. Culture

Aspergillus species grow within a few days on most media at room temperature. Species are identified according to the morphology of their conidial structures.

4. Serology

The ID test for precipitins to A fumigatus is positive in over 80% of patients with **aspergilloma** and **allergic forms of aspergillosis**, antibody tests are not helpful.

REFERENCES

Bhatina, V. N. and Mohapatra, L. N. (1969). Experimental Aspergillosis in mice, Part I: Pathogenic potential of *Aspergillus fumigatus, Aspergillus flavus* and *Aspergillus niger. Mycosen* **12**, 651-654.

Bhatia, V. N. and Mohapatra, L. N. (1970). Experimental aspergillosis in mice, Part II: Enhanced susceptibility of the cortisone treated mice to infection with *Aspergillus fumigatus, Aspergillus flavus* and *Aspergillus niger. Mycosen* **13**, 105-110.

Chitnis, V. R. and Deshpande, C. K. (1967). Disseminated aspergillosis, a case report. *Jour. post. grad. Med.* **13**, 131.

Gupta, I. M. and Viswanathan, R. (1963). Secondary aspergilloma. *J. ind. med. Assoc.* **40**, 514.

Mittal, O. P., Agarwala, Mi. C., Samuel, K. C., Prasad, R. and Singh, K. N. (1968). Aspergillus mycetoma in an old tuberculous cavity. *Ind. Journ. Surg.* **30**, 539-543.

Mohapatra, L. N. (1969). Systemic mycoses in India. *Bull. A. I. I. M. S.*, **3**, 7-19.

Randhawa, H. S. *et al.* Quoted by Mohapatra (1969).

Rapper, K. B. and Fennell, D. I. "The genus *Aspergillus*" Williams and Wilkins Co., Baltimore, 1965.

Reddy, D. J., Srinivasa Rao, K., Indira, C. and Showramma, A. (1965). Pulmonary and gastric aspergillosis. *Journ. ind. med. Assoc.* 45, 384-386.

Reinhardt, K. "*Das Mycetom.*" Ferdinand Enke Verl. Stuttgart, 1967.

Sandhu, D. K., Sharma, V. N., Sandhu, R. S., Damodaran, V. N. and Randhawa, H. S. (1966). Bronchopulmonary aspergilloma *Ind. Jour. chest. Dis.* **8**, 198-204.

26

Opportunistic Fungi Pneumocystis Carinii

Pneumocystis carinii causes Plasma cell pneumonia; in immunocompromised patients it is often seening dissemination is rare. Until recently, P. carinii was though to be a protozoan, but molecular biologic studies have proved that it is a fungus with a close relationship to Ascomycotina. P. carinii is present in the lungs of mammals (cats, dogs, ferrets, mice, rats, rabbits) but rarely causes disease unless the host is immunosuppressed. Prior to the introduction of effective chemoprohylactic regimens, it was a major cause of death among AIDS patients. Chemoprophylaxis has resulted in a dramatic decrease in the incidence of *P. carinii* pneumonia with increase in infections primarily in the bone marrow, lymph nodes and spleen.

The morphological forms of P. carinii are :

(1) thin-walled **TROPHOZOITES** and (2) cysts, which are thick-walled, spherical to elliptifal (4-6 um), and contain four to eight nuclei. **CYSTS ARE STAINED WITH SILVER STAIN (GMS) TOLUIDINE BLUE**, and **CALCOFLUOR WHITE**. In most clinical specimens, the trophozoites and cysts are present in a tight mass which reflects the mode of growth in the host.

P. carinii is an extracellular pathogen. Pulmonary growth is limited to the **SURFACTANT LAYER ABOVE ALVEOLAR EPITHELIUM**. It cause infiltration of the alveolar speces with plasma cells called **INTERSTITIAL PLASMA CELL PNEUMONIA.**

PLASMA CELLS ARE ABSENT IN AIDS-RELATED *P.CARINII* PNEUMONIA

Clinically, cyanosis is not uncommon.

Diagnostic / Investigation of *P. carinii* (pneumonia) includes specimens of broncho-alveolar lavage, sputum examination and lung biopsy. Stained sputum samples show the presence of **CYSTS AND TROPHOZOITES**. Other stains are Giemsa, toluidine blue, Gomoni's methenamine silver, and calcofluor white.

A SPECIFIC MONOCLONAL ANTIBODY is available for **direct fluorescent examination** of specimens. ***P.carinii*** cannot be cultured in artificial culture media. Serology to establish the prevalence of infection is done for academic interest.

Only in immunosuppression and in immunocompromised subjects, ***P. carinii*** causes the pneumonic disease. According to the serological evidence most of the individuals are infected in **early childhood. Cell-mediated immunity** (CMI) offers resistance to the infection. In AIDS patients often have significiant antibody titres and ***P. carinii*** pneumonia is manifested when the **CD4 + lymphocyte** count drops below 400/uL (cu. micron).

Acute pneumocystis pneumonia is treated with TMP-SMZ (trimethoprim-sulfamethoxazole) or TMS and/or pentamidine isethionate. Prophylaxis can be achieved with dailty TMP-SMZ and/or pentamidine inhaler (aerosolised).

P. carinii may be an obligate member of the normal flora. Persons at risk are provided with chemoprophylaxis. The mode of infection is not clear.

OTHER OPPORTUNISTIC MYCOSES

Opportunistic mycoses are **candidiasis, aspergillosis** and **mucormycosis**. But growing numbers of highly compromised patients are affected by pathogenic opportunistic fungi. Most disastrous infections are caused by Alternaria, Bipolaris, Curvalaria, Fusarium, Paecilomyces and species of many other fungi. Increasing use of antifungal antibiotics is a great contributory factor leads to the selection of resistant fungal species and strains as well.

REFFERENCES

Ampel NM (1996) : Emerging diseases issues and fungal pathogens associated with HIV infection. *Emerg Infect Dis.* **2**, 109.

Know-Chung Kj, Bennett JE : *Medical Mycology.* Lea & Febiger, 1992.

Masur H (1992) : Prevention and treatment of pneumocystis pneumonia. *N Eng J Med.* **327**, 1853.

Minamoto GY (1997) : Rosenberg AS : Fungal infections in patients with acquired immunodeficiency syndrome. *Med Clin North Am* **81**, 381.

Rippon JW : *Medical Mycology* : *The Pathogenic Fungi and the Pathogenic Actinomycetes*, 3rd ed. Saunders, 1988.

Smulian AG, Walzer PD (1992) : The biology of *Pneumocystis carinii*. ***Crit. Rev. Microbiol.*** **18**, 191.

Su TH. Martin WJ, II (1994) Pathogenesis and host response in *Pneumocystis carinii* pneumonia. *Annu, Rev. Med.* **45**, 261.

Wheat LJ : Endemic mycoses in AIDS (1995) : A clinical review. Clin Microbiol Rev. **8**, 146.

27

Cladosporiosis

(Systemic Mycosis)

Definition and Aetiology

A relatively rare mycosis caused by ***Cladosporium trichoides*** *(bantianum)* causing lesions of the central nervous system which are fatal within a few months if they cannot be excised surgically.

C. trichoides **is a neurotropic fungus,** though cutaneous abscesses have been reported. It is closely related to ***C. carrionii***, an agent of **Chromoblastomycosis**.

Badar (1965) mentioned eight mycologically proved cases of cerebral cladosporiosis already recognised in different parts of the world. Meanwhile, more cases were recorded,

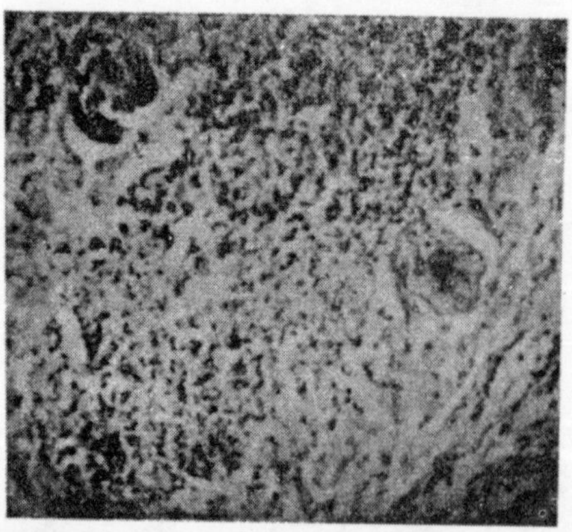

Fig. 27.1 : Abscess with central necrotic portion with cellular debris and leucocytes. Wall was poorly defined lined by numerous multinucleated giant cells and histiocytes. H & E; × 320

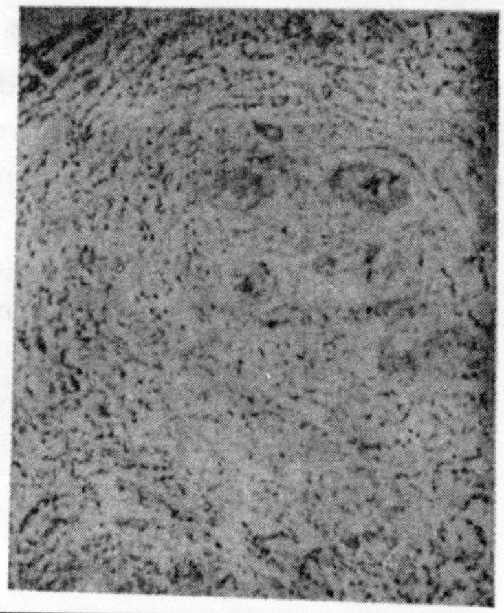

Fig. 27.2 : Gliosis of the surrounding area with thin walled blood vessels H & E; × 180

including two instances from India (Bagchi *et al.*, 1962; and Desai *et al.*, 1966).

The infection is supposed to gain entrance through the lungs.

Pathology

Lesions consist of **circumscribed, firm masses of granulation tissue in one of the hemispheres** of the cerebrum. They may be of several centimetres in diameter and some were easily shelled out. ***The cut surface*** shows a greyish appearance. Haemorrhage, circumscribed necrotic areas, or purulent tissue destruction may be seen. **The lesion may be mistaken for an abscess.**

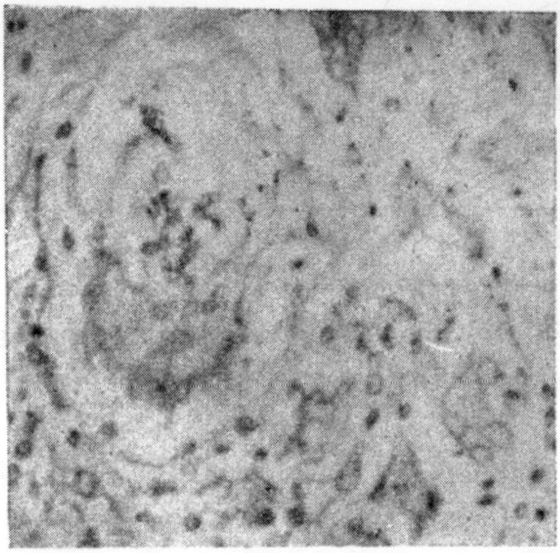

Fig. 27.3 : Granulomatous lesion of the brain produced by ***Cladosporium trichoides***. Periodic acid Schiff stain : PAS positive hyphae along with budding forms inside and outside the giant cells.
(Courtesy of the senior author Dr. A. Bagchi, *J. ind. med. Assoc.* 38(11): 602, 1962)

Clinical symptoms are related to the site of lesions and consist in **epileptiform seizures, headache, vomiting, and fever**.

HISTOPATHOLOGY

The picture is characterised by presence of microabscesses, with granulomatous tissue in the periphery consisting of numerous multinuclear giant cells, histiocytes and epithelioid granulomas. The inflammatory infiltrate consists of polymorphs and plasma cells. The periphery shows gliosis. Within the giant cells or in the pus there are hyphal or spherical forms of fungus with relatively short segments, not unlike candida. But the fungus in the non-necrotic areas posesses a strongly pigmented brown wall. Cultural differentiation from other pigmented fungi which may also cause brain lesions, is therefore, necessary.

Excellent microphotographs of gross and microscopic characteristics of cladosporiosis may be referred to in Emmons *et al.* (1964).

Desai (1968) has pointed out that **three different tissue phases of *C. trichoides*** may be seen:

1. **Budding chlamydospores in giant cells.**
2. **Dividing, branching mycelia in** the **abscess cavity** and,
3. **Buding spores of echinucleate bodies** as described by Symmers.

CLADOSPORIUM MANSONI

Taxonomy

The fungus has been placed in the genus *Cladosporium or Torula*. Dodge (1985) placed it in the genus Dematium Person, 1974, but the latter name is not followed by other mycologists.

The genus: ***Cladosporium* Link, 1816 belongs to the family: *Cladosporium* Soccardo, 1886; suborder: Blastosporinae; order: Thallosporales, the class: Fungi Imperfecti (Deutero Mycosis).**

For Cladosposium werneckii Horta, 1921, see Emmons page 127.

REFERENCES

Bagchi, A., Aikat, B. K. and Barua, D. (1962). Granulomatous lesion of the brain produced by *Cladosporium trichoides*. *Journ. ind. med. Assoc.* **38**, 602-604.

Desai, S. C., Bhatikar, M. L. and Mehta, R. S. (1966). Cerebral Chromoblastomycosis *Cladosporium trichoides* (bantianum), part II (mycopathologic lesions of brain and skin involvement). *Neurology, India* **14**, 6-18.

28

ZYGOMYCOSIS (Phycomycetes)

(Including Mucormycosis Subcutaneous/Systemic & Opportunistic Mycosis)

Infections with Zygomycosis (phycomycetes) are generally caused by the following organisms :

Zygomycosis (Phycomycetes)	
Families	**Genera**
(a) MUCORACEAE	ABSIDIA MUCOR RHIZOPUS
(b) ENTOMOPH-THORACEAE	BASIDIOBOLYS ENTOMOPHTHORA
(c) MORTIERELLACEAE	MORTIERELIA
CLASSIFICATION UNCERTAIN	*Hyphomyces destruens*

Clinically, there are at least three distinct manifestations of phycomycosis: (1) Systemic **Zygomycosis (Phycomycosis)**. (2) Subcutaneous **Zygomycosis (Phycomycosis)**. (3) **Rhino-entromophthoromycosis**.

Details about characteristic difference of the three types of phycomycosis are summarised in the table 27.1.

SYSTEMIC PHYCOMYCOSIS

(MUCORMYCOSIS)

History : A case of pulmonary phycomycosis is first reported in 1874 by Sluyter followed by Paltauf's classical description of a gastro-intestinal phycomycosis (1885) under the name of "mycosis mucorina". Since then a few sporadic cases have been seen in all parts of the world, but the majority of them have been published only in recent years. Straatsma *et al.* (1962) collected 183 cases from the literature, including presentation of 33 new cases from the records of the AFIP in Washington.

Aetiology : The historical term "mucormycosis" has been replaced in recent years by "**systemic Zygomycosis (phycomycosis)**" because species of the **genus "mucor"** are only occasionally involved. Most frequently *Absidia corymbifera* or Rhizopus are found to be the aetiological agents. **These organisms are ubiquitous and frequently present in decaying vegetation, soil and dung.**

Distribution : The infection has no geographic limitation but has been more frequently reported from areas where good laboratory facilities exist and from centres in which cytostatic treatment is used and where debilitating diseases accumulate.

Cases from India include a pulmonary mucormycosis diagnosed at post mortem examination in a 5-year old boy (Balasubrahmanyam and Chaudhury 1963, and a rhino-cerebral phycomycosis due to *Rhizopus oryzae* in a 16-year old male with uncontrolled diabetes (Gover *et al.*, 1966). Grueber has seen a post mortem case of **intestinal "mucormycosis"** in Ludhiana in a 42-year old man. A similar case was reported by Madhavan and Reddy

PLATE 28.I

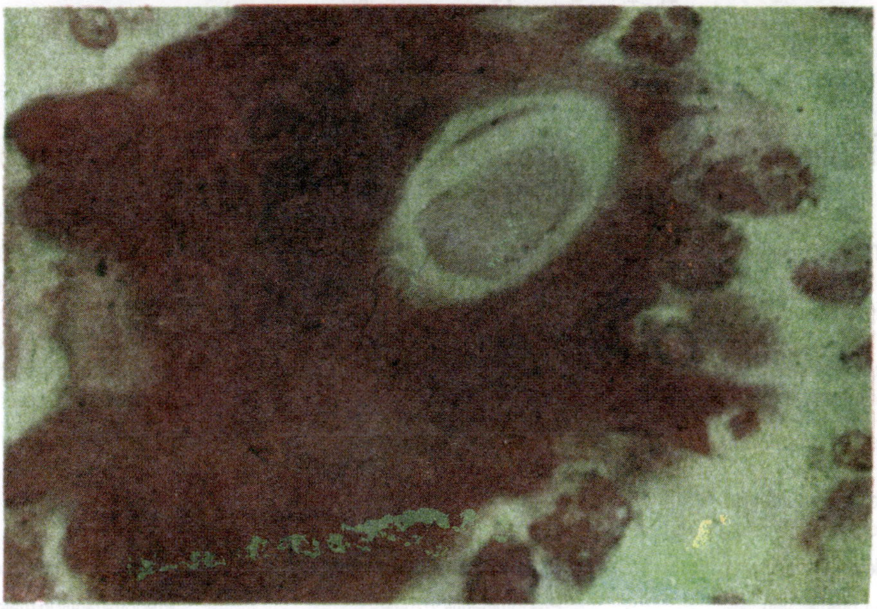

Subcutaneous phycomycetes. Note eosinophilic perihyphal space of phycomyces hyphae. PAS; × 1000

(1969). Another case of **systemic phycomycosis** in a 45-year old diabetic woman has just been documented in our laboratory. The patient had a large rapidly developing coin lesion in her left lung (**Pl. 28.I, upper left**) and rhino-ocular cerebral lesions. Rhizopus sp. was isolated from her sputum, from a purulent lesion in her hard palate and was shown to be present histologically in necrotic biopsy material from her paranasal sinuses (see **Pl. 28.I**, Figs. upper right & lower ones).

One case was clinically diagnosed by Dey (1986) around Kolkata area.

Epidemiology

In systemic Zygomycosis (phycomycosis) there appears to be no predilection of sex or race, since it is **primarily and "opportunistic" infection**, which is linked **with** an **impaired immune mechanism**, caused by debilitating conditions. Most commonly the patients suffer **from severe, uncontrolled diabetes**, from **malignant tumours, leukaemia, malnutrition, chronic renal or hepatic disease, or extensive burns**. It appears that acidosis which as a common denominator in diabetes, uraemia and severe diarrhoea, serves as a pacemaker for systemic phycomycosis. Other important factors are prolonged therapy with corticosteroids, with cystostatic drugs or radiation. Portals of entry are most frequently the lungs, but also the gastro-intestinal tract, the orbital conjunctiva, the respiratory mucosa of the nose and the paranasal sinuses. From these locations haematogenous dissemination to the central nervous system is frequent.

Clinical Features

According to the portals of entry, the following clinical types may be distinguished, though one has to be aware that dissemination often occurs rapidly and will not infrequently lead to a fatal end within days.

Landauer and Newcomer (1962) suggested (1) ***craniofacial type*** (also oculocerebral, rhino-oculo-cerebral or faciocerebral type), (2) ***thoracic* (or pulmonary *Zygomycosis* (*phycomycosis*),** (3) ***abdominal-peptic Zygomycosis (phycomycosis)*** **(gastro-intestinal), and (4) a *dermal type*.**

1. ***Cranio-facial*** Zygomycosis (*Phycomycosis) :* This type is associated with severe diabetes in about 3/4 of all cases

Table 28.1 : Diseases Caused By Zygomycosis (Phycomycetes) Charateristic Clinical Differences of The Three Types of Zygomycosis (Phycomycoses)

	Entomophthoromycosis Definition/Infection caused by Entomophthorales in otherwise healthy individuals	
(1) **SYSTEMIC PHYCOMYCOSIS**	(2) **SUBCUTANEOUS ZYGOMYCOSIS (PHYCOMYCOSIS)**	(3) **RHINO-ENTOMOPHTHORO-MYCOSIS** (or rhino-phycomycosis)
DEFINITION : Opportunistic infection caused by mucorales. Occurring in the presence of lowered host resistance, e.g., diabetes, malignancy, cytostatic therapy, etc.	DEFINITION : Subcutaneous infection with entomophthoraceae usually in individuals below the age of twenty	DEFINITION : Infection of the mucosae and the submucous tissue of the oropharyngeal and nasal mucosa and paranasal sinuses with Entomophthora sp. (? only)
(a) Predisposing factors : diabetes, malignancy, cytostatic therapy, uraemia, immunological disorders	In otherwise "healthy" individuals	In otherwise "healthy" individuals
(b) Middle or older age	Usually below the age of 20	Usually above the age of 20
(c) No sex predilection	Predominantly in males	Predominantly in males
(d) Rapid course if untreated	Chronic course, spontaneous regression possible but rare	Chronic course
(e) Almost invariably fatal	Responds to therapy, benign	Usually responds to therapy
(f) Systemic or generalised	Localised in subcutaneous tissue	Localised in respiratory and oral mucosa, lips
(g) Worldwise distribution	Almost exclusively in tropical and subtropical countries	In tropical and subtropical countries
	HISTOLOGICAL DIFFERENCE	
(a) Invasion of blood vessels	Usually no invasion of blood vessels	Usually no invasion of blood vessels
(b) Acute inflammatory response with abscesses and necrosis	Subacute and chronic granulomatous inflammatory response, necrosis possible. Characteristic eosinophilic perihyphal substance, eosinophil and plasma cell infiltrates	Subacute and chronic granulomatous response as in subcutaneous phycomycosis. Foreign body, giant cells, epithelioid cells, eosinophils and plasma cells present.

(Wilson and Plunkett, 1965), and usually accompanied by severe acidosis. Depending on the site of the infection, there may be swelling and necrosis of the mucosae of the paranasal sinuses with clouding and haziness in the Roentgenogram (**Pl. 28.I, Fig. upper left**, showing a corresponding appearance in Aspergillosis). Bony structures, palate, and the wall of the orbit may be involved. Proptosis, ophthalmoplegia, fixation of the pupil and cranial nerve involvement may be found as the disease progresses. Headache, blindness, signs of intracranial pressure develop suddenly and rapidly lead to coma and death. In all such cases ample tissue material from the brain should be taken at the post mortem examination for culture before fixation of the organ.

Plate 28.I A

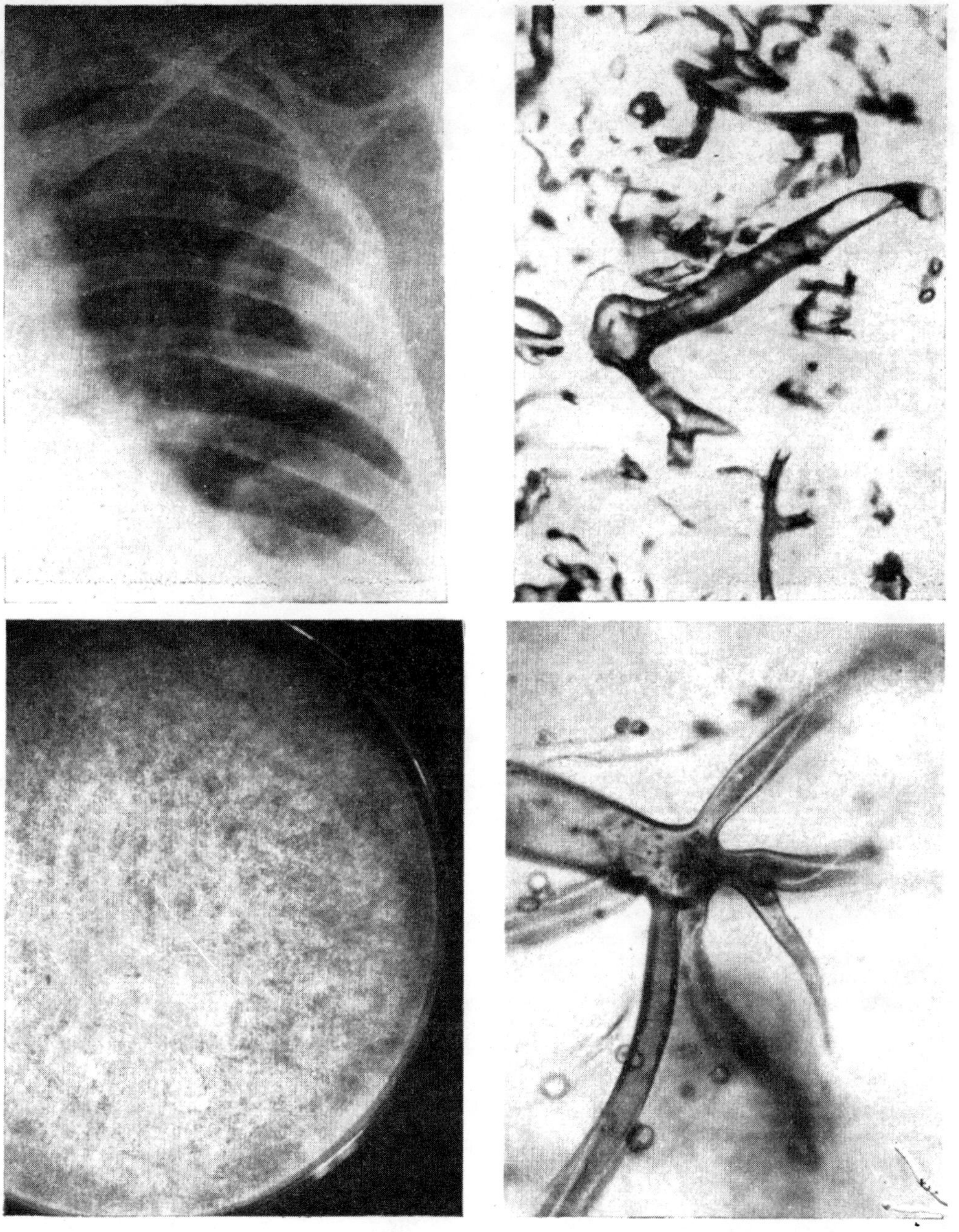

Figs. Upper left to Lower right are all the same case, a 45 year old diabetic woman with pulmonary and rhino-cerebral phycomycosis.

Figs. Upper left — Rapidly spreading pulmonary lesion due to phycomycosis with fungus ball in centre

Figs. Upper right — Phycomyces (*Rhizopus* sp.) hyphae in necrotic tissue from maxillary sinus. Weigert's H & E; ca. 520X

Figs. Lower left—5-day old *Rhizopus* colony on Sabouraud's agar, isolate from a destructive lesion of the hard plate.

Figs. Lower right — Culture mount of *Rhizopus* sp. showing diagnostic rhizoids; ca. 520X

PLATE 28.II

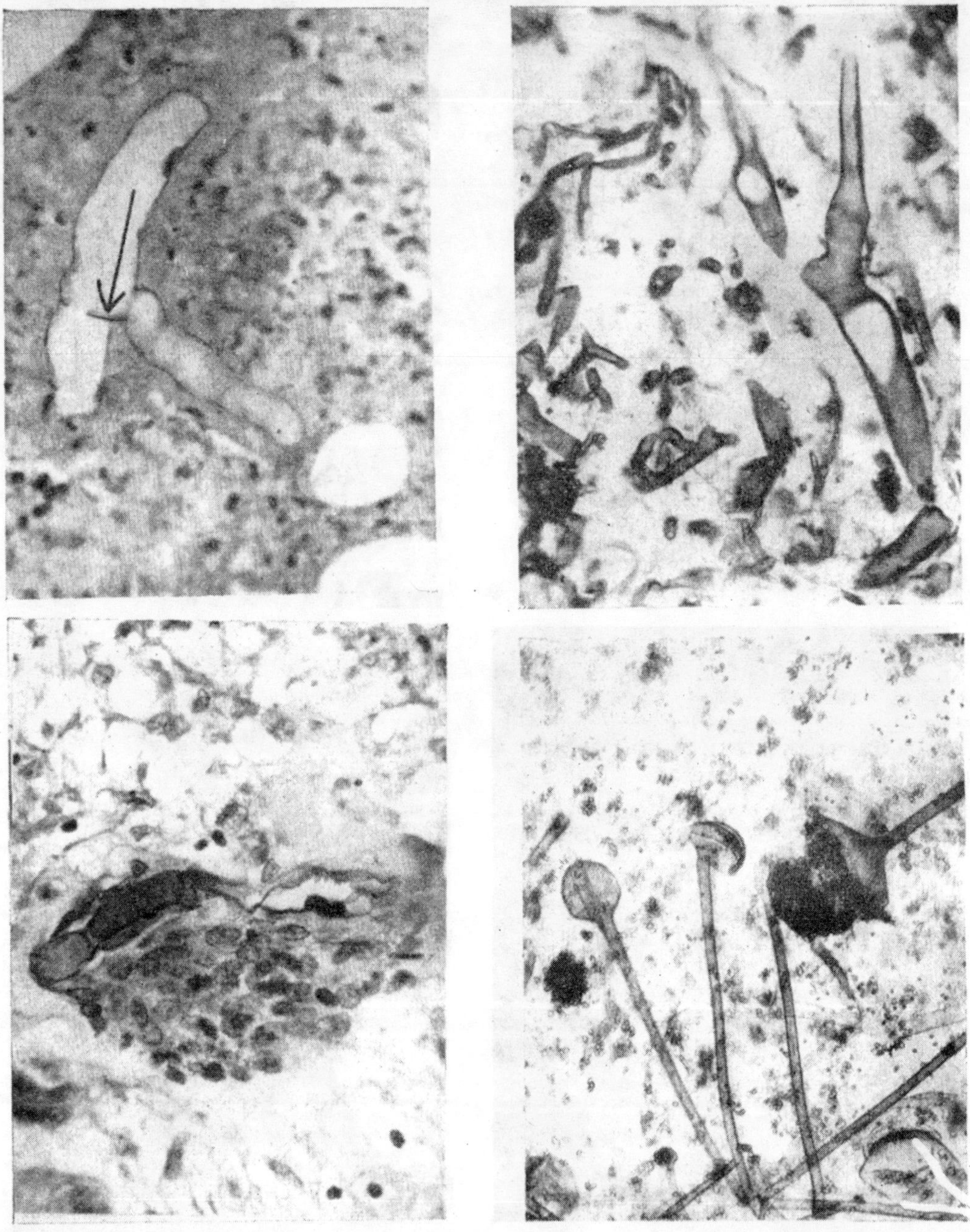

Fig. Upper left — Branching phycomyces hyphae in necrotic tissue. Pseudoseptum (arrow) is due to folding of mycelium. PAS, 520 X

Fig. Upper right — Width of hyphae varies in systemic phycomycosis (mucormycosis). Tissue section H & E, 520 X

Fig. Lower left — Collapsed and crumbling hyphae with adjacent foreign body giant cell (from a case of subcutaneous phycomycosis). PAS; 520 X

Fig. Lower right — Culture mount of *Rhizopus* sp. with ruptured sporangium at right, showing mushroom shaped columella.

Lactophenol cotton blue, 120 X

2. *Thoracic phycomycosis* is caused by inhalation or aspiration of infective material. It resembles clinically a progressive broncho-pneumonia, or isolated pulmonary lesions may rapidly develop (**Pl. 28.I, Fig. upper left**). Occasionally signs of pulmonary embolism and infarction are super imposed due to rapid invasion of blood vessels by the fungus with subsequent thrombosis and necrosis of dependent areas. Pericardial phycomycosis and phycomycotic myocardial infarction have been observed.

3. *Gastro-intestinal phycomycoses*

   They are probably lumen infections in most cases, or they may be caused by haematogenous spread in some. All segments of the **gastrointestinal tract can be involved**. Symptoms are variable and depend on the site of lesions and progress of the disease.

4. *Dermal phycomycosis*

   The skin is very rarely involved in opportunistic "**mucormycosis**". Lesions consist of dermatitis and ulcers and are different from subcutaneous phycomycosis, as already shown in the Table.

Histopathology

The hyphae are usually **easily recognisable in H & E sections**. However, sometimes they are not well identifiable in fresh lesions with cellular granulation tissue. Weigert's Iron Haematoxylin stains hyphae much more distinctly black blue than ordinary haematoxylin and can be of help in their identification. **PAS and silver methenamine stains are helpful but hyphae do not** always **stain well**, probably depending on their vitality.

Characteristics of hyphae in histologic sections : (1) *Variable* width from 8-20 microns (broader than hyphomyces, like aspergillus, **Pl. 28.I, Fig. upper right**).

(2) **Septa are absent** or very rarely seen. However, pseudosepta due to crinkling and folding can resemble septa **Pl. 28.II, Fig. upper left**).

(3) **The walls are fre**quently **not** quite **parallel** and the **width** *varies* from **small tubular to broad flask-shaped or club-like structures (Pl. 28.II, Fig. upper right**).

(4) Hyphae are frequently folded, collapsed, 'crumpled' together in contrast to hyphomyces (aspergillus, etc.) in which these phenomena are less conspicuous or absent (**Pl. 28.II, Fig. lower left**).

(5) **Eosinophilic perihyphal substance is absent in rapidly progressive cases in contrast to subacute and chronic cases of subcutaneous Zygomycosis (phycomycosis).**

TISSUE REACTIONS

There are three main characteristic features:

(1) **Polymorphonuclear reaction with pus formation,**

(2) **necrosis and**

(3) **rapid vascular invasion with thrombosis and subsequent secondary necrosis.** Eosinophils and plasma cells as well as giant cells may be present, but are not a characteristic feature of "**mucormycosis**". Successful therapy may alter the "typical" histological picture considerably and it may shift toward a granulomatous reaction.

Culture

Growth of Zygomycosis (phycomyces) sp. must be interpreted with caution. Patients with bronchiectasis may cough up spores of these fungi for several days after exposure to them in a dusty environment (Emmons *et al.*, 1963). Culture plates are also not infrequently contaminated by the ubiquitous spores or Zygomycosis (phycomycetes).

Systemic phycomycosis is most commonly caused by the two genera ***Rhizopus*** and ***Absidia***; less frequently by *Mucor*. All three grow rapidly **at 26 to 37°C on Sabouraud's medium** and fill a 10 cm diameter Petri dish **within 4-5 days**, first growing in a white, fluffy network on

PLATE 28.III

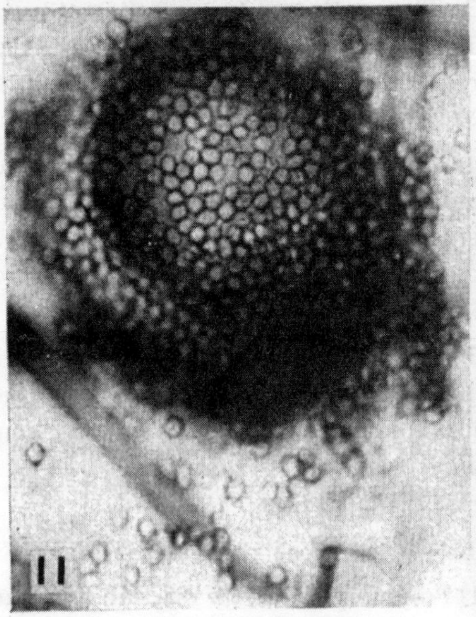

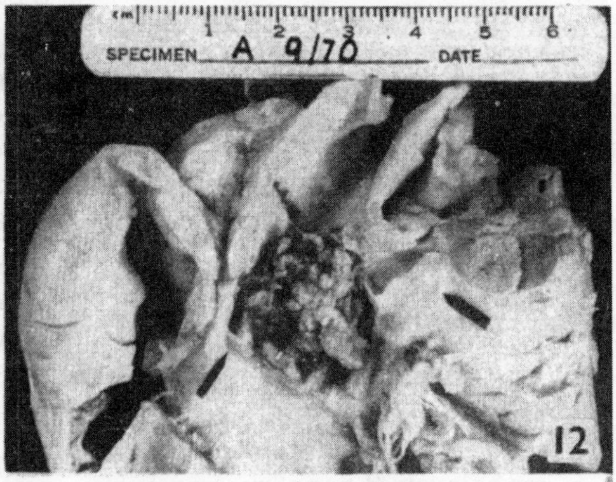

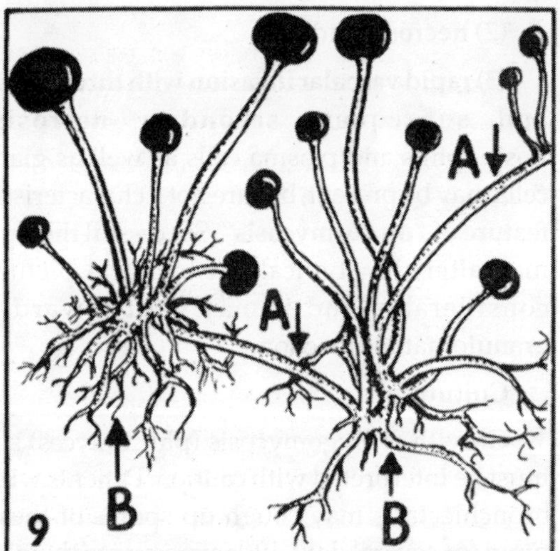

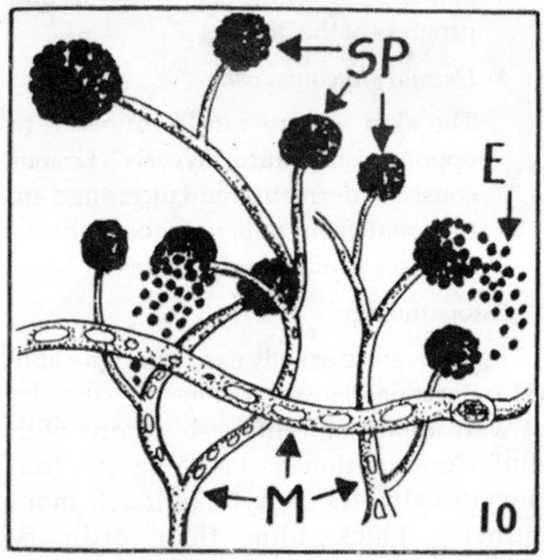

Fig. 9 : ***Rhizopus*** sp. : A = Stolons (runners); B = Rhizoids

Fig. 10 : ***Mucor*** sp. : Sp = Sporangia; E = Ruptured sporangium

Fig. 11 : Culture mount of ***Rhizopus*** sp. showing sporangiospores 520 X

Fig. 12 : Zygomycosis (Phycomycotic) endocarditis. This case is from a dog with cusp replacement of the aortic valve with a formalin preserved homograft cusp. The dog died suddenly 3 weeks after operation, a post-mortem finding. Both arrows pointing to healthy cusps.

hyphae (**Pl. 28.I, Fig. lower left**) which may turn grey after the 5th day due to the sporangia which can be recognised as tiny black dots. Some species are light brown or greyish from the beginning. The diagnostic structures of *Rhizopus* and *Absidia* are called rhizoids, resembling branching roots (**Pl. 28.I, Fig. lower right and Pl. 28.III, Fig. 9**). Long sporangiophores branch from the stolons (runners), originating opposite the rhizoids in *Rhizopus* (**Pl. 28.III, Fig. 9**), but between the rhizoids in ***Absidia. Mucor*** has **no rhizoids but sporangiophores** similar to those of the other two genera (**Pl. 28.III, Fig. 10**). The sporangia of all three genera consist of globular or pear shaped, easily ruptured structures, containing sporangiospores (**Pl. 28.III, Fig. 11**). After dehiscence of the spores, the columella of Rhizopus are mushroom-shaped and rest on a cup-shaped expansion called apophysis (see **Pl. 28.II, Fig. lower right**).

For a more detailed description of the morphology, the book "Medical Mycology" by Emmons, Binford and Utz may be consulted.

Therapy

Most important is the treatment of the underlying condition, usually diabetes.

Management includes supportive measures and if possible surgical excision of necrotic or infected tissue.

Antibacterial antibiotics and steroids should be avoided.

Previously, potassium iodide therapy in high doses was given with little success. To **Amphotericin therapy, systemic phycomycosis responds favourably**. It may be supported by potassium iodide.

By such measure life may be prolonged, but most patients cannot be saved.

SUBCUTANEOUS ZYGOMYCOSIS (PHYCOMYCOSIS) AND RHINO-ENTOMOPHTHOROMYCOSIS

Both infections are closely related (for details see Table). The matter has been discussed in more detail by Clark (1968) and by Grueber (1969).

In 1965, Lie and co-workers described for the first time a distinct subcutaneous mycosis in Indonesian children, mostly boys. Until 1961 more cases were seen in Indonesia, but soon further evidence was found of this disease in the Cameroons, Sudan, Uganda, Ghana, Senegal, Ivory Coast, Nigeria and also in India. (For complete literature, see Clark, 1968). All eight Indian cases up to 1968 have been summarised by Tunnell *et al.* (1968). Since then, another case has been diagnosed in a 5 year old boy from Chhatarpur, M.P. (Grueber, 1969).

Only Symmers (1962) found evidence of this infection in temperate climates, detecting a phycomycosis in a 13 year old girl in London. The patient, had, however, probably acquired the infection in Indonesia. Later, however, Symmers (1966) observed a second case of phycomycosis in a patient who had never been out of Britain. This probably represents the only genuine case reported from a country with a temperate climate. Bridges *et al.* (1961, 1962) saw similar infections in horses in Texas.

Epidemiology

The infection appears to gain entrance through small lesions of the skin. The most likely mode of infection in subcutaneous phycomycosis is that soil infected with droppings from lizards or insects is rubbed into the skin of persons squatting or lying on the ground. The overwhelming majority of the patients with subcutaneous phycomycosis are males below the age of 20, but Entomo-phthoromycosis mostly affects men over 20 years of age.

SUBCUTANEOUS ZYGOMYCOSIS (PHYCOMYCOSIS)

Clinical features : Lesions consist in hard swellings of the subcutaneous tissue of the buttocks, scrotum, legs (**Pl. 28.V Fig. upper left**), arms, and also of the chest, shoulders and, less frequently, the face. The overlying skin is smooth and does not ulcerate (unless there is secondary

PLATE 28.IV

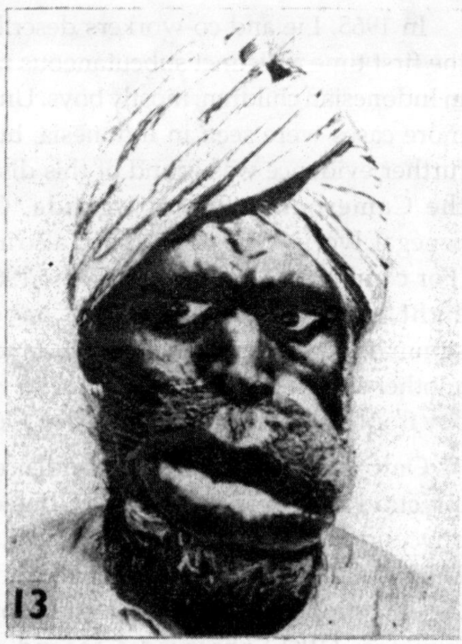

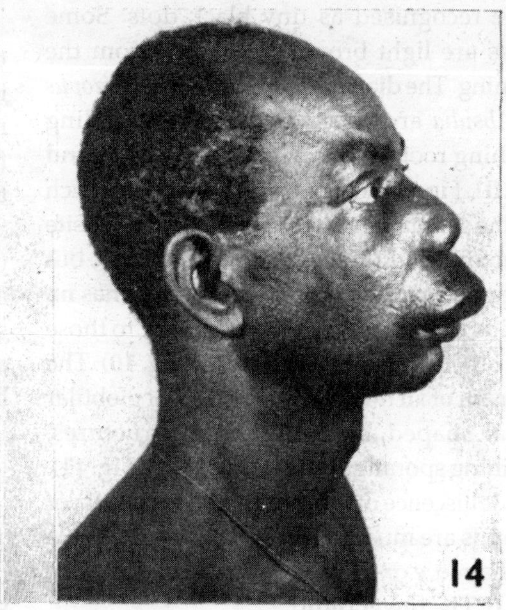

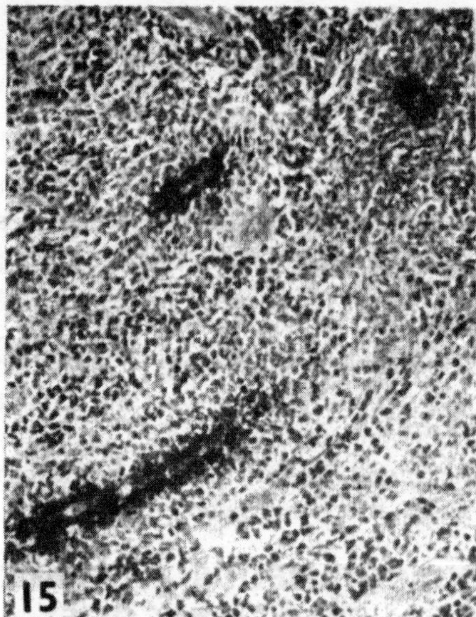

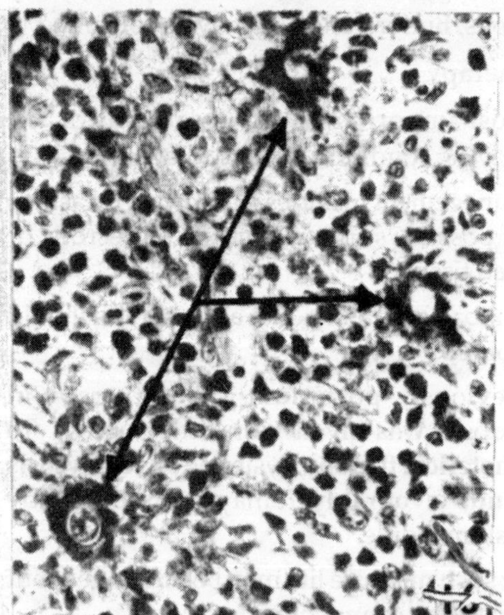

Fig. 13 : Subcutaneous Zygomycosis (phycomycosis) of lips (From *J.C.M. Assoc.* 1. 41; 284, 1966)

Fig. 14 : Nigerian patient suffering from Rhino-entomophthoromycosis (Courtesy Dr. B. M. Clark and Prof. D. D. Martinson, Univ. Coll. Hosp. Ibadan, Nigeria).

Fig. 15 : Longitudinal sections of **Zygomycosis** (*phycomyces)* hyphae in tissue; from the case in Fig. 13 H & E; 140 X

Fig. 16 : Transverse sections of **Zygomycosis** (*phycomyces)phycomyces* hyphae in tissue; from the case in Fig. 13 H & E; 560 X

PLATE 28.V

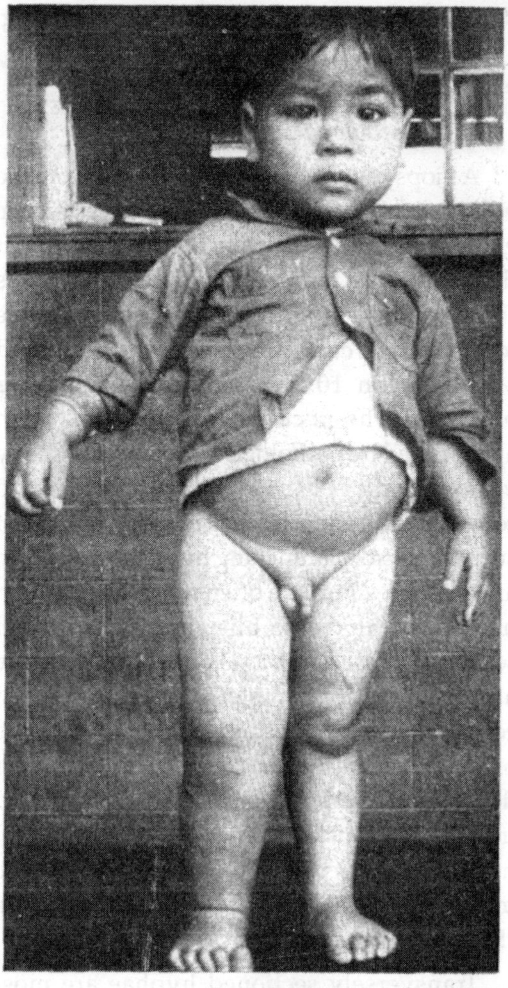

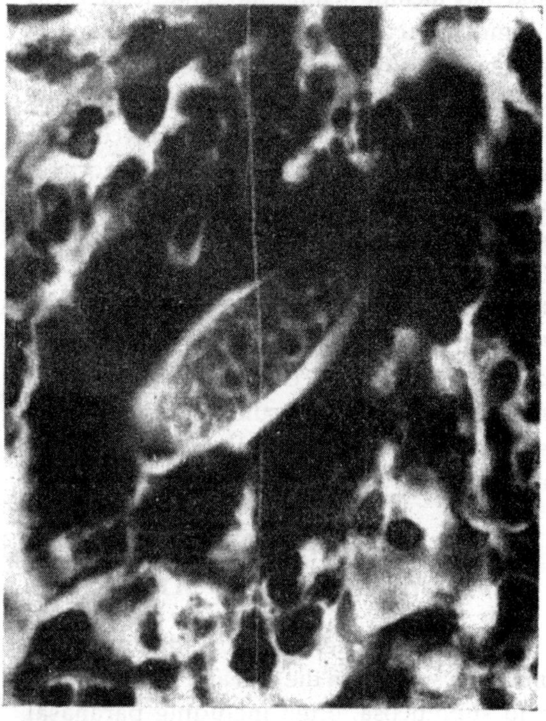

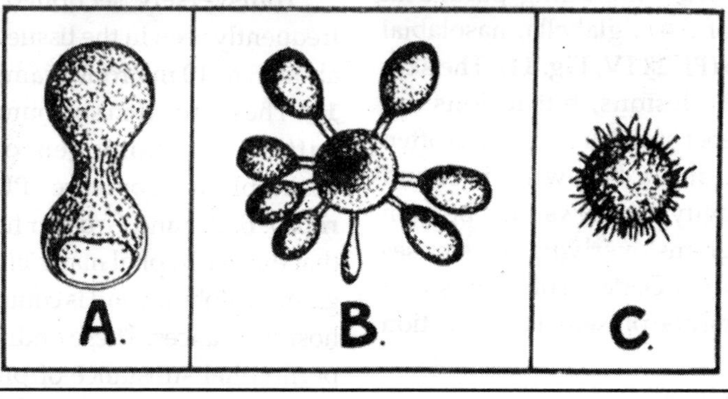

Fig. Upper right —Eosinophilia perihyphal substance in tissue section of subcutaneous phycomycosis. H & E; ca. 900 X

Fig. Lower—Culture mount of conidia of ***Entomophthora coronata*** A—single replican of conidium; B—multiple replication of **conidium** with formation of **conidiophores**, bearing secondary conidia.; C—**conidium** with hair-like radiating appendages (degeneration form?)

Fig. Upper left—Subcutaneous phycomycosis in a one year old boy (from N. Tunnel *et al.* 1968 *h. ind. med. Assoc.* **51**, 355)

infection). The size varies from small nodules of one to two centimeters in diameter to large areas, occasionally **involving adjacent structures like scrotum, neck, and lips**. There is usually a sharp edge between infiltrated and healthy tissue. However, satellite nodules near the edge may indicate slow progression. Subjective complaints are usually relatively few and may include itching, moderate tenderness or mechanical impediment. If the face is involved, disfigurement may be considerable (**Pl. 28.IV, Fig. 13**). The *differential diagnosis* includes elephantiasis, but in phycomycosis the limb distal to the lesion is perfectly normal, which distinguishes the condition from elephantiasis (Burkitt *et al.*, 1962). Histologically, lesions are most often confused with tuberculosis or simply diagnosed as a "granuloma". The case published by Tunnell *et al.* (1968) was unsuccessfully treated for 9 months with antituberculous drugs, until a second biopsy was examined and the true nature of the disease was appreciated.

RHINO-ENTOMOPHTHOROMYCOSIS

This most frequently affects adult men. The infection spreads in the oral and nasal mucosa and submucosa, often including paranasal sinuses, dorsum of nose, glabella, nasolabial fold and upper lip (**Pl. 28.IV, Fig. 14**). The skin is mobile over the lesions, but lesions are immobile over deeper structures. There is often nasal obstruction and a "growth" is found inside the nasal cavity. Shadows may be seen in the Roentgenograms overlying the sinuses or oro-nasal structures. Oedema of the lips, and proptosis of eyes may be present if the infection if widespread.

The differential diagnosis includes Rhinosporidiosis, which usually does not cause so much oedema and firm swelling of adjacent tissue. The infection may be further misinterpreted as a nasal polyposis, fibro-osseous dysplasia, rhinoscleroma, "goundou", a malignant growth or a lymphangioma, depending on the amount of stasis and swelling.

Laboratory Investigations

Histopathology

A biopsy should preferably be taker from the edge of subcutaneous lesions, because the growing fungus in this area shows least distortion. The **biopsy** should be so generous that part of the tissue can be sent in sterile saline for **mycologic culture**. The second part should be placed in 10% formaline solution for histology. The peculiar histology has been appreciated by pathologists in Africa for many years (Burkitt *et al.*, 1962), but they failed to recognise the fungi in the tissue. This is not surprising since branching hyphae are seldom and serial sections are often required to obtain not only transverse but also longitudinal sections of hyphae (**Pl. 28.IV, Fig. 15**). They have a characteristic diameter of 8-25 micron in contrast to **hyphomyces (aspergillus)** which are only four to five microns wide in tissue, expect degenerated forms which may be vesicular and up to 20 microns. Zygomycosis (Phycomyces) hyphae are *nonseptate*, though septation may be mimicked by folding of degenerate hyphae (**Pl. 28.II, Fig. upper left**).

Transversely sectioned hyphae are most frequently seen in the tissue as empty "gaps" of about 8 to 10 microns diameter (**Pl. 28.IV, Fig. 16**). These are often surrounded by an irregular cuffing of homogeneous or granular eosinophilic substance (**Pl. 28.V, Fig. upper right**). Bader and Grueber have recently shown that the perihyphal material absorbs antihuman gamma globulin and is composed of fungal and host substances. These findings suggest that the perihyphal substance of phycomyces may be considered as an immune precipitate; not a "capsult". It is characteristic of subcutaneous phycomycosis and is present as a valuable diagnostic aid. **Hyphae stain pale blue with haematoxylin eosin stain. Silver methenamine and PAS staining methods are helpful** in demonstrating hyphae but staining depends on vitality of the hyphae and other factors.

TISSUE REACTION

The general picture is that of a granulomatous lesion with a variable proportion of eosinophils, plasma cells and areas with foreign body giant cell formation. A scattering of eosinophils is almost always present. Giant cells are found in areas of degeneration of hyphae and may include degenerating, disfigured hyphal remains (**Pl. 28.II, Fig. lower left**).

Grueber has not found extensive necrosis in lesions of subcutaneous phycomycosis, though the picture may vary considerably and include areas of necrosis. The edges or necrotic areas should be searched for fungi. Occasionally **small eosinophilic abscesess or collections of neutrophils** are present. The periphery of the lesions shows a coarse fibroblastic granulation tissue and considerable lymphatic dilatation and obstruction may be associated with the inflammatory process.

The histological picture of *rhino-entomophthoromycosis* is virtually not different from that of subcutaneous phycomycosis. These two infections are distinguished clinically and by culture.

Culture

Tissue bits for culture should not be ground but cut with a sterile pair of scissors into small snippets.

BASIDIOBOLUS RANARUM (Eidam, 1887), frequently found in lesions of subcutaneous phycomycosis, grows on Sabouraud's agar in wax coloured or yellowish, radially folded, flat glabrous colonies of 1.5 cm diameter in the course of three or four days at room temperature. **Hyphae** are 8-20 microns in diameter and **produce numerous septa in culture** (in contrast to hyphae growing in tissue) particularly when spores develop. **Aerial hyphae, sporangiospores and chlamydospores are produced**. Characteristic **zygospores may be found after 8 days** or later. Zygospores **are spherical and thick walled, of 30-50 microns diameter**. They are formed after conjugation of two adjacent hyphal cells. Excellent photographs of these cells are reproduced in "Medical Mycology" by Emmons, Binford and Utz. Conidiophores bear a conidium each at the apex which is discharged in the fashion of a miniature water rocket by rupture of the turgid conidiophore.

The growth of ***ENTOMOPHTHORA CORONATA***, isolated from facial lesions, becomes visible on agar slopes after 3-4 days. Colonies are a first pale yellowish, flat, glabrous and with radial furrows, similar to colonies of *B. meristophorus*. As the colony matures, the surface becomes white and powdery, often forming a mat of growth with satellite colonies. ***Microscopically***, short branching hyphae with occasional septa, produce conidiophores which eject from their tips globose conidia of 10-20 microns diameter, which may, in turn, produce hyphae, secondary microconidia or, on ageing, a corona of fine radiating hair-like structures are characteristic of ***E. CORONATA*** (**Pl. 28.V, F. lower C**).

Therapy

In most cases of subcutaneous phycomycosis **oral iodide administration for two or three weeks potassium iodide 30 mg per kg body weight per day causes rapid disappearance of the symptoms**. In cases with secondary infection, alternation treatment with antibiotics may be necessary. Long standing cases and also rhinoentomophthoromycosis may require higher doses and more prolonged treatment with potassium iodide. Martinson and Clark suggested 3.0 g thrice daily in such cases. **Iodide treatment is so effective that no other measures are necessary**; if disfiguring swelling of the face, particularly of the lips, are present, **plastic surgery may be combined with** iodine therapy in order to achieve rapid results.

REFERENCES

Bader, G. and Grueber, H. L. E. Histochemical studies of tissue forms and perifungal substance phycomycosis. *Pathologia et Microbiologia*, in print.

Balasubrahmanya, M. and Chaudhuri, S. (1963). A case of pulmonary mucormycosis. *Ind. J. Path. Bact.* **6**, 60.

bridges, C. G. and Emmons, C. W. (1961). A Phycomycosis of horses caused by *Hyphomyces destruens. J. am. vet. med. Assoc.* **138**, 579-589.

Bridges, C. H., Romane, C. M. and Emmons, C. W. (1962). Phycomycosis of horses caused by *Entomophthora coronata. J. am. vet. med. Assoc.* **140**, 673-677.

Burkitt, D. P., Wilson, A. M. M. and Jeliffe, D. B. (1963). Subcutaneous phycomycosis. *East afr. med. J.* **40**, 34-38.

Clark, B. M. The Epidemiology of Phycomycosis. Ciba Foundation Symposium of Systemic Mycoses, 1968, p. 179-192. Editors: Wolstenholme, G. E. W. and Porter, R. J.; A. Churchill, London.

Grover, S., Naidu, A. and Junarkar, R. V. (1966). Rhinocerebral Phycomycosis (a case report). *Ind. J. Path. Bact.* **9**, 264.

Grueber, H. L. E. (1969). Rhino-Entomophthoromycosis. *J. Christ. med. Assoc. India* **44**, 20-24.

Landau, J. W. and Newcomer, V. D. (1962). Acute cerebral phycomycosis (mucormycosis). *J. Pediatr.* **61**, 363.

Lie-K-J, N-I, T. E., Pohan, A., Van Der Muelen, H. and Emmons, C. W. (1956). *Basidiobolus ranarum* as a cause of subcutaneous mycosis in Indonesia. *Arch. Derm. Syph.* **74**, 378-383.

Madhavan, M. and Reddy, D. J. (1969). Mucormycosis of Intestine. *Ind. J. Path. Bact.* **12**, 46-48.

Martinson, F. D. and Clark, B. M. (1967). Rhinophycomycosis Entomophthorae in Nigeria. *Am. J. trop. Med. Hyg.* **16**, 40-47.

Paltauf, A. (1885). Mycosis Mucornia, Ein Beitrag zur Kenntnis der Menschlichen Fadenpilzerkrankungen. *Arch. Path. Anat.* **102**, 543.

Straatsma, B. R., Zimmerman, L. E. and Gass, J. D. M. (1962). Phycomycosis. A clinicopathologic study of fifty-one cases. *Lab. Invest.* **11**, 963-985.

Symmers, W. St. Clair (1962). Histopathologic aspects of the pathogenesis of some opportunistic fungal infections, as exemplified in the pathology of aspergillosis and the phycomycetoses. *Lab. Invest.* **11**, 1073-1090.

Symmers, W. St. Clair (1966). Deep seated fungal infections currently seen in the histopathologic service of a medical school laboratory in Britain. *Am. J. clin. Path.* **46**, 514-537.

Tunnell, N., Grueber, H. L. E and Samuel, I. (1968). Subcutaneous Phycomycosis. *J. ind. med. Asoc.* **51**, 355-357.

Williams, A. Q. (1969). Pathology of Phycomycosis due to *Entomophthora* and *Basidiobolus* species. *Arch. Path.* **87**, 13-20.

Wilson, J. W. and Plunkett, O. A. The fungus diseases of man. University of California Press, Barkeley and Los Angeles, 1965.

29

CUTANEOUS MYCOSIS

Dermatophytes

· Dermatophytes from a closely related group of keratinophilic imperfect fungi, causing specific infection in man and animals. They infect the superficial keratinised horny layer of the skin, the hair and the nails and do not cause subcutaneous or systemic infection. In the parasitic form they show mycelial filaments in the skin, hair or nails but when old they may show arthrospores. When cultured in Sabouraud's glucose agar at room temperature, they show development of filamentous colonies with formation of ***Microsporum, Trichophyton, Epidermophyton*** **and** ***Keratinomyces.***

History and discovery of ringworm fungus

Ringworm has been known in India from time immemorial and reference of clinical symptoms has been found in old Ayurvedic literatures, where nothing has been mentioned definitely about the causative organisms.

Schoenlein (1893) discovered the first aetiological agent of the fungus disease in man causing favus. Remak (1845) called it ***Achorion schoenleinii***. causing smallspore ringworm of the scalp (hair) and Malmsten (1845) described ***Trichophyton tonsurans*** an aetiological agent of the hair of the scalp. Sabouraud (1907) described ***Epidermophyton*** *inguinale* from a case of Eczema marginatum which was later called ***E. floccosum***; whereas, Castellani (1910) reported *Endodermophyton concentricum* from a case of tinea imbricata or Tokelau ringworm. Sabouraud in his book on 'Les Teigne' described at least 45 species of dermatophytes. The species where differentiated on the basis of difference of clinical lesions from which they were isolated, and also the character of gross colony morphology on Sabouraud's standard medium,. Later, critical studies of these dermatophytes in natural or other media have discarded the genus *Achorion* and ***Endodermophyton*** and these genera have been included into the genus ***Trichophyton***. This genus now contains 14 species; whereas, the genus ***Microsporum*** contains 7 species and ***Epidermophyton*** contains a single species.

According to Sabouraud (1910), Remak in 1937 discovered fungus in the scutulum of a case of favus but he failed to realise that the fungus was responsible for the disease. In 1839, Schoenlein in Germany demonstrated fungus in favus scutula and thus the mycotic nature of **"dry pustule" or favus** was established. Lebert (1845) named the fungus which causes Favus as *Oidium scyoenleini* now called and in the same year Remak established the genus Achorion for it. Since then the fungus of favus was called ***Achorion schoenleinii*** for a long time till it was named ***Trichophyton schoenleinii*** (Lebert) Langeron & Milochevitch, 1930.

In 1941, Gurby while working in Paris published his first paper "Sur une vegetation qui constitute la rrai teigne" giving an accurate description of the fungus causing favus. In 1942, he published his second paper on the parasitic ringworm fungus later called Ectothrix Trichophyton by Sabouraud. In his third paper in 1843, he described small spored **Microsporum infection** of the hair of children and the fungus was called *Microsporum audouinii* after the name of ***Hean Victor Audouin*** (1797-1841), a distinguished entomologist, in honour of his work on Mascadrine disease in silkworm. **Malmsten** (1845) described ***Trichophyton (Trichomycosis) tonsurans*** as an aetiological agent for ringworm of the scalp. In 1948, he called it *T. tonsurans*. **Charles Robin (1847)** discovered the fungus now called *Trichophyton mentagrophytes*.

Sabouraud started the systematic studies of dermatophytosis after 1890 and published a series of papers. **Sabouraud (1907) described *Epidermophyton inguinale*** found in eczema marginatum of Hebra; whereas, Castellani (1910) reported *Endodermophyton concentricum* from **tinea imbricata** or **Tokelau ringworm.**

Sabouraud (1910) in his classic and memorable treaties "Les Teignes" reported 45 species of dermatophytes giving an account of 500 ringworm cases. More than 100 species of trichophyton alone are now listed in the literature. Sabouraud's work did not help to clarify the taxonomy of the Dermatophytes. The species were distinguished by him not only on the basis of (1) **appearance of lesions** from which they were isolated, (2) **gross appearance of colony characters on Sabouraud's proof (standard) medium** but also on (3) microscopic appearance of fruiting organisms. **Critical studies of these dermatophytes** (Ajello, 1968) **have reduced them into 4 genera**, e.g., **Keratinomyces, Microsporum, Trichophyton and Endodermophyton** have been discarded and incorporated in the genus trichophyton which now contains 15 valid species, while Microsporum contains 7 species, **Epidermophyton contains a single species** and **Keratinomyces also contains a single species.**

The works of Lucille K. Georg and Ajello are worth recording in the last decade. The former author has opened a new avenue for the study of trichophytons on the basis of their physiological requirement of vitamin and amino acids incorporated in a basal medium of casein and ammonium.

Ajello and Georg (1957) studied *in vitro* **hair cultures for differentiating atypical isolates of *T. mentagrophytes*** and ***T. rubrum*** as the aberrant form of these two dermatophytes are different to identify.

Vanbreuseghem (1952) made a discovery of ***keratinophilic* property of the dermatophytes** after their isolation from the soil by using "hair bait" on the moistened soil and he was able to isolate ***Keratinomyces ajelloi***. Dawson and Gentles (1959) using this hair-soil culture technique reported *K. ajelloi* producing its perfect phase and placed it in the *Gymnoascaceae* group as ***Arthroderma uncinatum***. It should be remembered that Nannizzi in 1926 put *Trichophyton mentagrophytes* in the Gymnoascaceae group when he cultivated this in natural media like feather,. leather, hair and other animal substrates. He considered the formation of these organs as **pycnidia corresponding to peridia** of the **gymnoascaceae**. Pycnidia differed from peridia in the fact that, while general morphology of both was similar, they did not contain asci but contained spores similar to aleuriospores and called pycnospores. In 1927, Nannizzi similarly put *Microsporum gypseum* on animal substrates, where true peridia were formed containing true asci. Thus the discussion found in the early literature (Matruchot and Dassonville, 1899) concerning the relationship of dermatophytes to *Gymnoascaceae* was fully substantiated.

In 1961, Dawson and Gentles described the perfect stage of *Microsporum nanum (Nannizzia obtusa)* ***and*** **Trichophyton terrestre (Arthroderma quadrifidum)**.

Therapy

In 1958 Gentles reported that the oral administration of the antibiotic griseofulvin cured experimental ringworm in guineapig and clinical trial in human cases of ringworm by the oral administration of the antibiotic cured resistant types of tinea capitis without any epilation and also effective in cases of onychomycosis.

Subsequently during first decade of this millenium (years) 2 efficacious medicines (chemotherapeutics) have gained mass popularity :

1. **TOPICAL IMIDZOLES** (Clotrimazole, miconazole)

2. **ORAL TERBINAFINE.**

Introduction

Since the time of Sabouraud (1910 or earlier) the classification of dermatophytes was based on clinical rather than mycological character as seen on a "proof" standard medium. The medical mycologists and dermatologists by this time have worked out cultural and microscopic characters of the cultures. They studied various ringworm fungi microscopically and grouped them in different genera on the basis of micro and macroconidia. **Nannizzi (1926, 1927)** first began to study **dermatophytes in moist soil in a methodical way** as their natural habitat and determined the perfect stage of dermatophytes by demonstration of pycnospores in soil. Thus he demonstrated that *M. gypseum* and *M. mentagrophytes* which form pycnidia in organic substance belonged to the family Gymnoascaceae in the class Ascomycetes.

Biochemical criteria are being considered for taxonomic classification of dermatophytes by Georg and Camp (1957) who used basal media of (1) **casein** and (2) **ammonium in media** as source of nitrogen, with glucose as carbon source and sulphate of magnesium, and phosphate of potassium as salts. They worked out the **trichophyton species** by adding thiamine to casein, inositol to casein, and both thiamine and inositol to casein and studied them as growth factors. It was noted that some strains of ***T. verrucosum*** (84%) required both thiamine and inositol as growth factors for their physiological growth but there are strains which require thiamine only (16%). Some strains (50%) of ***T. concentricum*** were *autotrophic* for physiological growth but the remaining 50 per cent of cultures required thiamine for their full fledged physiological growth, on the contrary ***T. schoenleinii*** strains are autotrophic for vitamins.

Again both granular and fluffy strains of *T. rubrum* and ***T. mentagrophytes*** are autotrophics for vitamins. On the other hand, different strains of *T. tonsurans* are stimulated by thiamine.

AMINO ACIDS

It was found out by adding histidine to NH_4NO_3 tube that ***T. megnini*** had complete requirements of histidine for a physiological growth and thus it can be differentiated from ***T. gallinae*** with which it is often confused.

Nicotinic acid : When cultured in casein medium supplemented with nicotinic acid ***T. equinum*** showed complete requirement for nicotinic acid. The study helped in the differentiation of the fungus from ***T. mentagrophytes*** which is autotrophic.

THIAMIN

When thiamin was added to casein, ***T. violaceum*** showed stimulation for a physiological growth. Thus it could be differentiated from ***T. ferrugineum*** which is autotrophic without a supplement.

In vitro ***hair culture for differentiation of species***. The *keratolytic* activity of dermatophytes has been utilised as a process of identification of these fungi. **The test is performed as follows: Short strands of human hair (1 cm long) are put in a Petri dish for autoclaving at 120°C for 10 minutes**. This is followed by addition of sterile water and 2 or 3 drops of 10 per cent sterile yeast extract solution. Plates are then inoculated with several small fragments of test cultures grown on Sabouraud's dextrose agar and incubated at 25ºC or room temperature for 4 weeks replenishing distilled water whenever necessary to maintain the moist condition.

For examination, remove hair segments from Petri dishes with sterile forceps at intervals, and placing them in a drop of lactophenol cotton blue and examine the coverslip preparation under the microscope to determine whether the hairs have been perforated by the fungus. Perforations develop in abundance after 10th to 14th day of incubation. Gentle heating of lactophenol cotton blue preparations helps the detection of perforation, as heat intensifies the staining of mycelia. The test is considered to be negative if perforations are not observed following the growth for 25 days.

NUTRITIONAL REQUIREMENTS

Practical methods of determining the nutritional requirement of dermatophytes, which can be carried out in a diagnostic laboratory has been worked out by Georg and Camp (1957). This is a physiological test which helps in classification of fungi.

For the test, the glass wares are scrupulously cleared. In this, a **casein glucose salt agar basal medium** is used to which thiamine, inositol or both, or nicotinic acid are used; whereas ammonium nitrate agar basal medium (0.15 per cent of NH_4NO_3) is used for the nutritional requirement of histidine.

For testing 2 ml of stock solutions, e.g., **thiamine solution (1 mg per cent in distilled water pH 4-5), inositol solution (250 mg per cent in distilled water) and L-histidine solution (150 mg per cent in distilled water) are mixed with basal media** mentioned above. These solutions are autoclaved at 120ºC for 10 minutes and preserved at 5ºC.

Inoculation of nutrition tubes and controls

Inoculum for testing may be taken from cultures grown on any isolation media, taking care that only a pin head size inoculum is taken in order to avoid carrying over an excess of non vitamin free isolation medium.

Casein-10 per cent hydrolysed, vitamin free (Nutritional Biohemical Co.) solution: 25 ml or **casamino acid** (vitamin free) -2.5 g, **glucose**-40.0 g, **$MgSO_4$** 0.1 g, **KH_2PO_4**-1.8 g, **agar** (Difco Bacto agar)-20 g and distilled water-100 ml. Dissolve by heating and distribute in 100 ml quantities in flasks and autoclave at 120ºC for 15 minutes; some tubes may be prepared as vitamin free controls and slanted. Other flasks may be kept for preparing different media containing vitamins as needed. **Nutritional medium** with numbers (1-7) **for the differentiation of Trichophyton species are manufactured by Difco.**

Nutritional pattern of Trichophyta as mentioned by Georg and Camp (1957) is shown in the following Tables.

Table 29.1 : Nutritional pattern of *T. Verrucosum*, *T. Schoenleinii* And *T. Concentricum*. This is a 4 Tube Test Incubated at 37°C And Read After 7—14 Days

Dermatophytes	Test Media: Casein	Casein-inositol	Casein-Thiamine	Casein-Thiamine & inositol
T. verrucosum				
84%	0	± *	0	4+ **
16%	0	0	4+	4+
T. schoenleinii	4+ ***	4+ ***	4+ ***	4+ ***
T. concentricum				
50%	4+ ***	4+ ***	4+ ***	
50%	2+	2+	4+	4+ ***

* ± Indicates trace of submerged growth about the inoculum.

** 4+ Indicates maximum growth for that series of tubes, growth in other tubes being judged by comparison.

*** Autotrophic for vitamins.

PLATE 29.I A

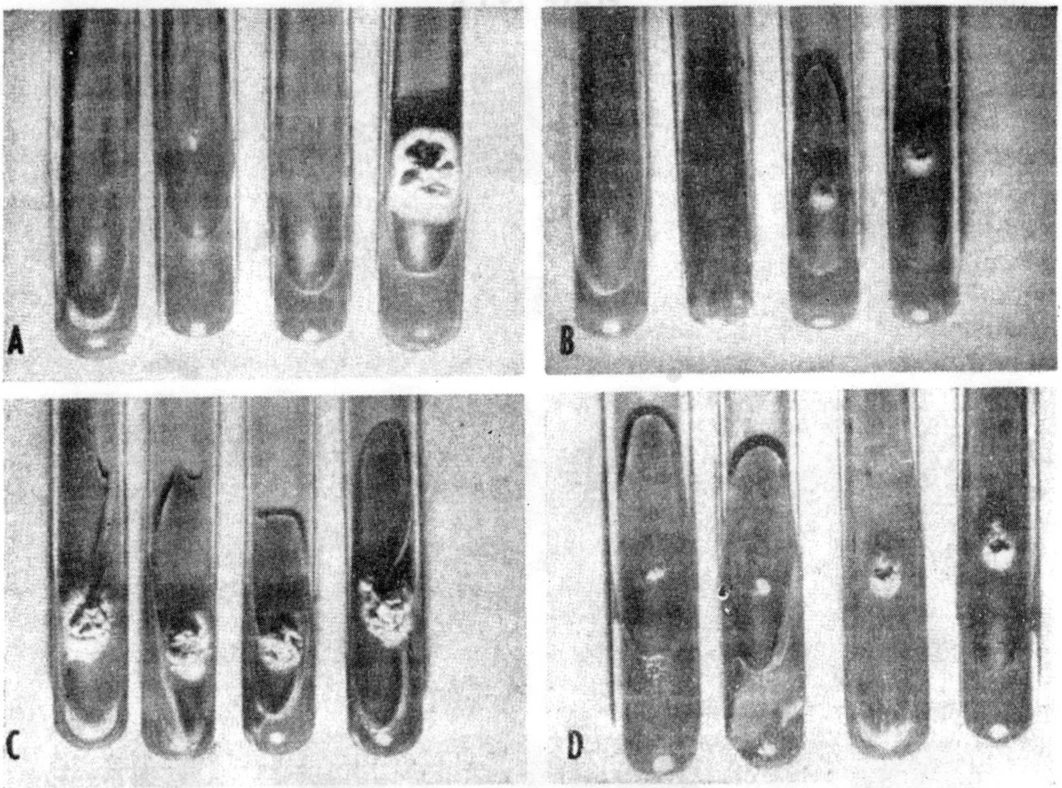

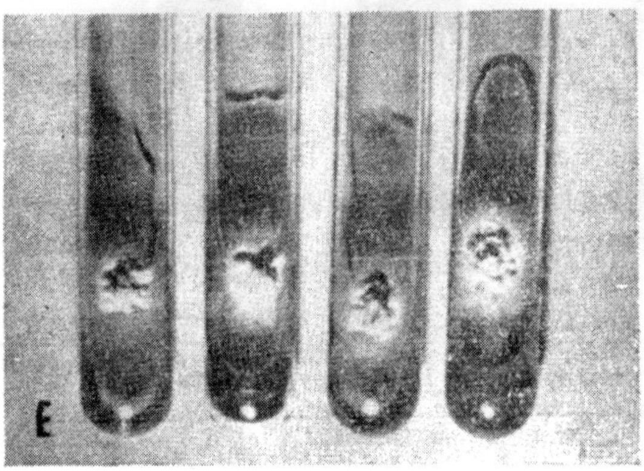

Four tube nutritional test : Left to right. (1) casein, (2) casein + inositol, (3) casein+thiamine, (4) casein+thiamine+inositol. **A. *Trichophyton verrucosum*** strains require both thiamine and inositol.

B. *Trichophyton verrucosum* strains requiring thiamine only.

C. *Trichophyton concentricum* strains are autotrophic for the vitamins.

D. *Trichophyton concentricum* strains are stimulated by thiamine.

E. *Trichophyton schoenleinii* strains are autotrophic for the vitamins. [Courtesy, George & Camp (1957)]. *J. Bact.* **74** (2), 113-121.

Table 29.2 : Nutritional Pattern of *T. tonsurans, T. mentagrophytes* And *T. rubrum*. This is a 2 Tubes Test Incubated at Room Temperature and Read After 7-10 Days

Dermatophytes	Test media	
	Casein	Casein-Thiamine
T. tonsurans	± to +	4+ *
T. mentagraphytes	4+ *	4+
T. rubrum	4+ **	4+ **

* Stimulated by thiamine. ** Autotrophic to both casein and thiamine.

PLATE 29.1 B

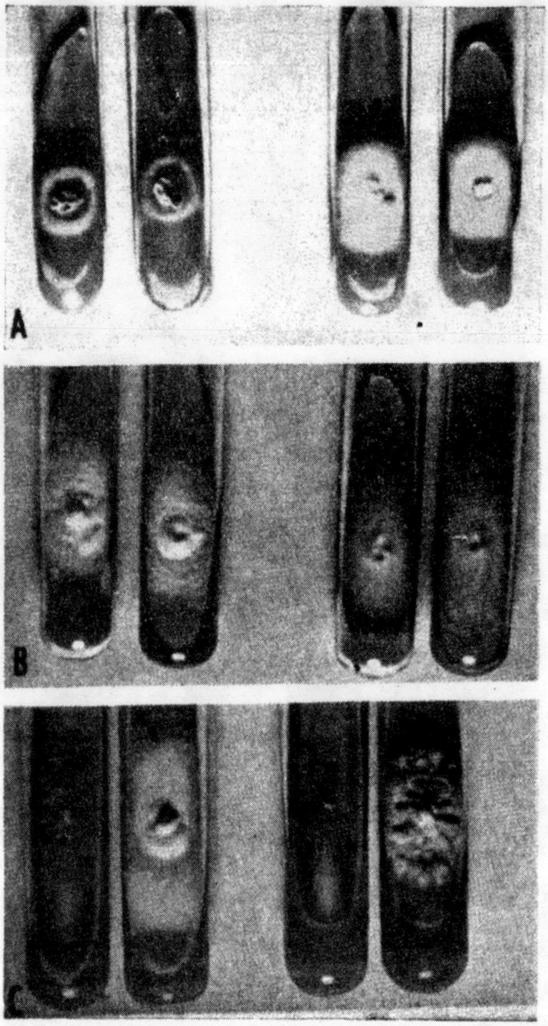

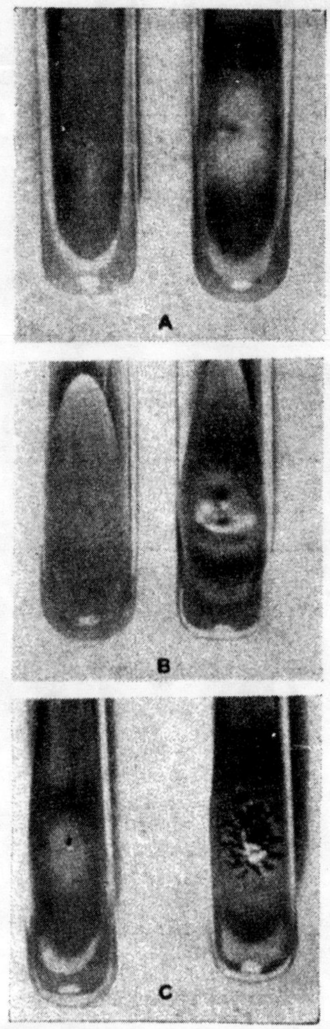

Fig. Left side : Two tube nutritional test: Left sided tube of each pair is the casein tube; right sided one is the casein+thiamine tube. **A. *Trichophyton* rubrum**; Left pair—granular strain; Right pair—fluffy strain.

B. ***Trichophyton mentagrophytes*:** Left pair—granular strain; Right pair—downy strain; both are autotrophic for thiamine.

C. ***Trichophyton tonsurans***; two different strains in both pair of tubes show stimulation by thiamine.

Fig. Right side : Two tube nutritional test: **A.** ***Trichophyton megninii.*** Left—NH_4NO_3 tube; Right—NH_4NO_3 + histidine tube showing complete requirement for histidine.

B. ***Trichophyton equinum***. Left—casein tube; Right—casein + nicotinic acid tube showing complete requirement for nicotinic acid.

C. ***Trichophyton violaceum***. Left—casein tube; Right—casein + thiamine tube showing stimulation of growth by thiamine.

Table 29.3 : Growth Pattern of *T. megninii* And *T. gallinae*

Dermatophyte	Test media	
	NH_4 NO_3	NH_4 NO_3 + histidine
T. megninii **	0	4+
T. gallinae *	4+	4+

* No other dermatophytes show regular requirement of histidine.

** *T. gallinae* is included for comparison as this species has been confused with *T. megninii* in the past.

Table 29.4 : Growth Pattern of *T. equinum as Compared With *T. mentagrophytes***

Dermatophytes	Test media	
	Casein	Casein + Nicotinic Acid
T. equinum *	0	4+
T. mentagrophytes **	4+	4+

* No other dermatophyte requires this nicotinic acid.

** *T. mentagrophytes* is often confused with *T. equinum* and hence included in this chart.

Table 29.5 : Growth Pattern of *T. ferrugineum* and *T. violaceum*

	Test media	
	Casein	Casein+ thiamine
T. ferrugineum	4+	4+
T. violaceum	±	4+

It is to be noted that ***T. violaceum*** has a similar nutritional requirement to that of ***T. tonsurans*** but it is not likely to be confused with it. ***T. violaceum*** grows extremely slowly even in the presence of thiamine and usually produces a glabrous colony with no conidia. ***T. tonsurans*** grows rapidly in presence of thiamine and shows many **microconidia** in the surface growth.

Diagnostic growth requirements of dermatophytes after George 1957.

Species	**Nutritional Requuirements**
Trichophyton equinum	Nicotinic acid
T. megninii	Histidine
T. tonsurans	Thiamine
T. verrucosum	Thiamine & inositol (84% of isolates)
T. verrucossum	Thiamine only (16% of isolates)

Tabble 29.6 : Complete List of Dermatophytes With Their Perfect Stages

A. Genus Microsporum

Small spores (3-5 microns), **hair fluorescent**	**Large spored** species (5-8 microns) **inside or outside the hair**
M. audouinii Gruby, 1843	*M. gypseum* (Bodin) Guiart and Grigorakis, 1928
M. cookei Ajello, 1959	*M. nanum* Guentes, 1956
M. canis Bodin 1902	*M. vanbreuseghemii* George *et al.*, 1962
M. distortum DiMenna and Marples, 1954	*M. fulvum*

B. Genus Keratinomyces

K. ajelloi Vanbreuseghem, 1952

C. Genus : Trichophyton ENDOTHRIX (HAIR NOT FLUORESCENT)

T. (crateriforme) tonsurans Malmsten, 1845	*T. gourvilii*
T. (achorion) violaceum Bodin 1902	*T. soudanense*
T. (purpureum) rubrum (Castellani) Sabouraud, 1911	*T. yaoundei*
	T. terrestre Durie and Frey (1957)

D. ECTOTHRIX MICROIDES Small spored

Micorides (3-5 micron spores, **HAIR FLUORESCENT TO WOOD'S LIGHT)**	*T. verrucosum* Georgem, 1950 Hair not fluorescent
	Ectothrix Macroides Large spored (Megaspores)
T. (gypseum) mentagrophytes (Robin) Blanchard, 1896 (Megaspores, 5-10 μ spores)	(Megaspores, 5-10 microns spores)
	T. gallinae
	T. equinum Gedeolst, 1902
	T. megninii (T. rosaccum)
	T. rubrum (rarely) Endthrix Trichophyta

E.Hyphae and air spaces in hair but **not spores**, dull fluorescence, and hair perforating scutulum: *T. schoenleinii*.

F.No invasion of hair: *Epidermophyton floccosum*, *T. (Endodermophyton) concentricum* (Castellani) Blanchard, 1896.

G.Perfect Stages

Imperfect stage	Perfect stage (ascigenous spores)
M. gypseum	*Nannizzia* incurvata
	N. gypsea
M. fulvum	*N. fulva*
M. nanum	N. obtusa
M. cookei	N. cajetana
M. vanbreuseghemii	N. grubyia
K. ajelloi	Arthroderma uncinatum
T. terrestre	*A. quadrifidum*

DERMATOPHYTOSIS IN INDIA

Infection

It depends on the seeds and the soil.

Seeds

The **anthropophilic fungi** cause infection by direct contact; whereas, **zoophilic strain** cause infection from animal sources. In this the infected materials are derived from animal hairs or feathers.

Soil

M. audouinii infection is specially in children in prepubertal age. It is quite likely that the gonadal secretions are quite different during this period than during and after puberty, when gonads and glandular secretions are different. Apocrine glands also come into full fledged activity at puberty. Sweat gland secretions which contain traces of **acetic, propionic and undecylenic acid** and similar aromatic substances by which the **infection is naturally prevented** are diluted and the intertriginous areas become susceptible to infection. Thus interdigital areas between fourth and fifth toe and the inguinal regions become highly susceptible to the infection with ringworm fungi. Once infection occurs these regions form a nidus, wherefrom the infection may spread to other parts of the body. This is why skin of the individual living in the highly humid tropical areas are liable to infection due to favourable temperature and moisture.

Epidemiology

Clinical study of 500 (longest series) suspected cases of dermatophytosis revealed about 250 cases which showed positive for evidences of fungus either by microscopic examination or culture or both (Mankoli and Kanvinde, 1969).

Incidence

Of the above 250 case *Tinea corporis* is commonest with incidence of 52.8 per cent and next is *Tinea cruris* in 34.4 per cent both being infection of glabrous skin.

Tinea capitis is prevalent in children and Dutta and Raman Rao (1970) isolated one each of ***T. rubrum***, ***T. violaceum*** and ***M. canis***. ***M. audouinii*** has not yet been found out in an Indian child as mentioned **before by Dey (1954)**. Tinea capitis is a problem in India and no case of ***M. audouinii*** infection has been reported as yet amongst the children of the soil (Dey, 1954), who studied tinea capitis in the School of Tropical Medicine, Calcutta from 1932 to 1950 and the following infections were encountered as shown in the following Table:

This is a representative figure when European and Anglo Indian Children were boarding in different Hill station Schools of India and attended the School of Tropical Medicine clinic, Calcutta. **T. capitis** is not common in India (Kalra *et al.* 1964). Desai (1961) reported 9.2 per cent of cases in his series to be those of **T. capitis**; whereas, Kalra (1964) reported it to be 3.1 per cent. Since then Mankodi and Kenvinde (1969) found in 15 (75%) cases.

Dey also found the same species causing infection amongst the Mohammadan children. It was equally prevalent amongst the European children as well (Table 29.9).

Table 29.7

Species	*No. of cases*	*Percentage*
Microsporum audouinii	94	47.47%
Trichophyton violaceum	81	40.91%
Trichophyton crateriforme (tonsurans)	7	3.54%
Trichophyton ferrugineum	8	4.04%
Trichophyton mentagrophytes	8	4.04%
Total	198	100%

Table 29.8

Tinea capitis	*T. rubrum*	*T. violaceum*	*T. tonsurans*	*T. mentagrophytes*
20	1	15	3	1

Age group

Kalra *et al.* (1964) reported 79.8 per cent to be in the age group of 2-30 years of age; whereas, Ghosh in the Calcutta School of Tropical Medicine, reported maximum incidence in the younger age group of 11 to 20 years. In the skin O. P. D. Again *T. corporis* is more common in the higher age group than *T. curis* which form the nidus of infection, might spread on the body in later life. Mankodi and Kanvinde (1963) observed that though persons of all age group are susceptible, it appears to become common in the adult above 20 years.

Species

India is a subcontinent with variable atmospheric temperature, altitude and humidity so the infective species also vary from place to place. It should also be remembered that the flora varies from time to time and place to place.

T. rubrum is the commonest dermatophyte and is shown in the Table 29.9, showing the prevalence of the disease in different parts of India.

Reports from various parts of India show that ***T. rubrum*** is the commonest dermatophytes of the glabrous skin in India (84.45 per cent) in Bombay (Dsai *et al.*, 1951). Prevalence of ***T. mentagrophytes*** is second in order of frequency in most of the series. The minimum incidence of ***T. violaceum*** is found in Bombay and Ahmedabad both being neighbouring states. Very high rate of *E. floccosum* (32.23 per cent) in the series by Ghosh in the Calcutta School of Tropical Medicine in 1968 can be explained in comparison to other places in India may be due to the change of flora because in the case of Panda's series to Orissa in 1967, only 1.96% were ***E. floccosum***. The highest incidence of *T. mentagrophytes* is recorded to be 31.73 per cent in Uttar Preadesh (Das Gupta *et al.*, 1959).

In conclusion while correlating clinical and mycological findings it appears that ***T. rubrum*** is chiefly responsible for ***T. corporis*** and ***T. cruris*** infections; whereas, ***T. violaceum*** is the chief causative agent for ***T. capitis***.

GENUS : MICROSPORUM

Gross : On Sabouraud's glucose agar at room temperature the colonies grow slowly, netted and furrowed, or **fast growing or powdery colony, tan or cinnamon brown in colour.**

Table 29.9 : Incidence of Dermatophytes (In Percentage) As Compared With Results From Different Parts of India

State/Place	*T. rubrum*	*T. mentagrophytes*	*T. violaceum*	*E. floccosum*
Bengal (Ghosh, 1948)	63.00	2.36	1.41	32.23
Uttar Pradesh (Gupta *et al.*, 1959)	56.73	31.73		1.92
Poona (Gokhale *et al.*, 1959)	72.38	17.16	2.24	2.24
Bombay (Desai *et al.*, 1961)	84.45	2.32	8.8	2.32
Kerala (Kurup *et al.* 1961)	68.30	10.00	.50	3.30
Dehli (Kalra *et al.*, 1964)	80.61	9.91	3.52	3.08
Warangal (Vasu, 1966)	63.83	16.86	4.81	8.30
Burla (Panda *et al.*, 1967)	78.43	10.78	—	1.96
Ahmedabad (present series)	84.50	1.50	8.50	2.00

Table 29.10 : Common Dermatophytes or Ringworm Fungi

Genus	*Species*
1. **MICROSPORUM**	*M. audouinii*
	M. canis
	M. gypseum
2. **TRICHOPHYTON**	
Rubrum group	*T. purpureum (T. rubrum)*
Crateriform group (**ENDOTHRIX**)	*T. tonsurans*
	T. sulphureum
	T. sabouraudii
GYPSEUM GROUP (ECTOTHRIX, SMALL SPORE)	*T. mentagrophytes*
ROSACEUM GROUP (ECTOTHRIX, LARGE SPORE)	*T. rosaceum*
Faviform group	*T. schoenleinii* and other varieties
	T. concentricum
	T. violaceum (Endothrix)
	T. discoides (**ECTOTHRIX, LARGE SPORE**)
3. **EPIDERMOPHYTON**	*E. floccosum*

Microcultures show numerous characteristic macroconidia 8 to 15 microns in width by 40-150 microns in length, spindle shaped, multicelled and rough thick-walled. The microconidia 2.5 to 4 by 3 to 5 microns, are in primary cultures, borne singly along the hyphae, sessile, or borne on short stalks. Raquet mycelia are common. Hyphae pectinate are seen in some. Many of microspora show perfect stage (see Table 29.6).

MICROSPORUM NANUM

and appears to be a pleomorphic or dwarf form of *M. gypseum*. It perforates hair *in vitro* and the perfect stages is known as *N. obtusa*. **Clinically, *M. naum*** causes **microsporic ectothrix ringworm** in human similar to more common ***M. gypseum***.

Epidemiology : Infection in man occurs; infection in pigs has been reported from several sources.

Cultural characters : The giant colony resembles to be one of pleomorphic ***M. gypseum***. The growth forms a fluffy, white, rapidly spreading colony with powdery buff surface due to the presence of a large numbers of macroconidia. The reverse may be reddish brown.

Microculture : Macroconidia are of ovoid or ellipsoidal, ovate and consist of one to three cells (one or two septa). They are relatively thin walled like *M. gypseum*. **Microconidia are clavate** and not specific for the species.

Perfect stage : *Nannizzia obtusa*.

GENUS : KERATINOMYCES

OIn Sabouraud's glucose agar at room temperature the colonies are powdery to fluffy cream to orange tan in colour.

Microcultures show *macroconidia* which are large smooth walled, cylindriform and multiseptate. ***Microconidia*** are seen **in some** strains. The genus contains one species, ***K. ajelloi***, infecting man and animals.

GENUS : TRICHOPHYTON

Gross

On Sabouraud's glucose agar at room temperature, the colonies are granular to powdery or glabrous covered with a velvety or waxy surface. Pigmentation varies from, light pink to red, purple, violet or brown, yellow or light buff colour.

Some glabrous species can be differentiated by their physiological study by nutritional requirements, and special media with these requirements show enhanced growth. Culture on 'hair baits' on moist soil may show development of perfect stage (Peridia formation) in some.

Microcultures show **microconidia** which are subspherical (*T. mentagrophytes*), pyriform or clavate (1.5 × 2 to 5 microns) distributed on sides of hyphae (*en thyreses*) or produced on conidiophores in cluster *en grappe*).

Macroconidia are characteristic but rare, appearing as thin walled, long, multiseptate and clavate (***T. mentagrophytes***), or cylindrical (***T. rubrum***). These are 4 to 6 microns by 10 to 50 microns in length, Racquet mycelium, nodular bodies, **coiled hyphae, parasitised** hair and **chlamydospores** are found in some species. They **invade hair of scalp and beard areas.** Some trichophyta are *endothrix* (human species), but others (often animal species) are *ectothrix*. ***T. concentricum*** does not attack hair at all. Some show small spores (3-5 microns), ***microides*** as in ***T. mentagrophytes***, others show large spores, *MEGASPORES*.

Perfect stage

T. terrestre Durie & Fry (1957) shows **perfect stage** and Dawson & Gentles (1961) called it **Arthroderma**.

PATHOGENESIS

It has already been stated that dermatophytes cause **superficial infections or dermatomycoses of keratinised parts** of the body, namely skin, hair and nails and they do not invade the deeper tissues or internal organs of man or experimental animals.

MICROSPOROSIS

M. audouinii *produces* scaly ringworm of the scalp, **clinically showing large scaly areas,** where the long stumps of affected hair broken at a distance from the surface, are covered with scales (*c.f.* **trichophytosis producing "blackdot" ringworm**). This is an anthropophilic species and does not cause inflammatory lesions. Hair stubs are surrounded by a sheath of spores and addition of chloroform gives Duckworth *test positive. M. canis* and *M. gypseum* being *zoophilic* species, inflammatory reactions resulting in a **BOGGY TUMOUR-LIKE MASS** or **"Kerion"**, resembling a pyoderma.

Source of Infection

Geographic classification of Dermatophytes is based on whether they are cosmopolitan or limited to some geographically limited areas. They are again divided into different groups according to their affinity for infection. There are some which spread from man and called ***anthropophilic***. Again there are other groups which spread from animal infection and called ***zoophilic***. There are some strains which are less adopted and cause marked inflammatory changes in hair producing *kerion* on the scalp or the beard area in skin. But there are other zoophilic species which are partially adopted to human beings producing less inflammatory changes. There are other species which are contracted to man or animal from soil. These are geophilic species.

A list of these fungi are quoted (modified from Ajello, Table 29.10).

TINEA PEDIS

The commonest infection is also known as athletes foot (dermatophytosis) caused by ***E. floccosum*** and ***T. rubrum***. It is an intractable chronic infection with acute exacerbation caused by secondary streptococci with acute inflammation and swelling, blister formation containing clear fluids. This is due to lymphangitis **spreading through cracks and fissures**. The acute condition subsides with proper treatment but the original infection continues as the part is macerated due to sweating. ***T. mentagrophytes*** may cause severe inflammatory change and may sensitise the skin by the fungus or its spores. There is then an **'ID' OR ALLERGIC REACTION**. In this condition

Table 29.11 : Geographic Classification of Dermatophytes According to Source of Infection (Modified From Ajello)

Cosmopolitan species				
Anthropophilic	*Zoophilic*	*Geophilic*	***ANTHROPOPHILIC***	***ZOOPHILIC***
E. floccosum	*M. canis*	*K. ajelloi* †	*M. distortum*	*M. distortum*
M. audouinii	*T. gallinae*	M. gypseum	*M. vanbreuseghemii*	*M. vanbreuseghemii*
	*T. mentagrophytes**		*T. concentricum* ††	*T. equinum*
*T. mentagrophytes**	*T. verrucosum*		*M. ferrugineum*	
T. rubrum			*T. gourvilli*	
T. schoenleinii			*T. megninii*	
T. tonsurans			*T. soudanense*	
T. violaceum			*Ty. yaoundei*	

* Equally prevalent in human and animals.

Two forms are known: (i) **granular**-usually **zoophilic** strains and may cause acute inflammatory lesions in human; (ii) **downy type** associated with human infection.

** Not reported from Asia & Africa.

† **A variant of *M. canis* with less pigment production**, common in rural areas of New Zealand and Australia. In USA traced from South American pet monkeys.

†† In tropical countries.

vesicles, indistinguishable from primary lesion may appear on other parts of the body particularly on the palms of the hands. Fungi are usually absent in these lesions and **trichophytin test is positive**. In this mycelia are carried in circulation from one part to another and cause hypersensitive reactions.

INFECTION OF THE NAIL

(Tinea unguinum)

This occurs from the margin of the nails of the fingers or toes. The **nail margin is brittle** and shows **a moth-eaten appearance**. There may be **hyperkeratosis of the nail** with white opaque appearance and on scraping these show the typical moth eaten appearance due to brittle and friable character. Paronychia is not common.

INFECTION OF BEARD AREAS

(Tinea barbae)

These are caused by animal species and associated with acute inflammatory changes of the hair follicles. Beard affections are caused by ***T. verrucosum, T. mentagrophytes*** and ***M. gypseum***.

INFECTION OF THE GLABROUS SKIN

This is caused by one of the following in order of frequency in the tropics: ***T. rubrum*** (66-70%), ***T. mentagrophytes, E. floccosum***.

They cause *Tinea glabrosa* and in the case of the latter two, the flora may vary or change from time to time in different endemic areas of India.

Adults are usually affected producing typical *ringed* or ***annular*** **lesions** studied with vesicles with a healing scaly centre and an actively spreading **erythematous vesiculo-pustular border**. Such infections are rare but may occur in children **secondary to infection on the scalp** due to dropping of the infective materia;. Being an infection of the horny layer, ***histopathology*** does not show any change in the skin except in the inflammatory lesions. Scraping shows the fungus in mycelial form.

RINGWORM OF THE SCALP

(*Trichophytosis*)

This infection of the scalp, *Tinea capitis* is specially seen in children and many of them heal up at puberty, if not cured during this period. This is true about microsporum infection. However, trichophytosis causes lesions which may not be spontaneously present at puberty but may continue to adult life. There are two types of *Tinea capitis*, namely, **Trichophytosis and Microsporum infections; trichophytosis causing "black dot" ringworm of the scalp** being caused ***by endothrix trichophyta*** (see Table 29.5A). The hairs are broken on the level of the scalp at heavy follicles with a black centre. Favous is a different type of trichophytosis produced by ***T. schoenleinii*** characterised by saucer shaped structures called scutula.

Ectothrix trichophytosis is caused by ***T. mentagrophytes*** producing inflammatory reaction, resulting in kerion formation; whereas, *megaspore* infection is caused by ***T. verrucosum*** and ***T. megninii***, and ***T. rubrum*** rarely produces such lesions.

GENUS : MICROSPORUM

MICROSPORUM AUDOUINII

Invasion of hair

M. audouinii causes infection of the hair and such hairs show dull fluorescent character.

Clinically, it causes **tinea capitis** in children below 12 years of age with long hair stumps buried in the scale and may infect the contiguous glabrous skin due to the falling of infective material (**Pl. 29.II**). The fungus is *anthropophilic* and cause no inflammatory changes unlike *M. canis* which is zoophilic and may cause inflammatory changes like kerion.

Epidemiology

The infection is prepubertal and restricted to man (**ANTHROPOPHILIC**). It has not yet ben found in an Indian even on surface inoculation with the fungus. All the cases studied in India were from European and Anglo-Indian population.

PLATE 29.II
MICROSPORUM AUDOUINI

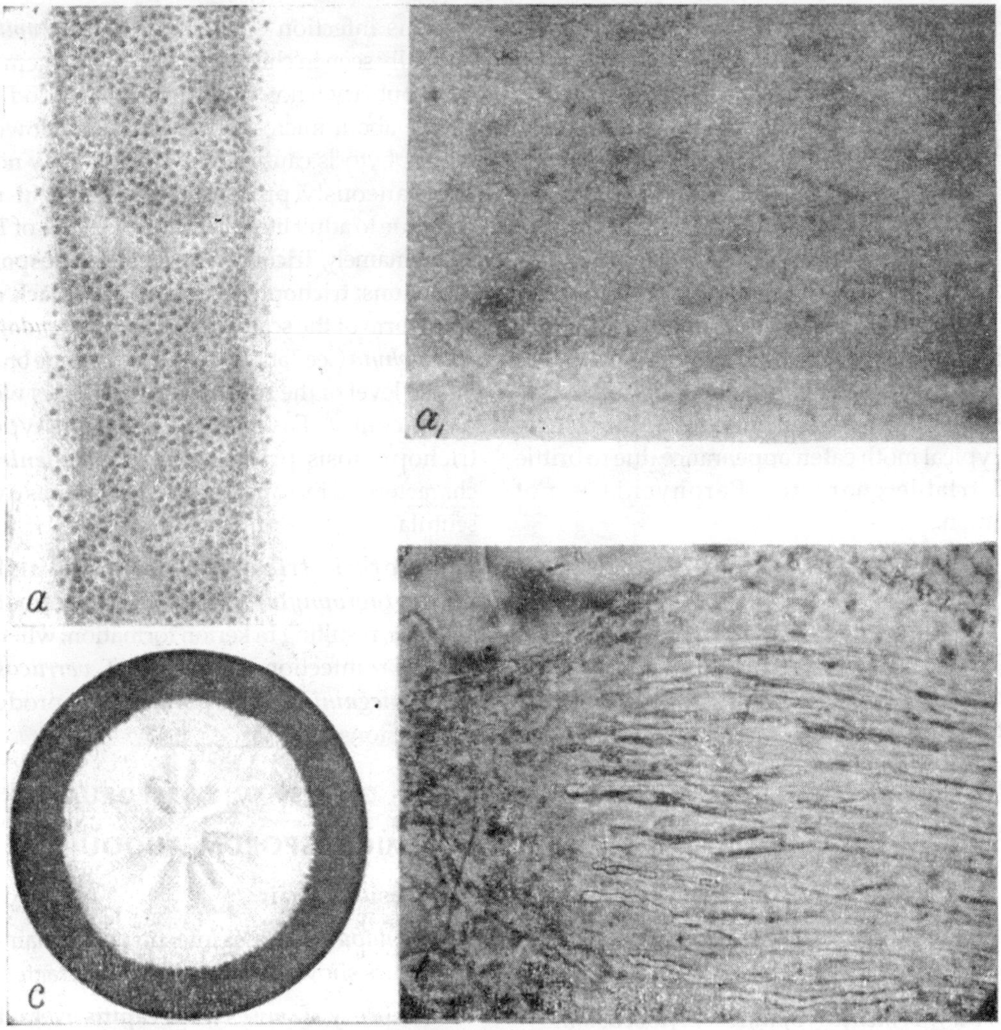

(a and b)—infected hair showing mosaic arrangement of small spores. (b) Adamson's fringe at the bulb of the hair. (c) Giant colony of ***M. audouinii***

Incidence

The incidence of this infection is 47.47 per cent (Dey, 1953).

Cultural characters

The colony appears on the third day after inoculation in *Sabouraud's glucose agar* with greyish white surface covered with small downy or mouse fur in texture. The central portion in a giant colony may be slightly raised or knob-like (**Pl. XLV**), and salmon pink or peach coloured pigment may appear under the colony specially in *potato-dextrose agar*. In contrast to *T. mentagrophytes* the colony of *M. audouinii* is less dense and more quickly spreading.

Microculture

M. audouinii shows fes *microconidia* (2.5 to 4 by 3 to 6 microns) but characteristic terminal chlamydospores which are characteristic are ill formed and diagnostic (**Fig. 29.1**).

PLATE 29.III

RINGWORM CAUSED BY *MICROSPORUM AUDOUINII*

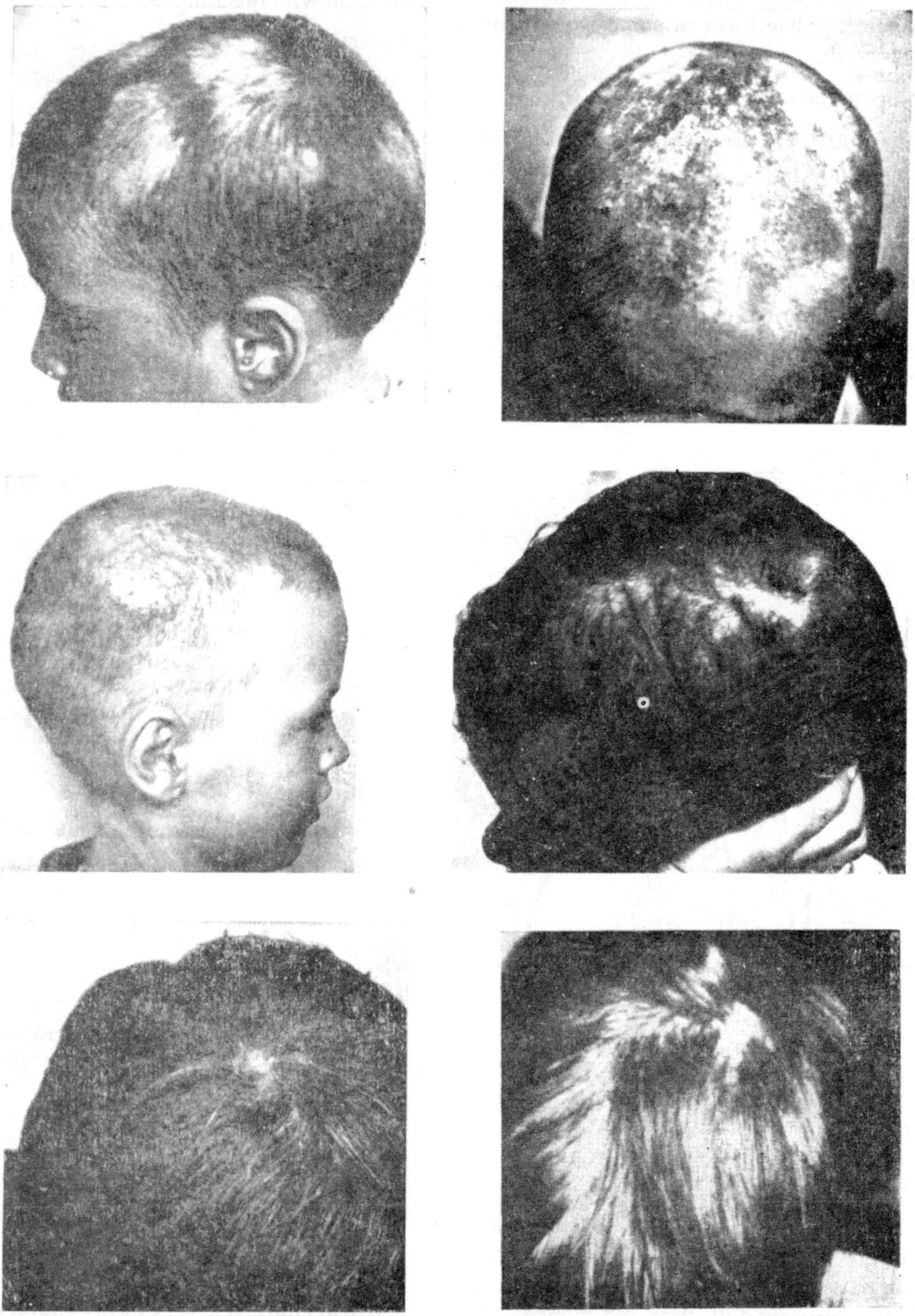

Scaly ringworm of the scalp caused by M. audouinii

Macroconidia are rare occasional species, show ***pluriseptate macroconidia*** which are smooth and poorly shaped or often distorted. Pectinate hyphae form in some isolates on routine media.

Nutritional requirements

The study of ***M. audouinii*** does not reveal any nutritional requirement but unlike *M. canis*, *M. audouinii* grows poorly on polished rice.

Variation

Isolates of *M. audouinii* have been obtained resembling the features of *M. canis* but not satisfactorily classified uptil now. The African form "langeronii" develop radial grooves and a tan colour.

Perfect stage

Not known as yet.

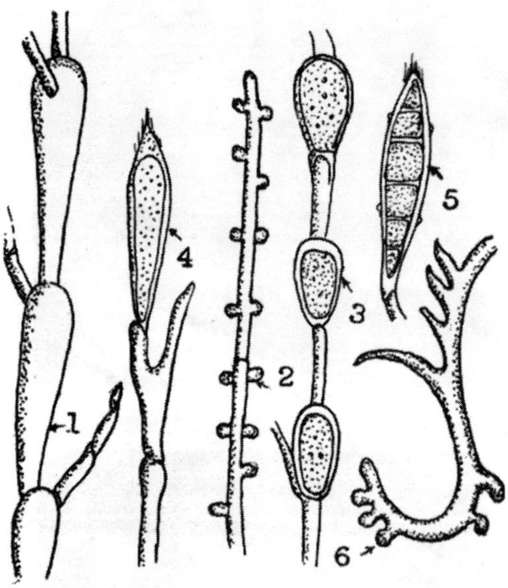

Fig. 29.1 : 1. Mycelial racquet, 2. Microconidia, 3. Chlamydospores, 4. Macroconidium without septa. 5. Septate macroconidium with hairs, 6. Hyphae pectinate

Woods's light

The infected hair stumps **fluoresce as bright, clear, green areas**. Usually all the hairs in a lesion are seen to be involved. In a ten days old culture it appears dull, clear and mouse-grey throughout.

Animal inoculation

Animals are resistant to infection after inoculation with this fungus.

Special media

Polished rice and potato dextrose agar are useful for differentiating ***M. audouinii*** from ***M. canis***, the former growing poorly in polished rice.

CHARACTERISTIC FEATURES

(1) ***Colony* characters** (see above), (2) **absence of *macroconidia*** except in rare strains, (3) **terminal *chlamydospores***, (4) **few distorted macroconidia**, (5) **failure to grow on *polished rice***.

MICROSPORUM COOKEI

(Ajello, 1959)

Natural invasion of hair is not known but perforates **hair baits** *in vitro*.

Clinically, *M. cookei* does not cause infection in man or animals.

Epidemiology : This is a ***geophilic species*** isolated from soil, like ***M. gypseum*** and the hair of small mammals.

Geographical distribution : The fungus has **worldwide distribution** but not yet reported from India.

Cultural characters

M. cookei shows a spreading colony with a characteristic deep grape red pigment. The surface is powdery and yellowish as greenish buff to dark brown in colour due to formation of *macroconidia*.

Microculture

The fungus shows thick celled echinulate *macroconidia* with thickened walls. *Microconidia* abundant and ovoid but not diagnostic.

Nutritional requirements : Not known.

Perfect stage

NANNIZZIA CAJETANA Ajello, 1961.

Special media : Not required.

Critical character : (1) **Colony characters**. (2) ***Macroconidia* with thick wall.**

MICROSPORUM CANIS

Synonyms : *M. lanosum, M. equinum, M. felineum, M. caninum, M. pseudolanosuml Sabouraudites lonosum, Sabouraudites lanatus, Closterosporia lanosa, M. stillianus, M. aurantiacum, M. simiae, M. obesum*, etc.

This often caused microsporosis with kerion formation. It has been recorded from Europe and America but not from India.

Direct examination

The infected hair is indistinguishable from that of *M. audouinii* (**Pl. 29.IV** *a*).

Cultural characters

In Sabouraud's medium it grows fast as a downy, fluffy growth with diffusion of yellowish pigmènt in the medium around the growth. In about 2 weeks the central part of the growth is depressed and the growth is buff tan

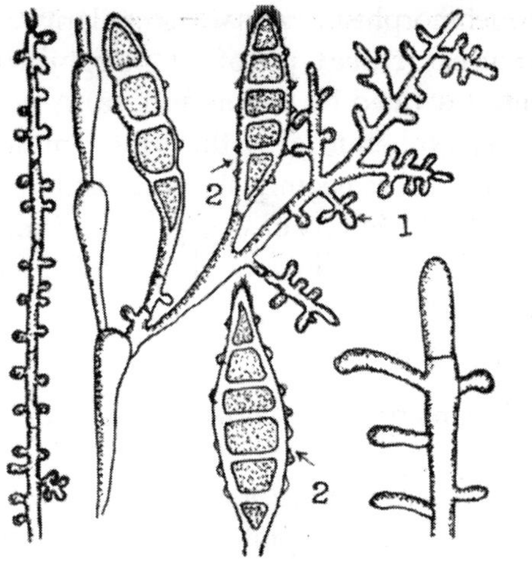

Fig. 29.2 : *M. canis.* 1. Microconidia, 2. Pluriseptate, tuberculated macroconidia

PLATE 29.IV
MICROSOPORUM CANIS

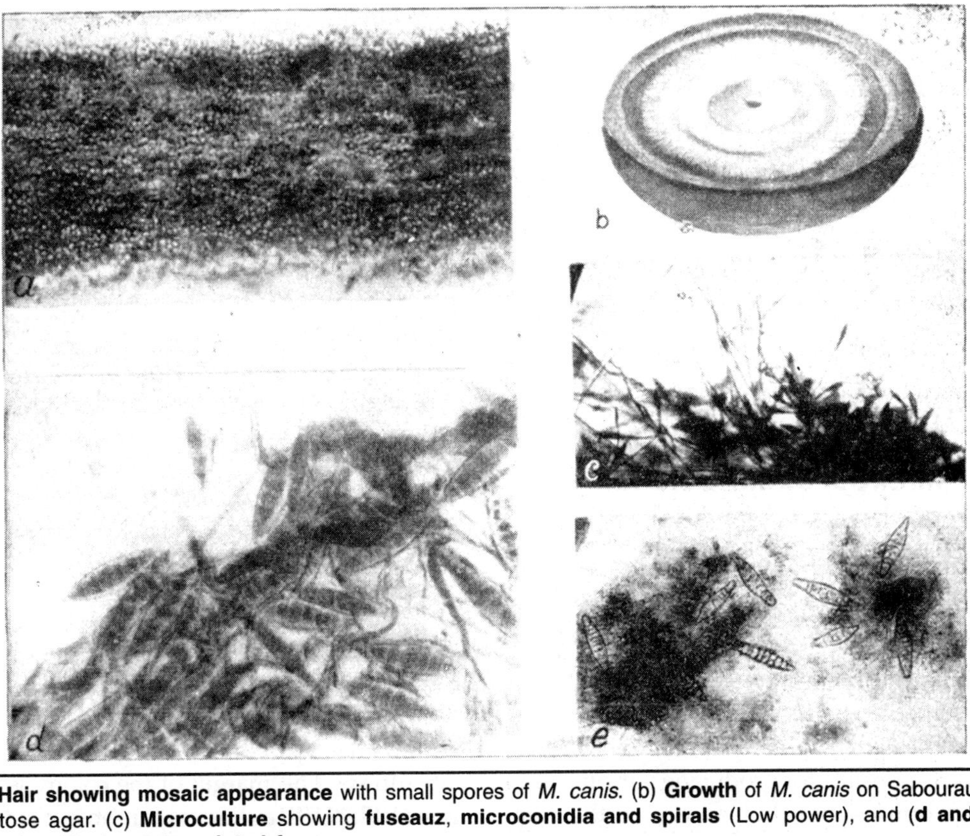

(a) **Hair showing mosaic appearance** with small spores of *M. canis.* (b) **Growth** of *M. canis* on Sabouraud's maltose agar. (c) **Microculture** showing **fuseauz, microconidia and spirals** (Low power), and (**d and e**) **showing Pluriseptate pointed fuseaux.**

in colour but at the same time covered with abundant wooly, aerial mycelia. (**Pl. 29.VII** *b*).

Pleomorphism starts with regularity after 4 to 5 weeks. Three types of pleomorphism have been described by Sabouraud. (a) A downy, white, exuberant growth, (b) an immersed growth with a smooth, glabrous, most, brownish surface, (c) a velvety or shaggy growth which is the most frequent and the stable form of pleomorphism.

Microculture

The microculture shows a large number of fusiform macroconidia tapering at the end. When young, they are non-septate but with age they become pluriseptate, showing usually 5 to 6 septa. At the tip of the fuseaux (macroconidia), fine hair like processes are often seen. Old macroconidia are tuberculated. Microconidia, mycelial racquets and hyphae pectinate are present (**Fig. 29.2**). Spirals are commonly seen in natural media like barley and wheat (**Pl. 29.VII** *c*).

Wood's light

The infected hairs appear bright, clear, green and resemble those of *M. audouinii*. After ten days colonies are bright and clear. Centre of the growth may be lavender-blue, salmon-pink or flesh-ochre in colour. **The mid-zone may be lavender blue** and the edge mouse grey.

Animal inoculation

Common laboratory animals are readily infected after inoculation with this fungus.

MICROSPORUM GYPSEUM

Gulart and Grigorakis, 1928.

Synonyms

Achorion gypseum, M. fulvum, M. flavescens, M. scorteum, Closterosporia fulva, M. xanthodes etc.

This produces ringworm of animals, like horses, monkeys, rabbits and mice and may be communicated to human particularly ***children***. This is responsible for producing various types of ringworm lesions in human, e.g., **erythemato-squamous ringworm, ringworm of the hair with kerion or favus**. In rabbits, it may produce favus (**Dey and Kakoti, 1955**).

Invasion of hair *causes* ***ECTOTHRIX SPORULATION*** causing dull fluorescence of hair and perforates hair *in vitro*.

Epidemiology

It is basically a saprophytic fungus, free-ling in the soil and is aptly called geophilic. It occasionally causes ringworm in man and animals and soil is the source of infections for both man and animals. It was first recorded in a rabbit by Dey & Kakoti in 1955 in Assam. They perforate hair baits *in vitro*.

Clinically, infection in human is occasional, in whom it may produce ***tinea corporis*** or ***tinea capitis***. The lesion is usually single or a few and may be inflammatory, bullous or impetiginous with rapid course, or resolution occurs. **In rabbits** it may form **scutula**.

Colony morphology

White downy colony, turning pink or buff with age and producing strawberry red pigment, diffusing into the medium. In old cultures, the centre is deep red.

Direct examination

The hair during the early infective stage shows spores arranged in beaded form but they may also show spores without any linear arrangement. The infected hair shows a sheath of **spores with rosaries**. When the infective stage is complete, the ***microscopic appearance*** of the hair is indistinguishable from that of other *microspora*. Lesions of the smooth skin show short chains of spores in the scale. ***FAVIC SCUTULA*** show mycelia and spores (**Pl. 29.VII** *a*).

Cultural characters

On Sabouraud's agar, the colony grows fast with brownish white, powdery, aerial mycelia often giving the appearance of suede. The cinnamon brown colour of the colony is characteristic. A central umbo almost in variability appears about 2 weeks after subculture. The margin of the colony is often fringed (**Pl. 29.VIII** *e*). Furrows are absent but there may be concentric rings in some cases.

PLATE 29.V

MICROSPORUM GYPSEUM
(From favus in a rabbit)

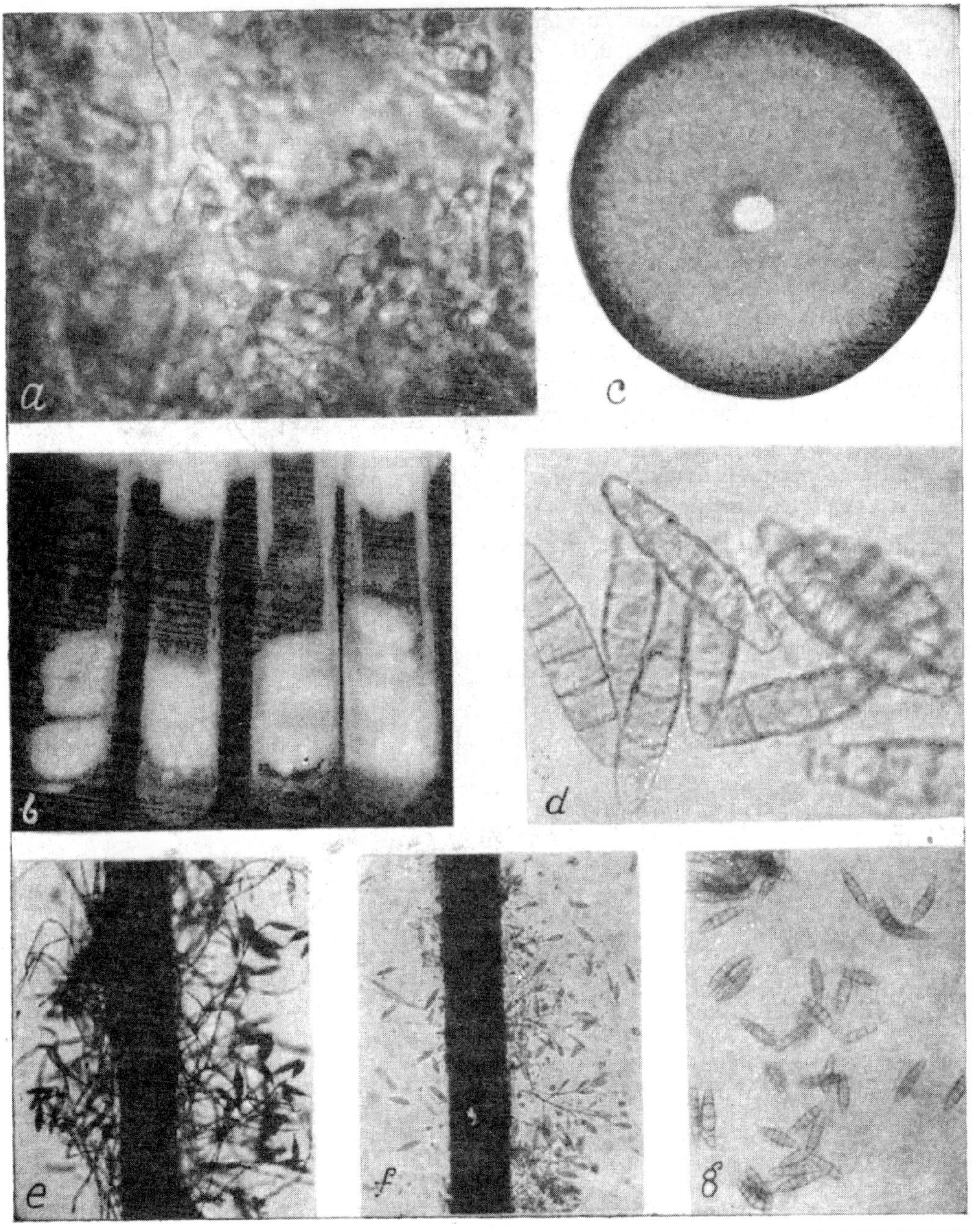

(**a**) Mycelia in the favic scutula, high power (50 × 12). (**b**) Growth in Sabouraud's maltose agar—primary culture and subcultures. (**c**) **Giant colony on Sabouraud's maltose agar**. (**d**) Macroconidia, high power, 50 × 12. (**e**) Hair is found naturally infected in the soil showing macroconidia. (**f**) Fungus growing from the infected hair, low power, 8 × 6. (**g**) Macroconidia under low power, 8 × 10.

Pleomorphism is characterised by growth of white sterile mycelia.

Microculture: Numerous macroconidia are produced. They are fusiform in nature, 8 to 12 by 15 to 50 μ. They are pluriseptate and often ellipsoid with thin walls, but tuberculated when old. (**Fig. 29.3** and **Pl. 29.VII** *a, c, f, g*). **Mycelial chlamydospores and microconidia** are often seen (**Fig. 29.3**).

Nutritional requirement: None known.

Variants: *M. gypseum* may give **two types of variants** which **form two types of perfect stages**, namely, ***N. incurvata*** and ***N. gypsea*** but **in the imperfect conidial forms** it is **difficult to distinguish these variants**. The strains related to *N. gypsea* produce more spreading and coarsely **granular colonies** with brighter surface colour and veins of deeper pigment under the colony. **Macroconidia** also appear to be somewhat **more plump**. A red pigmented form is sometimes seen isolated from soil or small mammals.

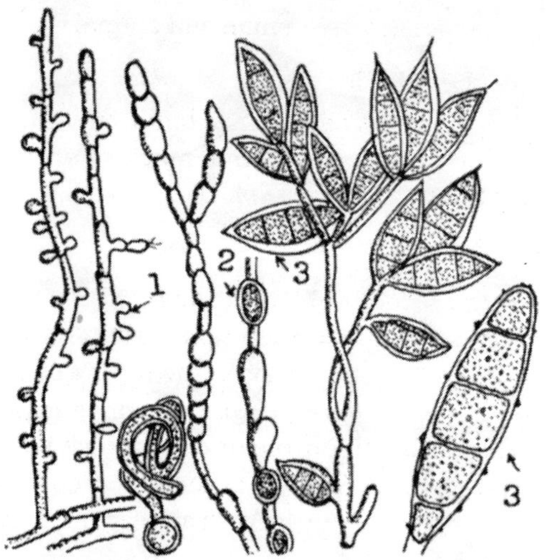

Fig. 29.3: *M. gypseum*, 1. Microconidia, 2. Chlamydospores. 3. Septate macroconidia, tuberculated when old.

The **imperfect conidial strains of** *M.* ***gypseum*** may give rise to ***N. incurvata*** and ***N. gypseum***.

Woods's light

The **infected hairs fluoresce light green** as in other microspora. The growth is dull and clear, cinnamon brown in colour throughout.

Animal inoculation

Young laboratory animals can be easily infected by *M. gypseum*.

MICROSPORUM GYPSEUM VAR. FULVUM

The *macroconidia* are longer and more clavate in form. They form in more open clusters rather than in future. The perfect stage is known as ***N. fulva***. This is considered to be a variant of ***M. gypseum*** rather than a new species as suggested by Stockdale (1963).

Critical characters

(1) **Typical macroconidia**, (2) **Colony characters**, (3) **Form asci** with culture grown from **single ascospores**.

Perfect Stage

Two perfect stages are known to be derived from ***M. gypseum*** namely, ***N. incurvata*** and ***N. gypseum*** often from soil. Rarely ascigerous form may be isolated from ringworm lesion. From this, it is often called ***N. fulva***.

Characterisitcs

(1) ***Macroconidia***, (2) ***colony*** **character**, (3) ***cultures*** **grown from single ascospores** of *N. incurvata* and ***N. incurvata***, or ***N. gypsea***.

MICROSPORUM DISTORTUM

This is better called ***Microsporum canis*** var. ***distortum*** being identical with and fluorescent like ***M. canis***.

Clinically, it produces a similar disease as caused by *M. canis*.

Epidemiology

It is an unusual cause of microsporic ***tinea capitis*** in man and often reported from United States, traced to contact with pet South American monkeys. Infection in dogs and other animals occur. It is considered to be a variant of ***M. canis***. No natural reservoir of the species is known as yet.

Colony characters : (1) **Growth** of New Zealand isolates is rapid but the strains isolated in United States are (2) **slow growing, dense**

and fuzzy. Colonies are flat with a tendency to develop radial grooves. Surface is velvety or fluffy, white to tan. (3) **Some strains have little aerial mycelium** and may appear waxy. Reserve side is yellowish-tan.

Microculture

Macroconidia* are numerous**, thick-walled with a rough surface (echinulate), markedly distorted in shape and often grossly bent. **Clavate *microconidia are often **abundant, pear-shaped and sessile.**

Variants are seen with various surface growths or none at all (see colony morphology) with variable pigment production.

Perfect stage

Not known.

Special media

Like *M. canis*, growth and sporulation occurs in polished rice (natural medium).

Special characteristics

Distorted macroconidia.

MICROSPORUM VANBREUSEGHEMII

Hair : The organism has got special affinity of perforating hair *in vitro*. It causes large spore *ECTOTHRIX* infection of hair which is **not fluorescent**.

Clinically, it is a rare causative organism of hair ringworm of human and animal.

Epidemiology

It has only been reported from United States.

Cultural character

It is a rapidly growing organism with cottony of fluffy surface growth which develops a lavender pink to deep-rose or buffy powdery surface due to *macroconidia*. The pigment on the reverse is light yellow or lacking.

Microculture

Microconidia are numerous, pyriform to ovate in shape, not diagnostic.

Nutritional requirement

None required.

Variants

(i) Fluffy forms are isolated and (ii) downy pleomorphism occurs readily.

Special media

Not required.

Perfect stage : ***Nannizzia grubyia***.

Critical characters

(1) **Colony morphology**. Macroconidia with rough thickened wall (Ajello), (2) **Perfect stage**.

GENUS : KERATINOMYCES

KERATINOMYCES AJELLOI

(Vanbreuseghem) Ajello, 1968

Invasion of hair

It is a rare cause of ringworm in man and animals and at best has the power of perforating hair *in vitro*.

Clinically

It causes disease in man and animal.

Epidemiology

It is found in soil mainly, but it may cause infection in man and animal much rarer than infection with ***M. gypseum***.

Colony morphology

It is a rapidly growing fungus. It may be flat or somewhat heaped up and folded. The surface is finely powdery or downy and cream to tan or orange tan in colour. Sometimes it may produce reddish purple or deep bluish balck colour on the reverse side of the colony. Pleomorphic downy growth appears rapidly.

Microculture

Macroconidia are numerous on the **powdery surface**. They are long, slender, smooth and cylindrofusiform with parallel walls tapering at each and composed of 8 to 12 cells.

The walls of macroconidia are wide (thicker than those of ***M. canis***) and have a smooth surface. They resemble those of ***E. flccosum***. **Macroconidia** are abundant in some strains.

Microconidia if at all present is not diagnostic. They are ovate to pyriform.

Perfect stage is ***Arthroderma uncinatum*** which has curved, rough, **peridial hyphae** with dumble-shaped cells.

Special media

Perfect stage obtained from mating of strain on soil baited with horse hair.

Variants

Strains with powdery surface and purple reverse side is the type in which pleomorphism occurs readily.

Critical characters

(1) **Buffy powdery surface of the *colony*,** (2) smooth walled, thick, **cigar-shaped *macroconidia*,** (3) ***morphology*** **of perfect stage.**

ENCOTHRIX TRICHOPHYTA SPECIES (CRATERIFORM GROUP)

TRICHOPHYTON TONSURANS

Malmsten, 1845.

Synonyms : *T. crateriforme* Sabouraud, 1902; *T. effractum* Sabouraud, 1910; *T. fumatum* Sabouraud, 1916; *T. regulare* Sabouraud, 1910; *T. Polygonum* Sabouraud, 1910; *T. umbilicatum* Sabouraud, 1910, etc.

The organism causes **endothrix trichophytosis or black-dot ringworm of the scalp** and this is generally seen to infect **children in France** and other parts of Europe. But **seven cases were recorded by Dey and Maplestone (1941)** in an epidemic in a European school **in Calcutta**. The lesions on the scalp are small patches, usually multiple, without any inflammatory changes. The hairs under the scales appear like letters 'S' or 'Z' or twisted spirally. It might infect the glabrous skin as an extension from the scalp; nails may be affected secondary to scalp infection. In smooth skin the lesions are circinate, non-inflammatory, with scale formation (tinea circinata).

Invasion of hair

It is an endothrix trichophyton forming chains of round spores inside the hair producing **"black dot" ringworm** or **"chicken skin" effect on the scalp**. The hair is scraped, as they cannot be plucked due to their brittleness.

Clinically, *T. tonsurans* primarily causes scalp ringworm but may cause *tinea corporis* and *capitis* in children **(Pl. 29.VI)**, occasionally onychomycosis or rarely *tinea pedis*. After attainment of puberty (12 years) there may not be spontaneous regression of *tinea capitis* and may even affect old women. ***Kerion*** **rarely occurs** (also see the following).

Two groups of fungi are responsible for this infection in India, namely, Microspora and Trichophyta. Out of 198 cases studied by the senior author in the Calcutta School of Tropical Medicine and in other places of India from 1932 to 1950, the following infections were encountered **(Dey, 1954)**.

Direct examination : In the hair, spores are large, seen in chains along the hair shaft **(Pl. 29.VII)**. Mycelial filaments are 4 to 5 μ in diameter and old mycelia show spores with

Table 29.13 : Difference Between Microsporum And Trichophyton Ringworm

	Microsporum (scaly) ringworm	Trichophyton (black-dot) ringworm
Patches on the scalp :	Only one or two **big scaly patches**.	**Multiple small "bklack-dot" patches**.
Hair stumps :	Stumps are longer and stick out in different directions. Greyish sheath forms at the root of the hair.	Stumps are very small and flush with the surface of the skin. No sheath is seen. They break easily at the time of epilation.
Microscoscopy :	Spores are small (3 μ) and give a **mosaic appearance**.	Spores are large (5 μ) and give a **beaded appearance** along the long axis of the hair.
Culture :	Shows characteristic growth of the species.	Shows characteristic growth of the species.

PLATE 29.VI

BLACK-DOT RINGWORM (TRICHOPHYTON TONSURANS)

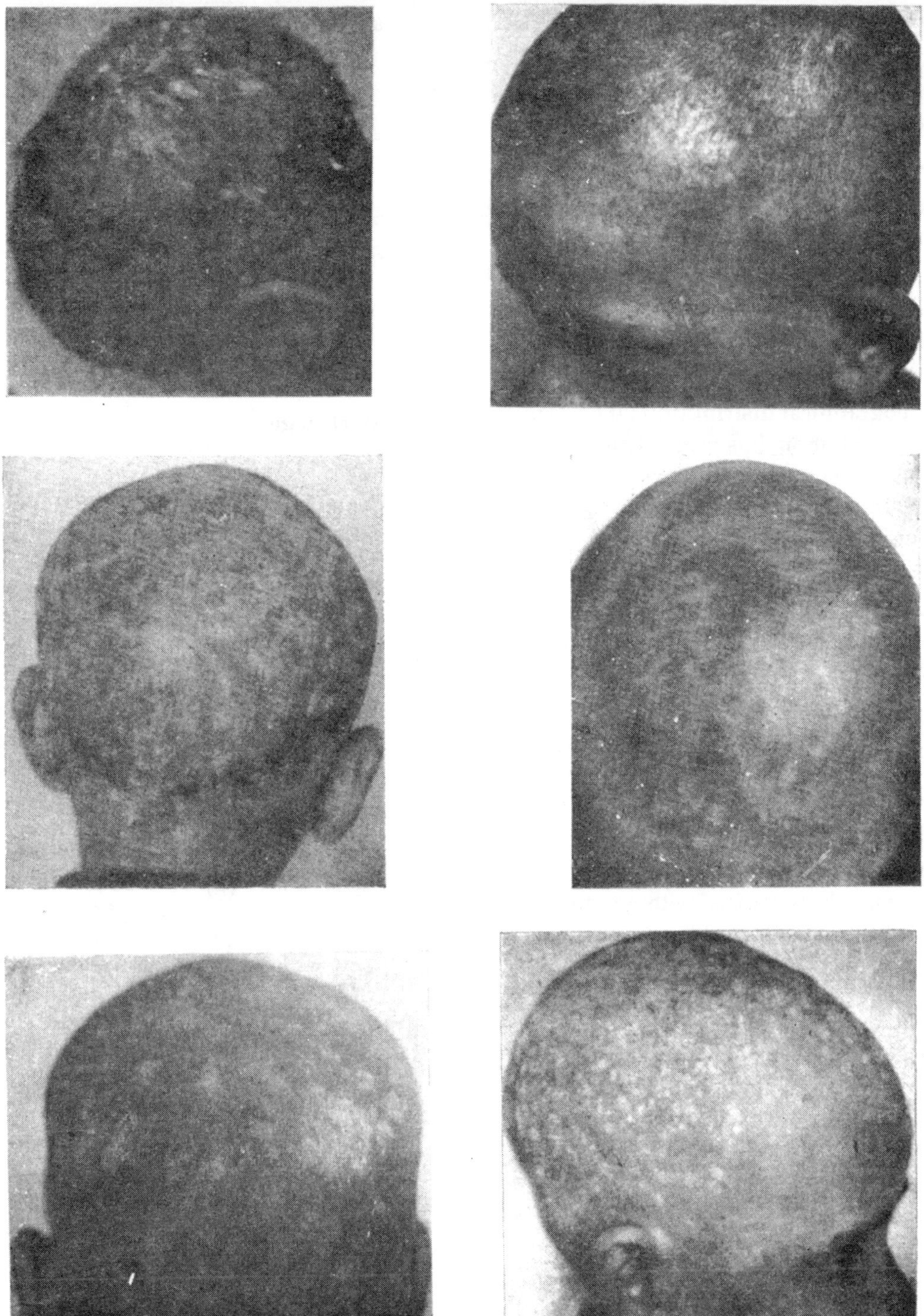

Black-dot ringworm of the scalp caused by *T. tonsurans*. Note the small, multiple areas of baldness caused by stumps of broken hair.

Species	No. of cases	Percentage
Microsporum audouinii	94	47.47%
Trichophyton violaceum	81	40.91%
Trichophyton crateriforme	7	3.54%
Trichophyton ferrugineum	8	4.04%
Trichophyton mentagrophytes	8	4.04%
	198	100%

square-cut ends. The spores are differentiated from those of other endothrix trichophyta, e.g., *T. accuminatum* or *T. violaceum* which are oval in shape.

Epidemiology : It infects only human subjects.

Geographical distribution : It has a world wide distribution. It is probably an old world species but spreading to Mexico, Canada and Latin American due to Latin immigrants. It is common in France and England.

Colony characters : On primary culture a flat powdery, yellowish colony with a reddish or yellowish back ground appears. This develops into a folded "*crateriform colony*" with a suede-like surface with buff or brownish pinkish tints. The under surface of the colony may show reddish brown or dark brown pigment diffusing into the medium of Sabouraud's agar. The young colonies often fluoresce green under *Wood's light.*

The growth is slow and grows a white or cream coloured, velvety surface with a central crater which develops with age. Pleomorphism is rare.

Microculture

Microcondia are most diagnostic **features** (Fig. 29.4). They are abundant and often large in size, the shape being that of tear drop or club shaped on open branching clusters. The individual microconidia has a tendency to enlarge forming clavate or balloon forms. *Microconidia* are thick-walled and less perfectly formed than those of *T. mentagrophytes*; arthrospores are frequent.

Nutritional requirements

Like *T. violaceum, T. tonsurans* grows better if thiamine is available.

Variants

T sulphureum variety is often distinguished specially in England characterised by a sulphur yellow colour of the colony. Differently folding types were encountered in France namely *crateriform, cerebriform, plicatile,* etc.

Perfect stage

Not known.

Special media

Thiamine may be incorporated to differentiate *T. tonsurans* from *T. rubrum, T. mentagrophytes* and *T. megninii.*

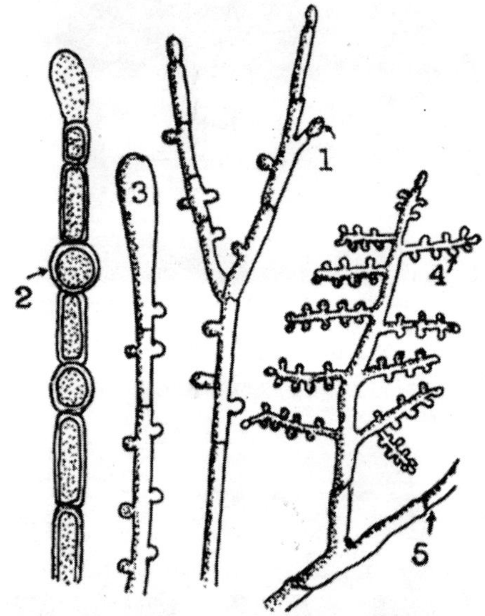

Fig. 29.4 : *Trchophyton tonsurans,* (1) Microconidia, (2) Chlamydospore, (3) Terminal club, (4) Microconidia on compound branches.

Critical characters

(1) *Colony* form, (2) *microconidia* and (3) *thiamine* requirement.

Plate 29.VII
TRICHOPHYTON TONSURANS

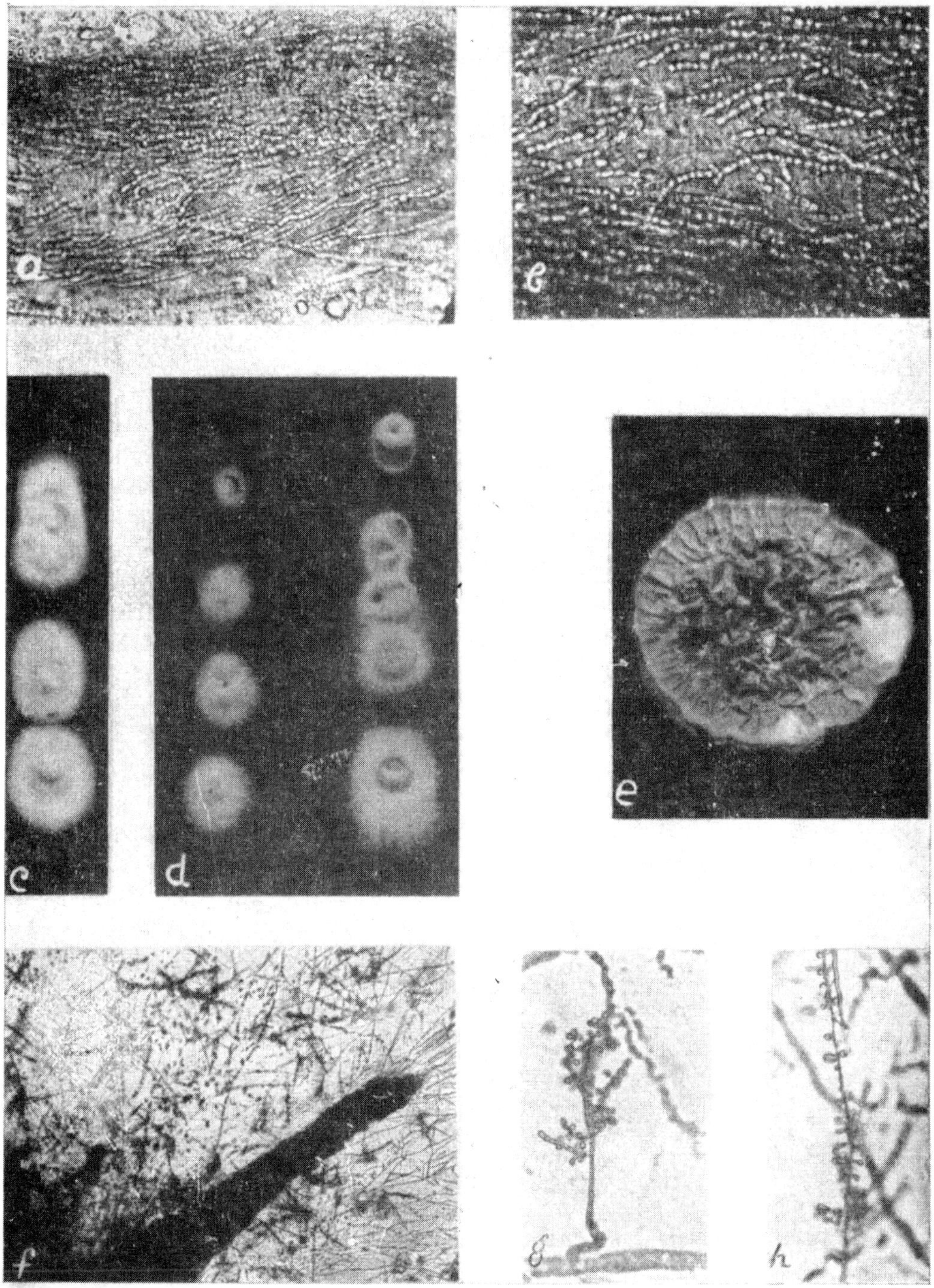

Hair showing endothrix trichophytosis caused by *T. tonsurans* : (**a**) 43 × 5, (b) 43 × 10; (**c** and **d**) Primary culture on Sabouraud's maltose agar; (**e**) Giant colony with a central crater; (**f**) Microculture from infected hair; (**g** and **h**) micronconidia on compound branches

FAVIFORM GROUP

TRICHOPHYTON SCHOENLEINII

Synonyms

Achorion schoenleinii Remak, 1845; *Oidium schoenleinii* Lebert, 1845.

Trichophyton schoenleinii causes favus of the scalp, smooth skin and nails in Europe. It is characterised by formation of scutula on the scalp with deposit of mudplaster like crust. When crusts are removed, a raw surface is exposed. On healing, scars are formed resulting in alopecia in patches. On the smooth skin, vesicular or ringworm lesions form followed by accumulation of favic material consisting of scales and fungus mixecd with secretion of a fatty nature. When the nails are affected, they are characterised by hypertrophy with heaped up favic material.

Invasion of hair

It is characteristic. Within the hair, air spaces are seen due to degeneration of hyphae; arthrospores are few or absent. Infected hairs remain intact (does not break) and it passes through saucer shaped scutula or godets. They collectively form mud plaster like masses which when removed show raw surface epithelium. Similar crusts may form in *T. mentagrophytes* or rarely in *T. violaceum* infections.

Clinically, the fungus produces favus, characterised by formation of scutula which collectively causes a mud plaster appearance due to deposit of crusts. On removal of these crusts a raw surface is exposed followed by scarring, atrophy and permanent loss of hair. Sometimes *T. schoenleinii* produces scaling *tinea capitis* which might be mistaken for *seborrhoea capitis* or dandruff. This fungus may produce infrequently *Tinea corporis* or rarely onychomycosis. Fatal brain abscesses have also been reported **Pl. 29.VIII**).

Epidemiology

The fungus causes anthropophilic infection causing deformity of the scalp.

Geographical distribution

The disease is widespread in Eurasia and North Africa, where **T. violaceum** also shares a prominent part amongst the deformity caused by *Tinea capitis*. The disease is less common in Western hemisphere. In India, it is rare but seen in amongst those races in Kashmir, Punjab and Rajasthan who wear turbans. These are endemic foci, for the disease may be seen in the same family in these endemic areas.

DIRECT EXAMINATION

Favic scutula show mycelia with masses of **branching hyphae** and often chlamydospores. Hair shows **endothrix infection** with spores or fragments of mycelia in the hair which often contain air spaces attached to its margin.

PLATE 29.VIII
TRICHOPHYTON SCHOENLEINII

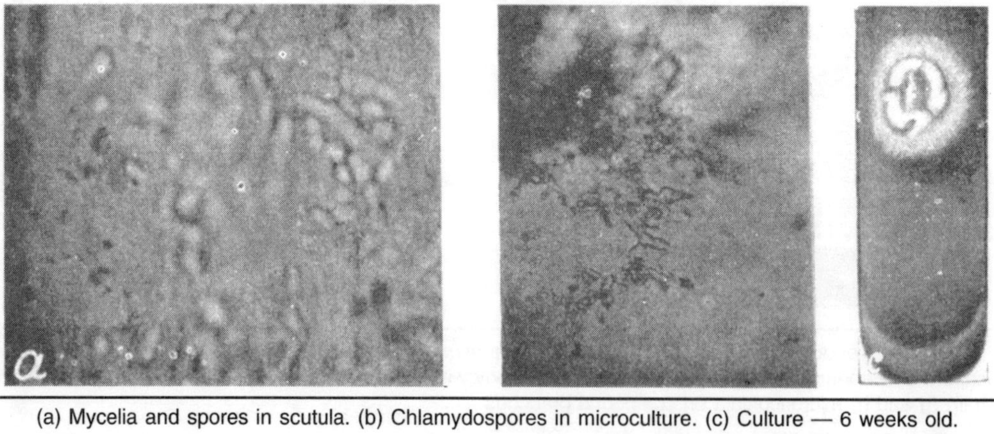

(a) Mycelia and spores in scutula. (b) Chlamydospores in microculture. (c) Culture — 6 weeks old.

PLATE 29.IX
TRICHOPHYTON ACTONI

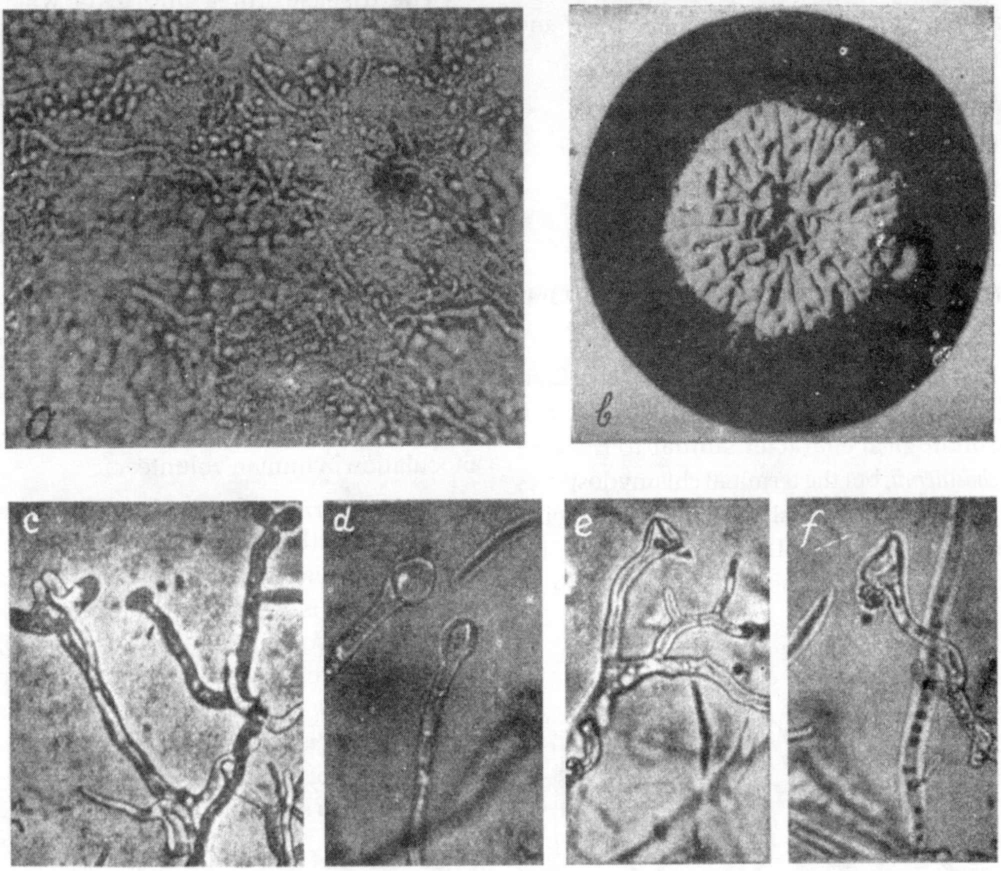

(a) Scutula showing mycelia and spores. (b) Giant colony. (c) Favic chandeliers.
(d) Oval and (e, f) Funnel-shaped terminal chlamydospores.

Culture : Characteristic growth appears in about 3 weeks. This shows a heaped up, convex, glabrous faviform, cream coloured growth having a cerebriform or honeycomb appearance (**Pl. 29.VIII**). Pleomorphism is uncommon.

New isolates on Sabouraud's agar are often glabrous or sometimes covered with fine white downs, with a convoluted surface of bees wax or plaster of paris. The colonies are raised and convex but a part of it is embedded in the agar at the margin of the growth often cracking the medium.

Microculture : Conidia either *macro* or *micro* are not seen; characteristically antler-like terminals with terminal branching swollen hyphal tips (nail head) are often seen looking like terminal chlamydospores (**Fig. 29.5**). They are often funnel shaped in *T. actoni*, isolated in India.

Nutritional requirements : Not known.

Varianst : *T. actoni* in India was isolated from Rajputana (**Dey & Mapl.estone, 1936**). Sometimes glabrous growth is covered with fine downs. These are common in isolates from Kashmir.

Wood's light : Infected hairs show greenish fluorescence. The culture gives a dull, olive-grey appearance throughout.

Animal inoculation : All the common laboratory animals may be infected.

Perfect stage : Not known.

Special media : Growth on Sabouraud's medium without thiamine distinguishes this species from *T. verrucosum, T. violaceum* and *T. concentricum.*

Critical characters

(1) **Typical glabrous growth;**

(2) **Fair type of antler hyphae;**

(3) **Growth in absence of thiamine.**

TRICHOPHYTON (ACHORION) ACTONI

(Dey and Maplestone, 1936)

Dey and Maplestone (1936) observed a favic trichophyton having a cultural and morphological character similar to those of *T. schoenleinii*, but the terminal chlamydospores are very often funnel shaped instead of being oval or round (**Pl. LII**). It is responsible for causing favus in Rajasthan and Kashmir in India.

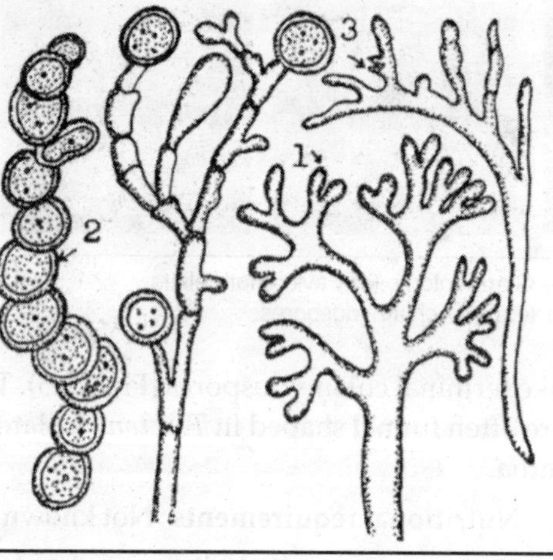

Fig. 29.5 : *Trichophyton schoenleinii* (a) Favic chandeliers; (2) **Intercalary chlamydospores**; (3) **Hyphae pectinate**

TRICHOPHYTON CONCENTRICUM

Blanchard, 1896

Synonyms : *Endodermophyton concentricum* Castellani, 1910; *Trichophyton Castellani* (Perry) Castellani, 1908; *Trichophyton mansoni* Castellani, 1905; *E. indicum* Castellani, 1911; *E. tropicale* Castellani, 1919; *E. roquettei* Fonseca, 1925.

History

The disease, tinea imbricata was first described by Dampier in "Voyage Round The World" published in 1789. Tilbury Fox described the disease in 1874 as **Tokelau ringworm** but he considered the fungus to be identical with that of European ringworm. Manson (1879-82) was the first to describe the disease in China in any detail and he gave the name **tinea imbricata**. He considered the fungus to be a non-cultivable trichophyton, to which Blanchard gave the name ***Trichophyton concentricum***. Castellani in Ceylon (1910-11) was the first to cultivate the fungus and proved it to be the causative organism by successful inoculation in human volunteers.

T. concentricum is recorded from **Oceania**, Asia, and **North, Central, and South America**. It is not present in the New World, North of Mexico, and it is absent from Europe and Africa. A supposed isolate of this species from Africa, when studied by Georg, proved to be ***T. verrucosum***. Pijper's report of this fungus in South Africa is erroneous, as the published descriptions of his isolate indicate that he was dealing with a saprophyte.

Invasion of hair

Not known to invade hair. They show **dull yellowish fluorescence.**

Clinically, the fungus causes **Tinea imbricata** (Tropical ringworm or Tokelau) on the glabrous skin and the lesion spreads with formation of concentric rings to cover the entire body on a gyrate pattern. The lesions are intensely pruritic. It is characterised by the presence of extensive tissue paper-like concentrically arranged scales (**Pl. 29.IX**).

Epidemiology

It is an anthropophilic species common in the tropics and often seen endemic amongst the aboriginal tribes.

Geographical distribution

The major endemic area is South Pacific Islands. The disease occurs commonly in the

PLATE 29.X
FAVUS (Scalp)

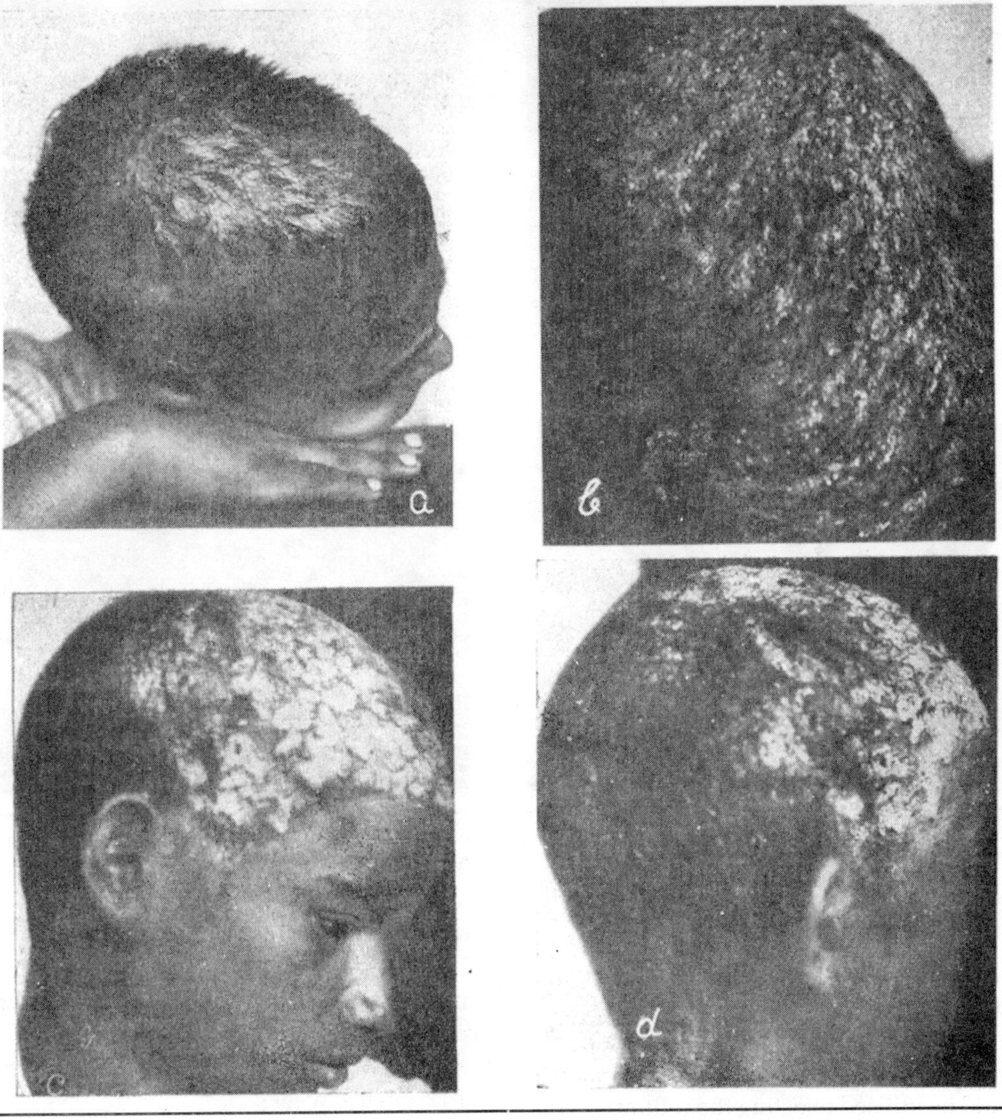

Favus caused by—(*a and b*) *T. schoenleinii*; (*c and d*) *T. actoni*

Malay Peninsula, South Pacific Islands, Southern China, **India**, **Ceylon**, South Africa, and **Central America**. **In India**, it is very common in the aboriginals and tribal population of Assam, residing in the submountainous regions, namely the Mikirs, Garos, Khasis, Ravas, Nagas and Aborrhs (Dey and Maplestone, 1941) and rarely in the planes. The disease was described by Acton and Ghosh (1934) in a Bengali inhabitant residing in the plane at the foot of Garo hills.

DIRECT EXAMINATION

Scales show septate intersecting mycelia (**Pl. 26.9**, a) easily stained with McGure's stain or seen under a coverslip preparation with sodium sulphide.

PLATE 29.XI
TRICHOPHYTON CONCENTRICUM

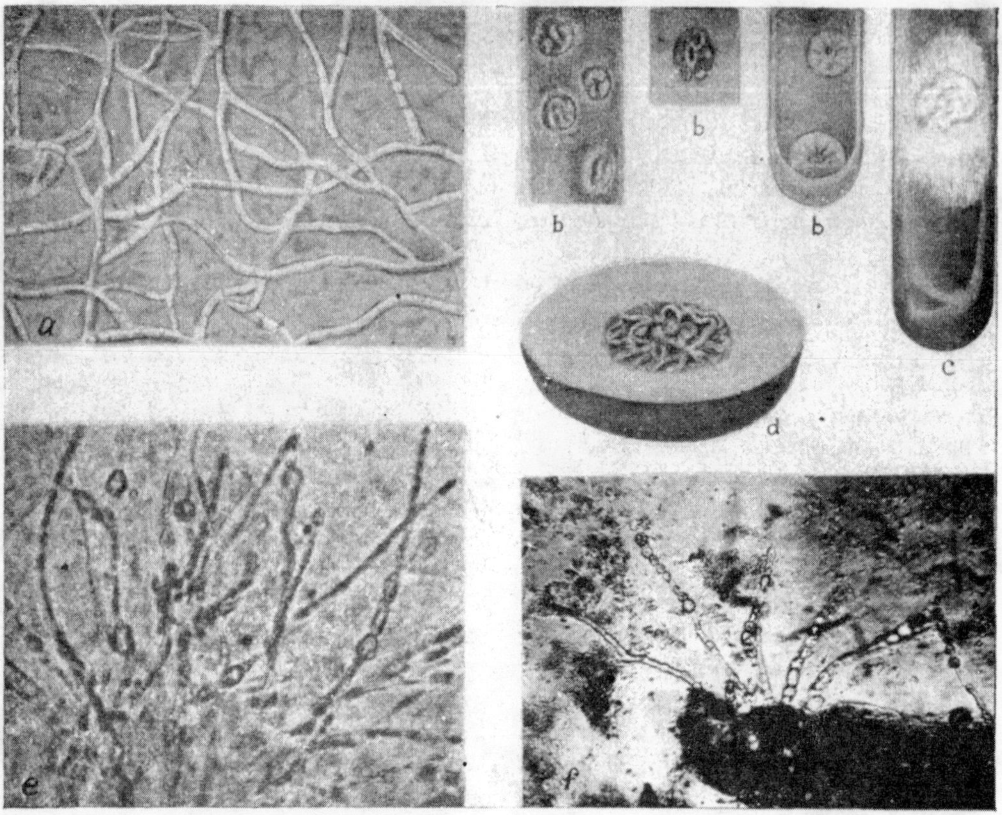

(a) Scales of tinea imbricata showing **mycelia**. (b) Primary cultures from scales. (c) Subculture. (d) Giant colony. (e) Microculture showing **chlamydospores**, (f) growth of the fungus in hanging-drop preparation from a scale; × 215

Cultural characters

Primary culture in solid medium is difficult owing to the fact that the redundant portion of the horny layer of the epidermis does not give any growth. Fresh material adhering to the skin should be utilised rejecting the redundant part sticking out.

For **culture**, treat the scales in alcohol for 5 minutes and inoculate tiny bits of infected material in glucose broth, Sabouraud's glucose agar and Czapek's synthetic agar. In **glucose broth**, the growth appears on the 3rd or 4th day and forms as a powder-puff ball of mycelia. On Sabouraud's glucose agar, visible growth occurs on the 4th or 5th day but in Czapek's synthetic medium, the growth is delayed till the 6th or 7th day. Subculture in Sabouraud's medium maintains the cultural character for a long time without pleomorphism. The growth is about 1 cm in 3 weeks. The character of the **growth is** ***cerebriform*** greyish brown colour, covered with white powder. The growth increases to the size of 1.5 cm in 6 weeks.

On glucose agar, the growth is raised, yellowish brown in colour. Under reduced oxygen tension in glucose agar, the colour develops to a salmon pink, when the tube is sealed with a rubber cap.

Microculture shows **branching hyphae without** ***micro*-or** ***macroconidia***. They show arthrospores, and mycelial agents are common on Sabouraud's glucose agar. Macroconidia are absent and microconidia are rudimentary (**Fig. 29.6**).

PLATE 29.XII
TINEA IMBRICATA
(*T. concentricum*)

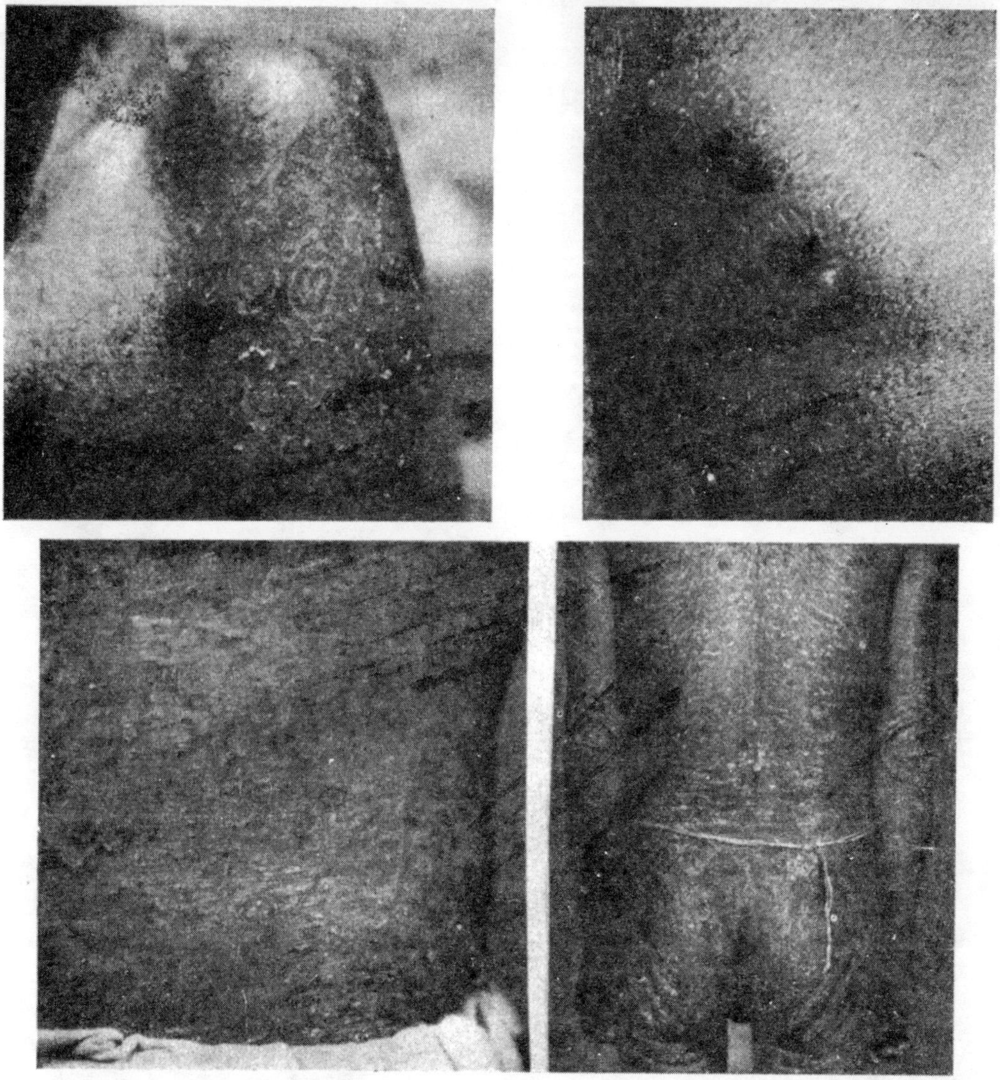

Concentric rings are seen in the upper left and parallel wavy lines in the lower left figure

Special nutritional requirements : About 50% strains are stimulated by thiamine.

Variants : The colour varies from whitish to brown. The whitish variety is usually seen after isolation due to white downs on the surface.

Perfect stage : Not known.

Special media

Response to thiamine in about 50% isolates occurs and this may be utilised as an aid to differentiate from *T. schoenleinii* and *T. verrucosum* with which it is often confused being glabrous growth. Ahearn observed on bean pod agar lateral macroconidia in absence of microconidia.

Critical characters

(1) **Characteristic lesion** in endemic areas, (2) **colony morphology**, (3) **response** of isolates **to thiamine**, (4) **lateral macroconidia on bean pod agar**.

PLATE 29.XIII
RINGWORM OF THE GLABROUS SKIN

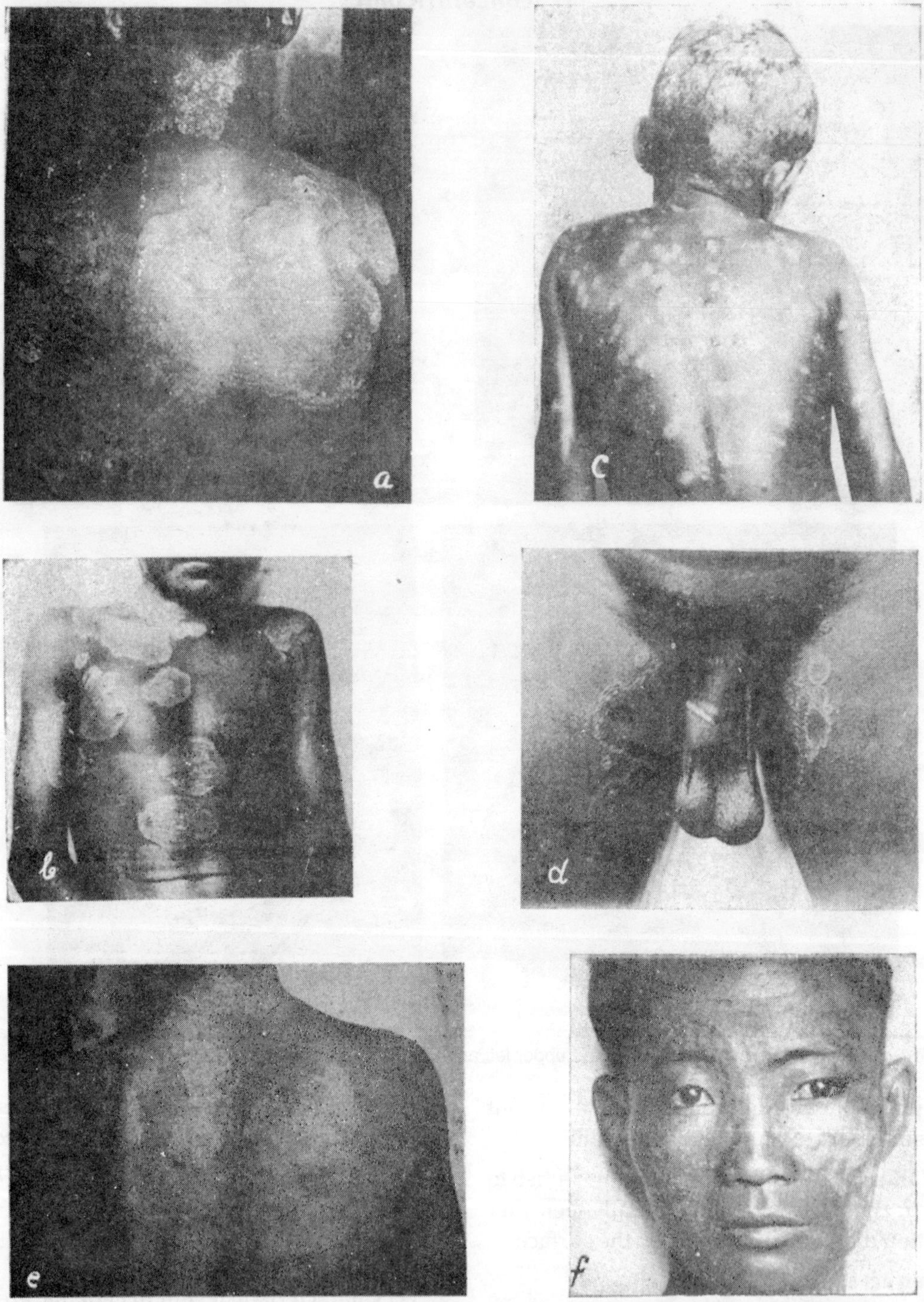

(a), (b), (e) and (f) caused by ***T. purpureum***. (c) Extension from ringworm of the scalp caused by ***T. violaceum***, (d) **Eczema marginatum of Hebra** caused by ***Epidermophyton floccosum***

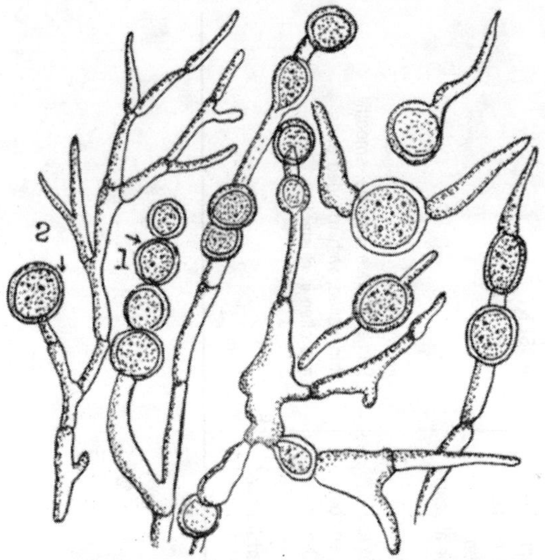

Fig. 29.6 : *T. concentricum*, (1) Intercalary chlamydospores; (2) Terminal chlamydospores

Wood's light : The culture shows a dull yellowish fluorescence.

Animal inoculation : Inoculations into human volunteers has been successful.

GENUS EPIDERMOPHYTON

EPIDERMOPHYTON FLOCCOSUM

(Harz) Langeron & Milochevitch, 1930.

Synonyms : *Acrothecium floccosum* Harz, 1871; *Trichothecium floccosum* Harz, 1870; *Epidermophyton inguinale* Sabouraud, 1910; *E. plicarum* Nicolau, 1913; *E. clypeiforme* MacCarthy, 1925; *Trichophyton inguinale* Sabouraud, 1907; *T. intertrigenis* Sabouraud, 1905; *T. cruris* Castellani, 1908; *e. cruris* Castellani and Chalmers, 1910.

It invades the skin producing a condition of eczema marginatum of Hebra or ringworm of the groin, commonly known as dhobie itch. But it may affect the glabrous skin of other parts of the body as well as the nail. It does not affect the hair.

Invasion of hair : Not known.

Clinically, it causes **tinea cruris, tinea corporis** (**Pl. 29.XI**) and **onychomycosis**.

Epidemiology : Infection occurs only in man and may be contracted from seats or bathing in a public bath or common showers.

Geographical distribution : World wide.

DIRECT EXAMINATION

Mycelia containing chains of spores are seen in a fairly large number. They are usually wavy in appearance (**Pl. 29.XII** *a*).

Colony character

The culture in a giant colony shows suede-kike structures on a folded glabrous growth. Olive green or yellow-olive colour is characteristic. Pleomorphism often appears in sectors characterised by white and densely fluffy downs. It is difficult to maintain growths as they become downy at room temperature.

Cultural characters

It is a slow growing fungus. In Sabouraud's glucose medium, the colour is lemon yellow olive-green with short, white, cottony duvet on the surface which are sterile. The centre of the **giant colony is folded with radiating furrows** (**Pl. 29.XII** *b*). Pleomorphism is characterised by formation of white duvet on the surface of the growth.

Microcultures : They show characteristic **blunt**, ***club-shaped macroconidia*** attached together (**Fig. 29.7**), often terminally in groups,

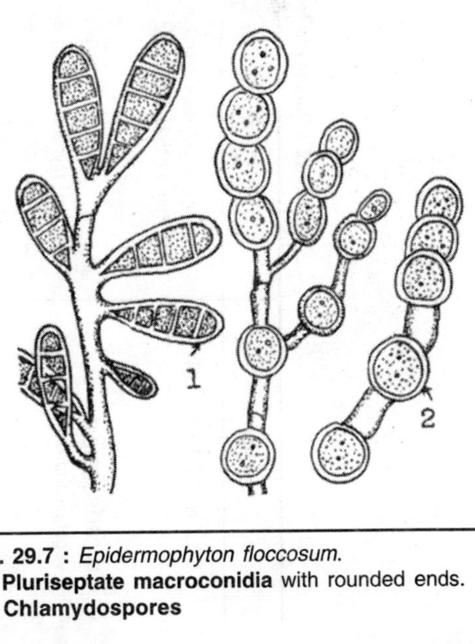

Fig. 29.7 : *Epidermophyton floccosum.*
(1) **Pluriseptate macroconidia** with rounded ends.
(2) **Chlamydospores**

Table 29.12 : Diagnostic features of Microsporum, Trichophyton And Epidermophyton Species

Description	*Microsporum*	*Trichophyton*	*Epidermophyton*
COLONY MORPHOLOGY	Cottony or powdery aerial mycelia, varying from white to buff colour	Cottony, glanular or grabrous (smooth and waxy), pigment may be present; loss of pigment on subcultures	Powdery or velvety with **radiating furrows** on the surface and lemon yellow in colour
MICROSCOPICAL Vegetative forms	Racquet mycelia, hyphae pectinate, chlamydospores well formed, spirals rare in species like *M. canis*	Racquet mycelia, chlamydospores, spiral hyphae, nodular organs, rarely hyphae pectinate. In faviform types—favic chandeliers and pectinate bodies in addition to racquets and chlamydospores	Occasionally racquets and frequently chalmydospores present
Microconidia	Small, clavate, 3 by 5 μ, borne along the hyphae, sessile or fixed on facets	Numerous small, clavate, subspherical or spherical, 2 by 4 μ, borne along the hyphae (thyrse) or in clusters (*en grappe*); they are rudimentary or absent in faviform group	**ABSENT**
Macroconidia	Large, spindle-shaped, thin-walled when young, but thick-walled and tuberculated when old; size 8 to 15 μ by 40 to 50 μ in length, pluriseptate, but often without septa in *M. audouinii*. They are pointed and contain fine hair at the tip	These are club shaped or cylindrical organs 4 to 6 μ by 10 to 50 μ , smooth, thin-walled, hyaline, containing 3 or 4 septa. **In faviform groups macroconidia are ABSENT**	Large, clavate, septate, smooth, thin-walled, with rounded ends and often in clusters

best seen in early glabrous olive green colonies. Older colonies of glabrous growth show chlamydospores. *MICROCONIDIA* **DO NOT FORM**.

Wood's light : The colony appears dark olive in colour.

Animal inoculation : Laboratory animals are not susceptible.

Nutritional requirement : Not known.

Variants : Colour variation is seen.

Perfect stage : Not known.

Special media : Not required.

Critical characters : (1) Characteristic colour of *Growth* is olive-green, (2) Characteristic blunt *macroconidia in* groups, (3) No microconidia.

TRICHOPHYTON FERRUGINEUM (TRICHOPHYTON CONTINUED)

Synonyms : *Microsporum ferrugineum* Ota, 1921; *M. japonicum* Dohi and Kambayashi, 1921; *M. aureum* Takeya, 1925; *M. orientale* Carol, 1928.

It causes ringworm of the hair in children in China. The infection was recorded in a European school boarding in India (Dey and Maplestone, 1939).

Epidemiology

It is common in China, Japan and Africa but recorded in India as well. In U. S. S. R. and Eastern European countries, the disease is spreading following relocation of oriental people since World War II. It is rare or unknown in western hemisphere except in Brazil, where a single case report suggests an endemic focus from an African emigrant.

DIRECT EXAMINATION

Infected hair shows spores having an arrangement like those of microsporum; they are about 3 μ in diameter.

Culture : The growth is slow, convex raised, glabrous and faviform with rusty colour and hence the name *ferrugineum*. Irregular radial furrows divide the growth into sectors.

The growth may be covered with duvet in the margin (**Pl. 29.XIII** *b*).

Microculture : According to Vanbreuseghem (1950) who studied 136 strains, *Microconidia* or *macroconidia* have not been observed (**Fig. 29.8**). The fungus is characterised by the presence of terminal chlamydospores; in length they are 20 to 30 μ with their ends pointed. **Atypical mycelial racquet and rudimentary *hyphae pectinate*** are often present. *Aleuriospores* are rare; when present they are sessile and atypical.(**Pl. 29.XII** *d*).

Nutritional requirements : None.

Variants : A flat, white, downy album variety may occur. Pleomorphic form may show white downy surface growth over the glabrous growth.

Perfect stage is not known.

Wood's light : The infected hairs do not show fluorescence.

Animal inoculation : Inoculation to laboratory animal are sometimes successful (**Pl. 29.XII** *d*).

RARE TRICHOPHYTA

The incidence of these trichophyton species are rare all over the World and some of them in my opinion should fall synonymous with other known species in absence of further data to

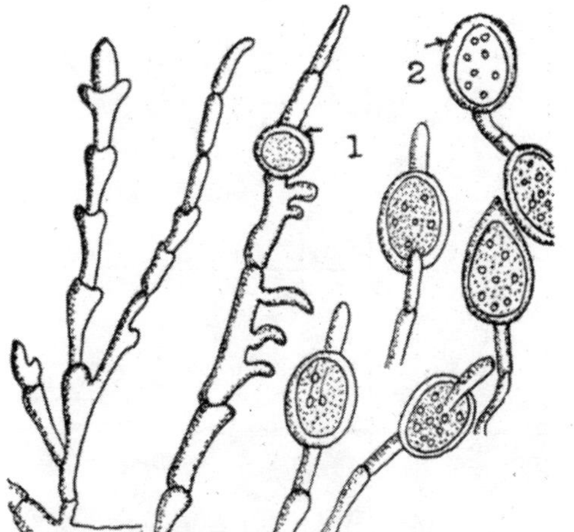

Fig. 29.8 : ***T. ferrugineum*. (1) Intercalary and (2) Terminal chlamydospores.**

PLATE 29.XIV
EPIDERMOPHYTON FLOCCOSUM

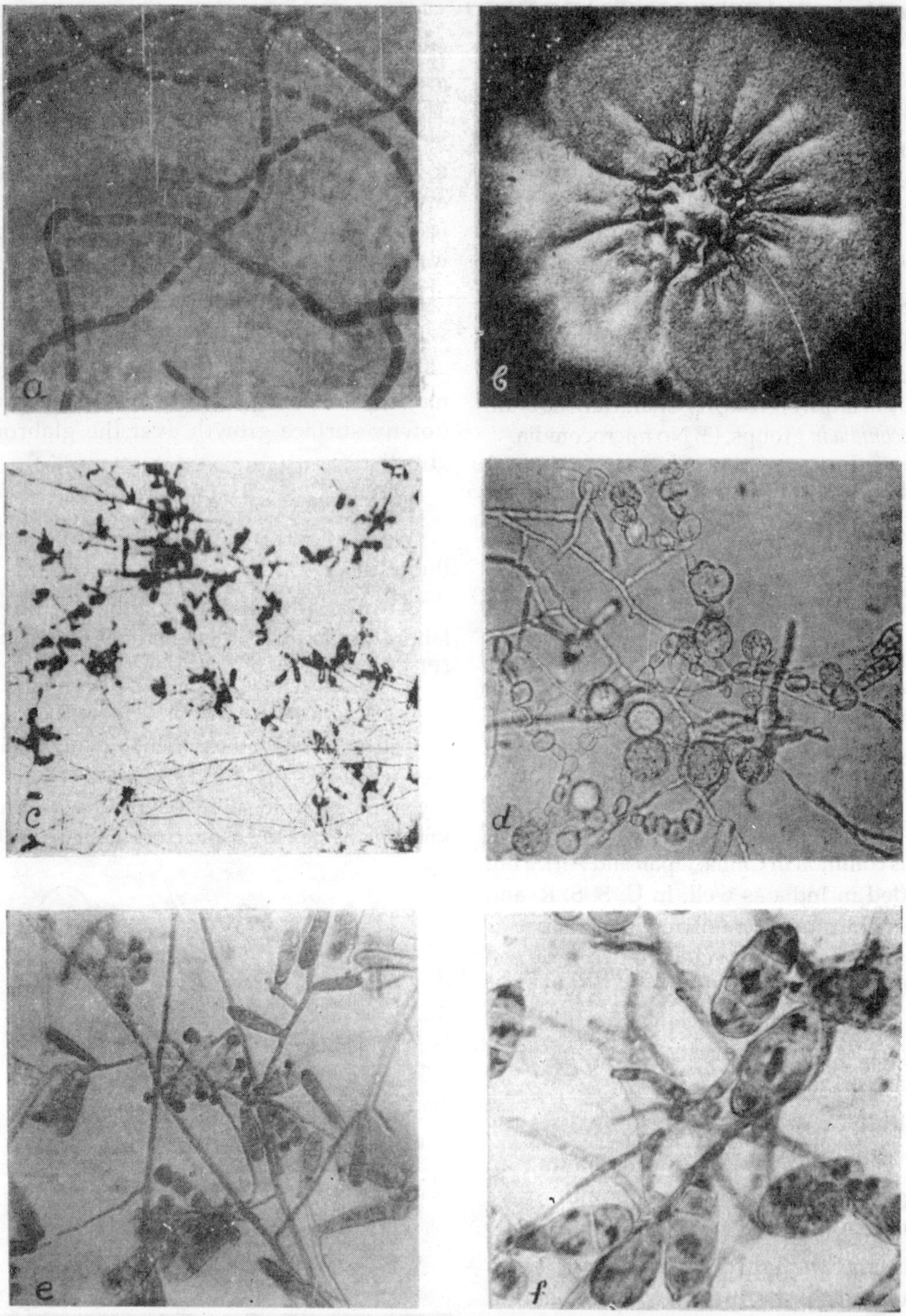

(a) Infected scale, (b) Growth on **Sabouraud's maltose agar**; (c) Macroconidia under low power; (d) **Chlamydospores** (high power 43 × 8). (**e and f**) **Macroconidia** under low power; (e) 43 × 5; (f) 43 × 12

PLATE 29.XV
TRICHOPHYTON FERRUGINEUM

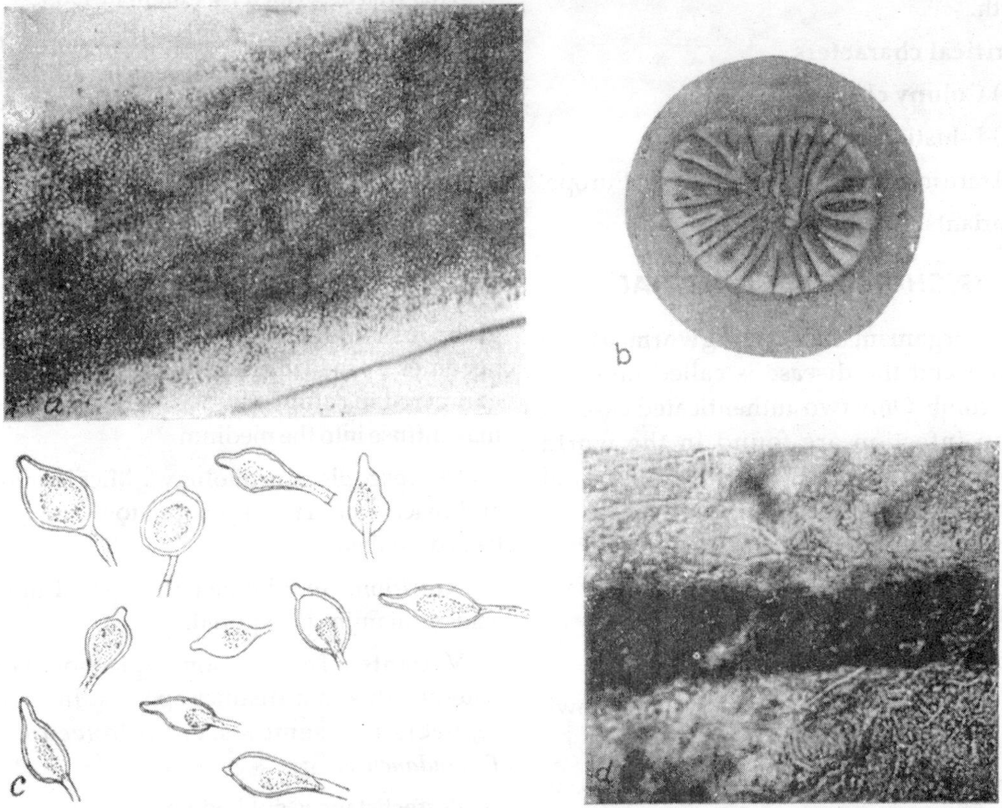

(a) Hair showing small spores with mosaic appearance; (b) Giant colony showing faviform character; (c) Macroconidia in microculture. (d) Hair of the experimental animal showing temporary infection

justify the establishment of a new valid species. This plurality of such species could be controlled. These species are : *T. soudanense*, *T. gourvilli* and *T. yaoundei*. It should be questioned, whether they are variants of *T. violaceum* of Europe or India. All these are endothrix trichophyta, show glabrous, slow growth of white to cream colour in the case of *T. yaoundei* but they may be glabrous or with surface velvety mycelia in case of *T. gourvilii*, the pigment varying from light lavender to a deep garnet red.

T. MEGNINII (T. rosaceum)

Invasion of hair : It causes *large spored ectothrix ringworm* infection in European countries.

Clinically, it causes ***tinea barbae***, ***tinea capitis*** and ***tinea corporis*** less commonly.

Epidemiology It is an anthropophilic species.

Geographical distribution

The fungus is found only in Europe and particularly seen in Portugal and Sardinia. It is not known in India or western hemisphere.

Colony morphology

It is typically rapidly growing suede or felt-like colony with pink surface and wide spaced radial grooves. Medium is stained red under the colony but not as intense as that of ***T. rubrum***.

Microscopic morphology

Small pyriform *microconidia*, but ***macroconidia*** which are seen **pencil shaped**, like those of ***T. rubrum*** are infringent.

Nutritional requirement : *T. megninii* requires L-histidine for the maximum growth.

Critical characters

(1) **Colony character**,

(2) **L-histidine requirement**,

(3) rare in other countries except in Europe.

Variants : Not known.

TRICHOPHYTON GALLINAE

The organism causes **ringworm of the chicken** and the disease is called favus or white comb. Only two authenticated cases of human infection are found in the world literature; one from France by Sabouraud (1910) and the other from Puerto-Rico by Torres and Georg (1956). Both the cases were in children and there was a contact with infected chicken.

Epidemiology

It causes ringworm in chicken and fowl, affecting mainly the comb and wattless.

Colony morphology : White downy colony forms turning pink or buff with age and producing strawberry red pigment, diffusing into the medium. In old cultures, the centre is deep red.

Microculture : ***Microconidia* are rare** in most of the cultures.

Macroconidia, which are clavate to cylindro-clavate are numerous. They have 1-20 cells, 15-50 by 6-8 microns, with smooth walls of medium thickness.

Variants : Pleomorphism develops and the **colonies become fluffy** with loss of colony and microscopic morphology and pigment production are characteristic.

Nutritional requirement : Not required.

Perfect stage : Not known.

Critical features are : (1) colony morphology with diffusing pigment and (2) microscopic morphology.

TRICHOPHYTON GOURVILLI

This was described by Cateni in 1933.

Clinically, this is an endothrix trichophyton like ***T. tonsurans*** or ***T. violaceum*** invading the interior of the hair.

Epidemiology

It causes endothrix trichophytosis in several African countries, e.g., Algeria, upper Malta and Chad.

Colony morphology : The **growth is glabrous**, folded and on subculture often becomes spreading and **lavender-pink to garnet red in colour**. A molasses brown colour may diffuse into the medium.

Microscopic morphology : Microconidia and macroconidia of Trichophyton type have been described.

Nutritional requirement : No special amino acid or thiamine is required.

Variants : There is some question as to whether this is a distinct species. In many respects it resembles ***T. violanceum*** or *T. soudanense*.

Perfect stage is not known.

TRICHOPHYTON SOUDANENSE

This species was first described by Joyeuk (1914).

The fungus caused endothrix trichophytosis of the scalp and has been isolated from **North African** countries and **Belgian Congo**. There is no record of its recovery from animal sources. It causes more inflammatory reaction than infection with ***T. Yaoundei***.

Clinically, ***T. soudanense*** causes endothrix trichophytosis which may be associated with tinea corporis spreading from the scalp.

Epidemiology

This is one of the commonest widespread endothrix trichophytosis in Africa, chiefly reported from West Africa (Algeria to Angola and into the Congo). It has been isolated from a child in Brazil, where it is endemic as a vestige of the slave trade. In Great Britain it has been reported by Calnan *et al.* (1952) and other countries receiving African immigrants.

Colony morphology

T. soudanense is a **slow growing trichophyton**, flat as slightly raised and folded, suede-like yellow, dried, apricot coloured colony, often with a submerged and fringed border. It becomes pleomorphic readily and does not survive refrigeration.

Microculture

Mycelia are characterised by formation of segments of **ARTHROSPORES** with rounded ends and reflexive branching at right angles. Rudimentary pyriform microconidia like those of ***T. violaceum*** occur along the hyphae but these are **not at all diagnostic**.

Nutritional requirement is not known and perfect stage has not been discovered.

The species is allied to ***T. violaceum*** in many respects.

TRICHOPHYTON YAOUNDEI

This endothrix trichophyton causing tinea capitis was recently described from French Cameronians in East Africa by Cochet and Duby-Dabois (1957), and Cochet *et al.* (1957). The name of the species was after an area of the Camerons known as Yaounde, where 241 cases where described amongst the native children. More recently Georg *et al.* (1963 described several strains of this fungus in the Mycology unit of communicable disease centre from clinical material received from the Republic of Congo.

Colony morphology

This is a show growing, glabrous, wet-loosing white to cream, heaped up, folded colonies and may be submerged in areas. The colony sooner or later, turns a deep tan or chocolate brown in colour, diffusing into the medium.

Microscopic morphology

Microconidia are absent or rare, but they are pyriform of trichophyton type. Typical macroconidia are absent. The short aerial hyphae may contain chlamydospore elements similar to those of ***T. violaceum***.

Nutritional requirement : None.

Variantis

T. yaoundei has a marked tendency to sector formation of the raised, glabrous growth, undergoes pleomorphic change of pigment which is often brownish. No two isolates are alike. Microscopic morphology is like that of *T. violaceum*. They are, therefore, in all probability a variant of *T. violaceum*.

Perfect stage is unknown.

Special media : The absence of requirement of thiamine is the only evidence to distinguish from ***T. verrucosum*** and glabrum variety of ***T. violaceum***.

OTHER IMPORTANT TRICHOPHYTA

TRICHOPHYTON VERRUCOSUM

(Bodin, 1902)

Synonyms : *Trichophyton discoides, Ectotrichophyton verrucosum* Bodin, 1902; *Ectotrichophyton ochraceum* Sabouraud, 1909; *Ectoptrichophyton album* Sabouraud, 1909; etc.

It primarily affects the cattle. It causes large-spored ectothrix trichophytosis with formation of kerion in children and ***Tinea barbae*** in adults. One case of *kerion* caused by this fungus has been recorded in India by **Dey & Kakoti (1955)**.

Horse hair is invaded with ectothrix *spores*. **Hiar is not fluorescent.**

Clinically, it produces highly inflammatory kerion-like condition on scalp, beard or exposed parts of the body, similar to **zoophilic *T. mentagrophytes***.

Epidemiology

The organism is the principal cause of ringworm in cattle. Infection in children occurs from direct contact or fomite causing kerion in them. It may cause *tinea barbae* in adult. The organism is ***zoophilic***.

Direct examination : Spores are seen in rows and are about **3 to 5 μ in diameter**. The mycelia

PLATE 29.XVI
TRICHOPHYTON VERRUCOSUM

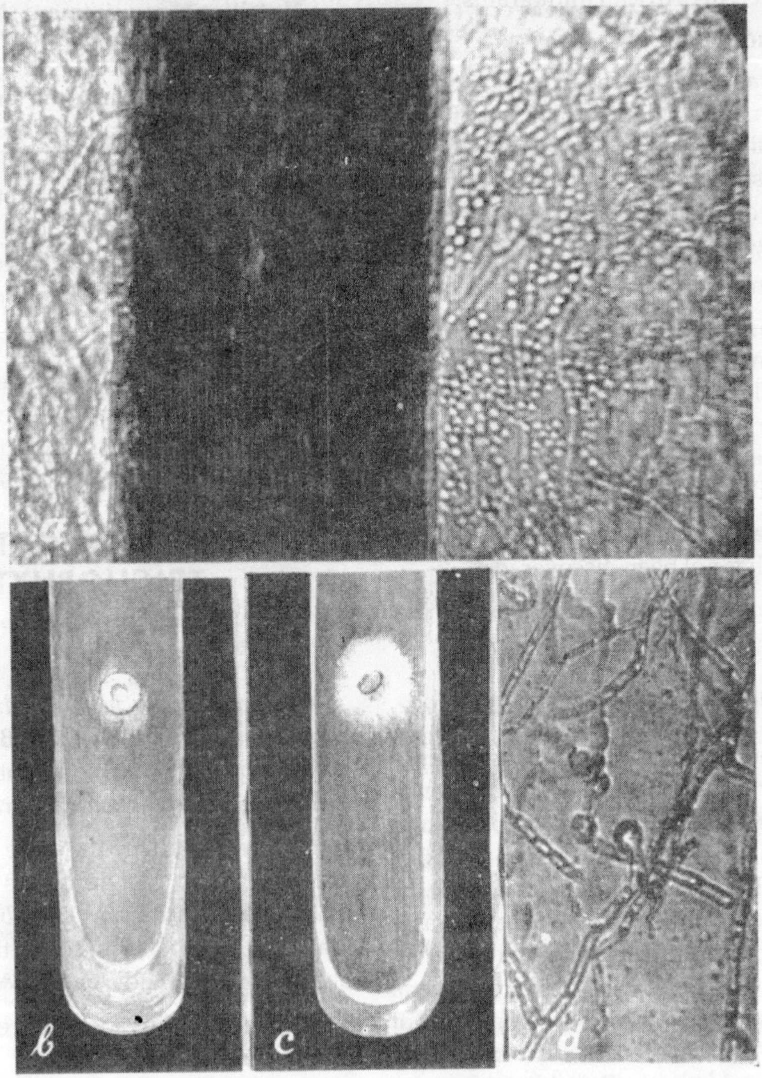

(a) **ECTOTHRIX TRICHOPHYTOSIS** of the hair. (b) Primary culture on Sabouraud's glucose agar (6 weeks old); (c) Subculture becoming pleomorphic; (d) Terminal chlamydospores in microculture.

often show a beaded appearance. No mycelia are seen inside the hair; (**Pl. 29.XIV** *a*).

Cultural characters : A blood agar base enriched with thiamine and inositol or yeast extract colonies are more rapidly spreading, flat and slightly downy. The **growth is enhanced at 37°C**. The fungus does not remain viable in the refrigeration.

The growth is convex, raised, smooth and waxy and faviform with irregular folds in the centre with radiating grooves at the periphery. It is characterised by the presence of a pigment when recently isolated (**Pl. 29.XIV**). But the pigment often disappears on sub-culture, and may be completely lost in subsequent subcultures. Pleomorphism which is characterised by the appearance of down on the surface of the glabrous growth is seen when subculture is delayed.

Microculture : On enriched media like blood agar enriched with thiamine, tear-shaped *microconidia* are seen and less frequently *macroconidia* which may have a rather characteristic rat-tail or string bean shape.

The mycelia show intercalary and terminal, oval or round chlamydospores (**Fig. 29.9, Pl. 29.XIV d**). Mycelial raquets are frequently present.

Nutritional requirement : All strains require thiamine and some require inositol. Unlike many dermatophytes, their **growth is enhanced at 37ºC**.

Special media : Absolute requirement for thiamine and sometimes inositol distinguishes this species from *T. schoenleinii* and other glabrous Trichophyta. Conidia also develop in the enriched media.

Perfect stage is not yet reported.

Variants : There are three principal variants, e.g., *album, ochraceum* and *discoides*. The white colony is called variety *album*; the usual growth like *T. schoenleinii*, yellow, glabrous colony is variety *ochraceum*; whereas, slightly downy, greyish white growth is variety *discoides*.

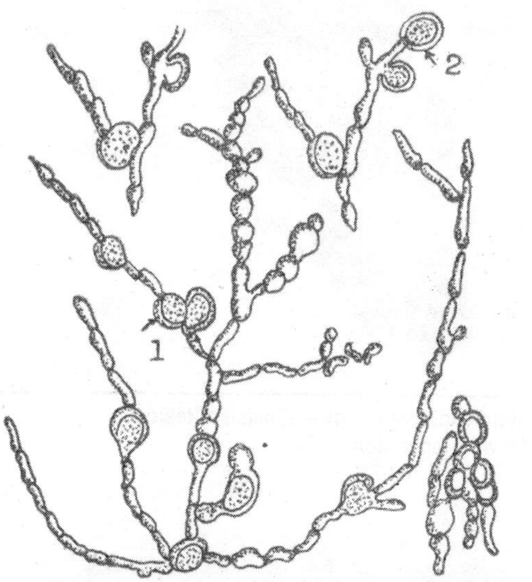

Fig. 29.9 : *Trichophyton verrucosum* with arthrospores and (1) Intercalary and (2) Terminal chlamydospores.

Critical characters : (a) Clinical lesion and history, (b) megathrix sporulation on hair, (c) colony form improves with thiamine and inositol, (d) grows very slowly, (e) growth improves at 37ºC.

TRICHOPHYTON TERRESTRE

Invasion of Hair : This does not invade hair but perforated hair baits *in vitro*.

Clinically, it does not cause any infection in man or animal but *in vitro* it perforates hair.

Epidermiology

This is a geophilic species isolated from soil and from the hair of small mammals. It has a world wide distribution.

Colony morphology

It forms a buff to yellow, granular, powdery colony resembling *T. mentagrophytes*. Reddish pigment may be present under the colony.

Microculture

Microconidia are large clavate or pedicled which may show transition to form cylindrical macroconidia. Conidia often stains more intensely with cotton blue than hyphae. They are characterised by flat base (sessile), spiral coils may be present in some cases.

Nutritional requirements : Not known.

Variants : A slow growing deeply pigmented form has been obtained from hedgehog quills.

Perfect stage : *ARTHRODERMA QUADRIFIDUM*.

Critical characters

(1) Colony characters,

(2) ***microconidia* square based** and show transition to macroconidia,

(3) **cylindrical** ***macroconidia***.

TRICHOPHYTON PURPUREUM

Synonyms : *Trichophyton "A"* Hodges, 1921. *Trichophyton "B"* Hodges, 1921, *Trichophyton purpureum* Bang, 1910; *Epidermophyton rubrum* Castellani, 1910; *E. pareneti* Castellani, *T. rubidum* Priestly, *T. plurizoniforme* MacCarthy, *T. multicolor* Maghalhaes and Neues, *E. salmoneum* Froilano de Mello, 1921.

PLATE 29.XVII
TINEA PEDIS

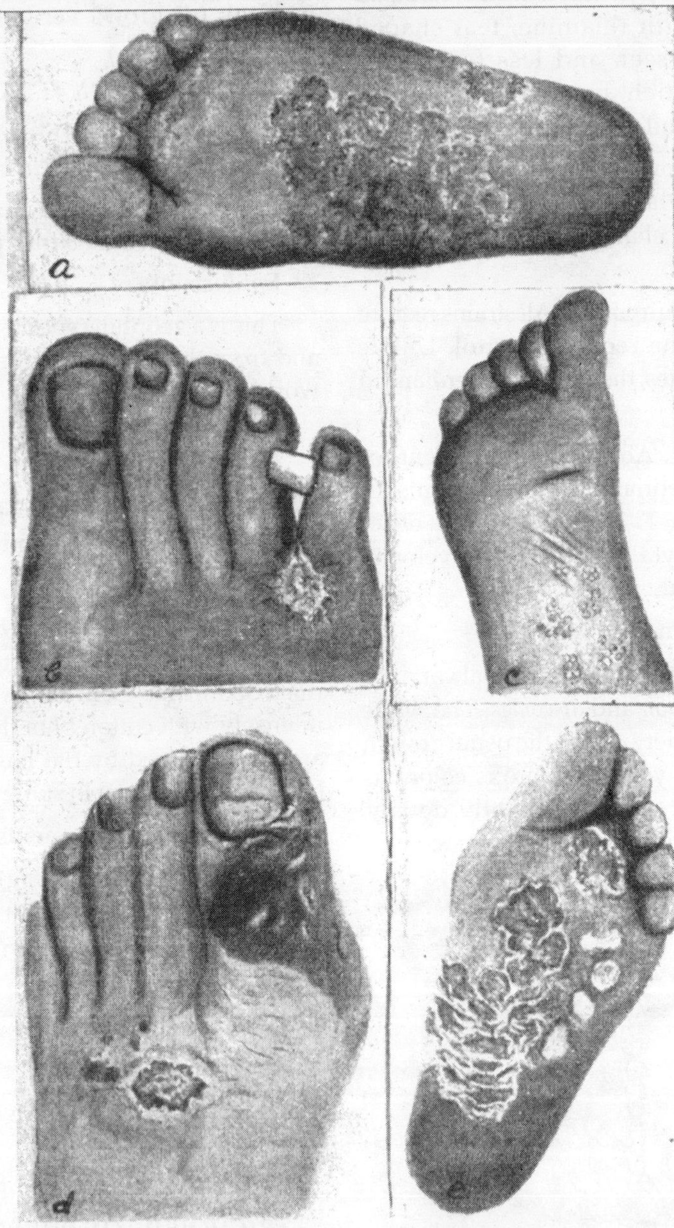

(a and e) — Hypertrophic patches. (b) — Interdigital lesion. (c) — Vesicular lesion. (d) Hypertrophic lesion with ulceration.

PLATE 29.XVIII
MELANOLEUCODERMA

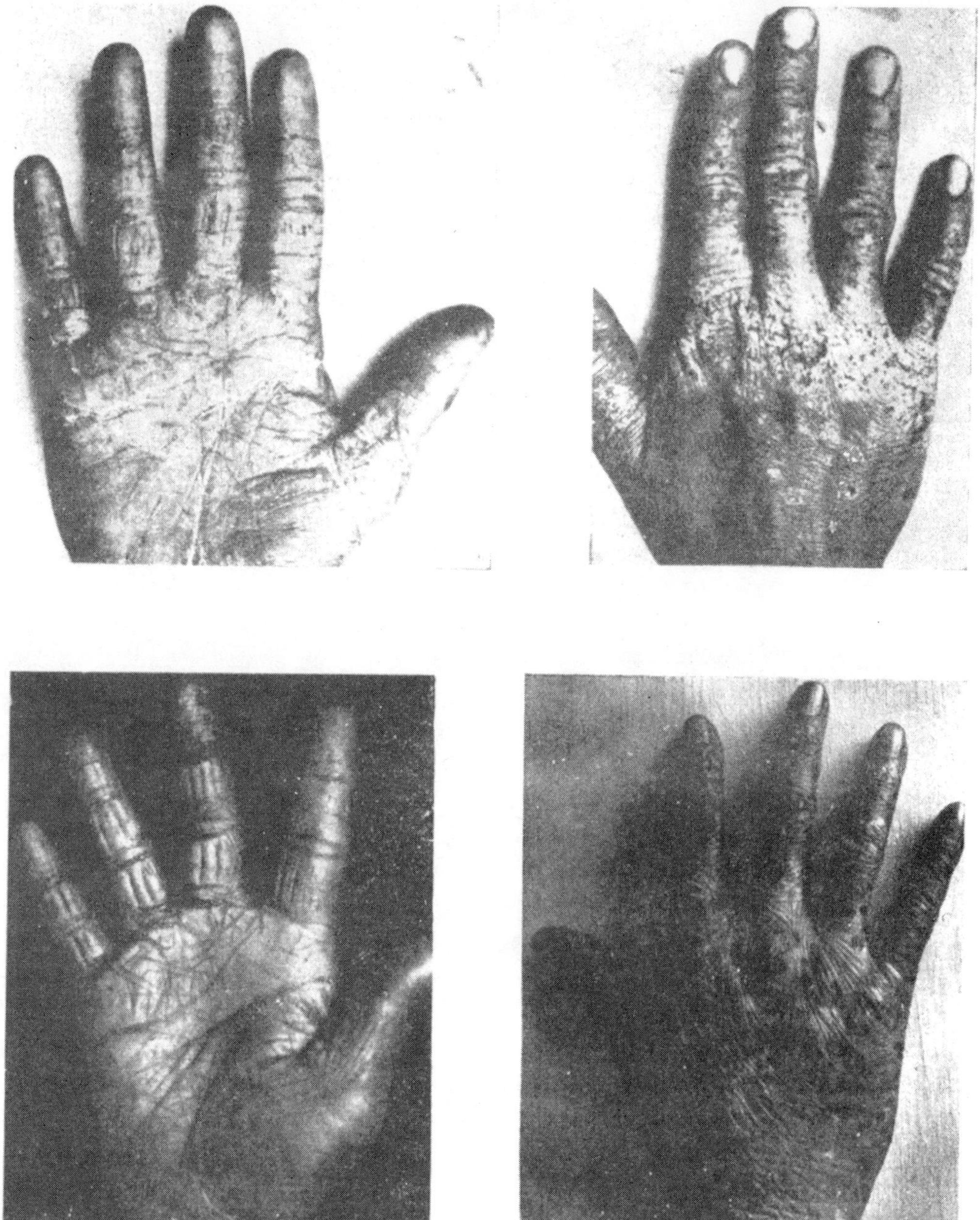

Melanoleucoderma of the hand. Note hyperkeratosis of the skin. *T. rubrum* was isolated from all of them.

PLATE 29.XIX
TRICHOPHYTON PURPUREUM

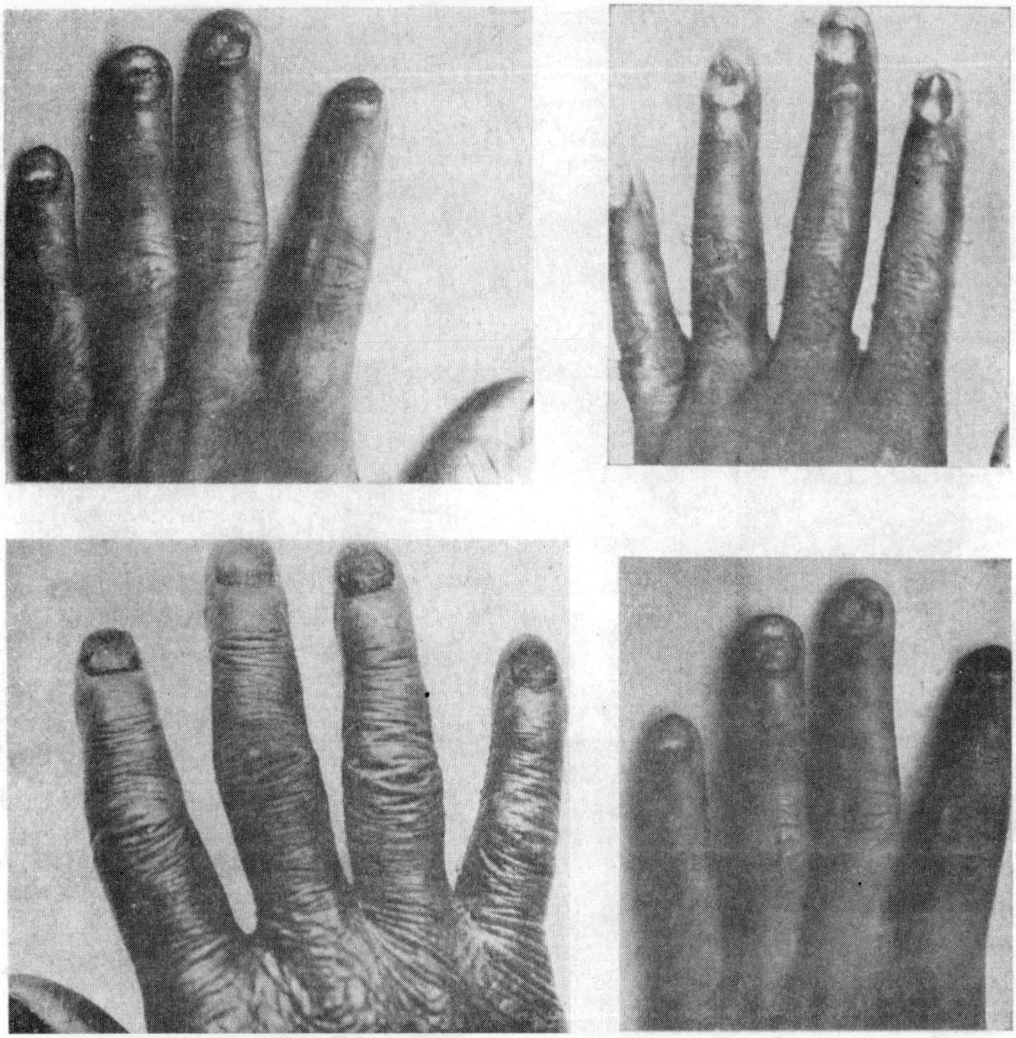

The affected nails show typical moth-eaten appearance. The skin on the dorsal aspect of the hand shows a condition of hyperkeratosis (lower left). The commonest organism in India is *T. purpureum (T. rubrum).*

It is the **commonest ringworm fungus in India** (**Dey, 1953**), China, Japan, and South of the United States and usually affects the skin and nails. In the skin it causes widespread ringworm of the body. It is associated with severe itchings, which may be confined to the areas of eruption. It is the commonest fungus that causes ringworm infection of the nail in India (Ghosh *et. al.* 1948).

Invasion of hair

It may cause ectothrix infection of the hair rather than endothrix infection. It has been mistaken with ***T. violaceum*** infection of hair follicles. The scalp infection is rare and fluorescence is absent.

Clinically, ***T. rubrum*** causes *tinea corporis, tinea cruris, tinea pedis* (**Pl. 29.XV**), ***Tinea manum*** **and onychomycosis** (**Pl. 29.XVI & 29.VII**). In *tinea cruris* or *tinea corporis* the lesion is chronic; whereas, in case of ***Tinea pedis***, the organism forms a nidus. The lesions are chronic in the body with inuginal region often red, indurated pigmented margin including the centre.

Granulomatous, follicular lesions (**MAJOCCHI'S GRANULOMA**) are sometimes seen involving the outer side of the leg of labourer and called Coolie's itch.

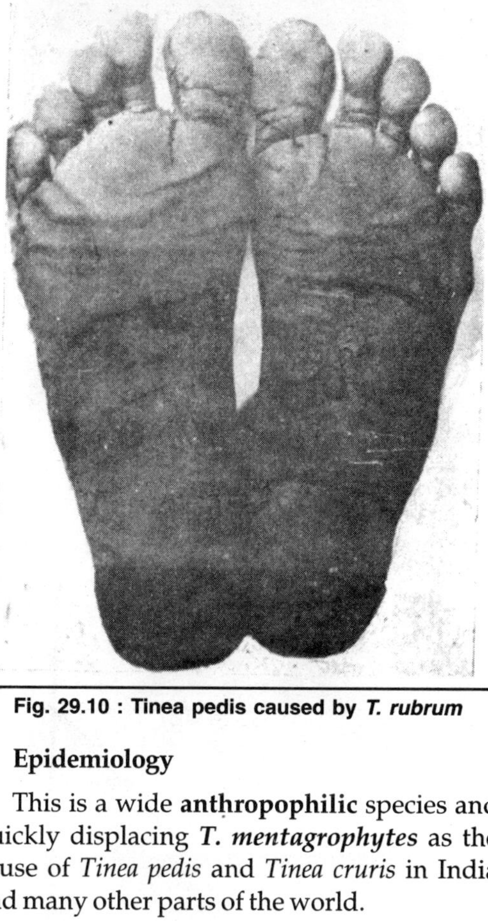

Fig. 29.10 : Tinea pedis caused by *T. rubrum*

Epidemiology

This is a wide **anthropophilic** species and quickly displacing ***T. mentagrophytes*** as the cause of *Tinea pedis* and *Tinea cruris* in India and many other parts of the world.

Geographical distribution

Cosmopolitan, particularly in the tropics.

Direct examination : Branching filaments of the fungus are present in the scales obtained specially from the margin of the lesion (**PL. 29.XVIII** *a*). The fungus filaments show septa at long intervals at the time of active vegetation in the monsoon, but in old lesions or chronic cases arthrospores are present. Nails show plenty of **ARTHROSPORES**.

Cultural characters

The growth is rapid. Cultures are variable and they produce deep red colour on Sabouraud's glucose agar. On Sabouraud's maltose agar the growths may be colourless or sometimes there is a shade of pink colour near the central portion than at the periphery. On **Sabouraud's glucose agar**, at least **two types of growth are seen on primary culture.**

(a) **Granular type of growth with red colour** is occasionally seen on primary culture. The centre of the growth may show short white duvet when the culture is old. On **subculture** they may become **fluffy with greenish white or white duvet**, suggesting a stage of degeneration. Pleomorphism is delayed in this type.

(b) **Downy culture with duvet** and the **red growth underneath**.

On maltose agar, the red colour is not marked and the growth is covered with fine, white downs and slight pink colour is seen near the centre of the growth. The surface of the culture shows a number of radiating furrows upto the periphery, superficial in character with some concentric rings; sometimes shallow radiating furrows are limited to the centre of the growth (**Pl. 29.XV** *b*, *c*, *d*, and *e*).

Microculture : In granular type of growth, ***macroconidia*** **are innumerable** with **a few *microconidia*** (**Fig. 29.10A**). In cottony or **fluffy growth**, macroconidia are few or absent but **microconidia**, which are pyriform in this case, are often produced along the **mycelium *en grappe***. Chlamydospores, mycelial racquets and nodular organs are also present. Spirals are characteristic.

Nutritional requirements : None.

Variants

(1) **Sporulating or granular from** and **nonsporulating downy forms** have been described above,

(2) **Colourless downy strains** may be mistaken for pleomorphic ***T. mentagrophytes***. Colour may not develop in some strain under alkaline condition, poor aeration and light, when colony may be yellow rather than red.

(3) **Rare isolates** form dark, **melanoid diffusing pigment**.

PLATE 29.XX
TRICHOPHYTON PURPUREUM

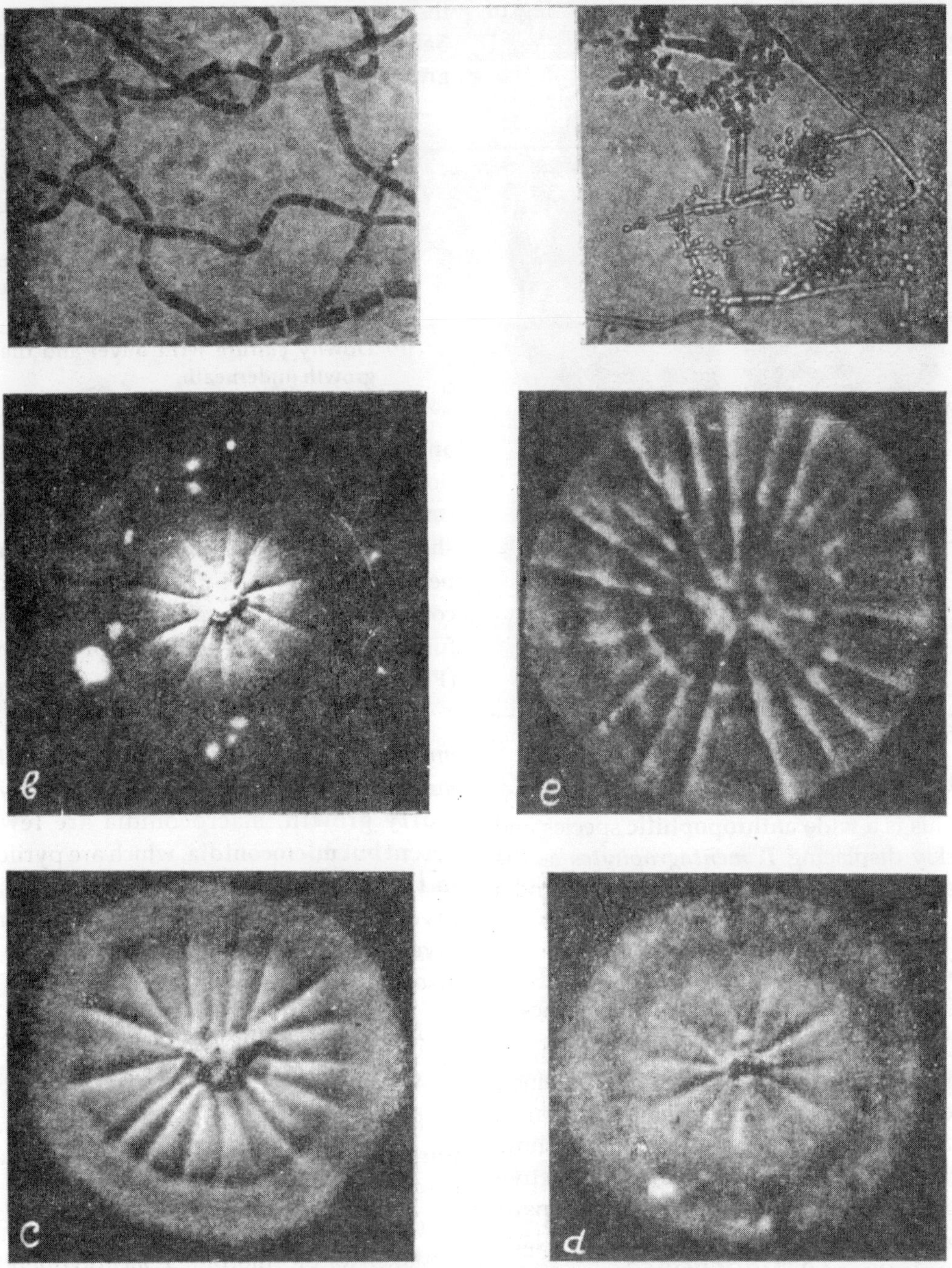

(a) *T. purpureum* in scales. (**b**, **c**, **d**, **e**). **Different giant colonies** of *T. purpureum* in Sabouraud's maltose agar. (f) Microconidia and macroconidia in microculture.

PLATE 29.XXI
TRICHOPHYTON VIOLACEUM BLACK DOT RINGWORM

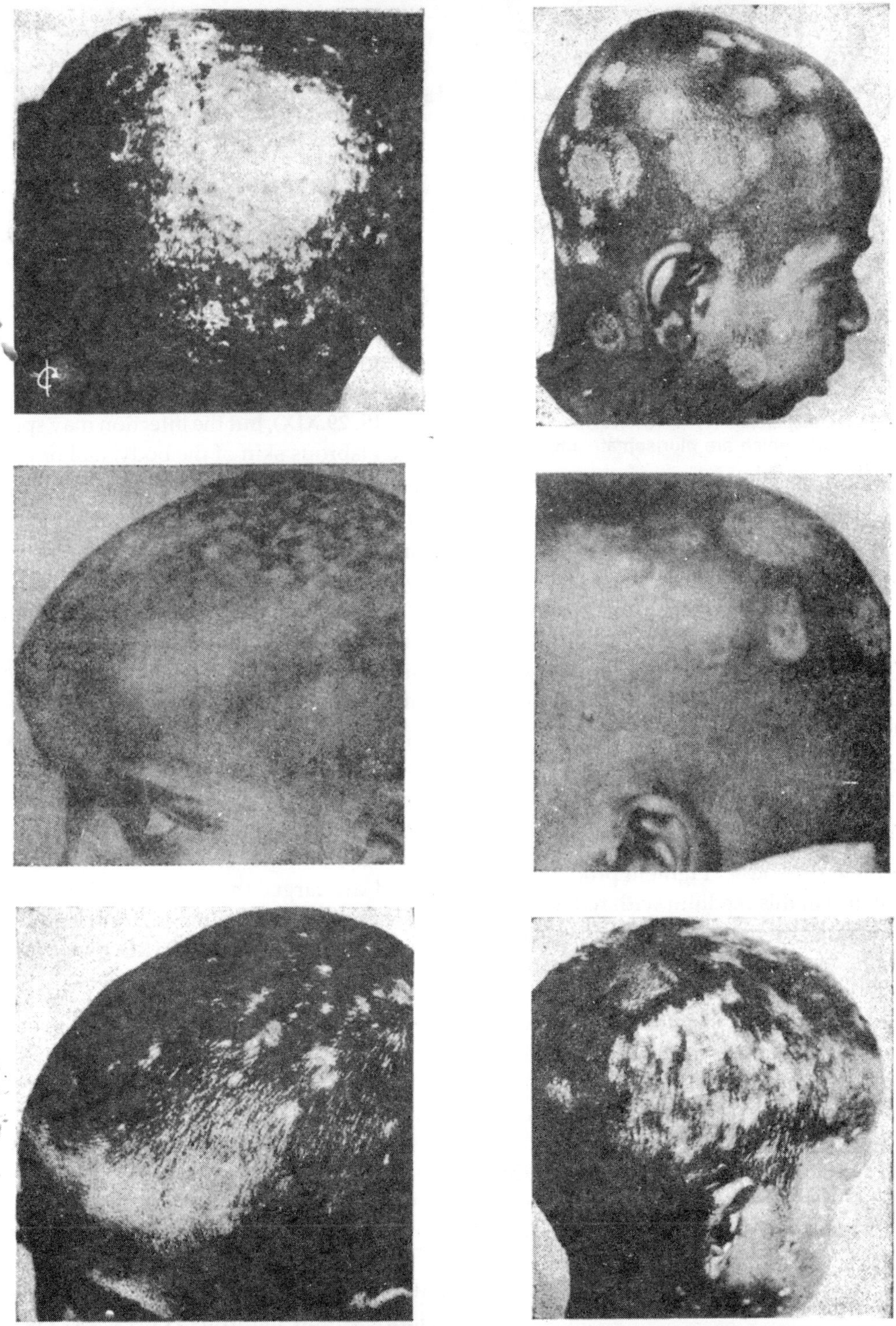

Multiple patches of **black-dot ringworm** of the scalp caused by ***T. violaceum***

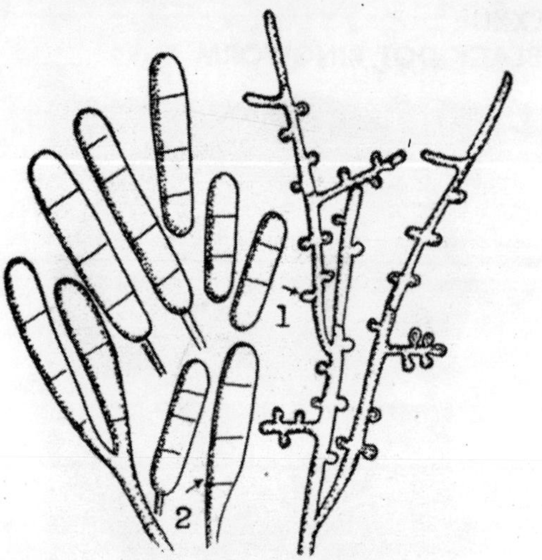

Fig. 2910A : *T. purpureum.* (1) Microconidia. (2) **Macroconidia which are pluriseptate and cylindrical**

Perfect stage : Not known.

Wood's light

In **young culture** the colour is **bright and light blue** throughout the colony but in the fluffy growth it is seen only along the periphery. In **older colonies a light blue border** is characteristic.

Animal inoculation : It is occasionally successful.

Special media

Potato-dextrose agar : Pigment production is very intense in this medium with reduction of fluffy surface growth and helps to distinguish ***T. rubrum*** from ***T. mentagrophytes***. *Meat infusion agar*. This stimulates production of few macroconidia in many fluffy strains. In absence of thiamine, *T. rubrum* grows well and serves to distinguish ***T. tonsurans*** and ***T. violaceum***. It does not require L-histidine and can thus be distinguished from ***T. megninii***. Hair baits *in vitro*. ***T. rubrum*** fails to penetrate hair and thus be distinguished from red pigmented ***T. mentagrophytes***, but red pigment of ***T. rubrum*** develops better on cornmeal dextrose agar.

Characteristic features

(1) Colony characters with pigment,

(2) cylindrical or **pencil shaped macroconidia**,

(3) **tear-shaped microconidia**,

(4) growth independent of thiamine and L-histidine,

(5) **failure to perforate hair** *in vitro*.

TRICHOPHYTON VIOLACEUM

Synonyms : *T. album* Sabouraud, 1909; *T. glabrum* Sabouraud, 1909; *Achorion violaceum* Bloch, 1911.

Clinically, the organism is responsible for causing endothrix trichophytosis of the hair characterised by black-dot ringworm of the scalp (**Pl. 29.XIX**), but the infection may spread to the glabrous skin of the body, nail or beard region.

Geographical distribution : It is an organism of tropical and subtropical regions, the most important organism to produce **ENDOTHRIX TRICHOPHYTOSIS of hair in India**, China, Eastern Europe, Russia and North Africa. It is relatively uncommon in the Western Hemisphere but endemic in Brazil.

Direct examination (**Pl. 29.XX** *a, b, c,d*). The stump of the hair, when examined under sodium sulphide solution, shows mycelia containing spores in rows and often giving a beaded appearance. The spores vary from 3 to 5μ and are larger than those of microspore. Mycelia are limited to the interior of the hair and not outside it (**ENDOTHRIX**).

Cultural Characters (**Pl. 29.XX, Figs. 1-4**). The organism grows slowly like other faviform trichophyta. The rate of growth is about 1.5 to 2 cm in 6 weeks. The growth is faviform and shows a glabrous surface with convolutions on the surface and radial groove marked near the periphery. The primary culture gives a deep violet appearance but the violet pigment fades on repeated subculture till there may be a central pigmented portion but the surrounding portion becomes colourless and glabrous (**Pl. 29.XX, Fig. 4**). Pleomorphism is rare.

PLATE 29.XXII
TRICHOPHYTON VIOLACEUM

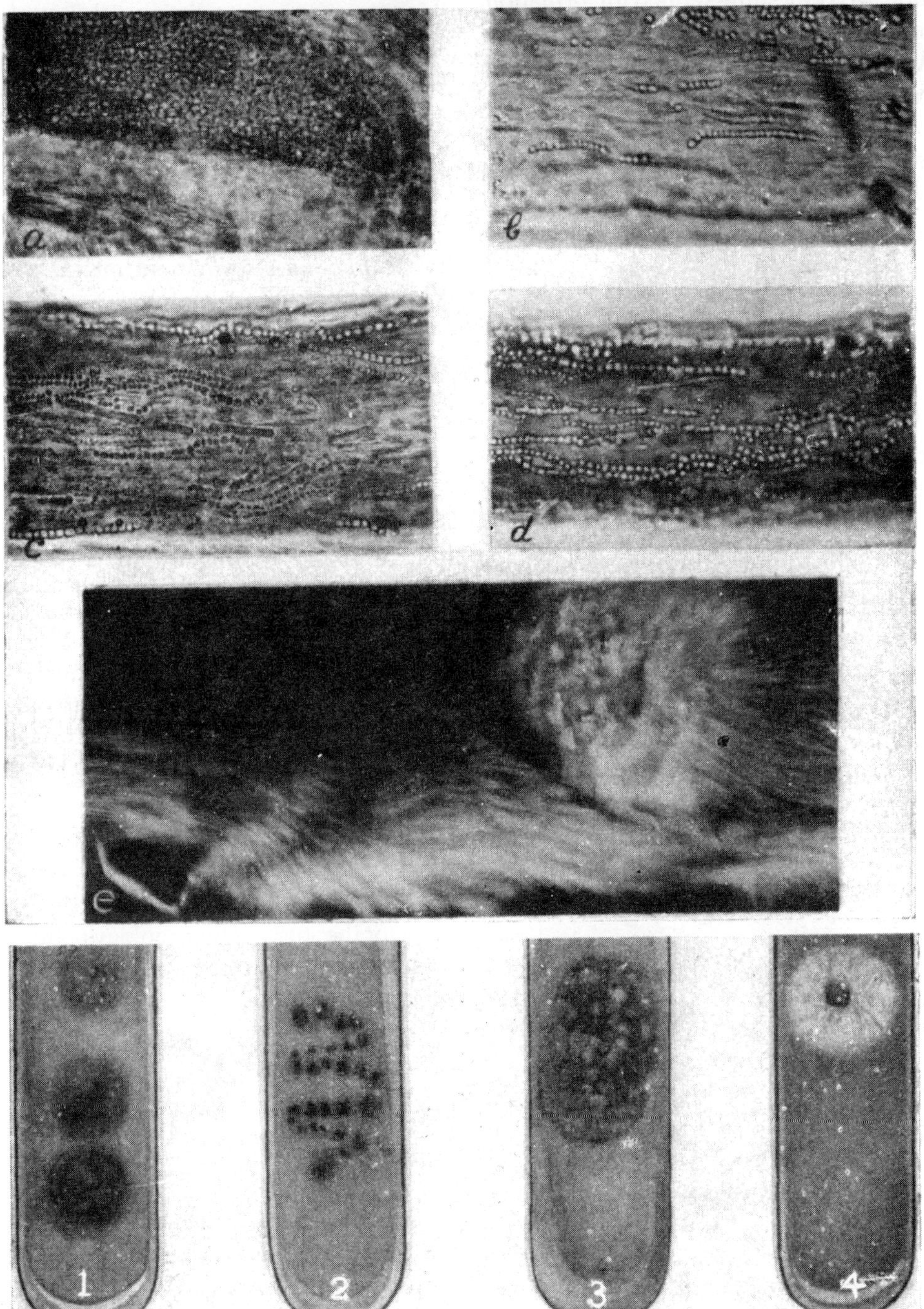

(a) Infection of interior of the hair; (c-d) Beaded appearance of the fungus infecting the hair; (c) Favic lesion in monkey. Isolation in Sabouraud's proof medium from (1) Hair, (2) Pus in the hair follicle, (3) Subculture, (4) Loss of pigment by repeated subculture in the same medium

Microculture (Fig. 29.11) : The hanging drop culture does not show any typical microconidia formation. **Mycelia show septa** with formation of numerous **ARTHROSPORES AND CHLAMYDOSPORES** which may be **intercalary or terminal**, like favic nails. Mycelia show irregular branching, and racquets are often seen but **ALEURIOSPORES** are either rudimentary or absent.

Nutritional requirements

Growth of strains of ***T. violaceum*** like *T. tonsurans* are partially dependent on thiamine and growth is found to be minimal without thiamine.

Perfect stage : Not known.

Variants : Sectoring and pleomorphism are often seen. Nonpigmented forms in subcultures are called ***T. glabrum***.

Wood's light : **Fluorescence** is **not seen** in the case of *T. violaceum* infection.

Animal inoculation

Animal inoculation on the skin of guineapigs and rabbits were found to be doubtful but the **inoculation in monkey** was **successful 10 days after inoculation** and in one of the two monkeys inoculated by the senior author, the **lesion was faviform** in character.

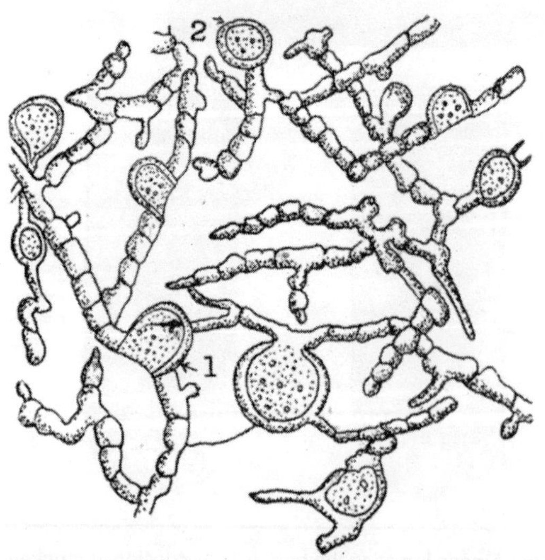

Fig. 29.11 : ***T. violaceum***. **Arthrospores** are plenty with: (1) **Intercalary** and (2) **Terminal chlamydospores**

Special enrichment media

Demonstration of partial requirements for thiamine may help to distinguish ***T. violaceum*** from other dermatophytes, like ***T. ferrugineum*** of similar morphology.

Characteristic features : (1) *ENDOTHRIX* hair invasion,

(2) **purple *glabrous growth* on primary isolation**,

(3) ***thiamine* requirement**, mand

(4) ***microculture* with or without thiamine**.

GROWTH OF DERMATOPHYTES IN HAIR

In *vitro* penetration of hair has helped the mycologists to differentiate dermatophytes. The manner of invasion of the hair *in vivo* by dermatophytes is rather different from natural infection when they grow in isolated hairs *in vitro*.

The *in vitro* study can be done in two ways as stated below.

(1) **PLUCK THE INFECTED HAIR CLEANED AND STERILISED WITH ETHER AND ALCOHOL** from the patient, keep it on a coverslip and invest it in a well slide preparation containing a drop of water, and the coverslip is sealed with paraffin. This is often done for the study of morphology of the fungus under normal condition.

(2) Human or animal hair is placed in a Petri dish, autoclaved, sterilised and put the hair in sterile water. Hairs are inoculated with

(a) arthrospores from infected hair or (b) fragments from a culture of dermatophyte to be studied.

The invasion of hair is not like one seen *in vivo* but the **fungus grows in the hair substrate** as in any artificial culture medium with development of **macroconidia and microconidia**. Here the **hair substrate is utilised as the source of nutrition**. Each species of dermatophyte grows in the hair in a manner in which it attacks the hair by the fungus secreted

PLATE 29.XXIII

LESIONS CAUSED BY TRICHOPHYTON MENTAGROPHYTES

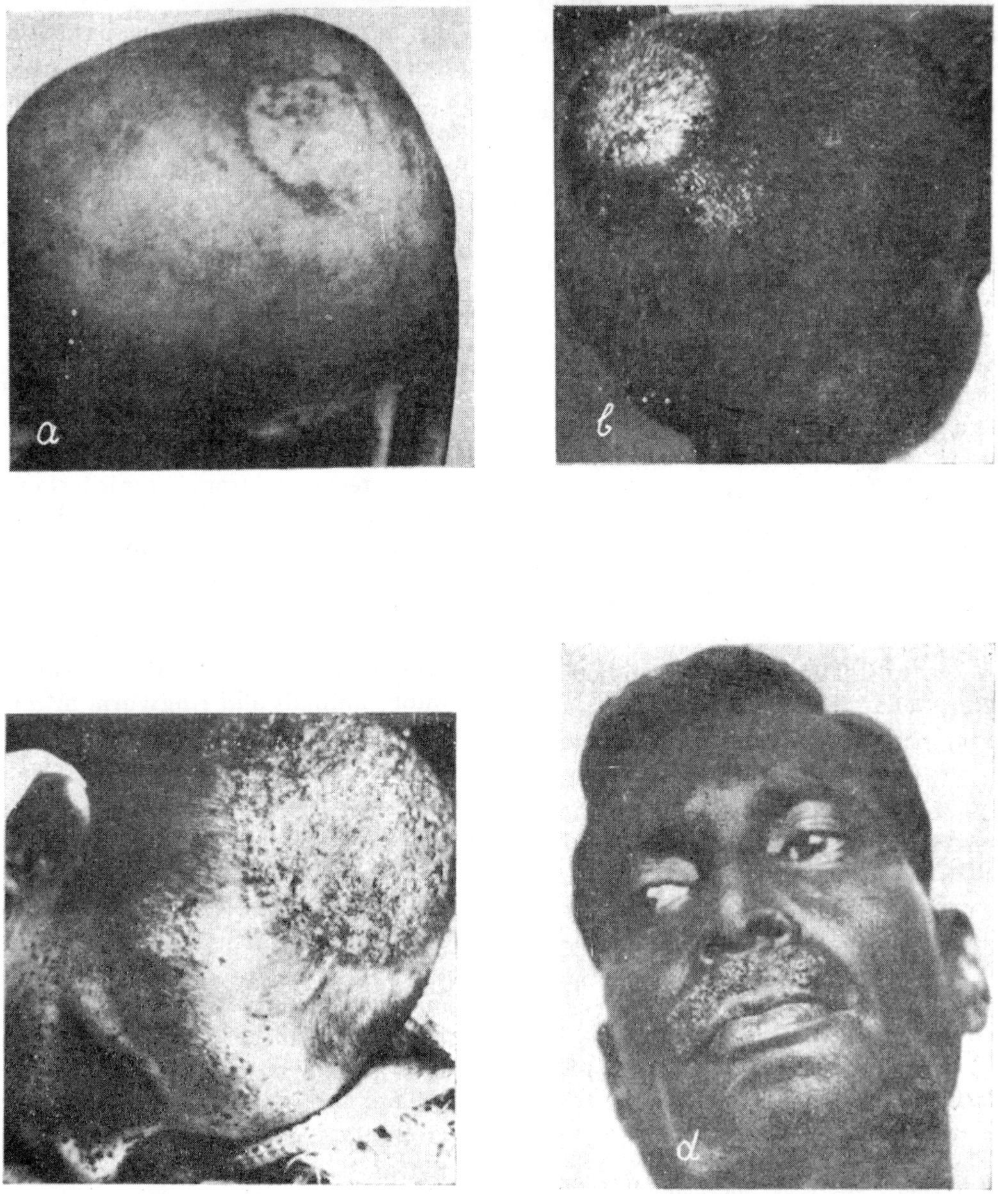

(a), (b) and (c) Kerion of the scalp, (d) Tinea barbae

enzymic activity. Some dermatophytes produce perforation in a manner of wedge, makes holes and tunnel bottles as a result of **enzyme secreted by the dermatophyte**, dissolving the keratin layer of the hair; whereas, other **dermatophyte species erode the surface of the hair** without perforation.

(a) The **fungus grows under the cuticle** lifting the cuticle away from the cortex,

(b) In some, the **cortical cells** are **loosened** and the **hair is eroded by the fungus**.

(c) In others, **penetration develops at right angles** to the hair staff and digest their way through the substance. All the species of **microsporum**, *Keratinomyces ajelloi* of the hair (Davidson and Gregory, 1934; Vanbreuseghem, 1959, page 150, in Daniels, '53) and some trichophyton species except ***T. rubrum*** are *keratinophilic* and perforate hair *in vitro*. ***M. audouinii*** perforates the hair slowly and it is imperceptible and difficult to demonstrate. *M. canis, M. gypseum* and *K. ajelloi* perforate readily in large number.

Barlow and Chattaway (1955) studied *in vitro* attack by dermatophytes of normal and chemically modified hair, ***M. audouinii***, ***T. rubrum*** and ***E. floccosum*** do not penetrate normal hair, but ***M. canis*** **does it partially but they grow on altered hair in which cross linkage between keratin molecule is broken** and they show a more marked and rapid attack when compared with controls with normal hair. In a similar way the attack was significantly less, when these fungi were grown on hair as to increase cross linkages on the side chains.

TRICHOPHYTON MENTAGROPHYTES

(Robin) Blanchard, 1896

Synonyms : *Microsporum mentagrophytes* Robin, 1958; *Achorion quinckaenum* Blanchard, 1896, *Trichophyton gypseum*, Bodin, 1902; *T. equinum* Gedoelst, 1902; *T. granulosum* Sabouraud, 1909; *T. radiolatum* Sabouraud, 1910; *T. asteroides* Sabouraud, 1910; *T. lacticolor* Sabouraud. 1910; *T. niveum* Sabouraud, 1910; *T. radians* Sabouraud, 1910; *T. denticulatum* Sabouraud, 1910; *T. interdigitale* Priestly, 1917; *Trichophyton* "C" Hodges, 1921, *Trichophyton kaufman-wolf* Ota, 1922; *T. pedis* Ota, 1922, etc.

INVASION OF HAIR

This is an **ectothrix** trichophyton with **small or moderate sized spores** on the surface of the hair. Affected **hair is not fluorescent**. The **hair perforates the hair baits** *in vitro*.

CLINICALLY

The fungus is found in different stages of evolution in different ringworm conditions. The fungus causes ***tinea capitis, tinea barbae*** **(Pl. 29.XX)**, ***tinea corporis and tinea cruris, tinea pedis*** and also ***onychomycosis***. The zoophilic fungus is inflammatory on the scalp and beard; whereas, **anthropophilic ones produce non-inflammatory changes**. The **anthropophilic strains are recovered from chronic lesions of *tinea pedis* and *tinea cruris*** but the **granular form** may be **recovered from pompholyx of feet**.

Epidemiology

The **granular from** is chiefly ***zoophilic*** being the **causes of *epizoothics* of ringworm** in laboratory animals and **ringworm of horse** (U. S. A.); in this form it is highly infectious and may be transferred to man often producing solitary lesions of ***tinea capitis*** or ***tinea barbae***. ***T. mentagrophytes*** var. *quinckaenum* may form favus like scutulum in mice.

The ***anthropophilic fluffy*** **form** of ***T. mentagrophytes*** often causes ***tinea pedis*** or ***tinea cruris***, but from **podopompholyx inflammatory granular form is** often **recovered**. It is difficult for the anthropophilic form to produce successful inoculation of the laboratory animals but sometimes they produce milder lesion without invasion of the hair. The ***anthropophilic*** **form may be transformed into *zoophilic* form by serial transfer on the skin of guineapig**.

DIRECT EXAMINATION

Hair shows the character of an **ECTOTHRIX TRICHOPHYTOSIS** and the fungi are seen outside the hair. The spores are small (**Pl. 29.XXI** ***a***) like those of microspora. In vesicles or macerated

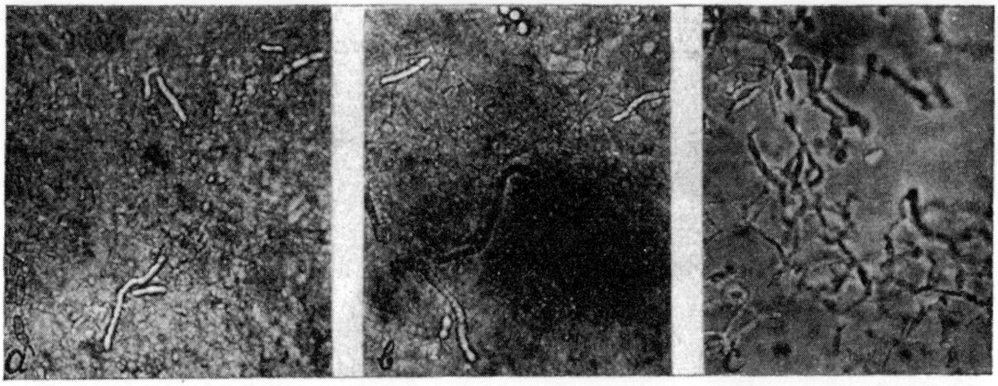

Fig. 29.12 : (a and b) Anthropophilic type with shreds of mycelia in the skin of blisters; (c) Mosaic fungus in the blister skin without any true mycelia; this is due to deposits of cholesterol

skin, ***T. interdigitale*** appears as short mycelia, often with shreds with little branching (**Fig. 29.12**).

Cultural character

The ***zoophilic type*** on Sabouraud's agar produces a flat, rapidly growing **flat colony with a granular surface**, yellowish buff in colour. Sometimes there may be powders sprinkled in **concentric rings (Pl. 29.XXI** *b*) and the margin may be fringed. They contain **clusters of microconidia**. The reverse is dark tanned but some strains may show red pigment on the underside of the colony.

The anthropophilic type of Sabouraud agar often shows a fluffy downy growth rich in microconidia but macroconidia are also present. They are white to cream coloured but older colonies become pink tinted. When transferred to potato dextrose agar, a characteristic stellate powdery colony appears. Those isolated from *tinea pedis* may show either fluffy or a mixed character often in sectors but those isolated from the inflammatory lesions of ***podopompholyx*** may show a **mixed or a pure zoophilic granular character**.

All strains of *T. mentagrophytes* rapidly tend to be pleomorphic, with the development of fluffy character and without any sporulation. **Nonpleomorphic colonies under Wood's light fluoresce blue-violet.**

Micoculture : *T. mentagrophytes* shows numerous macroconidia with either 3 or 4 septa. Microconida are almost always ***EN GRAPPE*** and also borne laterally. Microconidia are characteristically round in this case. Spirals are typical. **Chlamydospores, mycelial racquets, pectinate hyphae and nodular organs are often seen (Fig. 29.13)**.

In *anthropophilic type*, round microconidia are often seen in large cluster. Macroconidia and spirals are usually seen immediately after isolation but they gradually disappear in subsequent subcultures. Chalmydospores, mycelial racquet and nodular organs are usually present.

Nutritional requirement : None.

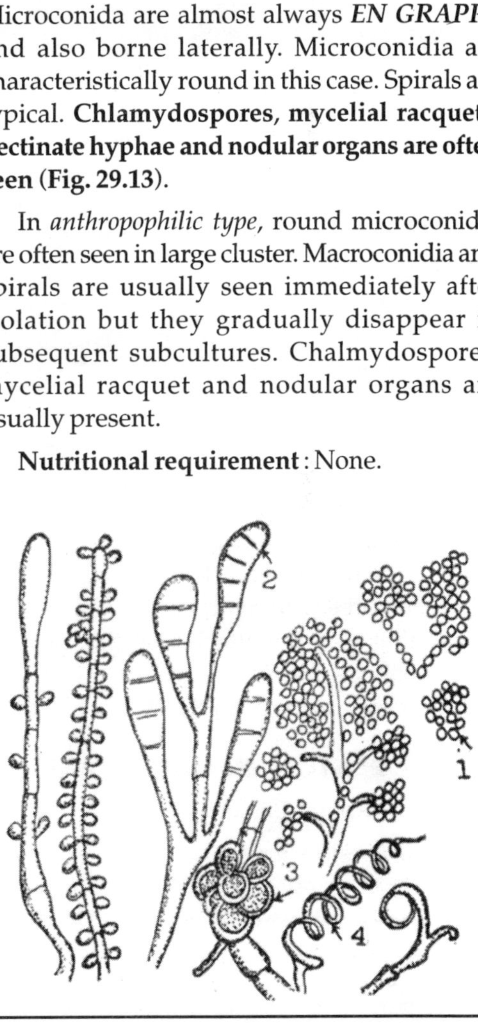

Fig. 29.13 : Trichophyton mentagrophytes. (1) **Microconidia en grappe.** (2) Macroconidia with a few septa, (3) Nodular organ, (4) Spirals

PLATE 29.XXIV
TRICHOPHYTON MENTAGBOPHYTES

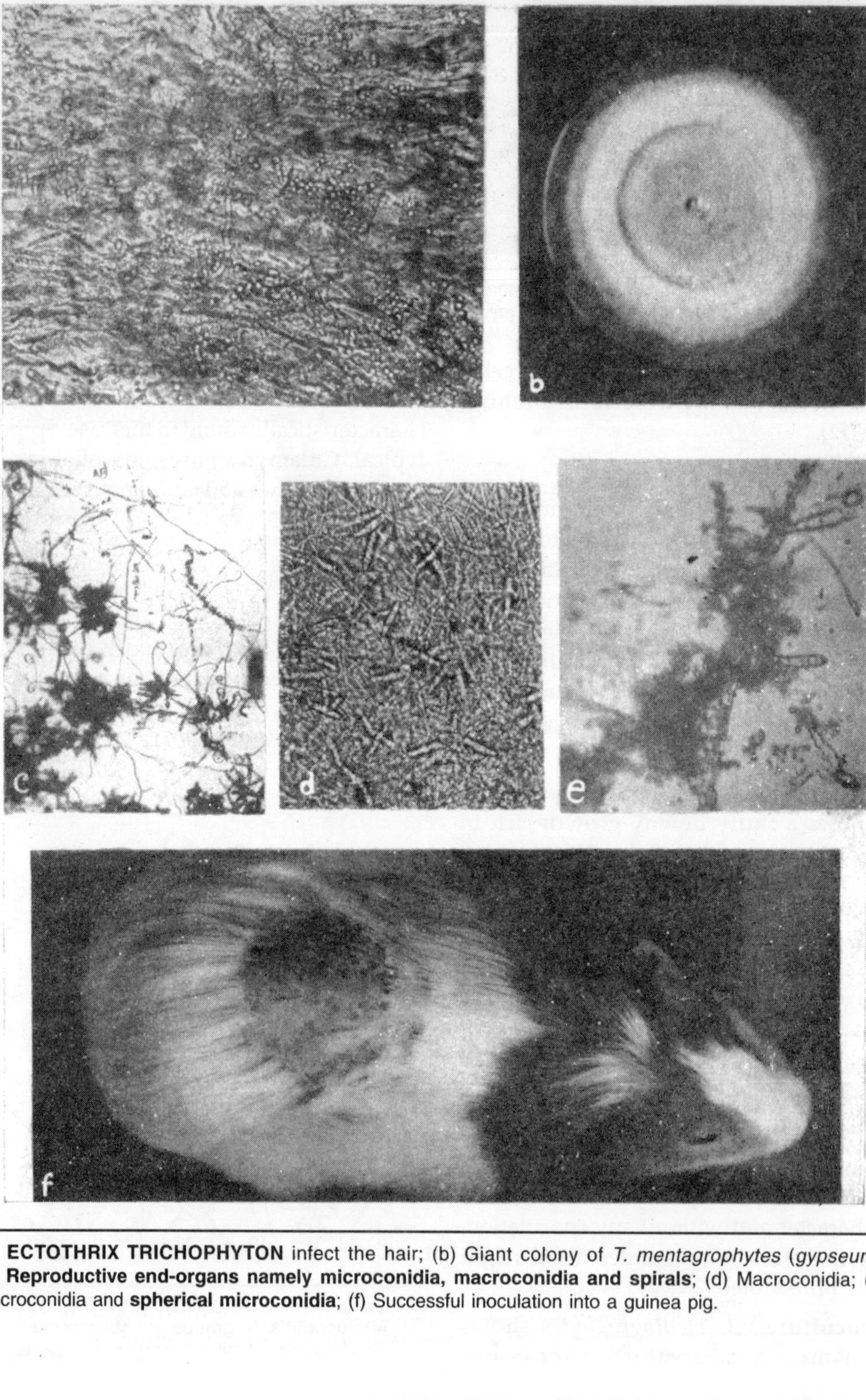

(a) **ECTOTHRIX TRICHOPHYTON** infect the hair; (b) Giant colony of *T. mentagrophytes* (*gypseum*); (c) **Reproductive end-organs namely microconidia, macroconidia and spirals**; (d) Macroconidia; (e) Macroconidia and **spherical microconidia**; (f) Successful inoculation into a guinea pig.

Variants : These are -

(1) **GRANULAR ZOOPHILIC TYPE** producing inflammatory lesions,

(2) **anthropophilic fluffy *pedis type*** often isolated **from *Tinea cruris*** or *tinea pedis,*

(3) ***var. quinckaenum*** having relatively few conidia have been isolated from mouse favus,

(4) ***var. erinacei,*** isolated from hedging and human infections in countries where hedgehogs are found. In this case, microconidia are clavate rather than spherical, spirals are absent and little growth occurs at pH 4.

Perfect stage : Not known.

Wood's light : In **ECTOTHRIX TRICHOPHYTOSIS OF THE SCALP OR BREAD AREAS, fluorescence is absent**; the skin or nails also do not show any fluorescence.

The **granular type of growth is typically bright** and **shows concentric bands of colour. It is bluish-violet** in the centre and fawn colour at the periphery. ***T. niveum*** shows a uniform yellow tone.

Animal inoculation

The granular type of growth of *zoophilic type* can be inoculated successfully into guinea pigs (**Pl. 29.XXI** *f*), cats, rabbits, but the fully anthropophilic type does not show the pathogenicity for these animals.

Special media

Potato dextrose agar. This is useful to differentiate fluffy growth *of* ***T. mentagrophytes*** from that of ***T. rubrum***. Most strains of *T. rubrum* develop a deep rose pigment contrasting with stellate powdery colonies of ***T. mentagrophytes***.

Corn meal agar. Pigment of dark brown or red strains of ***T. mentagrophytes*** does not develop in this.

Autoclaved hair. When grown on hair, ***T. mentagrophytes*** perforates hair; whereas, ***T. rubrum*** does not.

Critical Characters

(1) ***Colony*** morphology-typical, granular or fluffy colony;

(2) well-formed *macroconidia,*

(3) typical **round *microconidia,***

(4) *spirals* in the presence of typical microconidia (absent in var. *erinacei*),

(5) ability to **perforate *hair***.

Immunity

Extraction of a polysaccharide from mycelium of ***T. mentagrophytes*** was done by McNallesal, 1961. This is said to be antigenic for chinchillas that produced appreciable amount of antibodies.

TRICHOPHYTON MENTAGROPHYTES VAR. EQUINUM

(Rare Trichophyta)

Hair : The organism produces **large ectothrix spores in chain**. The infection is rare or unknown in human.

This was considered to be synonymous with ***T. mentagrophytes*** (now considered to be a separate species) but due to the requirement of nicotinic acid for its growth, it is now considered to be a separate species.

Epidemiology

It causes ringworm in horses, but ***T. mentagrophytes*** which commonly causes ringworm in horses caused some confusion of the two species. It has not been reported from Africa or Australia.

Cultural characters

The colony is like that of fluffy type of ***T. mentagrophytes*** with a deep yellow pigment underneath the colony but as the growth becomes old it tends to become orange brown. The edge of the colony is often ragged with furrows on the surface when compared to ***T. mentagrophytes***. There is no growth on vitamin from media.

Microculture

The ***microconidia*** are often **pyriform** and ***macroconidia*** are less frequently seen as **clavate** (club shaped), as in ***Trichophyton mentagrophytes*** types. These are thin with smooth walls. *Nodular organs* often characterise the yellow pigmented nodular variety of ***T. mentagrophytes*** which are few.

Nutritional requirement

The fungus requires *nicotinic acid* in the medium for its growth. It grows selectively on horse hair in *vitro* and will not grow on human hair. But it is not proved, however, that it is due to its niacin requirement.

Special media : Demonstration of requirement for niacin is specific and diagnostic Sabouraud's agar with yeast extract gives typical growth.

Perfect stage : Not known.

T. equinum is considered to be a pleomorphic form of ***T. mentagrophytes*** which has degraded further from the fluffy form but has still the power of infecting horse hair. This has lost the power of penetrating human hair.

Granular form is fluffy, anthropophilic; equinum form is zoophilic.

LABORATORY DIAGNOSIS OF DERMATOPHYTOSIS (RINGWORM)

Diagnosis is best made by :

(a) demonstration of **fungus in the hair (the skin or the nails)**,

(b) study of culture preferably of **giant colonies in Sabouraud's** proof or glucose agar **media** and

(c) study of **microcultures**.

The *Wood's light* is invaluable in locating and determining the extent of infection of the scalp. Hairs infected with ***M. audouinii*** and ***M. canis*** fluoresce when the light is held close to the scalp in a partially dark room. Hairs epilated from such infected areas of the scalp are examined or skin scraped from the erythematous lesions or it is obtained from the roofs of vesicles.

Nail is scraped from the discoloured friable areas of infected parts.

Hair, skin or nails are to be examined under the microscope **in a *keratolytic solution* like 10 per cent KOH**. Nails take a longer time for keratolysis and slight heating facilitates the observation. **Dermatophytes in the skin or nails appear as branching filaments or hyphae often with arthrospores.**

In hair, microsporosis shows Duckworth's *chloroform test* positive. In this test, on a few epilated hair stumps on a slide, add **a few drops of chloroform and the microsporum sheath** of the hair **becomes frosted** and prominent. In the hair, the appearance of the fungi in KOH preparation allows to some extent a generic determination. Microsporum species form dense spores, sheath on and around the hair stump with the spores crowded and giving a mosaic pattern (**Pl. 29.II & 29.IV**).

Species of Trichophyton show **parallel rows of large spores inside the hair in *ENDOTHRIX* SPECIES** and **small or large spores outside the hair stump in ECTOTHRIX CALLED *MICROIDES*,** or ***MEGASPORES*** respectively.

Although the appearance of the infected hair may allow identification of the genus, the **species are identified by specific character seen in gross and microcultures.**

CULTURES are routinely made by inoculating:

(a) **Sabouraud's glucose agar slants** with 3 or 4 fragments of infected material, each seeded at 1/2 to 1 inch apart,

(b) **Sabouraud's glucose agar** medium slants **with** the incorporation of **chloramphenicol (0.5 mg/ml) and cycloheximide (0.5 mg/ml)** to the medium. This facilitates greatly percentage of positive cultures by eliminating bacterial and fast growing fungus contaminants. All culture tubes must be maintained for 3 to 6 weeks at room temperature before being discarded as negative.

Growth should be studied in microcultures for development of **macro and microconidia** or **in** natural media like **moist soil with hair baits for** detection of **perforation of hair and development of the perfect stage**.

In studying microcultures, drop method on inverted Petri dish culture of **Dr. Curry of Manchester University** may be followed by inoculating hair or scales, etc., on those drops as a **primary culture** (see Ch. 4).

DERMATOPHYTIDES

(ID OR THE IDE LESIONS)

Dermatophytides are eruptions due to dissemination of mycelia, spores or any **allergen produced by the fungus from a primary focus of infection in the body, followed by sensitisation of the skin**.

Various **clinical types** have been described, namely, **papular, vesicular, pustular, eczematoid** or rarely of the type of **frank erythema multiforme**. The commoner is the **lichenoid type** in which groups of **conical, follicular papules appear symmetrically on the trunk and extremities. Fungi** are **present in the primary focus** but **CANNOT BE DEMONSTRATED IN** THE **IDE LESIONS**.

Dermatophytides usually occur in acute lesions caused by ringworm and **SENSITISATION** can be **DEMONSTRATED** by a positive **reaction to trichophytin. Epidermophytides** may be produced by *E. floccosum* **in the form of POMPHOLYX OF THE HAND** originating **from a primary focus in the foot**.

TREATMENT

The treatment of dermatophytides is **symptomatic**. They **disappear spontaneously when the primary focus is treated**.

THERAPY

SEE EARLIER THIS CHAPTER AND CHAPTER 8.

RNGWORM OF THE SCALP

Prophylasis

Ringworm of the scalp is a **highly contagious** disease and is extremely common amongst school children. **Epidemics** may break out **in boarding schools**. All children should have careful examination of the scalp before admission into the school and also periodically when they join the school after vacations. Help of **Wood's light is essential for this**. As soon as any infection is detected, the child should be isolated and put under treatment. Hats, hair brushes, combs, etc., of patients and all contacts should be sterilised but when this is not possible these articles used by the patient should be burnt out. The patient should wear a tightfitting linen cap through out the course of treatment. He or she should not attend the school till completely cured. In a boarding school the child should be segregated. **Protective and preventive shampoo containing selenium sulphide twice monthly is useful.**

Treatment

The principle of treatment is to get rid of the infected hairs by a process of **epilation**. Epilation is followed by application of mild antiseptic ointment which contains non-irritant, fungicidal agents. The ointment layer acts on the spores of the fungus and prevent the spread of infection by hair. The ointment should not be strong enough to cause too much irritation or inflammation of the affected part. Ointments usually used are Hydrarg ammon 10 gr. Hydrarg Ointment Baccz? 5 per cent ammoniated mercury, or 1 per cent thymol or 0.1 per cent **phenyl mercuric nitrate** or an ointment containing **0.5 per cent each of iodine crystal** and **cinnamon oil**. Irritant applications like croton oil, turpentine or chryserobin are hazardous and not useful. CLOTRIMAZOLE & other Azole compound in Lotion form is the best method of treatment. Never apply any ointment on the scalp to avoid boldness.

Loose hairs may be best removed by shampoos or a liquid soap. Rubbing too vigorously or too frequently may break the hair instead of making them loose. In infections with fungi of animal origin like ***T. mentagrophytes*** or ***M. lanosum*, inflammatory reaction is severe** and as a result there is a tendency of **suppuration and spontaneous cure by scarring of the hair follicles**. In these cases, prognosis is good and all that is required is the application of a mild antiseptic ointement.

ANTISEPTIC COMBINED WITH HORMONE THERAPY : Gonadotropic fungicidal paints containing thymol and hormones were tried by Ghosh and Dey (1945) in 23 cases of ringworm of the scalp combined with local application of **cinnamon oil** in **tincture iodi mitis** or 0.1 per cent application of **Phenyl mercuric nitrate**. The use of the **gonadotropic hormone has been limited**, apprehending that precocious puberty might be initiated specially in girls. Oestrogenic hormone like **stilboestrol** is given in doses of 1 mg. three times daily for 20 days or until the child exhibits symptoms of toxicity such as inflammation of the breast, vomiting, etc. It has been found that combined treatment is better than either hormone or antiseptic lotion alone. Those who are interested are referred to the original paper.

EPILATION

This may be rendered in one of the following ways. The hair should be cut short and the scalp should be examined with Wood's filter from time to time for the progress of the case.

MANUAL EPILATION

This is feasible when the patch is small and single. The infected hairs as revealed under Wood's light in a dark room, are pulled out with a pair of epilation forceps. The process is repeated twice weekly and the patches are covered with the antiseptic ointment after epilation. On the other days, the patient can apply soap or shampoo after which the ointment is applied. The patient should wear a tight linen cap which should be boiled every day.

EPILATION BY IRRITANT DRUGS

Local application of irritant drugs produces acute inflammation and loosening of the hairs which are then easily pulled out; e.g. croton oil or 2 per cent chryserobin in acetone. But in such applications there is danger of scarring and permanent baldness. Suppuration must, however, be stopped before the application of X-ray treatment.

EPILATION BY THALLIUM ACETATE

Thallium acetate administered orally in therapeutic doses, loosens hair so that all the hairs can be easily pulled out after 18 to 21 days. The drug has the disadvantage that the **therapeutic and toxic doses are close to each other**, so that there is always a risk or using it. The **therapeutic dose is 8 mg per kg body weight** given a a single dose in children upto 10 years. The **toxic dose is 8.5 mg per kg body weight**.

Mode of administration

The child must be accurately weighed in a nude state and the weight is recorded. The dose is calculated according to the body weight and thallium acetate is administered as a single dose in a cupful of sweetened water. Hair begins to fall after 3 weeks. The therapeutic action of the drug is due to the selective action through the sympathetic trophic nerves. Common **toxic symptoms are pain in the muscles and joints, convulsions of choreic nature, albuminuria, gastrointestinal disturbances**, etc. Adhesive plaster is applied all over the scalp on the 18th day after the draught and complete epilation is produced by taking off the plaster cap on a subsequent date. **FUNGICIDAL REMEDIES (ANTIFUNGAL CHEMOTHERAPEUTICS)** should then be applied to eradicate the infection thoroughly and completely before regrowth of the hair which commences after four weeks.

EPILATION BY X-RAYS

Total epilation with X-rays, followed by the application of antiseptics, is the standard method of treatment. **For X-ray epilation**, the method of **Adamson-Kienbock is followed**. In this the scalp is divided into five areas, namely frontal, vertical, occipital, and one over each ear and approximately a dose of 370 units of X-rays are administered to each area. This treatment does not kill the fungus, but their papillae are devitalised temporarily and within 21 days hairs fall out including the infected ones containing spores and the fungus. The scalp still remains infected and antiseptic ointment as mentioned before should be applied daily for this purpose along with the application of soap and wash.

The cure is such cases should be carefully observed and examined under Wood's light so that no area remains untreated or becomes newly infected.

Contraindications of X-rays

Impetigenous, pustular lesions should be treated and cured before the X-ray exposure; otherwise the lesion may spread on the whole scalp. **KERION CASES SHOULD NOT BE TREATED WITH X-RAY**.

Danger of X-ray

A permanent baldness may occur after an over-exposure. This should always be done by experts with an indemnity bond from the guardian of the patient. Ulcers may be aggravated and may cause deep scars with permanent baldness.

Treatment of Kerion

Thick crusts are removed with starch and boric poultrice daily for several days **till the scalp is clean**. Loose hairs are removed with a pair of epilation forceps. When the part is fairly clean, lotio triple dye is painted twice a day on the patch.

THE FUNDAMENT PRINCIPLES OF TREATMENT OF RINGWORM INFECTIONS WAS BEEN GROSSLY SIMPLIFIED, CONSISTING OF

I. TOPICAL IMIDAZOLE APPLICATION

II. ORAL TERBINAFINE

III. ORAL GRISEOFULVIN FOR 6 MONTHS IN RINGWORM OF THE NAIL AND SCALP

TINEA BARBAE

Treatment

In a very early case when the infection is limited to the skin and has not affected the hair follicles local application of Whitfield's ointment or application of tincture iodi mitis is quite effective to remove the infection. When the hair follicles are affected, a preliminary application of hot boric compress or starch boric **poultice** will help removal of the crust. Manual epilation is easy as the hairs are loose in the widely gaping hair follicles. When the part is clean, triple dye or tincture iodi mitis may be applied twice daily. **Phenyl mercuric nitrate preparation** (0.1 per cent) or 5 per cent ammoniated mercury may be applied at night. Sodium propionate (10 per cent solution in alcohol) may be substituted for ammoniated mercury ointment. **In persistent cases X-ray epilation of the affected parts is advised**. At least two microscopic examinations should be negative before the patient is declared to be cured.

(SEE ABOVE TINEA BARBAE)

RINGWORM OF THE GLABROUS SKIN

Prophylaxis

The disease is apt to recur and for this reason prophylaxis should be strictly observed. The nidus in the **interdigital and crural regions** should be treated properly and radically cured. Walking bare-footed is strictly avoided. As there is chance of contamination, underwears should be boiled every day or as frequently as possible. Public baths and comodes are sources of contaminations and these should be avoided as far as possible. In the rainy season shoes may become mouldy. The shoes should be disinfected with formalin spray at regular intervals and put in the sun. Socks should be changed and boiled or disinfected every day. Feet should be kept dry using sulphur and camphor powder; other flexural and intertriginous regions of the body should be similarly treated. Vinegar (3 per cent acetic acid) or powder containing propionic acid or undecylenic acid may be applied in the foot and intertriginous regions.

Treatment

Ringworm is a chronic disease but in the genitocrural and other regions where the skin is thin, acute exacerbation due to streptococcal infection is often seen. In the acute condition evaporating lotions, Goulard's lotion, lotio calamine of 1 : 4000 solution of potassium permanganate may be applied locally. But when the infection has further advanced, lotio acriflavine (1 : 5000 aqueous) may be applied on a gauze soaked with the lotion intermittently for two or three hours or four times a day. This is applied in such a way that the part is not sodden. When the oozing has stopped, unguentum zinc or borovaseline should be applied. Secondary infections in the interdigital part of the toes and cellulitis is best treated with application of Pot. permanganate lotion followed by application of lotio triple dye. When dry this may be followed by application of Castellani's paint.

In the chronic condition, unguentum Whitfield is the best application particularly on the thin, sensitive skin like those of scrotum, groin or face. In the groin as a routine measure, undecylenate powder (10 per cent zinc undecylenate) or sulphur and camphor powder should be applied in the day time twice or thrice daily and at night application of Whitfleld's ointment is advocated as a routine. Dithranol ointment cures tinea more quickly than other remedies but there is a great danger or irritation from the drug. For sensitive areas it should be used in a concentration of 0.5 per cent, but in other chronic or resistant cases it may be used in a concentration of 1 to 2 per cent. On the body,hands, feet and hypertrophic patches on the sole of the foot, ringworm paint is an excellent remedy. One application a day is quite effective; **unguentum Whitfield (Dipsalic) may be applied** in those areas at night. In areas of secondary infection causing **pyodermia**, **unguentum hydrarg. ammon. dil** may be applied to get rid of the secondary infection before Whitfield's ointment is applied on the part.

(SEE ABOVE TINEA BARBAE)

TINEA UNGUINUM (RINGWORM OF THE NAIL)

Treatment

No local treatment is satisfactory except avulsion of the nail under nitrous oxide anaesthesia. Thorium X varnish has been claimed to have given very good results. **Application of Castellani's paint twice daily**, when continued for about six months to 2 years may cures the condition is some cases.

This is an excellent remedy, but it will stain the nail Majenta. For this is the nail is scraped to such an extent that the affected tissue is almost completely removed before the paint is applied.

(SEE ALSO ABOVE TINEA BARBAE)

FAVUS (TINEA FAVOSA)

Prophylaxis

Favus is **HIGHLY CONTAGIOUS** and the complete segregation of the patient is imperative. Hats, turbans or other head-dresses should be burnt and tight-fitting washable cotton caps must be worn and boiled daily for disinfection, till the patient is completely cured.

Treatment

Scutula are removed with 10 per cent salicylic acid or 2 per cent mercuric oleate ointment. Epilation of the hair with X-rays is advisable and the scalp is then treated with thymol-iodine paint. Chances of baldness resulting from the disease are much greater than from X-ray treatment. If the nail is affected, it should be removed and the nail bed is treated as for ringworm of the nail. **Griseofulvin therapy** is advised in both scalp and nail conditions.

(SEE ALSO ABOVE TINEA BARBAE)

TINEA IMBRICATA

Treatment

No satisfactory treatment of tinea imbricata was known before. Strong keratolytics like 10 per cent chrysarobin or 5 per cent each of sulphur precipitate and salicylic acid in vaseline or 2 per cent derobin in an ointment base show a very good effect but results are temporary and lesions reappear within a short time. Castellani's paint has been recommended and this may be supplemented with Whitfield's ointment. A pigment containing resorcinol-dr. i, acid acetic-dr. i and tincture benzoin co. oz. 1 may be applied with some benefit. Ringworm paint is more effective but may irritate the tender skin.

At present successful treatment has come out (**SEE ABOVE TINEA BARBAE**, and Ch. 8)

REFERENCES

CLASSIFICATION

Dey, N. C. (1971). Proposal for a revised classification of dermatophytes. *Ind. J. Dermatol.* **16**(4), 1-19.

EPIDERMAPHYTON FLOCCOSUM

Dey, N. C. (1965). *Epidermophyton floccosum* infecting the lanugo hair. *Ind. Jour. Dermatol.* **10**(4), 150.

K. AJELLOII

Georg, L. K., Kaplan, W., Ajello, L. Williamson, W. M. and Tilden, E. B. (1959). The Parasitic Nature of the Soil fungus *Keratinomyces ajelloi. J. invest. Dermatol.* **32**, 539-544.

Dawson, C. D. and Gentles, J. C. (1961). The perfect States of *Keratinomyces ajelloi* Vanbreuseghem; *Trichophyton terrestre* Duris & Frey and *Microsporum nanum* Fuentes. *Sabouroudia* **1**, 49-57.

M. AUDOUINII

Kaplan, W. and Georg, L. K. (1957). Isolation of *Microsporum audouinii* from a dog. *J. invest. Dermatol.* **28**, 313-315.

Kligman, A. M. (1952). See under *M. canis*.

Kligman, A. M. (1955). See under *M. canis*.

Dey, N. C. (1953). A review of ringworm of the hair in India. *Ind. med. Gaz.* **88**, 194.

M. CANIS

Kaplan, W., Georg, L. K. and Ajello, L. (1958). Recent development in animal ringworm and their public health implications. *Ann. N. Y. Acad. Sci.* **70**, 636-649.

Kligman, A. M. (1952). The pathogenesis of tinea capitis due to *Microsporum audouinii* and *Microsporum canis*. 1. Gross observations following the inoculation of humans. *J. invest. Dermatol.* **18**, 231-245.

Kligman, A. M. (1955). The pathogenesis of tinea capitis due to *Microsporum audouinii* and *Microsporum canis*. II. Dynamics of host parasite relationship. *Arch. Dermatol.* **71**, 313-337.

M. DISTORTUM

Davidson, A. M., Dowding, E. S. and Buller, A. H. R. (1932). Hyphal fusions in dermatophytes. *Canad. Jour. Res.* **64**, 1.

Kaplan, W., Georg, L. K., Hendricles, L. and Leeper, R. A. (1954). Isolation of *Microsporum distortum* from Animals in the United States. *J. invest. Dermatol.* **28**, 449-453.

Marples, M. J. and Smith, J. M. B. (1962). Further observations on the ecology of *Microsporum distortum. New Zealand Med.* **61**, 539-544.

M. GYPSEUM

Dey, N. C., Kakoti, L. M. (1955). *Microsporum gypseum* in India. *Jour. ind. med. Assoc.* **25**, 160.

Stockdale, P. M. (1963). The *Microsporum gypseum Cowples* (*Nannizzia incurvata*, Stockdale, *N. gypsea* Nann. comb. nov., *N. fulva* sp. Nov.). *Sabouraudia* **3**, 114-126.

M. VANBREUSEGHEMII (*N. grubyia*)

Georg, L. K., Ajello, L., Friedman, L. and Brinkman, S. A. (1962). A new species of Microsporum pathogemic to man and animals. *Sabouraudia* **1**, 189-196.

MICROSPOROSIS AND TRICHOPHYTOSIS

Castellani, A. (1928). Fungi and fungus diseases. *Arch. Dermatol.* **17**, 354.

Catellani, A. and Chalmers, A. J. Mannual of tropical medicine; Baillier, Tindall and Cox, London, 1919.

Dawson, C. O. and Gentles, J. C. (1961). The perfect states of *Keratinomyces vanbreuseghemii, Trichophyton terrestre* Durie and Frey, *Microsporum nanum* Fuentes. *Sabouraudia* **1**, 49-57.

Dey, N. C. and Ghosh, L. M. (1945). Studies in ringworm, Part I. *Microsporum audouinii* infection in India. *Ind. med. Gaz.* **80**, 381. *Idem*. (1945). Studies in ringworm, Part II. Therapy of tinea capitis by gonadotropic hormone. *Ibid.* **80**, 381.

Dey, N. C. and Kakoti, L. M. (1956). Faviform ectothrix trichophytosis in India *Ind. Jour. Dermatol.* **1**, 3.

Ghosh. L. M., Dey N. C. and Panja, D. (1947). Studies in ringworm, Part IV. *Ind. med. Gaz.* **82**, 73.

Fuentes, C. A. (1956). A new Species of Microsporum. *Mycologia* **43**, 613-614.

Fuestes, C. A., Aboulafia, R. and Vidal, R. J. (1954). A dwarf form of *Microsporum gypseum. J. invest. Dermatol.* **23**, 51-57.

Lewis, G. N., Hopper, M. E. and Reiss, F. (1946). Ringworm of scalp: Clinical data on recent cases: Experiences with local endocrine therapy. *J. am. med. Assoc.* **132**, 62.

TRICHOPHYTON CONCENTRICUM

Acton, H. W. and Ghose, L. M. (1934). *Tinea imbricata* (Tokelau) in Bengal. *Ind. med. Gaz.* **69**, 426.

Dey, N. C. and Maplestone, P. A. (1942). *Tinea imbricata* in India. *Ind. J. med. Res.* **77**, 14.

T. FERRUGINEUM

Vanbreuseghem R. (1950). Etude de 136 souches de *Trichophyton ferrugineum* (Ota, 1921) Langeron *et al.* Milochevitch 1930 et da as variete blanche, isoles au Congo Belge. *Ann. Parasitol. Hum. et Comp.* 25, 485-492.

Vanbreuseghem, R. Mycoses of man and animals, transl. by J. Wilkinson, Sir Issac Pitman and Sons, Ltd., London, 1958, pp. 104-105.

T. MEGNINII, T. GALLINAE

Silva, M. and Benham, R. W. (1952). Nutritional studies of the dermatophytes with special reference to *Trichophyton megninii* Blanchard, 1896 comb. and *Trichophyton gallinae* (Megnin, comb. 1881) nov. *J. invest. Dermatol.* **18**, 453.

Tarres, G. and Georg, L. K. (1956). A human case of Trichophyton gallinae infection. Disease contracted from chickens. *Arch. Dermatol.* **74**, 191-197.

TRICHOPHYTON MENTAGROPHYTES

George, L. K. (1954). The relationship between the downy and granular forms of *Trichophyton mentagrophytes. J. invest. Dermatol.* **23**, 123-141.

Patridge, B. M. (1959). The use of the chorioallantoic membrane of the developing chick for culture of Dermatophytes – a modified technique. A preliminary report on its use for aerial passage. *J. invest. Dermatol.* **32**, 605-619.

T. MENTAGROPHYTES VAR. EQUI

Georg, L. K., Kaplan, W. and Camp, L. B. (1957). *Trichophyton equinum* – A re-evaluation of its Taxonomic Status. *J. invest. Dermatol.* **29**, 27-37.

T. RUBRUM (PURPUREUM)

Acton, H. W. and McGuire, C. (1927). Tinea cruris: Its manifestation, diagnosis and treatment. *Indian med. Gaz.* **72**, 419.

Dey, N. C. (1953). *Trichophyton purpureum* infection in India. *Ind. med. Gaz.* **88**, 525.

T. SCHOENLEINII

Dey, N. C., Maplestone, P. A. (1936). Favus in India. *Indian Jour. med. Res.* **23**, 687.

T. SOUDANENSE

Calnan, C. D., Djavahiszvill, N. and Hodgson, C. J. (1962). Trichophyton soudanense in Britain. *Brit. J. Dermatol.* **74**, 144-148.

Cochet, G., Doby-Dubois, N., Deblock, S., Doby, J. M. and Vaiva, C. (1957). Contribution a la connaissance des teignes infantiles due Cameroun. *Ann. Parasit. Hum. et Comp.* **32**, 580-589.

T. TONSURANS

Dey, N. C, Maplestone, P. A. (1941). *Trichophyton crateriforme in India. Ind. med. Gaz.* **76**, 410.

T. VERRUCOSUM, T. TERRESTRE

Durie, E. B. and Frey, D. (1957). A new species of trichophyton from New South Wales. *Mycologia* **49**, 401-411.

Marples, M. J. and Smith, J. M. B. (1962). *Trichophyton terrestre* as a resident in hedgehog skin. *Sabouraudia* **2**, 100-107.

Dawson, C. D. and Gentles, J. C. (1961). The perfect states of *Keratinomyces ajelloi*; Van breuseghem, *Trichophyton terrestre.* Durie and Frey and *Microsporum nanum Fuentes. Sabouraudia* **1**, *49-57.*

T. YAOUNDEI

Georg, L. B., Doupagne, P., Pattyn, S. R. and Neves, H. (1963). Trichophyton yaoundei, *a dermatophyte is indigenous to Africa.* **J. invest. Dermatol. 41**, 19-24.

Appendix A

Solvents And Stains For Direct Examination

Sodium sulphide solution

To prepare a saturated solution, put a few drops of distilled water to crystals of Merck's sodium sulphide. Take a volume of this solution; add an equal volume of rectified spirit or 90 per cent alcohol; on mixing turbidity appears. Add a few drops of distilled water till the turbidity disappears. The solution is then filtered with a filter paper. The **best result is obtained with a freshly prepared solution**. But for all practical purposes the solution can be used in a drop bottle for one month without any deterioration.

Potassium or sodium hydroxide solution

A 10 per cent solution of potassium or sodium hydroxide solution is used as a keratolytic agent. For nails, use a 40 per cent solution.

Sodium sulphide solution has been found to be the **best of all the keratolytic** agents and in the opinion of the authors, this is **better than potassium** or sodium **hydroxide solutions**.

MCGUIRE'S STAIN

TOLUIDIN BLUE	1 g
Alcohol	2 ml
Acetic acid	4 ml
Water	100 ml

CARBOL THIONIN

THIONIN	1 g
Phenol crystal	5 g
Alcohol	30 ml
Water	100 ml

BORAX METHYLENE BLUE

METHYLENE BLUE	2 g
Borax	5 g
Water	100 ml

This scales like those of tinea versicolor, erythrasma, pityriasis capitis may be placed in a slide and treated with a few drops of **McGuire's stain** or **carbol thionin for 2 minutes**. The stain is then mopped off with a piece of filter paper and a coverslip preparation is made with a drop of glycerine which gives a differentiating effect on the staining of the fungus. **For scales of pityriasis capitis, the scales are treated with ether or acetone for removal of a fat before they are treated with the stain**. Borax or any other polychrome (methylene blue) agent may give a satisfactory result.

AMMAN'S LACTOPHENOL *(Modified)*

Phenol crystal	20.00 g
Glycerine	20.00 g
Water	20.00 ml
LACTIC ACID (syrup)	20.00 g
Cotton blue	0.05 g

All the **ingredients except cotton blue are dissolved with gentle heating**. Cotton blue is then added and it dissolves in no time. **Acid fuchsin** may be **substituted for cotton blue** with an equally good result and it gives a red colour instead of a blue one. The **stain is** extremely **useful for the culture mount**. For this, hanging drop preparation from agar culture agar is teased, washed in water to remove traces of agar. The culture is then mounted in Amman's lactophenol in a coverslip preparation. Excess of the stain is removed by soaking with filter paper and the preparation is then sealed with cutex nail polish on all sides.

M. R. stain for the cytological study of ringworm fungi and parasitic protozoa (Mukherjea and Ray, 1957). *Bull. Calcutta School of trop. Med.* **5**, 176.

The staining solution is prepared as follows :

Solution 1

Sat. solution of mercuric chloride	64.00 ml
Absolute alcohol	32.00 ml
Glacial acetic acid	5.20 ml
Formalin (40% formaldehyde)	4.20 ml
THIONIN	0.04 g
AZURE 1	0.02 g

Solution 2

70 per cent alcohol	93.00 ml
Glacial acetic acid	7.00 ml
THIONIN	0.04 g
AZURE 1	0.06 g
Methyl green (chloroform extracted)	0.04 g
Eosin	0.24 g

The ingredients are mixed by shaking vigorously in a stoppered bottle and filtered through a filter paper.

Solutions 1 and 2 are mixed in equal proportion; filtered after 24 hours. This is a **fixative cum staining solution.**

Technique

Smears from culture of dermatophytes **or faeces are dipped in a jar containing the staining solution**. After **ten minutes** the slides are raised in tap water. The smears are then **dehydrated** by passing successively through **two changes of tertiary butyl alcohol 1 minute each**, cleared in toluol for 1 minute and xylol for 1 minute and **mounted** in DPX mount.

COLLECTION OF MATERIAL

(FOR RINGWORM, TINEA VERSICOLOR, etc.)

Equipments

(1) **Pairs of slides** wrapped in craft paper and sterilised in a hot air oven.

(2) **Ether and alcohol mixture** in equal volumes.

(3) **Swab of cotton** wool.

(4) **Paget's knife**. (5) **Epilation forceps**.

(6) **Spirit lamp**, etc.

The following points are **recorded in the register**, namely:- name, age, sex, date of collection, duration of illness, site from which the material is collected, clinical diagnosis, result of direct examination, culture report and remarks.

Scrapping from the glabrous skin

This is done for **direct examination of the scales and their culture**. The part to be scraped is cleaned and **indirectly for sterilised** with **a swab of ether and alcohol mixture**. The Paget's **knife** is dipped in ether alcohol mixture and **passed through the flame (bunsen burner** or spirit lamp) and thus the excess of spirit is burnt. The sterilised knife is used when cool. The scraping is usually done from the margin of the lesion without injuring the skin surface. This is done by **scraping with the blade**, kept **flat on the skin and not in an angular fashion**. The slide should be held in such a way that scales obtained by scraping may be received directly on the slides. When sufficient material is collected, the second slide is put on the **surface containing material on the** first **slide (where** the material was received). The slides are wrapped back in the craft paper. The serial number in the register is put down on the cover.

Collection of material from pompholyx

Boiled sago grain-like blisters on the **palms of the hand (Cheiropompholyx) or sole of the foot** (Podopompholyx) are due to ringworm infection or 'ID' **REACTION**. In the former case, ringworm fungus is present and the skin from the blister shows ringworm fungi, but in the latter instance, **no pathogenic organism can be detected. For collection** of **material from blisters, a pair of scissors is similarly sterilised**. The point of the scissors is introduced into the blister at its edge and the skin over the blister is cut at the periphery in a circular fashion holding the margin of the skin with a pair of dissecting forceps. The cut portion of the skin is received in between a pair of sterile slides. Skin from several blisters may be collected in this way for **direct examination and culture**.

Scraping from the anil

Nails are liable to be attacked by ringworm fungi. The affected **nail becomes opaque, brittle and hypertrophied**. Nails may also be **affected by favus**. Collection is made by shaving the nail bed with a Paget's knife or with a hard, safety razor blade. The material is received directly on the slide. Other procedures are the same as in the scraping of the glabrous skin.

Collection of infected hair

In ringworm of the scalp or beard area, **infected hairs are collected for difect examination and culture**. This is conveniently done with the help of a pair of epilation forceps. It should be emphasised that stumps of infected hair are small and the **epilation** in such a case **is not at all painful**. The fact should be borne in mind and if the epilation is painful, it should be presented that the epilated hair is healthy. **Hairs are quite loose in the widely gaping hair follicles in cases of kerion** of the **scalp and tinea barbae**.

CULTURE MEDIA

Sabouraud's proof medium

Maltose (brute de Chanut)	40 g
Agar-agar	23 g
Peptone (granules de Shassaing)	10 g
Distilled water	1000 ml
*p*H 5.6-6.2	

For the study of giant colonies, the above formula should be strictly followed because a slight difference in the composition of the medium or changing the brand of the ingredients may alter the character of the growth of the giant colony. In the tropics, the quantity of agar should be increased.

In Italy, Polacci and in Germany, Smidtz media are used as a substitute for Sabouraud's medium. In America, Wedman and Spring showed that Fairchild's Peptone could be substituted for Chassaing peptone without any difficulty or sacrifice in the final result. It was also demonstrated by Weidman and McMillan that crude American dextrose could act as a substitute for crude maltose de Chaunt used in the original Sabouraud's medium. It was later found that technical dextrose gave equally good results.

Wedman's Modification of Sabouraud's Medium

Agar (granular)	18 g
Dextrose (American crude)	40 g
Peptone (Fairchild)	10 g
Water	1000 ml
*p*H 5.6	

PREPARATION OF SABOURAUD'S PROOF MEDIUM

Mix the ingredient in a flask and allow the mixture to stand for half an hour. Then place the flask in the autoclave. Raise the temperature of the autoclave till a continuous jet of steam is coming out through the stop cock. The **stop cock is then closed** and the pressure is raised to 5 pounds and this is maintained for half an hour. The gas is then turned off and when the steam has fully escaped, the autoclave is opened. The solution is then filtered in a hot chamber and the pH is determined at 6.2 in a **Hellige or Lovibond comparator**. The medium is now distributed in test tubes and flasks (8 ml in a 6″ by 3/4″ test tubes or 75 ml in 250 ml Erlenmeyer flasks to give a depth of about 1.5 cm). The tubes are plunged with yellow cotton wool (the colour indicates Sabouraud's proof medium). Flasks and tubes are then autoclaved. For this, temperature of the autoclave is raised to 15 pounds, when the gas is turned off. The pressure gradually falls to nil. The steam is allowed to escape and the autoclave is then opened. The tubes are then slanted on a slightly inclined plane and allowed to cool without disturbance. The medium sets when cool. The tubes are used for routine work of isolation and subculture and the Erlenmeyer flasks are used for routine work for isolation and subculture and the Erlenmeyer flasks are used for culture of giant colonies.

Caution : As the reaction of the Sabouraud's medium is acidic, too much or prolonged heating hydrolyses the sugar and prevents gel formation of agar. The processing should be carried right through without allowing the medium to solidify at any stage before the slope is made. Preparation of slopes by melting the stock solid medium should always be avoided, as this has been found to be deliterious to gel formation and to the growth of the fungus.

SABOURAUD'S MEDIUM WITH GENTIAN VIOLET

This consists of Sabouraud's proof medium containing gentian violet in a proportion of 1 : 400,000 of the medium. For this one per cent solution of gentian violet is kept as a stock solution, and o.25 ml of this solution is added to one litre of Sabouraud's proof medium. This medium has been found to be very useful in preventing the growth but the growth is delayed in this medium and visible within five to seven days.

MALT EXTRACT AGAR MEDIUM (DIFCO)

for culture of *P. ovale* (Moore's method).

Maltose (Technical Difco)	12.75 g
Malt extract (Difco)	15.00 g
Dextrin (Difco)	2.75 g
Gycerine	2.35 g
Dipotassium phosphate	1.00 g
Ammonium chloride	0.78 g
Bacto peptone (Difco)	15.00 g
Bacto agar (Difco)	15.00 g
Water	1000.00 ml
*p*H 4.8	

Petroff's medium with gentian violet

500 g of ground level or weal is infused in 500 ml of 15 per cent solution of glycerine in water in a refrigerator. Next morning squeeze the meat in a sterile press and collect aseptically in a sterile beaker.

Mix **one part of meat juice** and **two parts of eggs by volume**. For this eggs are washed in soap and water and **kept immersed in 70 per cent alcohol for 10 minutes. Break the eggs in sterile beaker, mix and pass the mixture through sterile gauze.**

To the mixture of infusion and egg, add one per cent alcoholic solution of gentian violet to make a final dilution of 1 : 10,000 and mix the ingredients thoroughly. The mixture is distributed in screw-capped test tubes or bottles and sterilised by inspissation for three successive days.

PANJA'S MODIFICATION OF PETROFF'S MEDIUM for culture of *P. OVALE*, etc.

Meat infusion in 15 per cent glycerinated water	Equal parts
Content of one egg	

Genian violet-0.004 per cent; slope and inspissate. Meat infusion and egg are essential for the growth of *P. ovale*.

Blood agar and nutrient agar

These are prepared as in all routine bacteriological laboratories.

Corn meal agar : Corn meal 40 gm, agar 15-20 gm and water 1000 ml.

(a) Mix **corn meal and 500 ml of water** and steam in Koch's steriliser for about one hour. Filter through gauze and make up the volume.

(b) **Mix agar and melt in the autoclave (5 lb pressure for one hour)**. Mix (a) and (b); filter through lint; tube and sterilise in autoclave at 15 lbs pressure for 15 minutes.

N. B. *CANDIDA ALBICANS* is cultured in this medium to show production of **CHLAMYDOSPORES**.

CHLAMYDOSPORE AGAR (Nickerson and Man-kowski). This consists of a basal medium to which purified starch is added. The **BASAL MEDIUM** consists of ammonium sulphate-1 g, potassium dihydrogen phosphate-1 g, **biotin-5mμg**, Bacto agar-15 g; trypan blue-0.1 g and distilled water to make 1000 ml. Purified polysaccharide is sterilised separately by autoclaving or tyndallization) and added aseptically to sterilised basal medium in the proportion of 2 gm per 100 ml. **SLIDE CULTURES ARE PREPARED IN THIS MEDIUM** by taking inocula from young cultures grown on glucose peptone agar. Heavy inocula are put. The **cultures are incubated at 25°C.**

LITTMAN OXGALL AGAR

Peptone	10.00 g
Dextrose	10.00 g
Oxgall (dehydrated)	15.00 g
Agar	20.00 g
Crystal violet	0.01 g
Water	1000.00 ml

Autoclave at 12 pounds for 15 minutes. Cool to 46°C; add 30 units of streptomycin per cubic centimeter; pour into plates (15 ml in a 10 cm plate). 'Bacto' Littman oxygall agar should be prepared according to the direction supplied with the product.

The culture plate should contain enough medium to be incubated for 14 days at room temperature.

Strains of *C. albicans* which **fail to produce on Difco dehydrated cornmeal agar** may do so on freshly prepared cornmeal agar.

Purification of soluble starch

Add 5 to 7 gm of commercial soluble starch to 100 ml of boiling water. When the starch is dispersed and clear liquid is still hot, pour into an equal volume of saturated ammonium sulphate solution (76.0 g of ammonium sulphate to 100 ml of the solution at 25°C). After standing for 24 hours, the voluminous amorphous precipitate was filtered with suction, washed with cold distilled water until the filters gave a negative test with Nessler's reagent and then dried in vacuo. ***The starch solution should not contain any reducing sugar as determined by Fehling's reagent***. With iodine it gives a characteristic starch like reaction. Some soluble starch samples show a purple colour with iodine indicating a high content of dextrin.

Selective medium for isolation of pathogenic fungi

In this medium, Sabouraud's dextrose agar is distributed in 100 ml quantities into each of the Erlenmeyer flasks and sterilised in an autoclave as usual. The flasks are then allowed to **cool to 45°C** in a water bath and antibiotics to give the final concentration of Cycloheximide (Acti-dione, technical grade), penicillin and streptomycin are then added.

FORMULA OF THE SELECTIVE ISOLATION MEDIUM FOR PATHOGENIC FUNGI

Cycloheximide-Penicillin-Streptomycin Medium*

Sabouraud dextrose agar is prepared as usual and distributed into Erlenmeyer flasks-100 ml per flask. The medium is autoclaved and allowed to cool to 48°C in a water bath. The various antibiotics are added to give the following final concentrations :

Cycloheximide (Acti-dione)	0.5 mg per ml
Penicillin	20.0 units per ml
Streptomycin	40.0 units per ml

Stock Cycloheximide Solution

Place 250 mg crystalline **cycloheximide** (**Actidione**) in a 50 ml volumetric flask. Add 5 to 10 ml acetone to dissolve, then bring to volume with distilled water (the pH of the water must be previously adjusted to 6.5-6.8). Sterilise by filtration through a Seitz filter and store at 5°C (**Cycloheximide prepared in acid solution will remain stable at 5°C from 6-8 weeks**). Add 10 ml stock cycloheximide to 100 ml cooled medium to obtain 0.5 mg per ml.

Stock Streptomycin Solution

Streptomycin = 1,000,000 units/g

Dissolve 1 g in 20 ml of sterile physiological saline giving a solution containing 50,000 units/ml. Take 4 ml of this and add to 46 ml of sterile saline. This forms the stock solution containing 4.000 units/ml.

Add 1 ml of stock streptomycin solution to 100 ml of cooled medium to obtain 40 units/ml.

Stock Penicillin Solution

Add 20 ml sterile physiological saline to a vial containing 200,000 units of crystalline potassium penicillin, giving a concentration of 10,000 units/ml. Take 10 ml of this and add to 40 ml of saline giving a stock solution containing 2,000 units/ml.

Add 1 ml of stock penicillin solution to 100 ml of cooled medium to obtain 20 units/ml.

From stock solutions of Streptomycin (4000 units per ml) and Penicillin (2000 units per ml add 1 ml each to 100 ml of the cooled medium to give a final dilution of 40 and 20 units per ml respectively. This medium has been used for isolation of *COCCIDIOIDES IMMITIS*, dermatophytes from clinical materials, or from shoes and other extraneous sources.

MEDIUM ON A POLYSACCHARIDE BASE

(Langeron and Milochevitch)

Soluble starch	10 g
Maltose	10 g
Glucose	10 g
Peptone	10 g
Agar-agar	20 g
Water	1000 ml

Wheat or patoto starch has been found to be **better** substitutes than soluble starch.

CZAPEK'S SYNTHETIC AGAR

Dipotassium hydrogen phosphate	1.00 g
Potassium chloride	0.50 g
Magnesium sulphate, $7H_2O$	0.50 g
Ferrous sulphate	0.01 g
Agar-agar	20.00 g
Distilled water	1000.00 ml

Norris' Medium *(for culture of STREPTOMYCES KERATOLYTICA)*.

Soluble starch	2.00 g
Dipotassium hydrogen phosphate	0.50 g
Magnesium sulphate (hydrated)	0.25 g
Calcium chloride	0.50 g
Ferric chloride	0.01 g
Sodium nitrate	0.05 g
Asparagin	0.05 g
Agar-agar	20.00 g
Distilled water	1000.00 ml
pH 7 to 7.4	

Plates are prepared in Petri dishes from this medium.

***Acti-dione, trade name for cycloheximide,** distributed by **Upjohn Company, Kalamazoo, Michigan.**

A technical grade Acti-dione is available which is approximately one fourth the price of the therapeutic compound. This technical grade **Acti-dione is effective for use in the selective medium** as described above. This is distributed by Mycology Unit Laboratory Branch, Communicable Disease Center, Public Health Services **U. S. Department of Health, Education, and Welfare, Atlanta, Georgia.**

REFERENCES

Georg, L. K., Ajello, and Gordon, M. A. (1951). A selective medium for the isolation of Coccidioides immitis Science **114**, 387-389.

Georg, L. K. (1953). Use of a Cycloheximide medium for Isolation of Dermatophytes from Clinical Materials. *Arch. Dermatol. & Syph.* **67** (4), 355.

Ajello, L., Getz, M. E. (1954). Recovery of Dermatophytes from Shoes and Shower Stalls. *J. invest. Dermatol.* **22**, 17-21.

Appendix B

Common Therapeutic Recipes

LOTIONS

Lotio Acriflavine

Acriflavine	1 gr.
Aqua	ad 10 fl. oz.

To be applied on a gauze which should be kept soaked with the lotion 2 hrs. each time, in streptococcal infection of the skin.

Lotio Aluminium Acetate

Aluminium acetate	0.125 to 0.25 g
Water	1000.0 ml

To be applied as an evaporating lotion.

Lotio Calamine

Calamina preparata	90 gr.
Zinc oxide	60 gr.
Glycerine	20 ms.
Liq. calcis ad	1 fl. Oz.

To dab on the affected part thrice daily.

Lotio Calamine co.

Calamina preparata	60 gr.
Zinc oxide	60 gr.
Glycerine	60 ms.
Lotio plumbi subacet. (Fort)	20 ms.
Liq. carb. det.	10 ms.
Liq. calcis ad	1 fl. oz.

To dab on the affected part thrice daily. It is useful in acute weeping conditions.

Lotio Formalin Et Glycerine

Liq. formaldehyde	60 ms.
Glycerine	60 ms.
Aqua ad	1 fl. oz.

It is useful in *STREPTOMYCES KERATOLYTICA* infection of the hands and feet. To apply once at night.

For cracks and fissure's of the heel.

Lotio Eurisol

Eurisol	30 gr.
Spt. etheris	30 ms.
Oil ricini	10 ms.
Spt. vin. rect.	60 ms.
Spt. lavendulae	15 ms.
Spt. rosmarini	15 ms.
Aqua ad	1 oz.

To apply on the **scalp** twice daily. It is useful in **dandruff (seborrhoea capitis)**. It does not stain the hair of blondes as in the case of lotio resorcin.

Lotio Gentian Violet

Gentian violet	2 g
Alcohol	5 ml
Aqua add	100 ml

It is useful in **intertrigo** and **candidiasis**.

Lotio Potassium Permanganate

Pot. permanganate	0.05 g
Water	100.00 ml

Lotio Plumbi

Liq. plumbi subacet. (Forte)	100 ms.
Water	8 fl. oz.

To apply in case of **WEEPING ECZEMA**.

Lotio Resorcin

Resorcin	45 gr.
Oil ricini	10 ms.
Spt. etheris	30 ms.
Spt. lavender	15 ms.
Spt. rosemarini	15 ms.
Rose water ad	1 fl. oz.

To apply in **SEBORRHOEA CAPITIS**.

Lotio Resorcin Et Hg. Perchloride

Hg. perchloride ½ gr. is added to lotio resorcin

To apply on the scalp in **SEBORRHOEA CAPITIS**.

Lotio Sodium Thiosulphite

Sodium thiosulphite (hyposulphite)	48.0 g (20%)
Distilled water ad	240.0 ml

It is useful in **PITYRIASIS**.

Lotio Sulphuris Alba

Zinc sulphate	10 gr.
Potassium sulphurata	10 gr.
Glycerine	20 ms.
Water (or rose water)	1 fl. oz.

Lotio Sulphuris Flava

Sulphur (Precipitated)	60 gr.
Pulv. Tragacanth co	10 gr.
Water (or rose water)	1 fl. oz.

Lotio sulphuris flava and alba are efficacious in **TINEA VERSICOLOR, SEBORRHOEIC DERMATITIS, FLEXURAL ECZEMA AND ACNE VULGARIS**.

Lotio Triple Dye

Acriflavine	0.10 g
Brilliant green	0.25 g
Gentian violet	0.25 g
Water	100.00 ml

Useful in **INTERTRIGO, CANDIDIASIS, ATHLETE'S FOOT, TINEA BARBAE, SYCOSIS BARBAE**, etc.

LINIMENTS

Liniment Calamine

Calamina preparata	30 gr.
Zinc oxide	30 gr.
Phenol	3 gr.
Oil amygdale	4 dr.
Liq. calcis ad	1 fl. oz.

To apply on the patch of **ALLERGIC OR SEBORRHOEIC DERMATITIS, IRRITABLE SKIN**, etc.

Liniment Calamine With Ichthyol

Ichthyol	10 gr.
Liniment calamine	1 fl. oz.

Efficacious in **SEBORRHOEIC DERMATITIS**.

Liniment Calamine With Sulphur

Sulphur (Precipitated)	10 gr.
Liniment calamine ad	1 fl. oz.

PAINTS

Ringworm Paint

Resorcin	1 oz.
Acid salicylic	1 oz.
Acid carbolic	½ oz.
Acid acetic	1 fl. oz.
Glycerine	½ oz.
Phenyl mercuric acetate	5 gr.
Tincture benzoin co.	4 oz.

To apply on the **PERSISTENT PATCHES OF RINGWORM** of the body.

Pigment Resorcin Benzoin Co.

Resorcin	60 gr.
Tincture benzoin co. ad	1 fl. oz.

To apply on the ringworm patches including interdigital ringworm.

Pigment Fuchsin (CASTELLANI'S PAINT)

Basic fuchsin, sat. alcoholic sol.	10 ml
5 per cent aq. sol. Phenol	100 ml
Boric acid*	1 gm
Acetone**	5 ml
Resorcin†	10 g

CARBOL FUCHSIN :

Basic fuchsin, sat. alcoholic sol.	10 ml
Phenol	5 gm
Water	100 ml

Add* boric acid to filtered carbol fuchsin; after 2 hours add 5 ml of acetone** and 2 hours later add 10 g of resorcin†. Keep in a dark, stoppered bottle.

Apply on **ATHLETE'S** foot or **RINGWORM OF THE NAIL**.

Pigment Thymol, Cinnamon Oil And Iodine

Thymol	10 gr.
Cinnamon oil	10 gr.
Tincture iodi mitis ad	1 fl. oz.

To apply on the **RINGWORM OF THE SCALP AND BEARD** areas.

POWDERS

Desenex Powder

Zinc undecylenate	24 parts
Undecylenic acid	2 parts
Talcum powder	120 parts

To be dusted on areas of **RINGWORM** as a **PROPHYLACTIC POWDER.**

Mycostain powder (E. R. Squibb and Sons)

R. Nystain	1,500,000.00 units
Talcum base	15.00 G

It is useful in local lesion caused by fungi sensitive to **MYCOSTATIN.**

Pulv Sulphur Et Camphor

Sulphur (precipitated)	60 gr.
Camphor	60 gr.
Zinc oxide	60 gr.
Boric acid	60 gr.
Pulv. amylum or talcum ad	1 oz.

To be dispensed in a perforated powder can. To dust on the affected areas of **SEBORRHOEIC PATCHES, PRICKLY HEAT** and as a prophylactic in **INTERDIGITAL RINGWORM** infection.

Undecylenate Powder

Zinc undecylenate	20 parts
Undecylenic acid	2 parts
Pulv. talcum	78 parts

This should be used as a prophylactic in **TINEA CRURIS** and **INTERDIGITAL RINGWORM.**

Propionate Powder

Calcium propionate	15 parts
Zinc propionate	5 parts
Pulv. talcum	80 parts

To apply locally on the areas of **RINGWORM** as a **PROPHYLACTIC** powder.

SHAMPOO

Selsun (Pharmacia H. care)

Selenium disulphide	3.0 g (2.5%)
Emulsion base	120.0 ml

Spirit Aether Soap Shampoo

Sapo viridis	1 oz.
Aether	1 oz.
Alcohol or rectified spirit	1 oz.

To apply on the scalp by rubbing thoroughly and then to wash. This should be done twice weekly in seborrhoea capitis and psoriasis.

SPIRIT

Spt. Resorcin Et Hg. Perchloride

Hg. perchloride	1/4 gr.
Resorcin	15 gr.
Spirit vinum rect.	120 ms.
Water ad	1 fl. oz.

To be applied in **PRICKLY HEAT, SUMMER BOILS,** etc.

UNGUENTUM (Ointment)

Unguentum Dithranolis

Dithranol	0.5 g
Paraffin mole	99.5 g

To apply on patches of **RINGWORM AND PSORIASIS.**

The strength can be increased, if tolerated. Its action is like that of chryserobin but it is 3 or 4 times more active than it.

Unguentum Hg. Ammon Dil.

Hydrag. ammon	5 to 10 gr.
Hydrous ointment ad	1 oz.

Efficacious in **IMPETIGO.**

Unguentum Whitfield

Acid benzoic	25 gr.
Acid salicylic	15 gr.
Ol. cocos	4 dr.
Paraffin durum	2 dr.

Prepare the base first by melting and then add other ingredients.

To apply on the **RINGWORM OF THE GROIN** and other sensitive parts of the body.

Unguentum Iodi

Iodine (Crystal)	5 parts
Lanolin ad	100 parts

To apply locally as **ANTISEPTIC OR COUNTER-IRRITANT.**

Quinochlor Compound Ointment (Squibb)

Chlorohydroxyquinolin	0.50 parts
Benzoyl peroxide	10.00 parts
Aromatic oil	0.24 parts
Vehicle (inert) ad	1000.00 parts

To apply in cases of **SYCOSIS BARBAE** (Staphylococcal).

OR

Unguentum Aureomycin

used in Sycosis BAR BAE

3 per cent aureomycin in an water-soluble base.

OR

Penicillin Ointment B. P.

Pragmatar Ointment (Smith, Kline & French Labs.)

	gm. or ml.
R/Cetykl alcohol-coal tar distillate	0.3 (0.5%)
Near-colloidal sulfur	1.2 (2.0%)
Salicylic acid	1.2 (2.0%)
Oil-in-water emulsion base as needed to make	60.0 (2.0%)

Unguentum Sodium Propionate

Sodium propionate	16.4 parts
Propionic acid	3.6 parts
N-propyl alcohol	10.0 parts
Zinc stearate	5.0 parts
Carbo wax 4000	65.0 parts

For **PROPHYLAXIS OF RINGWORM.**

Desenex ointment

Zinc undecylenate	12 parts
Undecylenic acid	3 parts
Water soluble base	60 parts

For **PROPHYLAXIS OF RINGWORM.**

Sulphur-Salicylic ointment

Sulphur precipitate	1 g
Salicylic acid	1 g
Ointment base	100 g

For **SEBORRHOEIC GRANULOMA**

If required Liniment calamina may be added.

Appendix C

Stains And Histopathological Procedure

1. ALCIAN BLUE (SHORT'S METHOD) WITH OR WITHOUT HYALURONIDASE

Fixation : Neutral formalin 10%.

Technique : Sections are cut **at 6 microns** in paraffin embedding.

Reagents

(1) ***3% Acetic acid solution***

Glacial acetic acid	3 ml
Distilled water	97 ml

(2) ***Alcian blue solution***

Alcian blue	1 g
Acetic acid (3% sol)	100 ml

pH 2.5. Filter and add a few crystals of thymol.

(3) ***Kernechtrot solution***

Kernechtrot (Nuclear fast red)*	0.1 g
Aluminium sulphate (5% sol)	100 ml

Dissolve by heating; cool, filter and add a few crystals of thymol.

(4) ***Hyaluronidase solution***

Dissolve 1 vial hyaluronidase (150 TR Units) in 1 ml normal saline.

Staining Procedure

1. Deparaffinise sections through 2 changes each in xylol, absolute alcohol and 95 per cent alcohol.
2. Wash in tap water, then rinse in distilled water.

   N. B. **To test for hyaluronic acid, flood tissue with hyaluronidase solution** with the help of a syringe.

   Allow to stand for 1 hour in a Petri dish to minimise evaporation. Wash well with tap water.
3. Mordant in 3 per cent acetic acid solution for 3 minutes.
4. Stain in **Alcian blue** solution for 30 minutes.
5. Wash in running tap water for 10 minutes.
6. Rinse in distilled water.
7. **Counterstain in Kernechtrot** for 5 minutes.
8. Wash in running tap water for a minute.
9. Dehydrate with changes each in 95 per cent alcohol, absolute alcohol; clear in xylol with 3 changes and mount in DPX mount.

Results

ACID MUCOPOLYSACCHARIDES—BLUE.

NUCLEI—PINK.

When hyaluronidase is applied, acid mucopolysaccharide is digested; other elements remain same as in original stain.

2. FITE-FARACO STAIN, MODIFIED FOR NOCARDIA**

Fixation : Any well fixed tissue.

Sectioning : Paraffin sections at 4 to **6 microns**.

Solutions

Xylene-peanut Oil

Peanut oil	1 part
xylene	2 parts

1% Aqueous Sulphuric Acid

Sulphuric acid	1.0 ml
Tap water	99.0 ml

*Available at Roboz Surgical Instrument Co., Washington D. C.

**Modified by. S/Sgt. Jerry A. Coates, USAF, 1962, AFIP.

Carbol Fuchsin Solution (ZIEHL-NEELSEN)

Phenol crystals, melted	2.5 ml
Alcohol, absolute	5.0 ml
Basic fuchsin	0.5 g

Distilled water to make 50 ml. Filter before use. Keeps well at room temperature.

Working Methylene Blue Solution

Methylene Blue	0.5 g
Glacial acetic acid	0.5 ml
Tap water	100.0 ml

Staining Procedure

1. Deparaffinise sections with **xylene-peanut oil mixture, 2 changes** of 12 minutes each.
2. Drain, wipe off excess oil and blot to opacity. The residual oil helps to prevent shrinkage and injury of section.
3. **Stain in carbol fuchsin for 10 minutes**, (This is critical).
4. Wash in tap water for one to two minutes.
5. **Differentiate in 1% sulphuric acid in water** for approximately 5 to 10 minutes. Agitate to remove background as quickly as possible, this will maintain the redness of the organism which is very fragile. The overall **colour will be purplish pink**.
6. Wash in tap water.
7. **Counterstain** lightly **with working methylene blue**.
8. Rinse off excess methylene blue in tap water.
9. Blot and let stand for a few minutes to air dry.
10. Mount directly in Permount.

Results

Nocardia—red

Background—light blue.

N.B. **The Fite-Faraco technique demonstrates** certain strains of **Nocardia as long, thin, red structures on a pale blue background**. This technique, in the experience of the members of the staff of this installation, stands out as the most useful of the acid-fast procedures for the demonstration of these particular bacteria for the following reasons:

1. The protection afforded by the combination of peanut oil and xylene, which coats the filament capsule.
2. The elimination of alcohol from both the hydration and dehydration process, so that the oil coating applied around the filament through the above procedure is not dissolved.
3. The **decolorisation of the background with** a weak solution of **aqueous sulphuric acid (1%)** instead of the usual 1% acid alcohol is done.

3. GOMORI'S METHENAMINE-SILVER NITRATE (GMS) STAIN

(Grocott's application to fungi)

Fixation : Formalin 10 per cent.

Technique : Paraffin, celloidin, or frozen section.

Reagents

1. *5 per cent Chromic Acid Solution*

Chromic acid	5 g
Distilled water	100 ml

2. *5 per cent Silver Nitrate Solution*

Silver nitrate	5 g
Distilled water	100 ml

3. *3 PER CENT METHENAMINE SOLUTION*

Hexamethylene tetramine (methenamine)*	3 g
Distilled water	100 ml

4. *5 per cent Borax Solution*

Borax (Photographic grade)†	5 g
Distilled water	100 ml

5. *STOCK METHENAMINE SILVER NITRATE SOLUTION*

SILVER NITRATE (5 per cent sol.)	5 ml
Methenamine (3 per cent sol.)	100 ml

A white precipitate forms but dissolves immediately on shaking. Clear solution keeps well for months at refrigerator temperature.

6. *WORKING METHENAMINE-SILVER NITRATE SOLUTION*

Borax (5 per cent solution)	2 ml
Distilled water	25 ml
Mix and add	
METHENAMINE-SILVER NITRATE STOCK SOLUTION	25 ml

*To be obtained from Fischer Scientific Co. or Eastman Kodak Co.

† To be obtained from Eastman Kodak Co.

7. ***1 per cent Sodium Bisulphite Solution***

Sodium bisulphite	1 g
Distilled water	100 ml

8. ***0.1 per cent Gold Chloride Solution***

Gold chloride (1 per cent sol.)	10 ml
Distilled water	90 ml

This solution may be used repeatedly.

9. ***2 per cent Sodium Thiosulphate (Hypo) Solution***

Sodium thiosulphate	2 g
Distilled water	100 ml

Stock Light Green Solution

Light green, S. F. (yellow)††	0.2 g
Distilled water	100.0 ml
Glacial acetic acid	0.2 ml

Working Light Green Solution

Light green, stock solution	10.0 ml
Distilled water	50.0 ml

Staining Procedure

Run a control slide.

1. Deparaffinise sections through 2 changes each of xylene, absolute and 95% alcohols to distilled water as usual. Slides previously stained with most other stains may be used by removing cover glasses in xylol and running through alcohols to water. Subsequently chromic acid treatment will remove any remaining stain.
2. Oxidise in 5% chromic acid solution for 1 hour.
3. Wash in running tap water for a few seconds.
4. Rinse in 1% sodium bisulphite for 1 minute to remove any residual chromic acid.
5. Wash in tap water for 5 to 10 minutes.
6. Wash with 3 or 4 changes of distilled water.
7. **Place the working methenamine-silver nitrate solution in oven at 58° to 60°C for 30 to 60 minutes** until section turns **yellowish-brown**. Use paraffin coated forceps to remove slide from this solution. Dip slide in distilled water and **check for adequate silver impregnation under the microscope**. Fungi should be a dark-brown colour at this stage.

††Obtained from National Aniline Company.

8. Rinse in 6 changes of distilled water.
9. Tone in 0.1% gold chloride solution for 2 to 5 minutes.
10. Rinse in distilled water.
11. **Remove unreduced silver with 2% sodium thiosulphate (hypo) solution** for 2 to 5 minutes.
12. Wash thoroughly in tap water.
13. **Counterstain with working light green** for 30 to 45 seconds.
14. Dehydrate with 2 changes each of 95% alcohol, absolute alcohol; clear with 2 to 3 changes of xylene, and mount.

Results

FUNGI—SHARPLY DELINEATED IN BLACK.

MUCIN—TAUPE TO DARK GREY.

Inner parts of mycelia and hyphae—old rose.

Background—pale green.

N.B. There are several **important points** to be considered **in doing a GMS stain** as follows: The technique calls for incubation of slides in a 60°C oven for 30 to 60 minutes (the full time being preferable for all fungi, **but not for Nocardia**), or until sections turn yellowish-brown. In staining for filaments of Nocardia, it has been found that the incubation time for complete development may vary according to type and/or strain suspected to be present in the particular section. Therefore, if Nocardia is suspected, two slides should be processed, **one for 60 minutes** and **one for 90 minutes**, thereby assuring a positive demonstration of any filaments of Nocardia that may exist.

4. GRIDLEY'S STAIN FOR FUNGI

Fixation : Any well fixed tissue.

Technique : Cut paraffin sections **at 6 microns**.

Solutions

1. ***Chromic Acid Solution***

Chromic acid	4.00 g
Distilled water	100.00 ml

2. ***Coleman's Feulgen Reagent***

Bring 200 ml of distilled water to boil, remove from flame and add 1 g **BASIC**

FUCHSIN. When fuchsin is dissolved, cool and filter. Add 2 g sodium metabisulphite and 10 ml of normal hydrochloric acid. Let this bleach for 24 hours, and then add 0.5 g activated carbon (Norit); shake for about 1 minute, and filter through coarse paper. The filtrate should be colourless. Store in refrigerator.

3. *Normal Hydrochloric Acid*

Hydrochloric acid, sp. gr. 1.19	83.5 ml
Distilled water	916.5 ml

4. *Sodium Metabisulphite Solution*

Sodium metabisulphite	10.0 g
Distilled water	100.0 ml

5. *Sulphurous Rinse*

Sodium metabisulphite, (10% solution)	6.0 ml
Normal hydrochloric acid	5.0 ml
Distilled water	100.0 ml

6. *Aldehyde-Fuchsin Solution*

Basic fuchsin	1.0 g
Alcohol, 70%	200.0 ml
Paraldehyde	2.0 ml
Hydrochloric acid, concentrated	2.0 ml

Let it stand at room temperature for 3 days until solution turns deep blue. Keep in refrigerator. Filter and warm to room temperature before using.

7. *Metanil Yellow Solution (Counterstain)*

Metanil yellow	0.25 g
Distilled water	100.00 ml
Glacial acetic acid	2.00 drops

Staining Procedure

1. Xylene, absolute and 95% alcohols to distilled water.
2. **Place in 4% chromic acid for 1 hour (oxidiser).**
3. Wash in running tap water for 5 minutes.
4. **Place in Coleman's Feulgen reagent for 15 minutes.**
5. Rinse in 3 changes of **sulphurous acid rinse.**
6. Wash for 15 minutes in running tap water.
7. Place slides in **aldehyde-fuchsin solution** for 15 to 30 minutes.
8. Rinse off excess stain with 95% alcohol.
9. Rinse in tap water.
10. **Counterstain lightly with metanil yellow** solution **for 1 minute.**
11. Rinse in tap water.
12. Dehydrate through 95% and absolute alcohols.
13. Clear in xylene and mount.

Results

FUNGI, ELASTIC TISSUE AND MUCIN—DEEP ROSE TO PURPLE

Background—yellow.

N.B.

1. **Absolutely fresh Coleman's leuco-fuchsin solution** is a must for optimum results, especially for the demonstration of *HISTOPLASMA CAPSULATUM*.
2. Special attention must be given to the metanil yellow counterstain, since many of the small fungal bodies can be overshadowed if the counterstain is too dark. If this does happen, however, the excess stain can be removed by washing in running water.

5. PERIODIC ACID-SCHIFF REACTIONS (PAS)

Fixation : 10 per cent formalin or Zenker's solution.

Technique : Cut paraffin sections **at 6 microns.**

Solutions

1. *Coleman's Feulgen Reagent*

   Dissolve 1.0 g basic fuchsin in 200.0 ml hot distilled water. bring to boiling point. Cool and add 2.0 g potassium metabisulphite, and 10.0 ml normal hydrochloric acid. Let it bleach for 24 hours, then add 0.5 g activated carbon (Norit). Shake 1 minute and filter through coarse filter paper. Repeat filtration until solution is colourless. Store in refrigerator.

OR

2. *Schiff's Leuco-Fuchsin Solution*

   Dissolve 1.0 g basic fuchsin in 200.0 ml hot distilled water. Bring to boiling point. Cool to 50°C. Filter and add 20.0 ml normal hydrochloric acid. Cool further and add 1.0 g anhydrous sodium bisulphite, or sodium metabisulphite. Keep in the dark for 48 hours until solution becomes straw coloured. Store in refrigerator.

3. *Test for Schiff's Leuco-Fuchsin Solution*

Pour a few drops of Schiff's solution into 10 ml of 37 to 40% formaldehyde in a watch glass. If the solution turns reddish purple rapidly, it is good. If the reaction is delayed and the resultant colour deep blue-purple, the solution is breaking down.

4. ***0.5% PERIODIC ACID SOLUTION***

Periodic acid crystals	0.5 g
Distilled water	100.0 ml

5. ***Normal Hydrochloric Acid Solution***

Hydrochloric acid, conc. sp. gr. 1.19	83.5 ml
Distilled water	916.5 ml

6. ***0.2% Light Green Counterstain***

Light green crystals	0.2 g
Distilled water	100.0 ml
Glacial acetic acid	0.2 ml

OR

Harris's Haematoxylin

7. ***Diastase Solution***

Diastase	0.5 g
Distilled water, sterile	100.0 ml

Store in refrigerator; stays for 1 week.

Staining Procedure

1. Xylene.
2. Absolute alcohol.
3. Alcohol 95%.
4. If Zenker-fixed, remove mercury precipitate in iodine, wash in water and decolourise in **hypo** solution.
5. Rinse in distilled water.*
6. **PERIODIC ACID SOLUTION** for 5 minutes (oxidiser).
7. Rinse in distilled water.
8. **Place in Coleman's Feulgen** or Schiff's leuco-fuchsin for 15 minutes.
9. Place in running tap water for 10 minutes for **pink colour** to develop.
10. Stain in Harris' haematoxylin for 6 minutes, or light green counterstain for a few seconds. Light green is recommended for counterstaining sections in which fungi are to be demonstrated. Omit steps 11 through 15 if light green is used.
11. Rinse in tap water.
12. Differentiate in acid alcohol-3 to 10 quick dips.
13. Wash in tap water.
14. Dip in ammonia water **to blue sections**.
15. Wash in running tap water-10 minutes.
16. Alcohol, 95%.
17. Absolute alcohol, 2 changes.
18. Xylene, 2 changes.
19. Mount.

N.B. A solution of **5% aqueous Clorox bleach** will reduce overstaining by leuco-fuchsin. Running tap water or ammonia water will decolourise the light green.

Results

Glycogen, mucin, hyaluronic acid, reticulin, fibrin (of thrombi), colloid droplets, hyaline of arteriosclerosis, hyaline deposits in glomeruli, granular cells in the renal arterioles where preserved, most basement membranes, colloid of pituitary stalks and thyroid, amyloid infiltration and other elements may show **A POSITIVE REACTION–ROSE TO PURPLISH RED.**

NUCLEI—BLUE.

FUNGI—RED.

Background—pale green (with light green counterstaining).

REFERENCES

Faraco, J. (1938). *Rev. brasil. leprol.* **6**, 177.

Fite, G. L., Cambre, P. J., and Turner, M. H. (1947). Procedure for demonstrating lepra bacilli in paraffin sections. *Arch. Path.* **43**, 624-625.

Gridley, M. F. (1953). *Am. J. clin. Path.* **23**, 303-307.

Grocott, R. C. B. S. M. T. (ASCP), Technical Bulletin of the Registry of Medical Technologists, July 1955, Vol. **25**, No. 7, *American Journal of Clinical Pathology*, August, 1955, pages 975-979.

Kligman, A. M. Mescon, H., and de Lamater, E. D. (1951). The Hotchkiss-McManus stain for the histopathologic diagnosis of fungus disease. *Am. J. clin. Path.* **21**, 86-91.

Lillie, R. D. "*Histopathologic Tehnics and Practical Histochemistry*". The Blakiston Co., New York, 1954, page 381.

MacManus, J. F. A. (1949). *Stain Technology*, **23**, 99 (AFIP Modification).

Mallory, F. B. *Pathological Technique*, page 130. W. B. Saunders Co., Philadelphia, 1942.

Mowry, R. W. (1958). Improved Procedure for the Staining of Acidic Polysaccharides by Muller's Colloidal (Hydrous) Ferric Oxide and its Combination with the Feulgen and the Periodic Acid-Schiff Reactions. *Lab. Invest.* **7**, 566-576.

Robert W., Mowry, M.D. Department of Pathology, University of Alabama Medical Centrer, Brimingham 3, Alabama (Modified by AFIP) –

Steedman, H. F. (1950). Alcian Blue 8 GS; A New Stain for Mucin. *Quart. J. Micr. Sci.*, **91**, part 4, 477-479 (original reference).

Wade, W. H. (1952). "A modification using rectified turpentine instead of xylene and 20% aqueous sulfuric acid instead of acid alcohol." *Am. J. Path.* **28**, 157.

*If PAS reaction with digestion is desired, place sections in 0.5% diastase for 20 minutes. Saliva can also be used. Rinse in running tap water for 10 minutes, and wash in distilled water.

Index

M